STUDENT'S SOLUTIONS MANUAL

INTERMEDIATE ALGEBRA

SEVENTH EDITION

STUDENT'S SOLUTIONS MANUAL

INTERMEDIATE ALGEBRA
SEVENTH EDITION

Marvin L. Bittinger
Indiana University—Purdue University at Indianapolis

Mervin L. Keedy
Purdue University

Judith A. Penna

▲ **ADDISON-WESLEY PUBLISHING COMPANY**
Reading, Massachusetts • Menlo Park, California • New York
Don Mills, Ontario • Wokingham, England • Amsterdam • Bonn
Sydney • Singapore • Tokyo • Madrid • San Juan • Milan • Paris

Reprinted with corrections, December 1996.

Reproduced by Addison-Wesley from camera-ready copy supplied by the author.

Copyright © 1995 Addison-Wesley Publishing Company, Inc.

ISBN 0-201-59123-5

8 9 10 BAH 99 98

Table of Contents

The author thanks Patsy Hammond for her
excellent typing and Pam Smith for her careful
proofreading. Very special thanks are extended to
Mike Penna for sharing his expertise and for
his support.

STUDENT'S SOLUTIONS MANUAL

INTERMEDIATE ALGEBRA
SEVENTH EDITION

Chapter 1

Algebra and Real Numbers

Exercise Set 1.1

1. Substitute 1 for a and carry out the addition.
$$a + 2 = 1 + 2 = 3 \text{ m}$$
Substitute 5 for a and carry out the addition.
$$a + 2 = 5 + 2 = 7 \text{ m}$$
Substitute 9 for a and carry out the addition.
$$a + 2 = 9 + 2 = 11 \text{ m}$$

3. Substitute 6.5 for b and 15.4 for h and carry out the multiplication.
$$A = bh = 6.5 \times 15.4 = 100.1 \text{ cm}^2$$

5. Substitute 4 for z and carry out the multiplication.
$$83z = 83 \cdot 4 = 332$$

7. Substitute 24 for a and 8 for b and carry out the division.
$$\frac{a}{b} = \frac{24}{8} = 3$$

9. Substitute 36 for m and 4 for n and carry out the calculations.
$$\frac{m - n}{8} = \frac{36 - 4}{8} = \frac{32}{8} = 4$$

11. Substitute 9 for z and 2 for y and carry out the calculations.
$$\frac{5z}{y} = \frac{5 \cdot 9}{2} = \frac{45}{2}, \text{ or } 22\frac{1}{2}$$

13. Substitute \$1800 for P, 0.035 for r, and 2 for t and carry out the calculations.
$$I = Prt = \$1800(0.035)(2) = \$126$$

15. Phrase: 7 more than m
Algebraic expression: $7 + m$, or $m + 7$

17. Phrase: 11 less than c
Algebraic expression: $c - 11$

19. Phrase: 26 greater than q
Algebraic expression: $26 + q$, or $q + 26$

21. Phrase: z more than x
Algebraic expression: $x + z$, or $z + x$

23. Phrase: x less than y
Algebraic expression: $y - x$

25. Phrase: 28% of x
Algebraic expression: $28\%x$, or $0.28x$

27. Phrase: The sum of c and d
Algebraic expression: $c + d$, or $d + c$

29. Phrase: twice x
Algebraic expression: $2x$

31. Phrase: seven times t
Algebraic expression: $7t$

33. Phrase: The difference between 21 and a
Algebraic expression: $21 - a$

35. Phrase: 8 more than some number
Let y represent the number.
Algebraic expression: $8 + y$, or $y + 8$

37. Phrase: 54 less than some number
Let x represent the number.
Algebraic expression: $x - 54$

39. Phrase: 54 less some number
Let x represent the number.
Algebraic expression: $54 - x$

41. Phrase: A number p plus three times q
Algebraic expression: $p + 3q$

43. Distance = speed × time
$$d = rt$$

45. Substitute 2 for x and 4 for y and carry out the calculations.
$$\frac{y + x}{2} + \frac{3y}{x} = \frac{4 + 2}{2} + \frac{3 \cdot 4}{2} = \frac{6}{2} + \frac{12}{2} = 3 + 6 = 9$$

Exercise Set 1.2

In Exercises 1-5, consider the following numbers: -6, 0, 1, $-\frac{1}{2}$, -4, $\frac{7}{9}$, 12, $-\frac{6}{5}$, 3.45, $5\frac{1}{2}$, $\sqrt{3}$, $\sqrt{25}$, $-\frac{12}{3}$, $0.131331333133331\ldots$.

1. The natural numbers are numbers used for counting. The natural numbers in the list above are 1, 12, and $\sqrt{25}$ ($\sqrt{25} = 5$).

3. The rational numbers can be named as quotients of integers with nonzero divisors. The rational numbers in the list above are -6, 0, 1, $-\frac{1}{2}$, -4, $\frac{7}{9}$, 12, $-\frac{6}{5}$, 3.45 ($3.45 = \frac{345}{100}$), $5\frac{1}{2}$, $\sqrt{25}$ ($\sqrt{25} = 5$), and $-\frac{12}{3}$.

5. The real numbers consist of the rational numbers and the irrational numbers. All of the numbers in the list are real numbers.

In Exercises 7-11, consider the following numbers: $-\sqrt{5}$, -3.43, -11, 12, 0, $\dfrac{11}{34}$, $-\dfrac{7}{13}$, π, $-3.565665666566665\ldots$

7. The whole numbers consist of the natural numbers and 0. The whole numbers in the list above are 12 and 0.

9. The integers consist of the whole numbers and their opposites. The integers in the list above are -11, 12, and 0.

11. The irrational numbers are the numbers that are not rational. The irrational numbers in the list above are $-\sqrt{5}$, π, and $-3.565665666566665\ldots$.

13. We list the members of the set.
$\{m,a,t,h\}$

15. We list the members of the set.
$\{1,2,3,4,5,6,7,8,9,10,11,12\}$

17. We list the members of the set.
$\{2,4,6,8,\ldots\}$

19. We specify conditions by which we know whether a number is in the set.
$\{x|x$ is a whole number less than or equal to 5$\}$

21. We specify conditions by which we know whether a number is in the set.
$\left\{\dfrac{a}{b}\,\middle|\,a \text{ and } b \text{ are integers and } b \neq 0\right\}$

23. We specify conditions by which we know whether a number is in the set.
$\{x|x$ is a real number and $x > -3\}$, or $\{x|x > -3\}$

25. Since 13 is to the right of 0 on the number line, we have $13 > 0$.

27. Since -8 is to the left of 2 on the number line, we have $-8 < 2$.

29. Since -8 is to the left of 8, we have $-8 < 8$.

31. Since -8 is to the left of -3, we have $-8 < -3$.

33. Since -2 is to the right of -12, we have $-2 > -12$.

35. Since -9.9 is to the left of -2.2, we have $-9.9 < -2.2$.

37. Since 37.2 is to the right of -1.67, we have $37.2 > -1.67$.

39. We convert to decimal notation: $\dfrac{6}{13} = 0.461538\ldots$ and $\dfrac{13}{25} = 0.52$. Thus $\dfrac{6}{13} < \dfrac{13}{25}$.

41. $-8 > x$
The inequality $x < -8$ has the same meaning.

43. $-12 \leq y$
The inequality $y \geq -12$ has the same meaning.

45. $6 \leq -6$ False since neither $6 < -6$ nor $6 = -6$ is true.

47. $5 \geq -8$ True since $5 > -8$ is true.

49. $x < -2$
We shade all numbers less than -2. We indicate that -2 is not a solution by using an open circle at -2.

51. $x \leq -2$
We shade all the numbers to the left of -2 and use a solid circle at -2 to indicate that it is also a solution.

53. $x > -4$
We shade all the numbers to the right of -4 and use an open circle at -4 to indicate that it is not a solution.

55. $x \geq 2$
We shade all the numbers to the right of 2 and use a solid circle at 2 to indicate that it is also a solution.

57. The distance of -6 from 0 is 6, so $|-6| = 6$.

59. The distance of 28 from 0 is 28, so $|28| = 28$.

61. The distance of -35 from 0 is 35, so $|-35| = 35$.

63. The distance of $-\dfrac{2}{3}$ from 0 is $\dfrac{2}{3}$, so $\left|-\dfrac{2}{3}\right| = \dfrac{2}{3}$.

65. The distance of $\dfrac{0}{7}$, or 0, from 0 is 0, so $\left|\dfrac{0}{7}\right| = 0$.

67. We replace x by 8 and y by 20.
$8xy = 8 \cdot 8 \cdot 20 = 1280$

69. Recall that rational numbers can be named as quotients of integers with nonzero divisors. Five examples are 0, 10, -6, $-\dfrac{5}{8}$, and -987.6. Answers may vary.

71. $|-3| = 3$, so $|-3| \leq 5$.

73. $|-7| = 7$, so $|4| \leq |-7|$.

75. For comparison, we first write each number in decimal notation.

$$\frac{1}{11} = 0.090909\ldots$$

$$1.1\% = 0.011$$

$$\frac{2}{7} = 0.285714285714\ldots$$

$$0.3\% = 0.003$$

$$0.11 = 0.11$$

$$\frac{1}{8}\% = 0.00125 \qquad \left(\frac{1}{8} = 0.125\right)$$

$$0.009 = 0.009$$

$$\frac{99}{1000} = 0.099$$

$$0.286 = 0.286$$

$$\frac{1}{8} = 0.125$$

$$1\% = 0.01$$

$$\frac{9}{100} = 0.09$$

Then these rational numbers listed from least to greatest are $\frac{1}{8}\%$, 0.3%, 0.009, 1%, 1.1%, $\frac{9}{100}$, $\frac{1}{11}$, $\frac{99}{1000}$, 0.11, $\frac{1}{8}$, $\frac{2}{7}$, and 0.286.

Exercise Set 1.3

1. $-10 + (-18)$

The sum of two negative numbers is negative. We add their absolute values, $10 + 18 = 28$, and make the answer negative.

$$-10 + (-18) = -28$$

3. $7 + (-2)$

We find the difference of their absolute values, $7 - 2 = 5$. Since the positive number has the larger absolute value, the answer is positive.

$$7 + (-2) = 5$$

5. $-8 + (-8)$

The sum of two negative numbers is negative. We add the absolute values, $8 + 8 = 16$, and make the answer negative.

$$-8 + (-8) = -16$$

7. $7 + (-11)$

We find the difference of their absolute values, $11 - 7 = 4$. Since the negative number has the larger absolute value, the answer is negative.

$$7 + (-11) = -4$$

9. $-16 + 6$

We find the difference of their absolute values, $16 - 6 = 10$. Since the negative number has the larger absolute value, the answer is negative.

$$-16 + 6 = -10$$

11. $-26 + 0$

One number is 0. The sum is -26.

$$-26 + 0 = -26$$

13. $-8.4 + 9.6$

We find the difference of their absolute values, $9.6 - 8.4 = 1.2$. Since the positive number has the larger absolute value, the answer is positive.

$$-8.4 + 9.6 = 1.2$$

15. $-2.62 + (-6.24)$

The sum of two negative numbers is negative. We add the absolute values, $2.62 + 6.24 = 8.86$, and make the answer negative.

$$-2.62 + (-6.24) = -8.86$$

17. $-\frac{5}{9} + \frac{2}{9}$

We find the difference of their absolute values, $\frac{5}{9} - \frac{2}{9} = \frac{3}{9} = \frac{1}{3}$. Since the negative number has the larger absolute value, the answer is negative.

$$-\frac{5}{9} + \frac{2}{9} = -\frac{1}{3}$$

19. $-\frac{11}{12} + \left(-\frac{5}{12}\right)$

The sum of two negative numbers is negative. We add their absolute values, $\frac{11}{12} + \frac{5}{12} = \frac{16}{12}$, or $\frac{4}{3}$, and make the answer negative.

$$-\frac{11}{12} + \left(-\frac{5}{12}\right) = -\frac{4}{3}$$

21. $\frac{2}{5} + \left(-\frac{3}{10}\right)$

We find the difference of their absolute values.

$$\frac{2}{5} - \frac{3}{10} = \frac{2}{5} \cdot \frac{2}{2} - \frac{3}{10} = \frac{4}{10} - \frac{3}{10} = \frac{1}{10}$$

Since the positive number has the larger absolute value, the answer is positive.

$$\frac{2}{5} + \left(-\frac{3}{10}\right) = \frac{1}{10}$$

23. $-\frac{2}{5} + \frac{3}{4}$

We find the difference of their absolute values.

$$\frac{3}{4} - \frac{2}{5} = \frac{3}{4} \cdot \frac{5}{5} - \frac{2}{5} \cdot \frac{4}{4} = \frac{15}{20} - \frac{8}{20} = \frac{7}{20}$$

Since the positive number has the larger absolute value, the answer is positive.

$$-\frac{2}{5} + \frac{3}{4} = \frac{7}{20}$$

25. When $a = -4$, then $-a = -(-4) = 4$.

(The opposite, or additive inverse, of -4 is 4.)

27. When $a = -3.7$, then $-a = -(-3.7) = 3.7$.

(The opposite, or additive inverse, of -3.7 is 3.7.)

29. The opposite, or additive inverse, of 10 is -10, because $10 + (-10) = 0$.

31. The opposite, or additive inverse, of 0 is 0, because $0 + 0 = 0$.

33. $3 - 7 = 3 + (-7) = -4$

35. $-5 - 9 = -5 + (-9) = -14$

37. $23 - 23 = 23 + (-23) = 0$

39. $-23 - 23 = -23 + (-23) = -46$

41. $-6 - (-11) = -6 + 11 = 5$

43. $10 - (-5) = 10 + 5 = 15$

45. $15.8 - 27.4 = 15.8 + (-27.4) = -11.6$

47. $-18.01 - 11.24 = -18.01 + (-11.24) = -29.25$

49. $-\dfrac{21}{4} - \left(-\dfrac{7}{4}\right) = -\dfrac{21}{4} + \dfrac{7}{4} = -\dfrac{14}{4} = -\dfrac{7}{2}$

51. $-\dfrac{1}{3} - \left(-\dfrac{1}{12}\right) = -\dfrac{1}{3} + \dfrac{1}{12} = -\dfrac{4}{12} + \dfrac{1}{12} = -\dfrac{3}{12} = -\dfrac{1}{4}$

53. $-\dfrac{3}{4} - \dfrac{5}{6} = -\dfrac{3}{4} + \left(-\dfrac{5}{6}\right) = -\dfrac{9}{12} + \left(-\dfrac{10}{12}\right) = -\dfrac{19}{12}$

55. $\dfrac{1}{3} - \dfrac{4}{5} = \dfrac{1}{3} + \left(-\dfrac{4}{5}\right) = \dfrac{5}{15} + \left(-\dfrac{12}{15}\right) = -\dfrac{7}{15}$

57. $3(-7)$

The product of a positive number and a negative number is negative. We multiply their absolute values, $3 \cdot 7 = 21$, and make the answer negative.

$3(-7) = -21$

59. $-2 \cdot 4$

The product of a negative number and a positive number is negative. We multiply their absolute values, $2 \cdot 4 = 8$, and make the answer negative.

$-2 \cdot 4 = -8$

61. $-8(-3)$

The product of two negative numbers is positive. We multiply their absolute values, $8 \cdot 3 = 24$, and make the answer positive.

$-8(-3) = 24$

63. $-7 \cdot 16$

The product of a negative number and a positive number is negative. We multiply their absolute values, $7 \cdot 16 = 112$, and make the answer negative.

$-7 \cdot 16 = -112$

65. $-6(-5.7)$

The product of two negative numbers is positive. We multiply their absolute values, $6(5.7) = 34.2$, and make the answer positive.

$-6(-5.7) = 34.2$

67. $-\dfrac{3}{5} \cdot \dfrac{4}{7}$

The product of a negative number and a positive number is negative. We multiply their absolute values, $\dfrac{3}{5} \cdot \dfrac{4}{7} = \dfrac{12}{35}$, and make the answer negative.

$-\dfrac{3}{5} \cdot \dfrac{4}{7} = -\dfrac{12}{35}$

69. $-3\left(-\dfrac{2}{3}\right)$

The product of two negative numbers is positive. We multiply their absolute values, $3 \cdot \dfrac{2}{3} = \dfrac{6}{3} = 2$, and make the answer positive.

$-3\left(-\dfrac{2}{3}\right) = 2$

71. $-3(-4)(5)$

$= 12(5)$ The product of two negative numbers is positive.

$= 60$ The product of two positive numbers is positive.

73. $-4.2(-6.3)$

The product of two negative numbers is positive. We multiply the absolute values, $4.2(6.3) = 26.46$, and make the answer positive.

$-4.2(-6.3) = 26.46$

75. $-\dfrac{9}{11} \cdot \left(-\dfrac{11}{9}\right)$

The product of two negative numbers is positive. We multiply their absolute values, $\dfrac{9}{11} \cdot \dfrac{11}{9} = \dfrac{99}{99} = 1$, and make the answer positive.

$-\dfrac{9}{11} \cdot \left(-\dfrac{11}{9}\right) = 1$

77. $-\dfrac{2}{3} \cdot \left(-\dfrac{2}{3}\right) \cdot \left(-\dfrac{2}{3}\right)$

$= \dfrac{4}{9} \cdot \left(-\dfrac{2}{3}\right)$ The product of two negative numbers is positive.

$= -\dfrac{8}{27}$ The product of a positive number and a negative number is negative.

79. When a negative number is divided by a positive number, the answer is negative.

$\dfrac{-8}{4} = -2$

81. When a positive number is divided by a negative number, the answer is negative.

$\dfrac{56}{-8} = -7$

83. When a negative number is divided by a negative number, the answer is positive.

$$-77 \div (-11) = \frac{-77}{-11} = 7$$

85. When a negative number is divided by a negative number, the answer is positive.

$$\frac{-5.4}{-18} = \frac{5.4}{18} = 0.3$$

87. $\frac{5}{0}$ Undefined: Division by 0.

89. $\frac{0}{32} = 0$ because $0 \cdot 32 = 0$.

91. $\frac{9}{y-y}$ Undefined: $y - y = 0$ for any y.

93. The reciprocal of $\frac{3}{4}$ is $\frac{4}{3}$, because $\frac{3}{4} \cdot \frac{4}{3} = 1$.

95. The reciprocal of $-\frac{7}{8}$ is $-\frac{8}{7}$, because $-\frac{7}{8} \cdot \left(-\frac{8}{7}\right) = 1$.

97. The reciprocal of 25 is $\frac{1}{25}$, because $25 \cdot \frac{1}{25} = 1$.

99. The reciprocal of $-\frac{a}{b}$ is $-\frac{b}{a}$, because $-\frac{a}{b} \cdot \left(-\frac{b}{a}\right) = 1$.

101. $\frac{2}{7} \div \left(-\frac{11}{3}\right) = \frac{2}{7} \cdot \left(-\frac{3}{11}\right) = -\frac{6}{77}$

103. $-\frac{10}{3} \div -\frac{2}{15} = -\frac{10}{3} \cdot \left(-\frac{15}{2}\right) = \frac{150}{6}$, or 25

105. $18.6 \div (-3.1) = \frac{18.6}{-3.1} = -\frac{18.6}{3.1} = -6$

107. $(-75.5) \div (-15.1) = \frac{-75.5}{-15.1} = \frac{75.5}{15.1} = 5$

109. $-48 \div 0.4 = \frac{-48}{0.4} = -\frac{48}{0.4} = -120$

111. $\frac{3}{4} \div \left(-\frac{2}{3}\right) = \frac{3}{4} \cdot \left(-\frac{3}{2}\right) = -\frac{9}{8}$

113. $-\frac{5}{4} \div \left(-\frac{3}{4}\right) = -\frac{5}{4} \cdot \left(-\frac{4}{3}\right) = \frac{20}{12} = \frac{5}{3}$

115. $-\frac{2}{3} \div \left(-\frac{4}{9}\right) = -\frac{2}{3} \cdot \left(-\frac{9}{4}\right) = \frac{18}{12} = \frac{3}{2}$

117. $-\frac{3}{8} \div \left(-\frac{8}{3}\right) = -\frac{3}{8} \cdot \left(-\frac{3}{8}\right) = \frac{9}{64}$

119. $-6.6 \div 3.3 = \frac{-6.6}{3.3} = -2$

121. $\frac{-12}{-13} = \frac{12}{13}$, or $0.\overline{923076}$

123. $\frac{48.6}{-30} = \frac{16.2}{-10} = -1.62$

125. $\frac{-9}{17-17}$ Undefined: $17 - 17 = 0$

127. $\frac{2}{3}$:   The opposite of $\frac{2}{3}$ is $-\frac{2}{3}$, because
$$\frac{2}{3} + \left(-\frac{2}{3}\right) = 0.$$
The reciprocal of $\frac{2}{3}$ is $\frac{3}{2}$, because
$$\frac{2}{3} \cdot \frac{3}{2} = 1.$$

$-\frac{5}{4}$:   The opposite of $-\frac{5}{4}$ is $\frac{5}{4}$, because $-\frac{5}{4} + \frac{5}{4} = 0$.

The reciprocal of $-\frac{5}{4}$ is $-\frac{4}{5}$, because
$$-\frac{5}{4} \cdot \left(-\frac{4}{5}\right) = 1.$$

0 :   The opposite of 0 is 0, because $0 + 0 = 0$.

The reciprocal of 0 does not exist. (Only nonzero numbers have reciprocals.)

1 :   The opposite of 1 is -1, because $1 + (-1) = 0$.

The reciprocal of 1 is 1, because $1 \cdot 1 = 1$.

-4.5 :   The opposite of -4.5 is 4.5, because $-4.5 + 4.5 = 0$.

The reciprocal of -4.5 is $-\frac{1}{4.5}$, because

$-4.5 \cdot \left(-\frac{1}{4.5}\right) = 1$. (Note that

$-\frac{1}{4.5} = -\frac{1}{4.5} \cdot \frac{10}{10} = -\frac{10}{45}$, or $-\frac{2}{9}$.)

$x, x \neq 0$:   The opposite of x is $-x$, because $x + (-x) = 0$.

The reciprocal of x, $x \neq 0$, is $\frac{1}{x}$,

because $x \cdot \frac{1}{x} = 1$.

129. Phrase: Eight less than some number

Let x represent the number.

Algebraic expression: $x - 8$

131. Phrase: Eight more than three times a number

Let y represent the number.

Algebraic expression: $3y + 8$

133. Substitute 2 for x and 7 for y and carry out the addition.

$$x + y = 2 + 7 = 9$$

135.
$$\frac{1}{r_1} + \frac{1}{r_2}$$
$$= \frac{1}{12} + \frac{1}{6} \quad \text{Substituting 12 for } r_1 \text{ and 6 for } r_2$$
$$= \frac{1}{12} + \frac{2}{12}$$
$$= \frac{3}{12}, \text{ or } \frac{1}{4}$$

The conductance is $\frac{1}{4}$.

Exercise Set 1.4

1. Since $10 = 5 \cdot 2$, we multiply by $\frac{2}{2}$:

$$\frac{3x}{5} \cdot \frac{2}{2} = \frac{6x}{10}$$

3. Since $4x = 4 \cdot x$, we multiply by $\frac{x}{x}$:

$$\frac{3}{4} \cdot \frac{x}{x} = \frac{3x}{4x}$$

5. $\dfrac{25x}{15x} = \dfrac{5 \cdot 5x}{3 \cdot 5x}$ We look for the largest common factor of the numerator and denominator and factor each.

$$= \frac{5}{3} \cdot \frac{5x}{5x} \quad \text{Factoring the expression}$$

$$= \frac{5}{3} \cdot 1 \quad \left(\frac{5x}{5x} = 1\right)$$

$$= \frac{5}{3}$$

7. $\dfrac{100}{25x} = \dfrac{4 \cdot 25}{x \cdot 25}$ Factoring numerator and denominator

$$= \frac{4}{x} \cdot \frac{25}{25} \quad \text{Factoring the expression}$$

$$= \frac{4}{x} \cdot 1 \quad \left(\frac{25}{25} = 1\right)$$

$$= \frac{4}{x}$$

9. $w + 3 = 3 + w$ Commutative law of addition

11. $rt = tr$ Commutative law of multiplication

13. $4 + cd = cd + 4$ Commutative law of addition

or

$4 + cd = cd + 4 = dc + 4$ Commutative laws of addition and multiplication

or

$4 + cd = 4 + dc$ Commutative law of multiplication

15. $yz + x = x + yz$ Commutative law of addition

or

$yz + x = x + yz = x + zy$ Commutative laws of addition and multiplication

or

$yz + x = zy + x$ Commutative law of multiplication

17. $m + (n + 2) = (m + n) + 2$ Associative law of addition

19. $(7 \cdot x) \cdot y = 7 \cdot (x \cdot y)$ Associative law of multiplication

21. $(a + b) + 8 = a + (b + 8)$ Associative law

$= a + (8 + b)$ Commutative law

$(a + b) + 8 = a + (b + 8)$ Associative law

$= a + (8 + b)$ Commutative law

$= (a + 8) + b$ Associative law

$(a + b) + 8 = (b + a) + 8$ Commutative law

$= b + (a + 8)$ Associative law

Other answers are possible.

23. $7 \cdot (a \cdot b) = 7 \cdot (b \cdot a)$ Commutative law

$= (7 \cdot b) \cdot a$ Associative law

$7 \cdot (a \cdot b) = (7 \cdot a) \cdot b$ Associative law

$= b \cdot (7 \cdot a)$ Commutative law

$= b \cdot (a \cdot 7)$ Commutative law

$7 \cdot (a \cdot b) = 7 \cdot (b \cdot a)$ Commutative law

$= (b \cdot a) \cdot 7$ Commutative law

Other answers are possible.

25. Substitute -2 for x and 3 for y.

$$xy + x = -2 \cdot 3 + (-2) = -6 + (-2) = -8$$

27. Substitute -2 for x, 3 for y, and -4 for z.

$$(x + y)z = (-2 + 3)(-4) = 1(-4) = -4$$

29. Substitute -2 for x, 3 for y, and -4 for z.

$$xy - xz = -2 \cdot 3 - (-2)(-4) = -6 - 8 = -14$$

31. Substitute -4 for z.

$$5z + 3 = 5(-4) + 3 = -20 + 3 = -17$$

33. Substitute 120 for P, 6% for r, and 1 for t.

$$P(1 + rt) = 120(1 + 6\% \cdot 1)$$

$$= 120(1 + 0.06)$$

$$= 120(1.06)$$

$$= 127.2$$

The value of the account is \$127.20.

35. $4(a + 1) = 4 \cdot a + 4 \cdot 1$

$$= 4a + 4$$

37. $8(x - y) = 8 \cdot x - 8 \cdot y$

$$= 8x - 8y$$

39. $-5(2a + 3b) = -5 \cdot 2a + (-5) \cdot 3b$

$$= -10a - 15b$$

41. $2a(b - c + d) = 2a \cdot b - 2a \cdot c + 2a \cdot d$

$$= 2ab - 2ac + 2ad$$

43. $2\pi r(h + 1) = 2\pi r \cdot h + 2\pi r \cdot 1$

$$= 2\pi rh + 2\pi r$$

45. $\frac{1}{2}h(a+b) = \frac{1}{2}h \cdot a + \frac{1}{2}h \cdot b$

$\qquad = \frac{1}{2}ha + \frac{1}{2}hb$

47. $4a - 5b + 6 = 4a + (-5b) + 6$

The terms are $4a$, $-5b$, and 6.

49. $2x - 3y - 2z = 2x + (-3y) + (-2z)$

The terms are $2x$, $-3y$, and $-2z$.

51. $24x + 24y = 24 \cdot x + 24 \cdot y$

$\qquad = 24(x + y)$

53. $7p - 7 = 7 \cdot p - 7 \cdot 1$

$\qquad = 7(p - 1)$

55. $7x - 21 = 7 \cdot x - 7 \cdot 3$

$\qquad = 7(x - 3)$

57. $xy + x = x \cdot y + x \cdot 1$

$\qquad = x(y + 1)$

59. $2x - 2y + 2z = 2 \cdot x - 2 \cdot y + 2 \cdot z$

$\qquad = 2(x - y + z)$

61. $3x + 6y - 3 = 3 \cdot x + 3 \cdot 2y - 3 \cdot 1$

$\qquad = 3(x + 2y - 1)$

63. $ab + ac - ad = a \cdot b + a \cdot c - a \cdot d$

$\qquad = a(b + c - d)$

65. $\frac{1}{4}\pi rr + \frac{1}{4}\pi rs = \frac{1}{4}\pi r \cdot r + \frac{1}{4}\pi r \cdot s$

$\qquad = \frac{1}{4}\pi r(r + s)$

67. $7x + 5x = (7 + 5)x$

$\qquad = 12x$

69. $8b - 11b = (8 - 11)b$

$\qquad = -3b$

71. $14y + y = 14y + 1y$

$\qquad = (14 + 1)y$

$\qquad = 15y$

73. $12a - a = 12a - 1a$

$\qquad = (12 - 1)a$

$\qquad = 11a$

75. $t - 9t = 1t - 9t$

$\qquad = (1 - 9)t$

$\qquad = -8t$

77. $5x - 3x + 8x = (5 - 3 + 8)x$

$\qquad = 10x$

79. $3x - 5y + 8x = (3 + 8)x - 5y$

$\qquad = 11x - 5y$

81. $3c + 8d - 7c + 4d = (3 - 7)c + (8 + 4)d$

$\qquad = -4c + 12d$

83. $4x - 7 + 18x + 25 = (4 + 18)x + (-7 + 25)$

$\qquad = 22x + 18$

85. $\qquad 1.3x + 1.4y - 0.11x - 0.47y$

$= (1.3 - 0.11)x + (1.4 - 0.47)y$

$= 1.19x + 0.93y$

87. $\qquad \frac{2}{3}a + \frac{5}{6}b - 27 - \frac{4}{5}a - \frac{7}{6}b$

$= \left(\frac{2}{3} - \frac{4}{5}\right)a + \left(\frac{5}{6} - \frac{7}{6}\right)b - 27$

$= \left(\frac{10}{15} - \frac{12}{15}\right)a + \left(-\frac{2}{6}\right)b - 27$

$= -\frac{2}{15}a - \frac{1}{3}b - 27$

89. $P = 2l + 2w$

$P = 2 \cdot l + 2 \cdot w$

$P = 2(l + w)$

91. $-(-2c) = -1(-2c)$

$\qquad = [-1(-2)]c$

$\qquad = 2c$

93. $-(b + 4) = -1(b + 4)$

$\qquad = -1 \cdot b + (-1)4$

$\qquad = -b - 4$

95. $-(b - 3) = -1(b - 3)$

$\qquad = -1 \cdot b - (-1) \cdot 3$

$\qquad = -b + [-(-1)3]$

$\qquad = -b + 3$, or $3 - b$

97. $-(t - y) = -1(t - y)$

$\qquad = -1 \cdot t - (-1) \cdot y$

$\qquad = -t + [-(-1)y]$

$\qquad = -t + y$, or $y - t$

99. $-(x + y + z) = -1(x + y + z)$

$\qquad = -1 \cdot x + (-1) \cdot y + (-1) \cdot z$

$\qquad = -x - y - z$

101. $\qquad -(8x - 6y + 13)$

$= -8x + 6y - 13$ Changing the sign of every term inside parentheses

103. $\qquad -(-2c + 5d - 3e + 4f)$

$= 2c - 5d + 3e - 4f$ Changing the sign of every term inside parentheses

105.

$$-\left(-1.2x + 56.7y - 34z - \frac{1}{4}\right)$$

$$= 1.2x - 56.7y + 34z + \frac{1}{4} \quad \text{Changing the sign of every term inside parentheses}$$

107. $a + (2a + 5) = a + 2a + 5$
$$= 3a + 5$$

109. $4m - (3m - 1) = 4m - 3m + 1$
$$= m + 1$$

111. $5d - 9 - (7 - 4d) = 5d - 9 - 7 + 4d$
$$= 9d - 16$$

113.

$$-2(x + 3) - 5(x - 4)$$
$$= -2(x + 3) + [-5(x - 4)]$$
$$= -2x - 6 + [-5x + 20]$$
$$= -2x - 6 - 5x + 20$$
$$= -7x + 14$$

115.

$$5x - 7(2x - 3) - 4$$
$$= 5x + [-7(2x - 3)] - 4$$
$$= 5x + [-14x + 21] - 4$$
$$= 5x - 14x + 21 - 4$$
$$= -9x + 17$$

117.

$$8x - (-3y + 7) + (9x - 11)$$
$$= 8x + 3y - 7 + 9x - 11$$
$$= 17x + 3y - 18$$

119.

$$\frac{1}{4}(24x - 8) - \frac{1}{2}(-8x + 6) - 14$$
$$= \frac{1}{4}(24x - 8) + \left[-\frac{1}{2}(-8x + 6)\right] - 14$$
$$= 6x - 2 + [4x - 3] - 14$$
$$= 6x - 2 + 4x - 3 - 14$$
$$= 10x - 19$$

121. $-12 - (-19) = -12 + 19 = 7$

123.

$$-\frac{11}{5} - \left(-\frac{17}{10}\right)$$

$$= -\frac{11}{5} + \frac{17}{10}$$

$$= -\frac{22}{10} + \frac{17}{10}$$

$$= -\frac{5}{10}$$

$$= -\frac{1}{2}$$

125. The signs are different, so the product is negative.
$$-45(20) = -900$$

127. The signs are the same, so the product is positive.
$$-45(-90) = 4050$$

129. $\frac{1}{2} \cdot 0 = 0, \ \frac{0}{2} = 0$ \quad Substituting 0

$\frac{1}{2}(-4) = -2, \ \frac{-4}{2} = -2$ \quad Substituting -4

$\frac{1}{2}(12) = 6, \ \frac{12}{2} = 6$ \quad Substituting 12

The expressions seem to be equivalent.

131. $\frac{5 \cdot 0}{9} = 0, \ \frac{5}{9 \cdot 0}$ is undefined

The expressions are not equivalent.

Exercise Set 1.5

1. $\underbrace{4 \cdot 4 \cdot 4 \cdot 4 \cdot 4}_{5 \text{ factors}} = 4^5$

3. $\underbrace{5 \cdot 5 \cdot 5 \cdot 5 \cdot 5 \cdot 5}_{6 \text{ factors}} = 5^6$

5. $\underbrace{m \cdot m \cdot m \cdot m}_{4 \text{ factors}} = m^4$

7. $\underbrace{5b \cdot 5b \cdot 5b \cdot 5b}_{4 \text{ factors}} = (5b)^4$

9. There are 2 factors of 5, 3 factors of c, and 4 factors of d.
$5 \cdot 5 \cdot c \cdot c \cdot c \cdot d \cdot d \cdot d \cdot d = 5^2 c^3 d^4$

11. $2^7 = \underbrace{2 \cdot 2 \cdot 2 \cdot 2 \cdot 2 \cdot 2 \cdot 2}_{7 \text{ factors}}$, or 128

13. $(-2)^5 = \underbrace{(-2) \cdot (-2) \cdot (-2) \cdot (-2) \cdot (-2)}_{5 \text{ factors}}$, or -32

15. $w^4 = \underbrace{w \cdot w \cdot w \cdot w}_{4 \text{ factors}}$

17. $(-4b)^3 = \underbrace{(-4b) \cdot (-4b) \cdot (-4b)}_{3 \text{ factors}}$, or $-64bbb$

19. $(ab)^4 = \underbrace{ab \cdot ab \cdot ab \cdot ab}_{4 \text{ factors}}$

21. $5^1 = 5$ \quad (For any number a, $a^1 = a$.)

23. $(3y)^0 = 1$ \quad (For any nonzero number a, $a^0 = 1$.)

25. $(\sqrt{6})^0 = 1$ \quad (For any nonzero number a, $a^0 = 1$.)

27. $\left(\frac{7}{8}\right)^1 = \frac{7}{8}$ \quad (For any number a, $a^1 = a$.)

29. $y^{-5} = \frac{1}{y^5}$

31. $\dfrac{1}{a^{-2}} = a^2$

33. $(-11)^{-1} = \dfrac{1}{(-11)^1}$

35. $\dfrac{1}{3^4} = 3^{-4}$

37. $\dfrac{1}{b^3} = b^{-3}$

39. $\dfrac{1}{(-16)^2} = (-16)^{-2}$

41.
$$\begin{aligned}[12 - 4(5-1)] &= [12 - 4(4)]\\ &= [12 - 16]\\ &= -4\end{aligned}$$

43.
$$\begin{aligned}9[8 - 7(5-2)] &= 9[8 - 7 \cdot 3]\\ &= 9[8 - 21]\\ &= 9[-13]\\ &= -117\end{aligned}$$

45.
$$\begin{aligned}[5(8-6) + 12] - [24 - (8-4)] &= [5 \cdot 2 + 12] - [24 - 4]\\ &= [10 + 12] - [24 - 4]\\ &= 22 - 20\\ &= 2\end{aligned}$$

47.
$$\begin{aligned}[64 \div (-4)] \div (-2) &= -16 \div (-2)\\ &= 8\end{aligned}$$

49.
$$\begin{aligned}19(-22) + 60 &= -418 + 60\\ &= -358\end{aligned}$$

51.
$$(5+7)^2 = 12^2 = 144$$
$$5^2 + 7^2 = 25 + 49 = 74$$

53.
$$\begin{aligned}2^3 + 2^4 - 20 \cdot 30 &= 8 + 16 - 600\\ &= 24 - 600\\ &= -576\end{aligned}$$

55.
$$\begin{aligned}5^3 + 36 \cdot 72 - (18 + 25 \cdot 4)\\ = 5^3 + 36 \cdot 72 - (18 + 100)\\ = 5^3 + 36 \cdot 72 - 118\\ = 125 + 36 \cdot 72 - 118\\ = 125 + 2592 - 118\\ = 2717 - 118\\ = 2599\end{aligned}$$

57.
$$\begin{aligned}(13 \cdot 2 - 8 \cdot 4)^2 &= (26 - 32)^2\\ &= (-6)^2\\ &= 36\end{aligned}$$

59.
$$\begin{aligned}4000 \cdot (1 + 0.12)^3 &= 4000(1.12)^3\\ &= 4000(1.404928)\\ &= 5619.712\end{aligned}$$

61.
$$\begin{aligned}(20 \cdot 4 + 13 \cdot 8)^2 - (39 \cdot 59)^3\\ = (80 + 104)^2 - (2301)^3\\ = 184^2 - 2301^3\\ \approx 33{,}856 - 12{,}182{,}876{,}900\\ (2301^3 \approx 12{,}182{,}876{,}900)\\ \approx -12{,}182{,}843{,}044\end{aligned}$$
(Answers may vary due to rounding.)

63.
$$\begin{aligned}18 - 2 \cdot 3 - 9 &= 18 - 6 - 9\\ &= 12 - 9\\ &= 3\end{aligned}$$

65.
$$\begin{aligned}(18 - 2 \cdot 3) - 9 &= (18 - 6) - 9\\ &= 12 - 9\\ &= 3\end{aligned}$$

67.
$$\begin{aligned}[24 \div (-3)] \div \left(-\dfrac{1}{2}\right)\\ = -8 \div \left(-\dfrac{1}{2}\right)\\ = -8 \cdot (-2)\\ = 16\end{aligned}$$

69.
$$\begin{aligned}15 \cdot (-24) + 50 &= -360 + 50\\ &= -310\end{aligned}$$

71.
$$\begin{aligned}4 \div (8 - 10)^2 + 1 &= 4 \div (-2)^2 + 1\\ &= 4 \div 4 + 1\\ &= 1 + 1\\ &= 2\end{aligned}$$

73.
$$\begin{aligned}6^3 + 25 \cdot 71 - (16 + 25 \cdot 4)\\ = 6^3 + 25 \cdot 71 - (16 + 100)\\ = 6^3 + 25 \cdot 71 - 116\\ = 216 + 25 \cdot 71 - 116\\ = 216 + 1775 - 116\\ = 1991 - 116\\ = 1875\end{aligned}$$

75.
$$\begin{aligned}5000 \cdot (1 + 0.16)^3 &= 5000 \cdot (1.16)^3\\ &= 5000(1.560896)\\ &= 7804.48\end{aligned}$$

77.
$$\begin{aligned}4 \cdot 5 - 2 \cdot 6 + 4 &= 20 - 12 + 4\\ &= 8 + 4\\ &= 12\end{aligned}$$

79.
$$\begin{aligned}4 \cdot (6 + 8)/(4 + 3) &= 4 \cdot 14/7\\ &= 56/7\\ &= 8\end{aligned}$$

81.
$$\begin{aligned}[2 \cdot (5 - 3)]^2 &= [2 \cdot 2]^2\\ &= 4^2\\ &= 16\end{aligned}$$

83. $8(-7) + 6(-5) = -56 - 30$
$$= -86$$

85. $19 - 5(-3) + 3 = 19 + 15 + 3$
$$= 34 + 3$$
$$= 37$$

87. $9 \div (-3) + 16 \div 8 = -3 + 2$
$$= -1$$

89. $7 + 10 - (-10 \div 2) = 7 + 10 - (-5)$
$$= 7 + 10 + 5$$
$$= 17 + 5$$
$$= 22$$

91. $5^2 - 8^2 = 25 - 64$
$$= -39$$

93. $20 + 4^3 \div (-8) = 20 + 64 \div (-8)$
$$= 20 + (-8)$$
$$= 12$$

95. $-7(3^4) + 18 = -7 \cdot 81 + 18$
$$= -567 + 18$$
$$= -549$$

97. $9[(8 - 11) - 13] = 9[-3 - 13]$
$$= 9[-16]$$
$$= -144$$

99. $256 \div (-32) \div (-4) = -8 \div (-4)$
$$= 2$$

101. $\dfrac{5^2 - |4^3 - 8|}{9^2 - 2^2 - 1^5} = \dfrac{5^2 - |64 - 8|}{81 - 4 - 1}$
$$= \dfrac{5^2 - |56|}{77 - 1}$$
$$= \dfrac{5^2 - 56}{77 - 1}$$
$$= \dfrac{25 - 56}{76}$$
$$= \dfrac{-31}{76}$$
$$= -\dfrac{31}{76}$$

103. $\dfrac{30(8 - 3) - 4(10 - 3)}{10|2 - 6| - 2(5 + 2)} = \dfrac{30 \cdot 5 - 4 \cdot 7}{10| - 4| - 2 \cdot 7}$
$$= \dfrac{150 - 28}{10 \cdot 4 - 2 \cdot 7}$$
$$= \dfrac{122}{40 - 14}$$
$$= \dfrac{122}{26}$$
$$= \dfrac{61}{13}$$

105. $7a - [9 - 3(5a - 2)] = 7a - [9 - 15a + 6]$
$$= 7a - [15 - 15a]$$
$$= 7a - 15 + 15a$$
$$= 22a - 15$$

107. $5\{-2 + 3[4 - 2(3 + 5)]\} = 5\{-2 + 3[4 - 2(8)]\}$
$$= 5\{-2 + 3[4 - 16]\}$$
$$= 5\{-2 + 3[-12]\}$$
$$= 5\{-2 - 36\}$$
$$= 5\{-38\}$$
$$= -190$$

109.
$$[10(x + 3) - 4] + [2(x - 1) + 6]$$
$$= [10x + 30 - 4] + [2x - 2 + 6]$$
$$= 10x + 26 + 2x + 4$$
$$= 12x + 30$$

111.
$$[7(x + 5) - 19] - [4(x - 6) + 10]$$
$$= [7x + 35 - 19] - [4x - 24 + 10]$$
$$= [7x + 16] - [4x - 14]$$
$$= 7x + 16 - 4x + 14$$
$$= 3x + 30$$

113.
$$3\{[7(x - 2) + 4] - [2(2x - 5) + 6]\}$$
$$= 3\{[7x - 14 + 4] - [4x - 10 + 6]\}$$
$$= 3\{[7x - 10] - [4x - 4]\}$$
$$= 3\{7x - 10 - 4x + 4\}$$
$$= 3\{3x - 6\}$$
$$= 9x - 18$$

115.
$$4\{[5(x - 3) + 2^2] - 3[2(x + 5) - 9^2]\}$$
$$= 4\{[5(x - 3) + 4] - 3[2(x + 5) - 81]\}$$
$$= 4\{[5x - 15 + 4] - 3[2x + 10 - 81]\}$$
$$= 4\{[5x - 11] - 3[2x - 71]\}$$
$$= 4\{5x - 11 - 6x + 213\}$$
$$= 4\{-x + 202\}$$
$$= -4x + 808$$

117.
$$2y + \{8[3(2y - 5) - (8y + 9)] + 6\}$$
$$= 2y + \{8[6y - 15 - 8y - 9] + 6\}$$
$$= 2y + \{8[-2y - 24] + 6\}$$
$$= 2y + \{-16y - 192 + 6\}$$
$$= 2y + \{-16y - 186\}$$
$$= 2y - 16y - 186$$
$$= -14y - 186$$

119. Substitute 78 for l and 36 for w.
$$P = 2l + 2w, \text{ or } 2(l + w)$$
$$2l + 2w = 2 \cdot 78 + 2 \cdot 36 = 156 + 72 = 228$$
$$2(l + w) = 2(78 + 36) = 2 \cdot 114 = 228$$
The perimeter is 228 ft.

$A = lw$

$A = 78 \cdot 36$, or 2808 ft^2.

The area is 2808 ft^2.

121. The distance of $-\frac{9}{7}$ from 0 is $\frac{9}{7}$, so $\left| -\frac{9}{7} \right| = \frac{9}{7}$.

123. $-4 \geq -4$ is true, because $-4 = -4$.

125. $(-x)^2 = (-x)(-x) = (-1)x(-1)x = (-1)(-1)x \cdot x = x^2$

Thus, $(-x)^2 = x^2$ is true for any real number x.

127.
$$[11(a-3) + 12a] - \{6[4(3b-7) - (9b+10)] + 11\}$$
$$= [11a - 33 + 12a] - \{6[12b - 28 - 9b - 10] + 11\}$$
$$= [23a - 33] - \{6[3b - 38] + 11\}$$
$$= 23a - 33 - \{18b - 228 + 11\}$$
$$= 23a - 33 - \{18b - 217\}$$
$$= 23a - 33 - 18b + 217$$
$$= 23a - 18b + 184$$

129.
$$z - \{2z + [3z - (4z + 5x) - 6z] + 7z\} - 8z$$
$$= z - \{2z + [3z - 4z - 5x - 6z] + 7z\} - 8z$$
$$= z - \{2z - 7z - 5x + 7z\} - 8z$$
$$= z - \{2z - 5x\} - 8z$$
$$= z - 2z + 5x - 8z$$
$$= -9z + 5x$$

131.
$$x - \{x + 1 - [x + 2 - (x - 3 - \{x + 4 -$$
$$[x - 5 + (x - 6)]\})]\}$$
$$= x - \{x + 1 - [x + 2 - (x - 3 - \{x + 4 - [2x - 11]\})]\}$$
$$= x - \{x + 1 - [x + 2 - (x - 3 - \{x + 4 - 2x + 11\})]\}$$
$$= x - \{x + 1 - [x + 2 - (x - 3 - \{-x + 15\})]\}$$
$$= x - \{x + 1 - [x + 2 - (x - 3 + x - 15)]\}$$
$$= x - \{x + 1 - [x + 2 - (2x - 18)]\}$$
$$= x - \{x + 1 - [x + 2 - 2x + 18]\}$$
$$= x - \{x + 1 - [-x + 20]\}$$
$$= x - \{x + 1 + x - 20\}$$
$$= x - \{2x - 19\}$$
$$= x - 2x + 19$$
$$= -x + 19$$

Exercise Set 1.6

1. $3^6 \cdot 3^3 = 3^{6+3} = 3^9$

3. $6^{-6} \cdot 6^2 = 6^{-6+2} = 6^{-4} = \dfrac{1}{6^4}$

5. $8^{-2} \cdot 8^{-4} = 8^{-2+(-4)} = 8^{-6} = \dfrac{1}{8^6}$

7. $b^2 \cdot b^{-5} = b^{2+(-5)} = b^{-3} = \dfrac{1}{b^3}$

9. $a^{-3} \cdot a^4 \cdot a^2 = a^{-3+4+2} = a^3$

11. $(2x)^3(3x)^2 = 8x^3 \cdot 9x^2$
$$= 8 \cdot 9 \cdot x^3 \cdot x^2$$
$$= 72x^{3+2}$$
$$= 72x^5$$

13. $(14m^2n^3)(-2m^3n^2) = 14 \cdot (-2) \cdot m^2 \cdot m^3 \cdot n^3 \cdot n^2$
$$= -28m^{2+3}n^{3+2}$$
$$= -28m^5n^5$$

15. $(-2x^{-3})(7x^{-8}) = -2 \cdot 7 \cdot x^{-3} \cdot x^{-8}$
$$= -14x^{-3+(-8)}$$
$$= -14x^{-11} = -\dfrac{14}{x^{11}}$$

17. $(15x^{4t})(7x^{-6t}) = 15 \cdot 7 \cdot x^{4t} \cdot x^{-6t}$
$$= 105x^{4t+(-6t)}$$
$$= 105x^{-2t}$$
$$= \dfrac{105}{x^{2t}}$$

19. $\dfrac{8^9}{8^2} = 8^{9-2} = 8^7$

21. $\dfrac{6^3}{6^{-2}} = 6^{3-(-2)} = 6^{3+2} = 6^5$

23. $\dfrac{10^{-3}}{10^6} = 10^{-3-6} = 10^{-3+(-6)} = 10^{-9} = \dfrac{1}{10^9}$

25. $\dfrac{9^{-4}}{9^{-6}} = 9^{-4-(-6)} = 9^{-4+6} = 9^2$

27. $\dfrac{x^{-4n}}{x^{6n}} = x^{-4n-6n} = x^{-10n} = \dfrac{1}{x^{10n}}$

29. $\dfrac{w^{-11q}}{w^{-6q}} = w^{-11q-(-6q)} = w^{-11q+6q} = w^{-5q} =$

$\dfrac{1}{w^{5q}}$

31. $\dfrac{a^3}{a^{-2}} = a^{3-(-2)} = a^{3+2} = a^5$

33. $\dfrac{9a^2}{(-3a)^2} = \dfrac{9a^2}{9a^2} = 1 \qquad [(-3a)(-3a) = 9a^2]$

35. $\dfrac{-24x^6y^7}{18x^{-3}y^9} = \dfrac{-24}{18}x^{6-(-3)}y^{7-9}$
$$= -\dfrac{24}{18}x^{6+3}y^{-2}$$
$$= -\dfrac{4}{3}x^9y^{-2} = -\dfrac{4x^9}{3y^2}$$

37. $\dfrac{-18x^{-2}y^3}{-12x^{-5}y^5} = \dfrac{-18}{-12}x^{-2-(-5)}y^{3-5}$
$$= \dfrac{18}{12}x^{-2+5}y^{-2}$$
$$= \dfrac{3}{2}x^3y^{-2} = \dfrac{3x^3}{2y^2}$$

39. $(4^3)^2 = 4^{3 \cdot 2} = 4^6$

41. $(8^4)^{-3} = 8^{4(-3)} = 8^{-12} = \dfrac{1}{8^{12}}$

43. $(6^{-4})^{-3} = 6^{-4(-3)} = 6^{12}$

45. $(5a^2b^2)^3 = 5^3(a^2)^3(b^2)^3$
$\qquad = 125a^{2 \cdot 3}b^{2 \cdot 3}$
$\qquad = 125a^6b^6$

47. $(-3x^3y^{-6})^{-2} = (-3)^{-2}(x^3)^{-2}(y^{-6})^{-2}$
$\qquad = \dfrac{1}{(-3)^2}x^{3(-2)}y^{-6(-2)}$
$\qquad = \dfrac{1}{9}x^{-6}y^{12}$
$\qquad = \dfrac{y^{12}}{9x^6}$

49. $(-6a^{-2}b^3c)^{-2} = (-6)^{-2}(a^{-2})^{-2}(b^3)^{-2}c^{-2}$
$\qquad = \dfrac{1}{(-6)^2}a^{-2(-2)}b^{3(-2)}c^{-2}$
$\qquad = \dfrac{1}{36}a^4b^{-6}c^{-2} = \dfrac{a^4}{36b^6c^2}$

51. $\left(\dfrac{4^{-3}}{3^4}\right)^3 = \dfrac{(4^{-3})^3}{(3^4)^3} = \dfrac{4^{-3 \cdot 3}}{3^{4 \cdot 3}} = \dfrac{4^{-9}}{3^{12}} = \dfrac{1}{4^9 \cdot 3^{12}}$

53. $\left(\dfrac{2x^3y^{-2}}{3y^{-3}}\right)^3 = \dfrac{(2x^3y^{-2})^3}{(3y^{-3})^3}$
$\qquad = \dfrac{2^3(x^3)^3(y^{-2})^3}{3^3(y^{-3})^3}$
$\qquad = \dfrac{8x^9y^{-6}}{27y^{-9}}$
$\qquad = \dfrac{8}{27}x^9y^{-6-(-9)}$
$\qquad = \dfrac{8}{27}x^9y^3$

55. $\left(\dfrac{125a^2b^{-3}}{5a^4b^{-2}}\right)^{-5} = \left(\dfrac{5a^4b^{-2}}{125a^2b^{-3}}\right)^5$
$\qquad = \left(\dfrac{a^{4-2}b^{-2-(-3)}}{25}\right)^5$
$\qquad = \left(\dfrac{a^2b}{25}\right)^5$
$\qquad = \dfrac{(a^2)^5(b)^5}{(25)^5} = \dfrac{a^{2 \cdot 5}b^{1 \cdot 5}}{25^{1 \cdot 5}}$
$\qquad = \dfrac{a^{10}b^5}{25^5}, \text{ or } \dfrac{a^{10}b^5}{5^{10}}$
$\qquad\qquad [25^5 = (5^2)^5 = 5^{10}]$

57. $\left(\dfrac{-6^5y^4z^{-5}}{2^{-2}y^{-2}z^3}\right)^6 = (-1 \cdot 6^5 \cdot 2^2y^{4-(-2)}z^{-5-3})^6$
$\qquad = (-1 \cdot 6^5 \cdot 2^2y^6z^{-8})^6$
$\qquad = (-1)^6(6^5)^6(2^2)^6(y^6)^6(z^{-8})^6$
$\qquad = 1 \cdot 6^{5 \cdot 6}2^{2 \cdot 6}y^{6 \cdot 6}z^{-8 \cdot 6}$
$\qquad = 6^{30}2^{12}y^{36}z^{-48}$
$\qquad = \dfrac{6^{30}2^{12}y^{36}}{z^{48}}$

59. $[(-2x^{-4}y^{-2})^{-3}]^{-2} = [(-2)^{-3}(x^{-4})^{-3}(y^{-2})^{-3}]^{-2}$
$\qquad = [(-2)^{-3}x^{12}y^6]^{-2}$
$\qquad = [(-2)^{-3}]^{-2}(x^{12})^{-2}(y^6)^{-2}$
$\qquad = (-2)^6x^{-24}y^{-12}$
$\qquad = \dfrac{64}{x^{24}y^{12}}$

61. $\left(\dfrac{3a^{-2}b}{5a^{-7}b^5}\right)^{-7} = \left(\dfrac{5a^{-7}b^5}{3a^{-2}b}\right)^7$
$\qquad = \left(\dfrac{5a^{-7-(-2)}b^{5-1}}{3}\right)^7$
$\qquad = \left(\dfrac{5a^{-5}b^4}{3}\right)^7$
$\qquad = \dfrac{5^7(a^{-5})^7(b^4)^7}{3^7}$
$\qquad = \dfrac{5^7a^{-35}b^{28}}{3^7}$
$\qquad = \dfrac{5^7b^{28}}{3^7a^{35}}$

63. $\dfrac{10^{2a+1}}{10^{a+1}} = 10^{2a+1-(a+1)} = 10^{2a+1-a-1} = 10^a$

65. $\dfrac{9a^{x-2}}{3a^{2x+2}} = \dfrac{9}{3} \cdot \dfrac{a^{x-2}}{a^{2x+2}} = 3a^{x-2-(2x+2)} =$
$3a^{x-2-2x-2} = 3a^{-x-4}$

67. $\dfrac{45x^{2a+4}y^{b+1}}{-9x^{a+3}y^{2+b}} = \dfrac{45}{-9} \cdot \dfrac{x^{2a+4}y^{b+1}}{x^{a+3}y^{2+b}} =$
$-5x^{2a+4-(a+3)}y^{b+1-(2+b)} = -5x^{2a+4-a-3}y^{b+1-2-b} =$
$-5x^{a+1}y^{-1}$

69. $(8^x)^{4y} = 8^{x \cdot 4y} = 8^{4xy}$

71. $(12^{3-a})^{2b} = 12^{(3-a)(2b)} = 12^{6b-2ab}$

73. $(5x^{a-1}y^{b+1})^{2c} = 5^{2c}x^{(a-1)(2c)}y^{(b+1)(2c)} =$
$5^{2c}x^{2ac-2c}y^{2bc+2c}, \text{ or } (5^2)^cx^{2ac-2c}y^{2bc+2c} =$
$25^cx^{2ac-2c}y^{2bc+2c}$

75. $\dfrac{4x^{2a+3}y^{2b-1}}{2x^{a+1}y^{b+1}} = \dfrac{4}{2} \cdot \dfrac{x^{2a+3}y^{2b-1}}{x^{a+1}y^{b+1}} =$
$2x^{2a+3-(a+1)}y^{2b-1-(b+1)} = 2x^{2a+3-a-1}y^{2b-1-b-1} =$
$2x^{a+2}y^{b-2}$

77. $4.\underbrace{7,000,000,000.}_{\text{10 places}}$

Large number, so the exponent is positive.

$47,000,000,000 = 4.7 \times 10^{10}$

79. $2.\underbrace{50,000,000.}_{\text{8 places}}$

Large number, so the exponent is positive.

$250,000,000 = 2.5 \times 10^8$

81. $0.\underbrace{00000001}_{\text{8 places}}.6$

Small number, so the exponent is negative.

$0.000000016 = 1.6 \times 10^{-8}$

83. $0.\underbrace{00000000007}_{\text{11 places}}.$

Small number, so the exponent is negative.

$0.00000000007 = 7 \times 10^{-11}$

85. $6.\underbrace{73000000}_{\text{8 places}}.$

Positive exponent so the number is large.

$6.73 \times 10^8 = 673,000,000$

87. $0.\underbrace{00006}_{\text{5 places}}.6 \text{ cm}$

Negative exponent, so the number is small.

$6.6 \times 10^{-5} \text{ cm} = 0.000066 \text{ cm}$

89. $1.\underbrace{000000000}_{\text{9 places}}.$

Positive exponent, so the number is large.

$1 \times 10^9 = 1,000,000,000$

91. $0.\underbrace{0000000008}_{\text{10 places}}.923$

Negative exponent, so the number is small.

$8.923 \times 10^{-10} = 0.0000000008923$

93. $(2.3 \times 10^6)(4.2 \times 10^{-11})$

$= (2.3 \times 4.2)(10^6 \times 10^{-11})$

$= 9.66 \times 10^{-5}$

95. $(2.34 \times 10^{-8})(5.7 \times 10^{-4})$

$= (2.34 \times 5.7)(10^{-8} \times 10^{-4})$

$= 13.338 \times 10^{-12}$

$= (1.3338 \times 10^1) \times 10^{-12}$

$= 1.3338 \times (10^1 \times 10^{-12})$

$= 1.3338 \times 10^{-11}$

97. $\dfrac{8.5 \times 10^8}{3.4 \times 10^5} = \dfrac{8.5}{3.4} \times \dfrac{10^8}{10^5}$

$\qquad = 2.5 \times 10^3$

99. $\dfrac{4.0 \times 10^{-6}}{8.0 \times 10^{-3}} = \dfrac{4.0}{8.0} \times \dfrac{10^{-6}}{10^{-3}}$

$\qquad = 0.5 \times 10^{-3}$

$\qquad = (5 \times 10^{-1}) \times 10^{-3}$

$\qquad = 5 \times (10^{-1} \times 10^{-3})$

$\qquad = 5 \times 10^{-4}$

101. 2000 yr

$= 2000 \text{ yr} \times \dfrac{365 \text{ days}}{1 \text{ yr}} \times \dfrac{24 \text{ hr}}{1 \text{ day}} \times \dfrac{60 \text{ min}}{1 \text{ hr}} \times \dfrac{60 \text{ sec}}{1 \text{ min}}$

$= 63,072,000,000 \text{ sec}$

$= 6.3072 \times 10^{10} \text{ sec}$

103. $13 \text{ wk} = 13 \text{ wk} \times \dfrac{1 \text{ yr}}{52 \text{ wk}} = \dfrac{13}{52} \text{ yr} = 0.25 \text{ yr}$

Now light travels 5.87×10^{14} mi in 100 yr, or

$\dfrac{5.87 \times 10^{14} \text{ mi}}{100 \text{ yr}}$, so the distance traveled in 13 weeks, or

0.25 yr is given by:

$\qquad \dfrac{5.87 \times 10^{14} \text{ mi}}{100 \text{ yr}} \times 0.25 \text{ yr}$

$= \dfrac{5.87 \times 10^{14} \text{ mi}}{10^2 \text{ yr}} \times 2.5 \times 10^{-1} \text{ yr}$

$= (5.87 \times 2.5) \times \left(\dfrac{10^{14} \times 10^{-1}}{10^2}\right) \text{ mi}$

$= 14.675 \times 10^{11} \text{ mi}$

$= (1.4675 \times 10) \times 10^{11} \text{ mi}$

$= 1.4675 \times 10^{12} \text{ mi}$

105. $3,000,000 \times 365$

$= (3 \times 10^6) \times (3.65 \times 10^2)$

$= (3 \times 3.65) \times (10^6 \times 10^2)$

$= 10.95 \times 10^8$

$= (1.095 \times 10) \times 10^8$

$= 1.095 \times 10^9$

1.095×10^9 gal of orange juice is consumed in this country in one year.

107.
$$\frac{10,000}{300,000} = \frac{10^4}{3 \times 10^5}$$
$$\approx 0.333 \times 10^{-1}$$
$$\approx (3.33 \times 10^{-1}) \times 10^{-1}$$
$$\approx 3.33 \times 10^{-2}$$

The part of the total number of words in the English language that an average person knows is about 3.33×10^{-2}.

109.
$$9x - (-4y + 8) + (10x - 12)$$
$$= 9x + 4y - 8 + 10x - 12$$
$$= 19x + 4y - 20$$

111.
$$4^2 + 30 \cdot 10 - 7^3 + 16 = 16 + 30 \cdot 10 - 343 + 16$$
$$= 16 + 300 - 343 + 16$$
$$= 316 - 343 + 16$$
$$= -27 + 16$$
$$= -11$$

113.
$$20 - 5 \cdot 4 - 8 = 20 - 20 - 8$$
$$= 0 - 8$$
$$= -8$$

115.
$$\frac{(2^{-2})^{-4} \times (2^3)^{-2}}{(2^{-2})^2 \cdot (2^5)^{-3}} = \frac{2^8 \times 2^{-6}}{2^{-4} \cdot 2^{-15}}$$
$$= \frac{2^{8+(-6)}}{2^{-4+(-15)}}$$
$$= \frac{2^2}{2^{-19}}$$
$$= 2^{2-(-19)}$$
$$= 2^{21}$$

117.
$$\left[\left(\frac{a^{-2}}{b^7} \right)^{-3} \cdot \left(\frac{a^4}{b^{-3}} \right)^2 \right]^{-1} = \left[\frac{(a^{-2})^{-3}}{(b^7)^{-3}} \cdot \frac{(a^4)^2}{(b^{-3})^2} \right]^{-1}$$
$$= \left[\frac{a^6}{b^{-21}} \cdot \frac{a^8}{b^{-6}} \right]^{-1}$$
$$= \left[\frac{a^{6+8}}{b^{-21+(-6)}} \right]^{-1}$$
$$= \left[\frac{a^{14}}{b^{-27}} \right]^{-1}$$
$$= \frac{a^{-14}}{b^{27}}$$
$$= \frac{1}{a^{14}b^{27}}$$

119.
$$\left[\frac{(2x^a y^b)^3}{(-2x^a y^b)^2} \right]^2 = \left[\frac{(2x^a y^b)^3}{(2x^a y^b)^2} \right]^2 \quad [(-2x^a y^b)^2 = (2x^a y^b)^2]$$
$$= [(2x^a y^b)^{3-2}]^2$$
$$= (2x^a y^b)^2$$
$$= 2^2 (x^a)^2 (y^b)^2$$
$$= 4x^{2a} y^{2b}$$

Chapter 2

Solving Equations and Inequalities

Exercise Set 2.1

1.

$\quad x + 23 = 40 \quad$ Writing the equation

$\dfrac{17 + 23 \quad | \quad 40}{\qquad 40 \quad | \qquad}$ Substituting 17 for x

TRUE

Since the left-hand and the right-hand sides are the same, 17 is a solution of the equation.

3.

$\quad 2x - 3 = -18 \quad$ Writing the equation

$\dfrac{2(-8) - 3 \quad | \quad -18}{}$ Substituting -8 for x

$-16 - 3$

$\qquad -19 \quad | \qquad$ FALSE

Since the left-hand and the right-hand sides are not the same, -8 is not a solution of the equation.

5.

$\quad \dfrac{-x}{9} = -2 \quad$ Writing the equation

$\dfrac{-45}{9} \quad | \quad -2 \quad$ Substituting

$\qquad -5 \quad | \qquad$ FALSE

Since the left-hand and the right-hand sides are not the same, 45 is not a solution of the equation.

7.

$\quad 2 - 3x = 21$

$\dfrac{2 - 3 \cdot 10 \quad | \quad 21}{}$ Substituting

$2 - 30$

$\qquad -28 \quad | \qquad$ FALSE

Since the left-hand and the right-hand sides are not the same, 10 is not a solution of the equation.

9.

$\quad 5x + 7 = 102$

$\dfrac{5 \cdot 19 + 7 \quad | \quad 102}{}$

$95 + 7$

$\qquad 102 \quad | \qquad$ TRUE

Since the left-hand and the right-hand sides are the same, 19 is a solution of the equation.

11.

$\quad 7(y - 1) = 84$

$\dfrac{7(-11 - 1) \quad | \quad 84}{}$

$7(-12)$

$\qquad -84 \quad | \qquad$ FALSE

Since the left-hand and the right-hand sides are not the same, -11 is not a solution of the equation.

13.

$\quad y + 6 = 13$

$\quad y + 6 - 6 = 13 - 6 \quad$ Subtracting 6 on both sides

$\quad y + 0 = 7 \qquad$ Simplifying

$\quad y = 7 \qquad$ Using the identity property of 0

Check: $\dfrac{y + 6 = 13}{}$

$\dfrac{7 + 6 \quad | \quad 13}{13 \quad | \quad}$ TRUE

The solution is 7.

15.

$\quad -20 = x - 12$

$\quad -20 + 12 = x - 12 + 12 \quad$ Adding 12 on both sides

$\quad -8 = x + 0 \qquad$ Simplifying

$\quad -8 = x \qquad$ Using the identity property of 0

Check: $\dfrac{-20 = x - 12}{}$

$\dfrac{-20 \quad | \quad -8 - 12}{\quad | \quad -20}$ TRUE

The solution is -8.

17.

$\quad -8 + x = 19$

$\quad 8 - 8 + x = 8 + 19 \qquad$ Adding 8

$\quad 0 + x = 27$

$\quad x = 27 \qquad$ Using the identity property of 0

Check: $\dfrac{-8 + x = 19}{}$

$\dfrac{-8 + 27 \quad | \quad 19}{19 \quad | \quad}$ TRUE

The solution is 27.

19.

$\quad -12 + z = -51$

$\quad 12 + (-12) + z = 12 + (-51) \qquad$ Adding 12

$\quad 0 + z = -39$

$\quad z = -39$

The number -39 checks, so it is the solution.

21.

$\quad p - 2.96 = 83.9$

$\quad p - 2.96 + 2.96 = 83.9 + 2.96$

$\quad p + 0 = 86.86$

$\quad p = 86.86$

The number 86.86 checks, so it is the solution.

23.
$$-\frac{3}{8} + x = -\frac{5}{24}$$

$$\frac{3}{8} + \left(-\frac{3}{8}\right) + x = \frac{3}{8} + \left(-\frac{5}{24}\right)$$

$$0 + x = \frac{3}{8} \cdot \frac{3}{3} + \left(-\frac{5}{24}\right)$$

$$x = \frac{9}{24} + \left(-\frac{5}{24}\right)$$

$$x = \frac{4}{24}$$

$$x = \frac{1}{6}$$

The number $\frac{1}{6}$ checks, so it is the solution.

25. $3x = 18$

$$\frac{3x}{3} = \frac{18}{3} \quad \text{Dividing by 3 on both sides}$$

$$1 \cdot x = \frac{18}{3} \quad \text{Simplifying}$$

$$x = 6$$

Check:
$$\begin{array}{c|c} \multicolumn{2}{c}{3x = 18} \\ \hline 3 \cdot 6 & 18 \\ 18 & \\ \end{array} \quad \text{TRUE}$$

The solution is 6.

27. $-11y = 44$

$$\frac{-11y}{-11} = \frac{44}{-11}$$

$$1 \cdot y = \frac{44}{-11}$$

$$y = -4$$

Check:
$$\begin{array}{c|c} \multicolumn{2}{c}{-11y = 44} \\ \hline -11(-4) & 44 \\ 44 & \\ \end{array} \quad \text{TRUE}$$

The solution is -4.

29.
$$\frac{-x}{7} = 21$$

$$\frac{1}{7}(-x) = 21$$

$$7 \cdot \frac{1}{7}(-x) = 7 \cdot 21 \quad \text{Multiplying by 7 on both sides}$$

$$-x = 147$$

$$-1 \cdot (-x) = -1 \cdot 147 \quad \text{Multiplying by } -1$$

$$x = -147$$

Check:
$$\begin{array}{c|c} \multicolumn{2}{c}{\frac{-x}{7} = 21} \\ \hline \frac{-(-147)}{7} & 21 \\ \frac{147}{7} & \\ 21 & \\ \end{array} \quad \text{TRUE}$$

The solution is -147.

31. $-96 = -3z$

$$\frac{-96}{-3} = \frac{-3z}{-3} \quad \text{Dividing by } -3$$

$$\frac{-96}{-3} = 1 \cdot z$$

$$32 = z$$

Check:
$$\begin{array}{c|c} \multicolumn{2}{c}{-96 = -3z} \\ \hline -96 & -3 \cdot 32 \\ & -96 \\ \end{array} \quad \text{TRUE}$$

The solution is 32.

33. $4.8y = -28.8$

$$\frac{4.8y}{4.8} = \frac{-28.8}{4.8}$$

$$1 \cdot y = -\frac{28.8}{4.8}$$

$$y = -6$$

The number -6 checks, so it is the solution.

35.
$$\frac{3}{2}t = -\frac{1}{4}$$

$$\frac{2}{3} \cdot \frac{3}{2}t = \frac{2}{3} \cdot \left(-\frac{1}{4}\right)$$

$$1 \cdot t = -\frac{2}{12}$$

$$t = -\frac{1}{6}$$

The number $-\frac{1}{6}$ checks, so it is the solution.

37. $6x - 15 = 45$

$$6x - 15 + 15 = 45 + 15 \quad \text{Adding 15}$$

$$6x = 60$$

$$\frac{6x}{6} = \frac{60}{6} \quad \text{Dividing by 6}$$

$$x = 10$$

Check:
$$\begin{array}{c|c} \multicolumn{2}{c}{6x - 15 = 45} \\ \hline 6 \cdot 10 - 15 & 45 \\ 60 - 15 & \\ 45 & \\ \end{array} \quad \text{TRUE}$$

The solution is 10.

39.
$$5x - 10 = 45$$
$$5x - 10 + 10 = 45 + 10$$
$$5x = 55$$
$$\frac{5x}{5} = \frac{55}{5}$$
$$x = 11$$

Check:

$5x - 10 = 45$	
$5 \cdot 11 - 10$	45
$55 - 10$	
45	TRUE

The solution is 11.

41.
$$9t + 4 = -104$$
$$9t + 4 - 4 = -104 - 4$$
$$9t = -108$$
$$\frac{9t}{9} = \frac{-108}{9}$$
$$t = -12$$

Check:

$9t + 4 = -104$	
$9(-12) + 4$	-104
$-108 + 4$	
-104	TRUE

The solution is -12.

43. $-\dfrac{7}{3}x + \dfrac{2}{3} = -18$, LCM is 3

$$3\left(-\frac{7}{3}x + \frac{2}{3}\right) = 3(-18) \quad \text{Multiplying by 3 to clear fractions}$$
$$-7x + 2 = -54$$
$$-7x = -56 \quad \text{Subtracting 2}$$
$$x = \frac{-56}{-7} \quad \text{Dividing by } -7$$
$$x = 8$$

The number 8 checks. It is the solution.

45. $\dfrac{6}{5}x + \dfrac{4}{10}x = \dfrac{32}{10}$, LCM is 10

$$10\left(\frac{6}{5}x + \frac{4}{10}x\right) = 10 \cdot \frac{32}{10} \quad \text{Multiplying by 10 to clear fractions}$$
$$12x + 4x = 32$$
$$16x = 32 \quad \text{Collecting like terms}$$
$$x = \frac{32}{16} \quad \text{Dividing by 16}$$
$$x = 2$$

The number 2 checks. It is the solution.

47.
$$0.9y - 0.7y = 4.2$$
$$10(0.9y - 0.7y) = 10(4.2) \quad \text{Multiplying by 10 to clear fractions}$$
$$9y - 7y = 42$$
$$2y = 42 \quad \text{Collecting like terms}$$
$$y = \frac{42}{2}$$
$$y = 21$$

The number 21 checks, so it is the solution.

49.
$$8x + 48 = 3x - 12$$
$$5x + 48 = -12 \quad \text{Subtracting } 3x$$
$$5x = -60 \quad \text{Subtracting 48}$$
$$x = \frac{-60}{5} \quad \text{Dividing by 5}$$
$$x = -12$$

The number -12 checks, so it is the solution.

51.
$$7y - 1 = 23 - 5y$$
$$12y - 1 = 23 \quad \text{Adding } 5y$$
$$12y = 24 \quad \text{Adding 1}$$
$$y = \frac{24}{12} \quad \text{Dividing by 12}$$
$$y = 2$$

The number 2 checks, so it is the solution.

53.
$$3x - 4 = 5 + 12x$$
$$-4 = 5 + 9x \quad \text{Subtracting } 3x$$
$$-9 = 9x \quad \text{Subtracting 5}$$
$$-1 = x \quad \text{Dividing by 9}$$

The number -1 checks, so it is the solution.

55.
$$5 - 4a = a - 13$$
$$5 = 5a - 13 \quad \text{Adding } 4a$$
$$18 = 5a \quad \text{Adding 13}$$
$$\frac{18}{5} = a \quad \text{Dividing by 5}$$

The number $\dfrac{18}{5}$ checks. It is the solution.

57.
$$3m - 7 = -7 - 4m - m$$
$$3m - 7 = -7 - 5m \quad \text{Collecting like terms}$$
$$3m = -5m \quad \text{Adding 7}$$
$$8m = 0 \quad \text{Adding } 5m$$
$$m = \frac{0}{8} \quad \text{Dividing by 8}$$
$$m = 0$$

The number 0 checks, so it is the solution.

59. $5x + 3 = 11 - 4x + x$

$5x + 3 = 11 - 3x$ Collecting like terms

$8x + 3 = 11$ Adding $3x$

$8x = 8$ Subtracting 3

$x = \dfrac{8}{8}$ Dividing by 8

$x = 1$

The number 1 checks, so it is the solution.

61. $-7 + 9x = 9x - 7$

$-7 = -7$ Subtracting $9x$

The equation $-7 = -7$ is true. Replacing x by any real number gives a true sentence. Thus, any real number is a solution.

63. $6y - 8 = 9 + 6y$

$-8 = 9$ Subtracting $6y$

The equation $-8 = 9$ is false. No matter what number we try for x we get a false sentence. Thus, the equation has no solution.

65. $2(x + 7) = 4x$

$2x + 14 = 4x$ Multiplying to remove parentheses

$14 = 2x$ Subtracting $2x$

$7 = x$ Dividing by 2

Check:
$$\begin{array}{c|c} 2(x+7) = 4x & \\ \hline 2(7+7) & 4 \cdot 7 \\ 2 \cdot 14 & 28 \\ 28 & \text{TRUE} \end{array}$$

The solution is 7.

67. $80 = 10(3t + 2)$

$80 = 30t + 20$

$60 = 30t$

$2 = t$

Check:
$$\begin{array}{c|c} 80 = 10(3t+2) & \\ \hline 80 & 10(3 \cdot 2 + 2) \\ & 10(6+2) \\ & 10 \cdot 8 \\ & 80 \qquad \text{TRUE} \end{array}$$

The solution is 2.

69. $180(n - 2) = 900$

$180n - 360 = 900$

$180n = 1260$

$n = 7$

Check:
$$\begin{array}{c|c} 180(n-2) = 900 & \\ \hline 180(7-2) & 900 \\ 180 \cdot 5 & \\ 900 & \text{TRUE} \end{array}$$

The solution is 7.

71. $5y - (2y - 10) = 25$

$5y - 2y + 10 = 25$

$3y + 10 = 25$

$3y = 15$

$y = 5$

Check:
$$\begin{array}{c|c} 5y - (2y-10) = 25 & \\ \hline 5 \cdot 5 - (2 \cdot 5 - 10) & 25 \\ 25 - (10 - 10) & \\ 25 - 0 & \\ 25 & \text{TRUE} \end{array}$$

The solution is 5.

73. $7(3x + 6) = 11 - (x + 2)$

$21x + 42 = 11 - x - 2$

$21x + 42 = 9 - x$

$22x + 42 = 9$

$22x = -33$

$x = \dfrac{-33}{22}$

$x = -\dfrac{3}{2}$

The number $-\dfrac{3}{2}$ checks, so it is the solution.

75. $\dfrac{1}{8}(16y + 8) - 17 = -\dfrac{1}{4}(8y - 16)$

$2y + 1 - 17 = -2y + 4$

$2y - 16 = -2y + 4$

$4y - 16 = 4$

$4y = 20$

$y = 5$

The number 5 checks, so it is the solution.

77. $3[5 - 3(4 - t)] - 2 = 5[3(5t - 4) + 8] - 26$

$3[5 - 12 + 3t] - 2 = 5[15t - 12 + 8] - 26$

$3[-7 + 3t] - 2 = 5[15t - 4] - 26$

$-21 + 9t - 2 = 75t - 20 - 26$

$9t - 23 = 75t - 46$

$-23 = 66t - 46$

$23 = 66t$

$\dfrac{23}{66} = t$

The number $\dfrac{23}{66}$ checks, so it is the solution.

79. $\dfrac{2}{3}\left(\dfrac{7}{8} + 4x\right) - \dfrac{5}{8} = \dfrac{3}{8}$

$$\dfrac{2}{3}\left(\dfrac{7}{8} + 4x\right) = 1$$

$$\dfrac{7}{12} + \dfrac{8}{3}x = 1$$

$$12\left(\dfrac{7}{12} + \dfrac{8}{3}x\right) = 12 \cdot 1$$

$$7 + 32x = 12$$

$$32x = 5$$

$$x = \dfrac{5}{32}$$

The number $\dfrac{5}{32}$ checks, so it is the solution.

81. $(6x^5 y^{-4})(-3x^{-3}y^{-7})$

$\quad = 6 \cdot (-3) \cdot x^5 \cdot x^{-3} \cdot y^{-4} \cdot y^{-7}$

$\quad = -18x^{5+(-3)}y^{-4+(-7)}$

$\quad = -18x^2 y^{-11}$

$\quad = -\dfrac{18x^2}{y^{11}}$

83. $-4(3x - 2y + z) = -4 \cdot 3x - 4(-2y) - 4 \cdot z$

$\qquad\qquad\qquad\quad = -12x + 8y - 4z$

85. $4x - 10y + 2 = 2 \cdot 2x - 2 \cdot 5y + 2 \cdot 1$

$\qquad\qquad\qquad = 2(2x - 5y + 1)$

87. $\dfrac{3x}{2} + \dfrac{5x}{3} - \dfrac{13x}{6} - \dfrac{2}{3} = \dfrac{5}{6}$,   LCM is 6

$$6\left(\dfrac{3x}{2} + \dfrac{5x}{3} - \dfrac{13x}{6} - \dfrac{2}{3}\right) = 6 \cdot \dfrac{5}{6}$$

        Multiplying by 6 to clear fractions

$$3 \cdot 3x + 2 \cdot 5x - 1 \cdot 13x - 2 \cdot 2 = 1 \cdot 5$$

$$9x + 10x - 13x - 4 = 5$$

$$6x - 4 = 5 \quad \text{Collecting like terms}$$

$$6x = 9 \quad \text{Adding 4}$$

$$x = \dfrac{9}{6} \quad \text{Dividing by 6}$$

$$x = \dfrac{3}{2}$$

The number $\dfrac{3}{2}$ checks, so it is the solution.

89. $2x - 4 - (x + 1) - 3(x - 2) = 6(2x - 3) - 3(6x - 1) - 8$

$\quad 2x - 4 - x - 1 - 3x + 6 = 12x - 18 - 18x + 3 - 8$

$\qquad\qquad\quad -2x + 1 = -6x - 23$

$\qquad\qquad\qquad 4x + 1 = -23$

$\qquad\qquad\qquad\quad 4x = -24$

$\qquad\qquad\qquad\quad\ x = -6$

The number -6 checks, so it is the solution.

Exercise Set 2.2

1. Familiarize. We let $x =$ the length of one piece of cable and $x + 2 =$ the length of the other piece. Then we draw a picture.

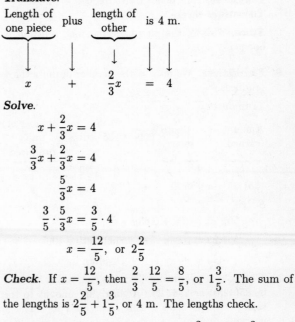

Translate.

| Length of one piece | plus | length of other | is 30. |

$$x + (x + 2) = 30$$

Solve.

$$x + (x + 2) = 30$$

$$2x + 2 = 30 \qquad \text{Collecting like terms}$$

$$2x = 28 \qquad \text{Subtracting 2}$$

$$x = 14 \qquad \text{Dividing by 2}$$

Check. If the length of one piece of cable is 14 ft long, then the length of the other is $14 + 2$, or 16 ft. The lengths of the pieces add up to 30 ft, so this checks.

State. One piece of cable is 14 ft long, and the other is 16 ft long.

3. Familiarize. We let $x =$ the length of the longer piece and $\dfrac{2}{3}x =$ the length of the shorter piece. Then we make a drawing.

Translate.

| Length of one piece | plus | length of other | is 4 m. |

$$x + \dfrac{2}{3}x = 4$$

Solve.

$$x + \dfrac{2}{3}x = 4$$

$$\dfrac{3}{3}x + \dfrac{2}{3}x = 4$$

$$\dfrac{5}{3}x = 4$$

$$\dfrac{3}{5} \cdot \dfrac{5}{3}x = \dfrac{3}{5} \cdot 4$$

$$x = \dfrac{12}{5}, \text{ or } 2\dfrac{2}{5}$$

Check. If $x = \dfrac{12}{5}$, then $\dfrac{2}{3} \cdot \dfrac{12}{5} = \dfrac{8}{5}$, or $1\dfrac{3}{5}$. The sum of the lengths is $2\dfrac{2}{5} + 1\dfrac{3}{5}$, or 4 m. The lengths check.

State. The lengths of the pieces are $2\dfrac{2}{5}$ m and $1\dfrac{3}{5}$ m.

5. *Familiarize.* Let x = the Burger King sales. Then $x + 475,000$ = the McDonalds sales.

Translate.

$$\underbrace{\text{Burger King sales}}_{x} \quad \underset{+}{\text{plus}} \quad \underbrace{\text{McDonalds sales}}_{x + 475,000} \quad \underset{=}{\text{are}} \quad \underset{2,525,000}{\$2,525,000.}$$

Solve.

$$x + x + 475,000 = 2,525,000$$
$$2x + 475,000 = 2,525,000$$
$$2x = 2,050,000$$
$$x = 1,025,000$$

Check. If $x = \$1,025,000$, then $x + \$475,000 =$ $\$1,500,000$. The total of these amounts is $\$2,525,000$. The answers check.

State. The average Burger King sales were $\$1,025,000$, and the average McDonalds sales were $\$1,500,000$.

7. *Familiarize.* We are asking "What percent of 18,000 is 14,040?" To translate, we let y = the number.

Translate.

$$\underbrace{\text{What percent}}_{y\%} \quad \underset{\times}{\text{of}} \quad \underset{18,000}{18,000} \quad \underset{=}{\text{is}} \quad \underset{14,040}{14,040?}$$

Solve.

$$y \times 0.01 \times 18,000 = 14,040 \quad \text{Substituting "} \times 0.01\text{" for "}\%\text{"}$$
$$180y = 14,040$$
$$\frac{180y}{180} = \frac{14,040}{180} \quad \text{Dividing by 180}$$
$$y = 78$$

Check. $78\% \cdot 18,000 = 0.78 \times 18,000 = 14,040$ The answer checks.

State. 78% of the pieces of junk mail sent out will be opened.

9. *Familiarize.* We let y = the number of units the worker produces.

Translate.

$$\underbrace{\text{Amount earned}}_{551.25} \underset{=}{\text{is}} \underbrace{\text{base salary}}_{190} \underset{+}{\text{plus}} \underset{0.85}{\$0.85} \underset{\cdot}{\text{times}} \underbrace{\text{number of units produced}}_{y}$$

Solve.

$$551.25 = 190 + 0.85y$$
$$361.25 = 0.85y \quad \text{Subtracting 190}$$
$$425 = y \quad \text{Dividing by 0.85}$$

Check. For producing 425 units the worker earns $\$0.85(425)$, or $\$361.25$. Adding this amount to the base salary, we get $\$190 + \$361.25$ or $\$551.25$. The answer checks.

State. 425 units were produced.

11. *Familiarize.* Let m = the number of minutes the call lasted. Then $m - 1$ minutes are charged at the 26¢ rate.

Translate.

$$\underbrace{\text{Cost of first minute}}_{0.30} \underset{+}{\text{plus}} \underset{0.26}{\$0.26} \underset{\cdot}{\text{times}} \underbrace{\text{number of additional minutes}}_{(m-1)} \underset{=}{\text{is}} \underset{11.74}{\$11.74.}$$

Solve.

$$0.30 + 0.26(m - 1) = 11.74$$
$$30 + 26(m - 1) = 1174 \quad \text{Multiplying by 100}$$
$$30 + 26m - 26 = 1174$$
$$26m + 4 = 1174$$
$$26m = 1170$$
$$m = 45$$

Check. If the call lasted 45 minutes, the charge was $\$0.30 + \$0.26(44) = \$0.30 + \$11.44 = \$11.74$. The answer checks.

State. The call was 45 minutes long.

13. *Familiarize.* Let x = the number.

Translate.

$$\underset{18 + 5}{18 + 5 \text{ times a number}} \quad \underset{\cdot}{} \quad \underset{x}{\text{a number}} \quad \underset{= 7}{\text{is } 7 \text{ times}} \quad \underset{\cdot}{} \quad \underset{x}{\text{the number.}}$$

Solve.

$$18 + 5x = 7x$$
$$18 = 2x$$
$$9 = x$$

Check. Five times 9 is 45. If we add 18, we get 63. This is $7 \cdot 9$. The answer checks.

State. The number is 9.

15. *Familiarize.* Let n = the number.

Translate.

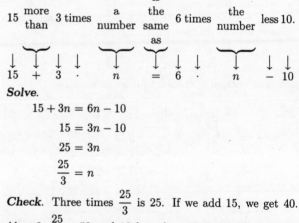

Solve.

$$15 + 3n = 6n - 10$$
$$15 = 3n - 10$$
$$25 = 3n$$
$$\frac{25}{3} = n$$

Check. Three times $\frac{25}{3}$ is 25. If we add 15, we get 40. Also, $6 \cdot \frac{25}{3} = 50$ and 10 less than 50 is 40. The answer checks.

State. The number is $\frac{25}{3}$.

17. *Familiarize*. We let q = the former price.

Translate.

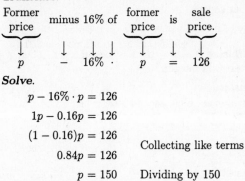

Former price minus 16% of former price is sale price.

$$p - 16\% \cdot p = 126$$

Solve.

$$p - 16\% \cdot p = 126$$
$$1p - 0.16p = 126$$
$$(1 - 0.16)p = 126 \qquad \text{Collecting like terms}$$
$$0.84p = 126$$
$$p = 150 \qquad \text{Dividing by 150}$$

Check. 16% of $150 is $24. Subtracting this price decrease from $150 (the former price), we get $126 (the sale price). This checks.

State. The former price was $150.

19. *Familiarize*. We let c = the original cost of the CD player.

Translate.

Original cost plus 12% of original cost is $504.

$$c + 12\% \cdot c = 504$$

Solve.

$$c + 12\% \cdot c = 504$$
$$1c + 0.12c = 504$$
$$(1 + 0.12)c = 504 \qquad \text{Collecting like terms}$$
$$1.12c = 504$$
$$c = 450$$

Check. 12% of $450 is $54. Adding this to $450 (the original cost), we get $504 (the amount that pays off the purchase). This checks.

State. The original cost of the CD player was $450.

21. *Familiarize*. We draw a picture. We let x = the measure of the first angle. The second angle is three times the first, so its measure is $3x$. The third angle is 12° less than twice the first, so its measure is $2x - 12$.

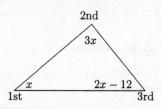

2nd

$3x$

x $2x - 12$

1st   3rd

The measures of the angles of a triangle add up to 180°.

Translate.

Measure of 1st angle + Measure of 2nd angle + Measure of 3rd angle is 180°

$$x + 3x + (2x - 12) = 180$$

Solve.

$$x + 3x + 2x - 12 = 180$$
$$6x - 12 = 180$$
$$6x = 192$$
$$x = 32$$

The angles will then have measures as follows:

1st angle: $x = 32°$

2nd angle: $3x = 3 \cdot 32$, or 96°

3rd angle: $2x - 12 = 2 \cdot 32 - 12$, or 52°

Check. The angle measures add up to 180° so they give the answer to the problem.

State. The angles are 32°, 96°, and 52°.

23. *Familiarize*. We let w = the width. Then $2w$ = the length. We draw a picture.

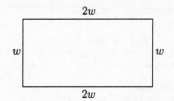

$2w$

w w

$2w$

Translate. Recall that the formula for the perimeter of a rectangle is $P = 2l + 2w$. We substitute $2w$ for l and 180 for P as follows:

$$2l + 2w = P$$
$$2 \cdot 2w + 2w = 180$$

Solve.

$$2 \cdot 2w + 2w = 180$$
$$4w + 2w = 180$$
$$6w = 180$$
$$w = 30$$

Then $w = 30$ (the width) and $2 \cdot w = 2 \cdot 30 = 60$ (the length).

Check. The length is twice the width. The perimeter is $2 \cdot 60 + 2 \cdot 30$, or 180. This checks.

State. The dimensions of the court are 60 ft and 30 ft.

25. *Familiarize*. We let x = the first integer. Then the next two are $x + 1$ and $x + 1 + 1$, or $x + 2$.

Translate.

First integer plus twice the second integer plus three times the third is 80.

$$x + 2(x + 1) + 3(x + 2) = 80$$

Solve.

$$x + 2(x + 1) + 3(x + 2) = 80$$
$$x + 2x + 2 + 3x + 6 = 80$$
$$6x + 8 = 80$$
$$6x = 72$$
$$x = 12$$

Check. If $x = 12$, then $x + 1 = 13$ and $x + 2 = 14$. The sum of 12 plus $2 \cdot 13$ plus $3 \cdot 14$ is $12 + 26 + 42$, or 80. The numbers check.

State. The consecutive integers are 12, 13, and 14.

27. Familiarize. We let $s =$ the old salary. Then $5\%s =$ the raise.

Translate.

Old salary plus raise is new salary.

$$s + 5\%s = 8610$$

Solve.

$$s + 5\%s = 8610$$
$$1s + 0.05s = 8610$$
$$(1 + 0.05)s = 8610$$
$$1.05s = 8610$$
$$s = 8200$$

Check. 5% of \$8200 is \$410 (the raise). Adding this to \$8200 (the old salary), we get \$8610 (the new salary). This checks.

State. The old salary was \$8200.

29. Familiarize. We let $x =$ the cost of room and board. Then $x + 906 =$ the tuition.

Translate.

Tuition plus room and board is \$7386.

$$(x + 906) + x = 7386$$

Solve.

$$(x + 906) + x = 7386$$
$$2x + 906 = 7386$$
$$2x = 6480$$
$$x = 3240$$

Then $x = 3240$ (the room and board) and $x + 906 = 4146$ (the tuition).

Check. The tuition is \$906 more than the room and board. The total cost for tuition plus room and board is $\$4146 + \$3240$, or \$7386. This checks.

State. The tuition is \$4146.

31. Familiarize. Let $x =$ the number.

Translate.

Eleven less than seven times a number is five more than six times the number.

$$7x - 11 = 6x + 5$$

Solve.

$$7x - 11 = 6x + 5$$
$$x - 11 = 5$$
$$x = 16$$

Check. Eleven less than seven times 16 is $7 \cdot 16 - 11$, or 101. Five more than six times 16 is $6 \cdot 16 + 5$, or 101. The number checks.

State. The number is 16.

33. Familiarize. Let $x =$ the first odd integer. Then $x + 2 =$ the next odd integer.

Translate.

First integer plus second integer is 137.

$$x + (x + 2) = 137$$

Solve.

$$x + x + 2 = 137$$
$$2x + 2 = 137$$
$$2x = 135$$
$$x = 67.5$$

Check and State. The number 67.5 is not an odd integer. There is no solution to the problem.

This problem could also be solved as follows. The sum of any two odd integers is always even. Thus, the sum can never be 137. There is no solution.

35. Familiarize. After the next test there will be six test scores. The average of the six scores is their sum divided by 6. We let $x =$ the next test score.

Translate.

The average of the six scores is 86%

$$\frac{92 + 84 + 78 + 86 + 90 + x}{6} = 86$$

Solve.

$$\frac{92 + 84 + 78 + 86 + 90 + x}{6} = 86$$

$$\frac{430 + x}{6} = 86 \qquad \text{Adding in the numerator}$$

$$430 + x = 516 \qquad \text{Multiplying by 6}$$

$$x = 86$$

Check. If the next test score is 86%, the average will be

$$\frac{92 + 84 + 78 + 86 + 90 + 86}{6} = \frac{516}{6}, \text{ or } 86\%. \text{ The answer}$$

checks.

State. The sixth score must be 86%.

37. Familiarize. Note that each odd integer is two more than the one preceding it. If we let $n =$ the first odd integer, then the second is 2 more than the first, or $n + 2$, and the third is 2 more than the second, or 4 more than the first, or $n + 4$. We are told that the sum of the first, twice the second, and three times the third is 70.

Translate.

First plus two times second plus three times third is 70.

$$n + 2(n + 2) + 3(n + 4) = 70$$

Solve.

$$n + 2(n + 2) + 3(n + 4) = 70$$
$$n + 2n + 4 + 3n + 12 = 70$$
$$6n + 16 = 70$$
$$6n = 54$$
$$n = 9$$

Check. The numbers are 9, 11, and 13. They are consecutive odd integers. Also, $9 + 2 \cdot 11 + 3 \cdot 13 = 9 + 22 + 39$, or 70. The numbers check.

State. The integers are 9, 11, and 13.

39. **Familiarize**. Let n represent the number of romance novels. Then $65\%n$, or $0.65n$, represents the number of science fiction novels; $46\%(0.65n)$, or $(0.46)(0.65n)$, or $0.299n$, represents the number of horror novels; and $17\%(0.299n)$, or $0.17(0.299n)$, or $0.05083n$, represents the number of mystery novels.

Translate.

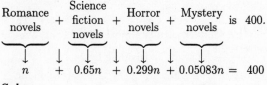

$$n + 0.65n + 0.299n + 0.05083n = 400$$

Solve.

$$n + 0.65n + 0.299n + 0.05083n = 400$$
$$1.99983n = 400$$
$$n \approx 200$$

Check. If the number of romance novels is 200, then the other novels are:

Science fiction: $0.65(200) = 130$

Horror: $0.299(200) \approx 60$

Mystery: $0.05083(200) \approx 10$

The total number of novels is $200 + 130 + 60 + 10$, or 400. The numbers check.

State. The professor has 130 science fiction novels.

41. **Familiarize**. Let $p =$ the total percent change, represented in decimal notation. We represent the original population as 1 (100% of the population). After a 20% increase the population is $1 + 0.2 \cdot 1$, or 1.2. When this new population is increased by 30%, we have $1.2 + 0.3(1.2)$, or $1.3(1.2)$. When this population is decreased by 20% we have $1.3(1.2) - 0.2[1.3(1.2)]$, or $0.8(1.3)(1.2)$.

Translate.

Percent of original population after changes	minus	100% of original population	is	total percent change
$0.8(1.3)(1.2)$	$-$	1	$=$	p

Solve.

$$0.8(1.3)(1.2) - 1 = p$$
$$1.248 - 1 = p \qquad \text{Multiplying}$$
$$0.248 = p$$
$$0.25 \approx p \qquad \text{Rounding}$$

Since p is a positive number, it represents an increase in population.

Check. We repeat the computations. The result checks.

State. The total percent change is a 25% increase.

43. **Familiarize**. We first make drawings. We let $x =$ the length of the shorter piece of wire. Then the length of the longer piece is $100 - x$.

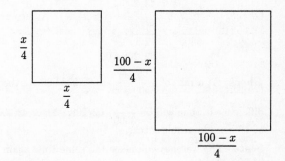

The squares have sides of length $\dfrac{x}{4}$ and $\dfrac{100 - x}{4}$.

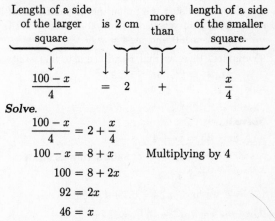

Translate.

Length of a side of the larger square	is	2 cm	more than	length of a side of the smaller square.
$\dfrac{100 - x}{4}$	$=$	2	$+$	$\dfrac{x}{4}$

Solve.

$$\frac{100 - x}{4} = 2 + \frac{x}{4}$$
$$100 - x = 8 + x \qquad \text{Multiplying by 4}$$
$$100 = 8 + 2x$$
$$92 = 2x$$
$$46 = x$$

Check. If one piece of wire is 46 cm, the other piece is $100 - 46$, or 54 cm. The side of one square is $\dfrac{46}{4}$, or $11\frac{1}{2}$ cm. The side of the other is $\dfrac{54}{4}$, or $13\frac{1}{2}$ cm, which is 2 cm greater than the side of the first square. The numbers check.

State. The wire should be cut into pieces of 46 cm and 54 cm.

45. *Familiarize.* Let $s =$ the number of seconds after which the watches will show the same time again. The difference in time between the two watches is

$$2.5\,\frac{\text{sec}}{\text{hr}} = 2.5\,\frac{\text{sec}}{\text{hr}} \times \frac{1\,\text{hr}}{60\,\text{min}} \times \frac{1\,\text{min}}{60\,\text{sec}} = \frac{2.5\,\text{sec}}{3600\,\text{sec}}.$$

The watches will show the same time again when the difference in time between them is

$$12\,\text{hr} = 12\,\text{hr} \times \frac{60\,\text{min}}{1\,\text{hr}} \times \frac{60\,\text{sec}}{1\,\text{min}} = 43{,}200\,\text{sec}.$$

Translate.

Difference in time, per 3600 sec	times	how many seconds	is	43,200 sec
$\dfrac{2.5}{3600}$	$\cdot$	s	$=$	$43{,}200$

Solve.

$$\frac{2.5}{3600}s = 43{,}200$$

$$\frac{3600}{2.5} \cdot \frac{2.5}{3600}s = \frac{3600}{2.5} \cdot 43{,}200$$

$$s = 62{,}208{,}000$$

Check. At a rate of $\dfrac{2.5\,\text{sec}}{3600\,\text{sec}}$, in 62,208,000 sec the difference in time will be $\dfrac{2.5}{3600} \cdot 62{,}208{,}000$, or 43,200 sec, or 12 hr. The result checks.

State. The watches will show the same time again after 62,208,000 sec.

47. *Familiarize.* Using the properties of parallel lines intersected by a transversal, we know that $m\angle 8 = m\angle 4$. Also $m\angle 4 + m\angle 2 = 180°$ and $m\angle 4 = m\angle 1$.

Translate. First we find x and use it to find $m\angle 4$.

$$\begin{array}{ccc} m\angle 8 & = & m\angle 4 \\ \downarrow & \downarrow & \downarrow \\ 5x + 25 & = & 8x + 4 \end{array}$$

Solve.

$$5x + 25 = 8x + 4$$
$$21 = 3x \qquad \text{Subtracting } 5x \text{ and } 4$$
$$7 = x$$

If $x = 7$, we have $m\angle 4 = 8 \cdot 7 + 4 = 60°$.

Then $m\angle 4 + m\angle 2 = 180°$, so

$$60° + m\angle 2 = 180°$$
$$m\angle 2 = 120°. \qquad \text{Subtracting } 60°$$

Also, $m\angle 4 = m\angle 1 = 60°$.

Check. We go over the computations. The results check.

State. $m\angle 2 = 120°$ and $m\angle 1 = 60°$.

Exercise Set 2.3

1. $A = lw$

$\dfrac{A}{l} = w \qquad$ Dividing by l

3. $W = EI$

$\dfrac{W}{E} = I \qquad$ Dividing by E

5. $d = rt$

$\dfrac{d}{t} = r \qquad$ Dividing by t

7. $I = Prt$

$\dfrac{I}{Pr} = t \qquad$ Dividing by Pr

9. $E = mc^2$

$\dfrac{E}{c^2} = m \qquad$ Dividing by c^2

11. $P = 2l + 2w$

$P - 2w = 2l \qquad$ Subtracting $2w$

$\dfrac{P - 2w}{2} = l \qquad$ Dividing by 2

or $\dfrac{P}{2} - w = l$

13. $c^2 = a^2 + b^2$

$c^2 - a^2 = b^2 \qquad$ Subtracting a^2

15. $Ax + By = C$

$By = C - Ax \qquad$ Subtracting Ax

$y = \dfrac{C - Ax}{B} \qquad$ Dividing by B

17. $A = \pi r^2$

$\dfrac{A}{\pi} = r^2 \qquad$ Dividing by π

19. $W = \dfrac{11}{2}(h - 40)$

$\dfrac{2}{11}W = h - 40 \qquad$ Multiplying by $\dfrac{2}{11}$

$\dfrac{2}{11}W + 40 = h \qquad$ Adding 40

21. $V = \dfrac{4}{3}\pi r^3$

$\dfrac{3V}{4\pi} = r^3 \qquad$ Multiplying by $\dfrac{3}{4\pi}$

23. $A = \dfrac{1}{2}h(c - d)$

$2A = h(c - d) \qquad$ Multiplying by 2

$2A = hc - hd$

$2A + hd = hc \qquad$ Adding hd

$\dfrac{2A + hd}{h} = c \qquad$ Dividing by h

or $\dfrac{2A}{h} + d = c$

25. $F = \dfrac{mv^2}{r}$

$F = m \cdot \dfrac{v^2}{r}$

$\dfrac{Fr}{v^2} = m \qquad$ Multiplying by $\dfrac{r}{v^2}$

27. Solve for a:

$P = 8a - 3ab$

$P = a(8 - 3b) \quad$ Factoring

$\dfrac{P}{8 - 3b} = a \qquad$ Dividing by $8 - 3b$

Solve for b:

$P = 8a - 3ab$

$P - 8a = -3ab \qquad$ Subtracting $8a$

$\dfrac{P - 8a}{-3a} = b, \text{ or} \qquad$ Dividing by $-3a$

$\dfrac{P}{-3a} + \dfrac{8}{3} = b$

29. $\dfrac{80}{-16} = -\dfrac{80}{16} = -5$

31. $-\dfrac{1}{2} \div \dfrac{1}{4} = -\dfrac{1}{2} \cdot \dfrac{4}{1} = -\dfrac{4}{2} = -2$

33. $-\dfrac{2}{3} \div \left(-\dfrac{5}{6}\right) = -\dfrac{2}{3} \cdot \left(-\dfrac{6}{5}\right) = \dfrac{2 \cdot 6}{3 \cdot 5} = \dfrac{2 \cdot 2 \cdot 3}{3 \cdot 5} = $

$\dfrac{2 \cdot 2 \cdot \cancel{3}}{\cancel{3} \cdot 5} = \dfrac{4}{5}$

35. $A = \pi rs + \pi r^2$

$A - \pi r^2 = \pi rs$

$\dfrac{A - \pi r^2}{\pi r} = s$

37. Solve for V_1:

$\dfrac{P_1 V_1}{T_1} = \dfrac{P_2 V_2}{T_2}$

$V_1 = \dfrac{T_1 P_2 V_2}{P_1 T_2} \quad$ Multiplying by $\dfrac{T_1}{P_1}$

Solve for P_2:

$\dfrac{P_1 V_1}{T_1} = \dfrac{P_2 V_2}{T_2}$

$\dfrac{P_1 V_1 T_2}{T_1 V_2} = P_2 \quad$ Multiplying by $\dfrac{T_2}{V_2}$

39. We substitute \$75 for P, \$3 for I, and 5%, or 0.05, for r in the formula $t = \dfrac{I}{Pr}$.

$t = \dfrac{I}{Pr}$

$t = \dfrac{\$3}{\$75(0.05)} \qquad$ Substituting

$t = 0.8$

It will take 0.8 yr.

1. $x - 2 \geq 6$

-4 : We substitute and get $-4 - 2 \geq 6$, or $-6 \geq 6$, a false sentence. Therefore, -4 is not a solution.

0 : We substitute and get $0 - 2 \geq 6$, or $-2 \geq 6$, a false sentence. Therefore, 0 is not a solution.

4 : We substitute and get $4 - 2 \geq 6$, or $2 \geq 6$, a false sentence. Therefore, 4 is not a solution.

8 : We substitute and get $8 - 2 \geq 6$, or $6 \geq 6$, a true sentence. Therefore, 8 is a solution.

3. $t - 6 > 2t - 5$

0 : We substitute and get $0 - 6 > 2 \cdot 0 - 5$, or $-6 > -5$, a false sentence. Therefore, 0 is not a solution.

4 : We substitute and get $4 - 6 > 2 \cdot 4 - 5$, or $-2 > 3$, a false sentence. Therefore, 4 is not a solution.

-3 : We substitute and get $-3 - 6 > 2(-3) - 5$, or $-9 > -11$, a true sentence. Therefore, -3 is a solution.

-1 : We substitute and get $-1 - 6 > 2(-1) - 5$, or $-7 > -7$, a false sentence. Therefore, -1 is not a solution.

5. $x + 2 > 1$

$x + 2 - 2 > 1 - 2 \qquad$ Subtracting 2

$x > -1$

The solution set is $\{x | x > -1\}$.

7. $y + 3 < 9$

$y + 3 - 3 < 9 - 3 \qquad$ Subtracting 3

$y < 6$

The solution set is $\{y | y < 6\}$.

9. $a - 9 \leq -31$

$a - 9 + 9 \leq -31 + 9 \qquad$ Adding 9

$a \leq -22$

The solution set is $\{a | a \leq -22\}$.

11. $t + 13 \geq 9$

$t + 13 - 13 \geq 9 - 13 \qquad$ Subtracting 13

$t \geq -4$

The solution set is $\{t | t \geq -4\}$.

13.
$$y - 8 > -14$$
$$y - 8 + 8 > -14 + 8 \qquad \text{Adding 8}$$
$$y > -6$$
The solution set is $\{y | y > -6\}$.

15.
$$x - 11 \leq -2$$
$$x - 11 + 11 \leq -2 + 11 \qquad \text{Adding 11}$$
$$x \leq 9$$
The solution set is $\{x | x \leq 9\}$.

17.
$$8x \geq 24$$
$$\frac{8x}{8} \geq \frac{24}{8} \qquad \text{Dividing by 8}$$
$$x \geq 3$$
The solution set is $\{x | x \geq 3\}$.

19.
$$0.3x < -18$$
$$\frac{0.3x}{0.3} < \frac{-18}{0.3} \qquad \text{Dividing by 0.3}$$
$$x < -60$$
The solution set is $\{x | x < -60\}$.

21.
$$-9x \geq -8.1$$
$$\frac{-9x}{-9} \leq \frac{-8.1}{-9} \qquad \text{Dividing by } -9 \text{ and reversing the inequality symbol}$$
$$x \leq 0.9$$
The solution set is $\{x | x \leq 0.9\}$.

23.
$$-\frac{3}{4}x \geq -\frac{5}{8}$$
$$-\frac{4}{3}\left(-\frac{3}{4}x\right) \leq -\frac{4}{3}\left(-\frac{5}{8}\right) \qquad \text{Multiplying by } -\frac{4}{3} \text{ and reversing the inequality symbol}$$
$$x \leq \frac{20}{24}$$
$$x \leq \frac{5}{6}$$
The solution set is $\left\{x \left| x \leq \frac{5}{6}\right.\right\}$.

25.
$$2x + 7 < 19$$
$$2x + 7 - 7 < 19 - 7 \qquad \text{Subtracting 7}$$
$$2x < 12$$
$$\frac{2x}{2} < \frac{12}{2} \qquad \text{Dividing by 2}$$
$$x < 6$$
The solution set is $\{x | x < 6\}$.

27.
$$5y + 2y \leq -21$$
$$7y \leq -21 \qquad \text{Collecting like terms}$$
$$\frac{7y}{7} \leq \frac{-21}{7} \qquad \text{Dividing by 7}$$
$$y \leq -3$$
The solution set is $\{y | y \leq -3\}$.

29.
$$2y - 7 < 5y - 9$$
$$-5y + 2y - 7 < -5y + 5y - 9 \qquad \text{Adding } -5y$$
$$-3y - 7 < -9$$
$$-3y - 7 + 7 < -9 + 7 \qquad \text{Adding 7}$$
$$-3y < -2$$
$$\frac{-3y}{-3} > \frac{-2}{-3} \qquad \text{Dividing by } -3 \text{ and reversing the inequality symbol}$$
$$y > \frac{2}{3}$$
The solution set is $\left\{y \left| y > \frac{2}{3}\right.\right\}$.

31.
$$0.4x + 5 \leq 1.2x - 4$$
$$-1.2x + 0.4x + 5 \leq -1.2x + 1.2x - 4 \qquad \text{Adding } -1.2x$$
$$-0.8x + 5 \leq -4$$
$$-0.8x + 5 - 5 \leq -4 - 5 \qquad \text{Subtracting 5}$$
$$-0.8x \leq -9$$
$$\frac{-0.8x}{-0.8} \geq \frac{-9}{-0.8} \qquad \text{Dividing by } -0.8 \text{ and reversing the inequality symbol}$$
$$x \geq 11.25, \text{ or } x \geq \frac{45}{4}$$
The solution set is $\left\{x \left| x \geq \frac{45}{4}\right.\right\}$.

33. $5x - \dfrac{1}{12} \le \dfrac{5}{12} + 4x$

$12\left(5x - \dfrac{1}{12}\right) \le 12\left(\dfrac{5}{12} + 4x\right)$ Clearing fractions

$60x - 1 \le 5 + 48x$

$60x - 1 - 48x \le 5 + 48x - 48x$ Subtracting $48x$

$12x - 1 \le 5$

$12x - 1 + 1 \le 5 + 1$ Adding 1

$12x \le 6$

$\dfrac{12x}{12} \le \dfrac{6}{12}$ Dividing by 12

$x \le \dfrac{1}{2}$

The solution set is $\left\{x \middle| x \le \dfrac{1}{2}\right\}$.

35. $4(4y - 3) \ge 9(2y + 7)$

$16y - 12 \ge 18y + 63$ Removing parentheses

$16y - 12 - 18y \ge 18y + 63 - 18y$ Subtracting $18y$

$-2y - 12 \ge 63$

$-2y - 12 + 12 \ge 63 + 12$ Adding 12

$-2y \ge 75$

$\dfrac{-2y}{-2} \le \dfrac{75}{-2}$ Dividing by -2 and reversing the inequality symbol

$y \le -\dfrac{75}{2}$

The solution set is $\left\{y \middle| y \le -\dfrac{75}{2}\right\}$.

37. $3(2 - 5x) + 2x < 2(4 + 2x)$

$6 - 15x + 2x < 8 + 4x$

$6 - 13x < 8 + 4x$ Collecting like terms

$6 - 17x < 8$ Subtracting $4x$

$-17x < 2$ Subtracting 6

$x > -\dfrac{2}{17}$ Dividing by -17 and reversing the inequality symbol

The solution set is $\left\{x \middle| x > -\dfrac{2}{17}\right\}$.

39. $5[3m - (m + 4)] > -2(m - 4)$

$5(3m - m - 4) > -2(m - 4)$

$5(2m - 4) > -2(m - 4)$

$10m - 20 > -2m + 8$

$12m - 20 > 8$ Adding $2m$

$12m > 28$ Adding 20

$m > \dfrac{28}{12}$

$m > \dfrac{7}{3}$

The solution set is $\left\{m \middle| m > \dfrac{7}{3}\right\}$.

41. $3(r - 6) + 2 > 4(r + 2) - 21$

$3r - 18 + 2 > 4r + 8 - 21$

$3r - 16 > 4r - 13$ Collecting like terms

$-r - 16 > -13$ Subtracting $4r$

$-r > 3$ Adding 16

$r < -3$ Multiplying by -1 and reversing the inequality symbol

The solution set is $\{r | r < -3\}$.

43. $19 - (2x + 3) \le 2(x + 3) + x$

$19 - 2x - 3 \le 2x + 6 + x$

$16 - 2x \le 3x + 6$ Collecting like terms

$16 - 5x \le 6$ Subtracting $3x$

$-5x \le -10$ Subtracting 16

$x \ge 2$ Dividing by -5 and reversing the inequality symbol

The solution set is $\{x | x \ge 2\}$.

45. $\dfrac{1}{4}(8y + 4) - 17 < -\dfrac{1}{2}(4y - 8)$

$2y + 1 - 17 < -2y + 4$

$2y - 16 < -2y + 4$ Collecting like terms

$4y - 16 < 4$ Adding $2y$

$4y < 20$ Adding 16

$y < 5$

The solution set is $\{y | y < 5\}$

47. $2[4 - 2(3 - x)] - 1 \ge 4[2(4x - 3) + 7] - 25$

$2[4 - 6 + 2x] - 1 \ge 4[8x - 6 + 7] - 25$

$2[-2 + 2x] - 1 \ge 4[8x + 1] - 25$

$-4 + 4x - 1 \ge 32x + 4 - 25$

$4x - 5 \ge 32x - 21$

$-28x - 5 \ge -21$

$-28x \ge -16$

$x \le \dfrac{-16}{-28}$ Dividing by -28 and reversing the inequality symbol

$x \le \dfrac{4}{7}$

The solution set is $\left\{x \middle| x \le \dfrac{4}{7}\right\}$.

49. $\dfrac{4}{5}(7x - 6) < 40$

$5 \cdot \dfrac{4}{5}(7x - 6) < 5 \cdot 40$ Clearing the fraction

$4(7x - 6) < 200$

$28x - 24 < 200$

$28x < 224$

$x < 8$

The solution set is $\{x | x < 8\}$.

51. $\frac{3}{4}(3 + 2x) + 1 \geq 13$

$4\left[\frac{3}{4}(3 + 2x) + 1\right] \geq 4 \cdot 13$ Clearing the fraction

$3(3 + 2x) + 4 \geq 52$

$9 + 6x + 4 \geq 52$

$6x + 13 \geq 52$

$6x \geq 39$

$x \geq \frac{39}{6}$, or $\frac{13}{2}$

The solution set is $\left\{x \middle| x \geq \frac{13}{2}\right\}$.

53. $\frac{3}{4}\left(3x - \frac{1}{2}\right) - \frac{2}{3} < \frac{1}{3}$

$\frac{9x}{4} - \frac{3}{8} - \frac{2}{3} < \frac{1}{3}$

$24\left(\frac{9x}{4} - \frac{3}{8} - \frac{2}{3}\right) < 24 \cdot \frac{1}{3}$ Clearing fractions

$54x - 9 - 16 < 8$

$54x - 25 < 8$

$54x < 33$

$x < \frac{33}{54}$, or $\frac{11}{18}$

The solution set is $\left\{x \middle| x < \frac{11}{18}\right\}$.

55. $0.7(3x + 6) \geq 1.1 - (x + 2)$

$10[0.7(3x + 6)] \geq 10[1.1 - (x + 2)]$ Clearing decimals

$7(3x + 6) \geq 11 - 10(x + 2)$

$21x + 42 \geq 11 - 10x - 20$

$21x + 42 \geq -9 - 10x$

$31x + 42 \geq -9$

$31x \geq -51$

$x \geq -\frac{51}{31}$

The solution set is $\left\{x \middle| x \geq -\frac{51}{31}\right\}$.

57. $a + (a - 3) \leq (a + 2) - (a + 1)$

$a + a - 3 \leq a + 2 - a - 1$

$2a - 3 \leq 1$

$2a \leq 4$

$a \leq 2$

The solution set is $\{a | a \leq 2\}$.

59. A number <u>is less than</u> 12.

$x < 12$

61. My cholesterol level L <u>is greater than or equal to</u> 185.

$L \geq 185$

63. The price of a compact disc <u>is at most</u> $18.95.

$p \leq \$18.95$

65. 24 minus 3 times a number <u>is less than</u> 16 plus the number.

$24 - 3x < 16 + x$

67. Fifteen times the sum of two numbers <u>is at least</u> 78.

$15(a + b) \geq 78$

69. *Familiarize.* We let $x =$ the number of miles traveled in a day. Then the total rental cost for a day is $30 + 0.20x$.

Translate. The total cost for a day must be less than or equal to $96. This translates to the following inequality:

$30 + 0.20x \leq 96$

Solve.

$0.20x \leq 66$ Adding -30

$x \leq \frac{66}{0.20}$ Dividing by 0.20

$x \leq 330$

Check. If you travel 330 miles, the total cost is $30 + 0.20(330)$, or $30 + 66 = \$96$. Any mileage less than 330 will also stay within the budget.

State. Mileage less than or equal to 330 miles allows you to stay within budget.

71. *Familiarize.* List the information in a table. Let $x =$ the score on the fourth test.

Test	Score
Test 1	89
Test 2	92
Test 3	95
Test 4	x
Total	360 or more

Translate. We can easily get an inequality from the table.

$89 + 92 + 95 + x \geq 360$

Solve.

$276 + x \geq 360$ Collecting like terms

$x \geq 84$ Adding -276

Check. If you get 84 on the fourth test, your total score will be $89 + 92 + 95 + 84$, or 360. Any higher score will also give you an A.

State. A score of 84 or better will give you an A.

73. *Familiarize.* We let $x =$ the ticket price. Then $300x =$ the total receipts from the ticket sales assuming 300 people will attend. The first band will play for $250 + 50\%(300x)$. The second band will play for $550.

Translate. For school profit to be greater when the first band plays, the amount the first band charges must be less than the amount the second band charges. We now have an inequality.

$250 + 50\%(300x) < 550$

Solve.

$250 + 0.5(300x) < 550$

$250 + 150x < 550$

$150x < 300$ Subtracting 250

$x < 2$ Dividing by 150

Check. For $x = \$2$, the total receipts are $300(2)$, or 600. The first band charges $250 + 50\%(600)$, or 550. The second band also charges 550. The school profit is the same using either band. For $x = \$1.99$, the total receipts are $300(1.99)$, or 597. The first band charges $250 + 50\%(597)$, or 548.50. This is less than the second band charges, so the first band produces more profit. For $x = \$2.01$, the total receipts are $300(2.01)$, or 603. The first band charges $250 + 50\%(603)$, or 551.50. This is more than the second band charges, so the second band produces more profit. For these values, the inequality $x < 2$ gives correct results.

State. The ticket price must be less than $2. The highest price, rounded to the nearest cent, less than $2 is $1.99.

75. Familiarize. List the given information in a table. Let p = the monthly lease payments for which it will cost less to lease the car than to buy the car.

Buying	Leasing
\$100 down	\$4000 down
\$350 per month for 36 months	\$p per month for 36 months
Total: \$100 + \$350(36)	Total: \$4000 + \$p(36)

Translate.

Amount spent buying car	is greater than	amount spent leasing car
$100 + 350(36)$	$>$	$4000 + p(36)$

Solve. We solve the inequality.

$$100 + 350(36) > 4000 + p(36)$$
$$100 + 12,600 > 4000 + 36p$$
$$12,700 > 4000 + 36p$$
$$8700 > 36p \qquad \text{Subtracting } 4000$$
$$241.6\overline{6} > p \qquad \text{Dividing by } 36$$

Check. The amount spent buying the car will be $100 + $350(36)$, or $12,700. We calculate the amount spent leasing for $p = \$241.6\overline{6}$, for some amount less than $241.6\overline{6}$, and for some amount greater than $241.6\overline{6}$.

For $p = \$241.6\overline{6}$: $\$4000 + \$241.6\overline{6}(36) = \$12,700$.

For $p = \$200$: $\$4000 + \$200(36) = \$11,200$.

For $p = \$250$: $\$4000 + \$250(36) = \$13,000$.

For a monthly payment less than $241.67 (rounding to the nearest cent), leasing costs less than buying. We cannot check all the possibilities, so we stop here.

State. For monthly payments less than $241.67, leasing costs less than buying.

77. Familiarize. We make a table of information.

Plan A: Monthly Income	Plan B: Monthly Income
\$500 salary	\$750 salary
4% of sales	5% of sales over \$8000
Total: 500 + 4% of sales	Total: 750 + 5% of sales over 8000

Translate. We write an inequality stating that the income from Plan B is greater than the income from Plan A. We let S = gross sales. Then $S - 8000$ = gross sales over 8000.

$$750 + 5\%(S - 8000) > 500 + 4\%S$$

Solve.

$$750 + 0.05S - 400 > 500 + 0.04S$$
$$350 + 0.05S > 500 + 0.04S$$
$$0.01S > 150$$
$$S > \frac{150}{0.01}$$
$$S > 15,000$$

Check. We calculate for $x = \$15,000$ and for some amount greater than $15,000 and some amount less than $15,000.

Plan A:	Plan B:
$500 + 4\%(15,000)$	$750 + 5\%(15,000 - 8000)$
$500 + 0.04(15,000)$	$750 + 0.05(7000)$
$500 + 600$	$750 + 350$
\$1100	\$1100

When $x = \$15,000$, the income from Plan A is equal to the income from Plan B.

Plan A:	Plan B:
$500 + 4\%(16,000)$	$750 + 5\%(16,000 - 8000)$
$500 + 0.04(16,000)$	$750 + 0.05(8000)$
$500 + 640$	$750 + 400$
\$1140	\$1150

When $x = \$16,000$, the income from Plan B is greater than the income from Plan A.

Plan A:	Plan B:
$500 + 4\%(14,000)$	$750 + 5\%(14,000 - 8000)$
$500 + 0.04(14,000)$	$750 + 0.05(6000)$
$500 + 560$	$750 + 300$
\$1060	\$1050

When $x = \$14,000$, the income from Plan B is less than the income from Plan A.

State. Plan B is better than Plan A when gross sales are greater than $15,000.

79. *Familiarize*. Plan A will pay the house painter $200 + 12n$, while Plan B will pay the painter $20n$.

Translate. We write an inequality stating that the income from the job under Plan A is greater than the income from the job under Plan B.

$$200 + 12n > 20n$$

Solve. We solve the inequality.

$$
\begin{aligned}
200 + 12n &> 20n \\
200 &> 8n \qquad \text{Subtracting } 12n \\
25 &> n \qquad \text{Dividing by 8}
\end{aligned}
$$

Check. We calculate for $n = 25$, for some number of hours less than 25, and for some number of hours greater than 25.

For $n = 25$:   Plan A gives $200 + \$12 \cdot 25 = \$500$.
            Plan B gives $\$20 \cdot 25 = \$500$.

For $n = 20$:   Plan A gives $\$200 + \$12 \cdot 20 = \$440$.
            Plan B gives $\$20 \cdot 20 = \$400$.

For $n = 30$:   Plan A gives $\$200 + \$12 \cdot 30 = \$560$.
            Plan B gives $\$20 \cdot 30 = \$600$.

For a value of n less than 25, Plan A is better than Plan B. We cannot check all possible values of n, so we stop here.

State. Plan A is better for values of n less than 25.

81. *Familiarize*. We want to find the values of s for which $I > 36$.

Translate. $2(s + 10) > 36$

Solve.

$$
\begin{aligned}
2s + 20 &> 36 \\
2s &> 16 \\
s &> 8
\end{aligned}
$$

Check. For $s = 8$, $I = 2(8 + 10) = 2 \cdot 18 = 36$. Then any U.S. size larger than 8 will give a size larger than 36 in Italy.

State. For U.S. dress sizes larger than 8, dress sizes in Italy will be larger than 36.

83. a) Substitute 0 for t and carry out the calculation.

$$
\begin{aligned}
N &= 0.733(0) + 8.398 \\
N &= 0 + 8.398 \\
N &= 8.398
\end{aligned}
$$

Each person drank 8.398 gal of bottled water in 1990.
Substitute 5 for t and carry out the calculation.

$$
\begin{aligned}
N &= 0.733(5) + 8.398 \\
N &= 3.665 + 8.398 \\
N &= 12.063
\end{aligned}
$$

Each person drank 12.063 gal of bottled water in 1995.

In 2000, $t = 2000 - 1990 = 10$. Substitute 10 for t and carry out the calculation.

$$
\begin{aligned}
N &= 0.733(10) + 8.398 \\
N &= 7.33 + 8.398 \\
N &= 15.728
\end{aligned}
$$

Each person will drink 15.728 gal of bottled water in 2000.

b) *Familiarize*. The amount of bottled water that each person drinks t years after 1990 is given by $0.733t + 8.398$.

Translate.

Amount drunk t years after 1990	is at least	15 gal
$\downarrow$	$\downarrow$	$\downarrow$
$0.733t + 8.398$	$\geq$	15

Solve. We solve the inequality.

$$
\begin{aligned}
0.733t + 8.398 &\geq 15 \\
0.733t &\geq 6.602 \qquad \text{Subtracting 8.398} \\
t &\geq 9 \qquad \text{Rounding}
\end{aligned}
$$

Check. We calculate for 9, for some number less than 9, and for some number greater than 9.

For $t = 9$: $0.733(9) + 8.398 \approx 15$.

For $t = 8$: $0.733(8) + 8.398 \approx 14.3$.

For $t = 10$: $0.733(10) + 8.398 \approx 15.7$.

For a value of t greater than or equal to 9, the number of gallons of bottled water each person drinks is at 15. We cannot check all the possible values of t, so we stop here.

State. Each person will drink at least 15 gal of bottled water 9 or more years after 1990, or for all years after 1999.

85. $10(3x - 7y - 4) = 10 \cdot 3x - 10 \cdot 7y - 10 \cdot 4$
$$ = 30x - 70y - 40$$

87. $30x - 70y - 40 = 10 \cdot 3x - 10 \cdot 7y - 10 \cdot 4$
$$ = 10(3x - 7y - 4)$$

89. $-2x + 6y - 8x - 10 + 3y - 40$
$$= (-2 - 8)x + (6 + 3)y + (-10 - 40)$$
$$= -10x + 9y - 50$$

91. a) *Familiarize*. We will use

$$S = 460 + 94p \quad \text{and} \quad D = 2000 - 60p.$$

Translate. Supply is to exceed demand, so we have

$$S > D, \text{ or}$$
$$460 + 94p > 2000 - 60p.$$

Solve. We solve the inequality.

$$
\begin{aligned}
460 + 94p &> 2000 - 60p \\
460 + 154p &> 2000 \qquad \text{Adding } 60p \\
154p &> 1540 \qquad \text{Subtracting 460} \\
p &> 10 \qquad \text{Dividing by 154}
\end{aligned}
$$

Check. We calculate for $p = 10$, for some value of p less than 10, and for some value of p greater than 10.

For $p = 10$: $S = 460 + 94 \cdot 10 = 1400$
$D = 2000 - 60 \cdot 10 = 1400$

For $p = 9$: $S = 460 + 94 \cdot 9 = 1306$
$D = 2000 - 60 \cdot 9 = 1460$

For $p = 11$: $S = 460 + 94 \cdot 11 = 1494$
$D = 2000 - 60 \cdot 11 = 1340$

For a value of p greater than 10, supply exceeds demand. We cannot check all possible values of p, so we stop here.

State. Supply exceeds demand for values of p greater than 10.

b) We have seen in part (a) that $D = S$ for $p = 10$, $S < D$ for a value of p less than 10, and $S > D$ for a value of p greater than 10. Since we cannot check all possible values of p, we stop here. Supply is less than demand for values of p less than 10.

93. True

95. No. Let $x = 2$. Then $x < 3$ is true, but $0 \cdot x < 0 \cdot 3$, or $0 < 0$, is false.

97. $x + 8 < 3 + x$

$8 < 3$ Subtracting x

We get a false inequality. Thus, the original inequality has no solution.

Exercise Set 2.5

1. $\{1, 3, 7, 12\} \cap \{3, 6, 9, 12\}$

The numbers 3 and 12 are common to the two sets, so the intersection is $\{3, 12\}$.

3. $\{2, 4, 6, 8\} \cap \{1, 3, 5\}$

There are no numbers common to the two sets, so the intersection is the empty set, $\emptyset$.

5. $\{8, 9, 10\} \cap \emptyset$

Since $\emptyset$ has no members, there are no numbers common to the two sets. Thus, the intersection is $\emptyset$.

7. $1 < x < 6$

The graph is the intersection of the graphs of $x > 1$ and $x < 6$.

9. $6 > -x \geq -2$

$-6 < x \leq 2$ Multiplying by -1

The graph is the intersection of the graphs of $x > -6$ and $x \leq 2$.

11. $-10 \leq 3x + 2$ *and* $3x + 2 < 17$

$-12 \leq 3x$ *and* $3x < 15$

$-4 \leq x$ *and* $x < 5$

The solution set is the intersection of the solution sets of the individual inequalities. The numbers common to both sets are those that are greater than or equal to -4 *and* less than 5. Thus the solution set is $\{x| -4 \leq x < 5\}$.

13. $3x + 7 \geq 4$ *and* $2x - 5 \geq -1$

$3x \geq -3$ *and* $2x \geq 4$

$x \geq -1$ *and* $x \geq 2$

$\{x|x \geq -1\} \cap \{x|x \geq 2\} = \{x|x \geq 2\}$

15. $4 - 3x \geq 10$ *and* $5x - 2 > 13$

$-3x \geq 6$ *and* $5x > 15$

$x \leq -2$ *and* $x > 3$

$\{x|x \leq -2\} \cap \{x|x > 3\} = \emptyset$

17. $-4 < x + 4 < 10$

$-4 - 4 < x + 4 - 4 < 10 - 4$ Subtracting 4

$-8 < x < 6$

The solution set is $\{x| -8 < x < 6\}$.

19. $1 < 3y + 4 \leq 19$

$1 - 4 < 3y + 4 - 4 \leq 19 - 4$ Subtracting 4

$-3 < 3y \leq 15$

$\dfrac{-3}{3} < \dfrac{3y}{3} \leq \dfrac{15}{3}$ Dividing by 3

$-1 < y \leq 5$

The solution set is $\{y| -1 < y \leq 5\}$.

21. $-10 \leq 3x - 5 \leq -1$

$-10 + 5 \leq 3x - 5 + 5 \leq -1 + 5$ Adding 5

$-5 \leq 3x \leq 4$

$\dfrac{-5}{3} \leq \dfrac{3x}{3} \leq \dfrac{4}{3}$ Dividing by 3

$-\dfrac{5}{3} \leq x \leq \dfrac{4}{3}$

The solution set is $\left\{x\left| -\dfrac{5}{3} \leq x \leq \dfrac{4}{3}\right.\right\}$.

23. $2 < x + 3 \leq 9$

$2 - 3 < x + 3 - 3 \leq 9 - 3$ Subtracting 3

$-1 < x \leq 6$

The solution set is $\{x| -1 < x \leq 6\}$.

25. $-6 \leq 2x - 3 < 6$

$-6 + 3 \leq 2x - 3 + 3 < 6 + 3$

$-3 \leq 2x < 9$

$\dfrac{-3}{2} \leq \dfrac{2x}{2} < \dfrac{9}{2}$

$-\dfrac{3}{2} \leq x < \dfrac{9}{2}$

The solution set is $\left\{x\left| -\dfrac{3}{2} \leq x < \dfrac{9}{2}\right.\right\}$.

27.

$$-\frac{1}{2} < \frac{1}{4}x - 3 \leq \frac{1}{2}$$

$$-\frac{1}{2} + 3 < \frac{1}{4}x - 3 + 3 \leq \frac{1}{2} + 3$$

$$\frac{5}{2} < \frac{1}{4}x \leq \frac{7}{2}$$

$$4 \cdot \frac{5}{2} < 4 \cdot \frac{1}{4}x \leq 4 \cdot \frac{7}{2}$$

$$10 < x \leq 14$$

The solution set is $\{x | 10 < x \leq 14\}$.

29.

$$-3 < \frac{2x - 5}{4} < 8$$

$$4(-3) < 4\left(\frac{2x - 5}{4}\right) < 4 \cdot 8$$

$$-12 < 2x - 5 < 32$$

$$-12 + 5 < 2x - 5 + 5 < 32 + 5$$

$$-7 < 2x < 37$$

$$\frac{-7}{2} < \frac{2x}{2} < \frac{37}{2}$$

$$-\frac{7}{2} < x < \frac{37}{2}$$

The solution set is $\left\{x \left| -\frac{7}{2} < x < \frac{37}{2}\right.\right\}$.

31. $\{4, 5, 6, 7, 8\} \cup \{1, 4, 6, 11\}$

The numbers in either or both sets are $1, 4, 5, 6, 7, 8$, and 11, so the union is $\{1, 4, 5, 6, 7, 8, 11\}$.

33. $\{6, 8, 12, 14\} \cup \{2, 7, 11\}$

The numbers in either or both sets are $2, 6, 7, 8, 11, 12$, and 14, so the union is $\{2, 6, 7, 8, 11, 12, 14\}$.

35. $\{4, 8, 11\} \cup \emptyset$

The numbers in either or both sets are $4, 8$, and 11, so the union is $\{4, 8, 11\}$.

37. $x < -2 \text{ or } x > 1$

The graph is the union of the graphs of $x < -2$ and $x > 1$.

39. $x \leq -3 \text{ or } x > 1$

The graph is the union of the graphs of $x \leq -3$ or $x > 1$.

41.

$$x + 7 < -2 \qquad\qquad or \qquad\qquad x + 7 > 2$$

$$x + 7 + (-7) < -2 + (-7) \quad or \quad x + 7 + (-7) > 2 + (-7)$$

$$x < -9 \qquad\qquad or \qquad\qquad x > -5$$

The solution set is $\{x | x < -9 \text{ or } x > -5\}$.

43.

$$2x - 8 \leq -3 \qquad or \qquad x - 8 \geq 3$$

$$2x - 8 + 8 \leq -3 + 8 \quad or \quad x - 8 + 8 \geq 3 + 8$$

$$2x \leq 5 \qquad or \qquad x \geq 11$$

$$\frac{2x}{2} \leq \frac{5}{2} \qquad or \qquad x \geq 11$$

$$x \leq \frac{5}{2} \qquad or \qquad x \geq 11$$

The solution set is $\left\{x \left| x \leq \frac{5}{2} \text{ or } x \geq 11\right.\right\}$.

45. $7x + 4 \geq -17 \text{ or } 6x + 5 \geq -7$

$$7x \geq -21 \text{ or } \qquad 6x \geq -12$$

$$x \geq -3 \text{ or } \qquad x \geq -2$$

The solution set is $\{x | x \geq -3\}$.

47.

$$7 > -4x + 5 \qquad or \qquad 10 \leq -4x + 5$$

$$7 - 5 > -4x + 5 - 5 \quad or \quad 10 - 5 \leq -4x + 5 - 5$$

$$2 > -4x \qquad or \qquad 5 \leq -4x$$

$$\frac{2}{-4} < \frac{-4x}{-4} \qquad or \qquad \frac{5}{-4} \geq \frac{-4x}{-4}$$

$$-\frac{1}{2} < x \qquad or \qquad -\frac{5}{4} \geq x$$

The solution set is $\left\{x \left| x \leq -\frac{5}{4} \text{ or } x > -\frac{1}{2}\right.\right\}$.

49. $3x - 7 > -10 \text{ or } 5x + 2 \leq 22$

$$3x > -3 \text{ or } \qquad 5x \leq 20$$

$$x > -1 \text{ or } \qquad x \leq 4$$

All real numbers are solutions.

51.

$$-2x - 2 < -6 \qquad or \qquad -2x - 2 > 6$$

$$-2x - 2 + 2 < -6 + 2 \quad or \quad -2x - 2 + 2 > 6 + 2$$

$$-2x < -4 \qquad or \qquad -2x > 8$$

$$\frac{-2x}{-2} > \frac{-4}{-2} \qquad or \qquad \frac{-2x}{-2} < \frac{8}{-2}$$

$$x > 2 \qquad or \qquad x < -4$$

The solution set is $\{x | x < -4 \text{ or } x > 2\}$.

53.

$$\frac{2}{3}x - 14 < -\frac{5}{6} \qquad or \qquad \frac{2}{3}x - 14 > \frac{5}{6}$$

$$6\left(\frac{2}{3}x - 14\right) < 6\left(-\frac{5}{6}\right) \quad or \quad 6\left(\frac{2}{3}x - 14\right) > 6 \cdot \frac{5}{6}$$

$$4x - 84 < -5 \qquad or \qquad 4x - 84 > 5$$

$$4x - 84 + 84 < -5 + 84 \quad or \quad 4x - 84 + 84 > 5 + 84$$

$$4x < 79 \qquad or \qquad 4x > 89$$

$$\frac{4x}{4} < \frac{79}{4} \qquad or \qquad \frac{4x}{4} > \frac{89}{4}$$

$$x < \frac{79}{4} \qquad or \qquad x > \frac{89}{4}$$

The solution set is $\left\{x \left| x < \frac{79}{4} \text{ or } x > \frac{89}{4}\right.\right\}$.

55. $\dfrac{2x-5}{6} \le -3 \qquad or \qquad \dfrac{2x-5}{6} \ge 4$

$6\left(\dfrac{2x-5}{6}\right) \le 6(-3) \quad or \quad 6\left(\dfrac{2x-5}{6}\right) \ge 6\cdot 4$

$2x-5 \le -18 \qquad or \qquad 2x-5 \ge 24$

$2x-5+5 \le -18+5 \quad or \quad 2x-5+5 \ge 24+5$

$2x \le -13 \qquad or \qquad 2x \ge 29$

$\dfrac{2x}{2} \le \dfrac{-13}{2} \qquad or \qquad \dfrac{2x}{2} \ge \dfrac{29}{2}$

$x \le -\dfrac{13}{2} \qquad or \qquad x \ge \dfrac{29}{2}$

The solution set is $\left\{x \Big| x \le -\dfrac{13}{2} \ or \ x \ge \dfrac{29}{2}\right\}$.

57. $(-2x^{-4}y^6)^5 = (-2)^5(x^{-4})^5(y^6)^5$

$\qquad = -32x^{-20}y^{30}$

$\qquad = -\dfrac{32y^{30}}{x^{20}}$

59. $\dfrac{-4a^5b^{-7}}{5a^{-12}b^8} = -\dfrac{4a^{5-(-12)}b^{-7-8}}{5}$

$\qquad = -\dfrac{4a^{17}b^{-15}}{5}$

$\qquad = -\dfrac{4a^{17}}{5b^{15}}$

61. $\left(\dfrac{56a^5b^{-6}}{28a^7b^{-8}}\right)^{-3} = (2a^{-2}b^2)^{-3}$

$\qquad = 2^{-3}a^{-2(-3)}b^{2(-3)}$

$\qquad = \dfrac{1}{2^3}a^6b^{-6}$

$\qquad = \dfrac{a^6}{8b^6}$

63. a) Substitute $\dfrac{5}{9}(F-32)$ for C in the given inequality.

$1063 \le \dfrac{5}{9}(F-32) < 2660$

$9\cdot 1063 \le 9\cdot \dfrac{5}{9}(F-32) < 9\cdot 2660$

$9567 \le 5(F-32) < 23,940$

$9567 \le 5F-160 < 23,940$

$9727 \le 5F < 24,100$

$1945.4 \le F < 4820$

The inequality for Fahrenheit temperatures is $1945.4° \le F < 4820°$.

b) Substitute $\dfrac{5}{9}(F-32)$ for C in the given inequality.

$960.8 \le \dfrac{5}{9}(F-32) < 2180$

$9(960.8) \le 9\cdot \dfrac{5}{9}(F-32) < 9\cdot 2180$

$8647.2 \le 5(F-32) < 19,620$

$8647.2 \le 5F-160 < 19,620$

$8807.2 \le 5F < 19,780$

$1761.44 \le F < 3956$

The inequality for Fahrenheit temperatures is $1761.44° \le F < 3956°$.

65. Find x such that $-8 < x-3 < 8$:

$-8 < x-3 < 8$

$-5 < x < 11$

The solution set is $\{x| -5 < x < 11\}$.

67. $x-10 < 5x+6 \le x+10$

$-10 < 4x+6 \le 10 \qquad$ Subtracting x

$-16 < 4x \le 4$

$-4 < x \le 1$

The solution set is $\{x| -4 < x \le 1\}$.

69. $-\dfrac{2}{15} \le \dfrac{2}{3}x - \dfrac{2}{5} \le \dfrac{2}{15}$

$-\dfrac{2}{15} \le \dfrac{2}{3}x - \dfrac{6}{15} \le \dfrac{2}{15}$

$\dfrac{4}{15} \le \dfrac{2}{3}x \le \dfrac{8}{15}$

$\dfrac{3}{2}\cdot \dfrac{4}{15} \le \dfrac{3}{2}\cdot \dfrac{2}{3}x \le \dfrac{3}{2}\cdot \dfrac{8}{15}$

$\dfrac{2}{5} \le x \le \dfrac{4}{5}$

The solution set is $\left\{x \Big| \dfrac{2}{5} \le x \le \dfrac{4}{5}\right\}$.

71. $3x < 4-5x < 5+3x$

$0 < 4-8x < 5 \qquad$ Subtracting $3x$

$-4 < -8x < 1$

$\dfrac{1}{2} > x > -\dfrac{1}{8}$

The solution set is $\left\{x \Big| -\dfrac{1}{8} < x < \dfrac{1}{2}\right\}$.

73. $x+4 < 2x-6 \le x+12$

$4 < x-6 \le 12 \qquad$ Subtracting x

$10 < x \le 18$

The solution set is $\{x| 10 < x \le 18\}$.

Exercise Set 2.6

1. $|5x| = |5| \cdot |x| = 5|x|$

3. $|7x^2| = |7| \cdot |x^2|$
$= 7|x^2|$
$= 7x^2$ Since x^2 is never negative

5. $|-2x^2| = |-2| \cdot |x^2|$
$= 2|x^2|$
$= 2x^2$ Since x^2 is never negative

7. $|-6y| = |-6| \cdot |y| = 6|y|$

9. $\left|\dfrac{-2}{x}\right| = \dfrac{|-2|}{|x|} = \dfrac{2}{|x|}$

11. $\left|\dfrac{x^2}{-y}\right| = \dfrac{|x^2|}{|-y|}$
$= \dfrac{x^2}{|-y|}$
$= \dfrac{x^2}{|y|}$ The absolute value of the opposite of a number is the same as the absolute value of the number.

13. $\left|\dfrac{-8x^2}{2x}\right| = |-4x| = |-4| \cdot |x| = 4|x|$

15. $|-8 - (-46)| = |38| = 38$, or
$|-46 - (-8)| = |-38| = 38$

17. $|36 - 17| = |19| = 19$, or
$|17 - 36| = |-19| = 19$

19. $|-3.9 - 2.4| = |-6.3| = 6.3$, or
$|2.4 - (-3.9)| = |6.3| = 6.3$

21. $|-5 - 0| = |-5| = 5$, or
$|0 - (-5)| = |5| = 5$

23. $|x| = 3$
$x = -3 \ or \ x = 3$ Absolute-value principle
The solution set is $\{-3, 3\}$.

25. $|x| = -3$
The absolute value of a number is always nonnegative. Therefore, the solution set is $\emptyset$.

27. $|q| = 0$
The only number whose absolute value is 0 is 0. The solution set is $\{0\}$.

29. $|x - 3| = 12$
$x - 3 = -12 \ or \ x - 3 = 12$ Absolute-value principle
$x = -9 \ or \ x = 15$
The solution set is $\{-9, 15\}$.

31. $|2x - 3| = 4$
$2x - 3 = -4 \ or \ 2x - 3 = 4$ Absolute-value principle
$2x = -1 \ or \ 2x = 7$
$x = -\dfrac{1}{2} \ or \ x = \dfrac{7}{2}$
The solution set is $\left\{-\dfrac{1}{2}, \dfrac{7}{2}\right\}$.

33. $|4x - 9| = 14$
$4x - 9 = -14 \ or \ 4x - 9 = 14$
$4x = -5 \ or \ 4x = 23$
$x = -\dfrac{5}{4} \ or \ x = \dfrac{23}{4}$
The solution set is $\left\{-\dfrac{5}{4}, \dfrac{23}{4}\right\}$.

35. $|x| + 7 = 18$
$|x| + 7 - 7 = 18 - 7$ Subtracting 7
$|x| = 11$
$x = -11 \ or \ x = 11$ Absolute-value principle
The solution set is $\{-11, 11\}$.

37. $574 = 283 + |t|$
$291 = |t|$ Subtracting 283
$t = -291 \ or \ t = 291$ Absolute-value principle
The solution set is $\{-291, 291\}$.

39. $|5x| = 40$
$5x = -40 \ or \ 5x = 40$
$x = -8 \ or \ x = 8$
The solution set is $\{-8, 8\}$.

41. $|3x| - 4 = 17$
$|3x| = 21$ Adding 4
$3x = -21 \ or \ 3x = 21$
$x = -7 \ or \ x = 7$
The solution set is $\{-7, 7\}$.

43. $7|w| - 3 = 11$
$7|w| = 14$ Adding 3
$|w| = 2$ Dividing by 7
$w = -2 \ or \ w = 2$ Absolute-value principle
The solution set is $\{-2, 2\}$.

45. $\left|\dfrac{2x - 1}{3}\right| = 5$
$\dfrac{2x - 1}{3} = -5 \ or \ \dfrac{2x - 1}{3} = 5$
$2x - 1 = -15 \ or \ 2x - 1 = 15$
$2x = -14 \ or \ 2x = 16$
$x = -7 \ or \ x = 8$
The solution set is $\{-7, 8\}$.

47. $|m+5|+9=16$

$\quad\quad |m+5| = 7 \quad\quad$ Subtracting 9

$\quad m+5 = -7 \quad or \quad m+5 = 7$
$\quad\quad m = -12 \quad or \quad\quad m = 2$

The solution set is $\{-12, 2\}$.

49. $10 - |2x-1| = 4$

$\quad\quad -|2x-1| = -6 \quad\quad$ Subtracting 10

$\quad\quad |2x-1| = 6 \quad\quad$ Multiplying by -1

$\quad 2x-1 = -6 \quad or \quad 2x-1 = 6$

$\quad\quad 2x = -5 \quad or \quad\quad 2x = 7$

$\quad\quad x = -\dfrac{5}{2} \quad or \quad\quad x = \dfrac{7}{2}$

The solution set is $\left\{-\dfrac{5}{2}, \dfrac{7}{2}\right\}$.

51. $|3x-4| = -2$

The absolute value of a number is always nonnegative. The solution set is $\emptyset$.

53. $\left|\dfrac{5}{9} + 3x\right| = \dfrac{1}{6}$

$\quad \dfrac{5}{9} + 3x = -\dfrac{1}{6} \quad or \quad \dfrac{5}{9} + 3x = \dfrac{1}{6}$

$\quad\quad 3x = -\dfrac{13}{18} \quad or \quad\quad 3x = -\dfrac{7}{18}$

$\quad\quad x = -\dfrac{13}{54} \quad or \quad\quad x = -\dfrac{7}{54}$

The solution set is $\left\{-\dfrac{13}{54}, -\dfrac{7}{54}\right\}$.

55. $|3x+4| = |x-7|$

$\quad 3x+4 = x-7 \quad or \quad 3x+4 = -(x-7)$

$\quad 2x+4 = -7 \quad or \quad 3x+4 = -x+7$

$\quad\quad 2x = -11 \quad or \quad 4x+4 = 7$

$\quad\quad x = -\dfrac{11}{2} \quad or \quad\quad 4x = 3$

$\quad\quad x = -\dfrac{11}{2} \quad or \quad\quad x = \dfrac{3}{4}$

The solution set is $\left\{-\dfrac{11}{2}, \dfrac{3}{4}\right\}$.

57. $|x+3| = |x-6|$

$\quad x+3 = x-6 \quad or \quad x+3 = -(x-6)$

$\quad\quad 3 = -6 \quad or \quad x+3 = -x+6$

$\quad\quad 3 = -6 \quad or \quad\quad 2x = 3$

$\quad\quad 3 = -6 \quad or \quad\quad x = \dfrac{3}{2}$

The first equation has no solution. The second equation has $\dfrac{3}{2}$ as a solution. There is only one solution of the original equation. The solution set is $\left\{\dfrac{3}{2}\right\}$.

59. $|2a+4| = |3a-1|$

$\quad 2a+4 = 3a-1 \quad or \quad 2a+4 = -(3a-1)$

$\quad -a+4 = -1 \quad or \quad 2a+4 = -3a+1$

$\quad\quad -a = -5 \quad or \quad 5a+4 = 1$

$\quad\quad a = 5 \quad or \quad\quad 5a = -3$

$\quad\quad a = 5 \quad or \quad\quad a = -\dfrac{3}{5}$

The solution set is $\left\{5, -\dfrac{3}{5}\right\}$.

61. $|y-3| = |3-y|$

$\quad y-3 = 3-y \quad or \quad y-3 = -(3-y)$

$\quad 2y-3 = 3 \quad or \quad y-3 = -3+y$

$\quad\quad 2y = 6 \quad or \quad\quad -3 = -3$

$\quad\quad y = 3 \quad\quad$ True for all real values of y

All real numbers are solutions.

63. $|5-p| = |p+8|$

$\quad 5-p = p+8 \quad or \quad 5-p = -(p+8)$

$\quad 5-2p = 8 \quad or \quad 5-p = -p-8$

$\quad\quad -2p = 3 \quad or \quad\quad 5 = -8$

$\quad\quad p = -\dfrac{3}{2} \quad\quad$ False

The solution set is $\left\{-\dfrac{3}{2}\right\}$.

65. $\left|\dfrac{2x-3}{6}\right| = \left|\dfrac{4-5x}{8}\right|$

$\quad \dfrac{2x-3}{6} = \dfrac{4-5x}{8} \quad or \quad \dfrac{2x-3}{6} = -\left(\dfrac{4-5x}{8}\right)$

$\quad 24\left(\dfrac{2x-3}{6}\right) = 24\left(\dfrac{4-5x}{8}\right) \quad or \quad \dfrac{2x-3}{6} = \dfrac{-4+5x}{8}$

$\quad 8x-12 = 12-15x \quad or \quad 24\left(\dfrac{2x-3}{6}\right) = 24\left(\dfrac{-4+5x}{8}\right)$

$\quad 23x-12 = 12 \quad or \quad 8x-12 = -12+15x$

$\quad\quad 23x = 24 \quad or \quad -7x-12 = -12$

$\quad\quad x = \dfrac{24}{23} \quad or \quad\quad -7x = 0$

$\quad\quad\quad\quad\quad\quad\quad\quad\quad\quad x = 0$

The solution set is $\left\{\dfrac{24}{23}, 0\right\}$.

67. $\left|\frac{1}{2}x - 5\right| = \left|\frac{1}{4}x + 3\right|$

$\frac{1}{2}x - 5 = \frac{1}{4}x + 3$ *or* $\frac{1}{2}x - 5 = -\left(\frac{1}{4}x + 3\right)$

$\frac{1}{4}x - 5 = 3$ *or* $\frac{1}{2}x - 5 = -\frac{1}{4}x - 3$

$\frac{1}{4}x = 8$ *or* $\frac{3}{4}x - 5 = -3$

$x = 32$ *or* $\frac{3}{4}x = 2$

$x = 32$ *or* $x = \frac{8}{3}$

The solution set is $\left\{32, \frac{8}{3}\right\}$.

69. $|x| < 3$

$-3 < x < 3$ Part (b)

The solution set is $\{x | -3 < x < 3\}$.

71. $|x| \geq 2$

$x \leq -2$ or $x \geq 2$ Part (c)

The solution set is $\{x | x \leq -2 \text{ or } x \geq 2\}$.

73. $|x - 1| < 1$

$-1 < x - 1 < 1$ Part (b)

$0 < x < 2$

The solution set is $\{x | 0 < x < 2\}$.

75. $|x + 4| \leq 1$

$-1 \leq x + 4 \leq 1$ Part (b)

$-5 \leq x \leq -3$ Subtracting 4

The solution set is $\{x | -5 \leq x \leq -3\}$.

77. $|2x - 3| \leq 4$

$-4 \leq 2x - 3 \leq 4$ Part (b)

$-1 \leq 2x \leq 7$ Adding 3

$-\frac{1}{2} \leq x \leq \frac{7}{2}$ Dividing by 2

The solution set is $\left\{x \middle| -\frac{1}{2} \leq x \leq \frac{7}{2}\right\}$.

79. $|2y - 7| > 10$

$2y - 7 < -10$ *or* $2y - 7 > 10$ Part (c)

$2y < -3$ *or* $2y > 17$ Adding 7

$y < -\frac{3}{2}$ *or* $y > \frac{17}{2}$ Dividing by 2

The solution set is $\left\{y \middle| y < -\frac{3}{2} \text{ or } y > \frac{17}{2}\right\}$.

81. $|4x - 9| \geq 14$

$4x - 9 \leq -14$ *or* $4x - 9 \geq 14$ Part (c)

$4x \leq -5$ *or* $4x \geq 23$

$x \leq -\frac{5}{4}$ *or* $x \geq \frac{23}{4}$

The solution set is $\left\{x \middle| x \leq -\frac{5}{4} \text{ or } x \geq \frac{23}{4}\right\}$.

83. $|y - 3| < 12$

$-12 < y - 3 < 12$ Part (b)

$-9 < y < 15$ Adding 3

The solution set is $\{y | -9 < y < 15\}$.

85. $|2x + 3| \leq 4$

$-4 \leq 2x + 3 \leq 4$ Part (b)

$-7 \leq 2x \leq 1$ Subtracting 3

$-\frac{7}{2} \leq x \leq \frac{1}{2}$ Dividing by 2

The solution set is $\left\{x \middle| -\frac{7}{2} \leq x \leq \frac{1}{2}\right\}$.

87. $|4 - 3y| > 8$

$4 - 3y < -8$ *or* $4 - 3y > 8$ Part (c)

$-3y < -12$ *or* $-3y > 4$ Subtracting 4

$y > 4$ *or* $y < -\frac{4}{3}$ Dividing by -3

The solution set is $\left\{y \middle| y < -\frac{4}{3} \text{ or } y > 4\right\}$.

89. $|9 - 4x| \geq 14$

$9 - 4x \leq -14$ *or* $9 - 4x \geq 14$ Part (c)

$-4x \leq -23$ *or* $-4x \geq 5$ Subtracting 9

$x \geq \frac{23}{4}$ *or* $x \leq -\frac{5}{4}$ Dividing by -4

The solution set is $\left\{x \middle| x \leq -\frac{5}{4} \text{ or } x \geq \frac{23}{4}\right\}$.

91. $|3 - 4x| < 21$

$-21 < 3 - 4x < 21$ Part (b)

$-24 < -4x < 18$ Subtracting 3

$6 > x > -\frac{9}{2}$ Dividing by -4 and simplifying

The solution set is $\left\{x \middle| 6 > x > -\frac{9}{2}\right\}$, or

$\left\{x \middle| -\frac{9}{2} < x < 6\right\}$.

93. $\left|\frac{1}{2} + 3x\right| \geq 12$

$\frac{1}{2} + 3x \leq -12$ *or* $\frac{1}{2} + 3x \geq 12$ Part (c)

$3x \leq -\frac{25}{2}$ *or* $3x \geq \frac{23}{2}$ Subtracting $\frac{1}{2}$

$x \leq -\frac{25}{6}$ *or* $x \geq \frac{23}{6}$ Dividing by 3

The solution set is $\left\{x \middle| x \leq -\frac{25}{6} \text{ or } x \geq \frac{23}{6}\right\}$.

95. $\left|\frac{x - 7}{3}\right| < 4$

$-4 < \frac{x - 7}{3} < 4$ Part (b)

$-12 < x - 7 < 12$ Multiplying by 3

$-5 < x < 19$ Adding 7

The solution set is $\{x | -5 < x < 19\}$.

97. $\left|\dfrac{2-5x}{4}\right| \geq \dfrac{2}{3}$

$\dfrac{2-5x}{4} \leq -\dfrac{2}{3}$ *or* $\dfrac{2-5x}{4} \geq \dfrac{2}{3}$ Part (c)

$2-5x \leq -\dfrac{8}{3}$ *or* $2-5x \geq \dfrac{8}{3}$ Multiplying by 4

$-5x \leq -\dfrac{14}{3}$ *or* $-5x \geq \dfrac{2}{3}$ Subtracting 2

$x \geq \dfrac{14}{15}$ *or* $x \leq -\dfrac{2}{15}$ Dividing by -5

The solution set is $\left\{x \Big| x \leq -\dfrac{2}{15} \ or \ x \geq \dfrac{14}{15}\right\}$.

99. $|m+5| + 9 \leq 16$

$|m+5| \leq 7$ Subtracting 9

$-7 \leq m+5 \leq 7$

$-12 \leq m \leq 2$

The solution set is $\{m| -12 \leq m \leq 2\}$.

101. $7 - |3-2x| \geq 5$

$-|3-2x| \geq -2$ Subtracting 7

$|3-2x| \leq 2$ Multiplying by -1

$-2 \leq 3-2x \leq 2$ Part (b)

$-5 \leq -2x \leq -1$ Subtracting 3

$\dfrac{5}{2} \geq x \geq \dfrac{1}{2}$ Dividing by -2

The solution set is $\left\{x \Big| \dfrac{5}{2} \geq x \geq \dfrac{1}{2}\right\}$, or $\left\{x \Big| \dfrac{1}{2} \leq x \leq \dfrac{5}{2}\right\}$.

103. $\left|\dfrac{2x-1}{0.0059}\right| \leq 1$

$-1 \leq \dfrac{2x-1}{0.0059} \leq 1$

$-0.0059 \leq 2x-1 \leq 0.0059$

$0.9941 \leq 2x \leq 1.0059$

$0.49705 \leq x \leq 0.50295$

The solution set is $\{x| 0.49705 \leq x \leq 0.50295\}$.

105. $-43.5 + (-5.8) = -49.3$

(Add absolute values and make the result negative.)

107. $-43.5(-5.8) = 252.3$

(Multiply absolute values and make the result positive.)

109. $-\dfrac{7}{8} \div \dfrac{3}{4} = -\dfrac{7}{8} \cdot \dfrac{4}{3}$

$= -\dfrac{7 \cdot 4}{8 \cdot 3}$

$= -\dfrac{7 \cdot 4}{2 \cdot 4 \cdot 3}$

$= -\dfrac{7 \cdot 4}{2 \cdot 4 \cdot 3}$

$= -\dfrac{7}{6}$

111. From the definition of absolute value, $|2x-5| = 2x-5$ only when $2x-5 \geq 0$. Solve $2x-5 \geq 0$.

$2x-5 \geq 0$

$2x \geq 5$

$x \geq \dfrac{5}{2}$

The solution set is $\left\{x \Big| x \geq \dfrac{5}{2}\right\}$.

113. $|x+5| = x+5$

From the definition of absolute value, $|x+5| = x+5$ only when $x+5 \geq 0$, or $x \geq -5$. The solution set is $\{x| x \geq -5\}$.

115. $|7x-2| = x+4$

From the definition of absolute value, we know $x+4 \geq 0$, or $x \geq -4$. So we have $x \geq -4$ and

$7x-2 = x+4$ *or* $7x-2 = -(x+4)$

$6x = 6$ *or* $7x-2 = -x-4$

$x = 1$ *or* $8x = -2$

$x = 1$ *or* $x = -\dfrac{1}{4}$

The solution set is $\left\{x \Big| x \geq -4 \ and \ x = 1 \ or \ x = -\dfrac{1}{4}\right\}$, or $\left\{1, -\dfrac{1}{4}\right\}$.

117. $|x-6| \leq -8$

From the definition of absolute value we know that $|x-6| \geq 0$. Thus $|x-6| \leq -8$ is false for all x. The solution set is $\emptyset$.

119. $|x+5| > x$

The inequality is true for all $x < 0$ (because absolute value must be nonnegative). The solution set in this case is $\{x| x < 0\}$. If $x = 0$, we have $|0+5| > 0$, which is true. The solution set in this case is $\{0\}$. If $x > 0$, we have the following:

$x+5 < -x$ *or* $x+5 > x$

$2x < -5$ *or* $5 > 0$

$x < -\dfrac{5}{2}$ *or* $5 > 0$

Although $x > 0$ and $x < -\dfrac{5}{2}$ yields no solution, $x > 0$ and $5 > 0$ (true for all x) yield the solution set $\{x| x > 0\}$ in this case. The solution set for the inequality is $\{x| x < 0\} \cup \{0\} \cup \{x| x > 0\}$, or all real numbers.

121. Using part (b), we find that $-3 < x < 3$ is equivalent to $|x| < 3$.

123. Using part (c), we find that $x \leq -6$ or or $x \geq 6$ is equivalent to $|x| \geq 6$

125. $x < -8$ *or* $x > 2$

$x+3 < -5$ *or* $x+3 > 5$ Adding 3

$|x+3| > 5$ Part (c)

Chapter 3

Graphs of Equations and Inequalities

Exercise Set 3.1

1. $A(4,1)$ is 4 units right and 1 unit up.

   $B(2,5)$ is 2 units right and 5 units up.

   $C(0,3)$ is 0 units left or right and 3 units up.

   $D(0,-5)$ is 0 units left or right and 5 units down.

   $E(6,0)$ is 6 units right and 0 units up or down.

   $F(-3,0)$ is 3 units left and 0 units up or down.

   $G(-2,-4)$ is 2 units left and 4 units down.

   $H(-5,1)$ is 5 units left and 1 unit up.

   $J(-6,6)$ is 6 units left and 6 units up.

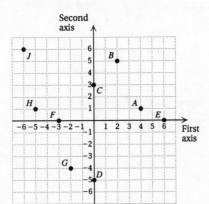

3.

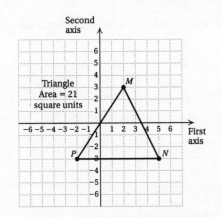

A triangle is formed. The area of a triangle is found by using the formula $A = \frac{1}{2}bh$. In this triangle the base and height are respectively 7 units and 6 units.

$$A = \frac{1}{2}bh = \frac{1}{2} \cdot 7 \cdot 6 = \frac{42}{2} = 21 \text{ square units}$$

5. We substitute 11 for x and 2 for y.

$$2x - 7y = 8$$

$2 \cdot 11 - 7 \cdot 2$	8
$22 - 14$	
8	

The equation becomes true; $(11,2)$ is a solution.

7. We substitute 3 for x and -2 for y.

$$2y - 5x = 8 + 3x$$

$2(-2) - 5 \cdot 3$	$8 + 3 \cdot 3$
$-4 - 15$	$8 + 6$
-19	14

The equation becomes false; $(3,-2)$ is not a solution.

9. We substitute 1 for p and 4 for q.

$$3p - q = 1$$

$3 \cdot 1 - 4$	1
$3 - 4$	
-1	

The equation becomes false; $(1,4)$ is not a solution.

11. Graph: $y = 5x$

    We choose any number for x and then determine y. We find several ordered pairs in this manner, plot them, and draw the line.

    When $x = 0$,　$y = 5 \cdot 0 = 0$.

    When $x = -1$, $y = 5(-1) = -5$.

    When $x = 1$,　$y = 5 \cdot 1 = 5$.

    When $x = \frac{1}{2}$,　$y = 5 \cdot \frac{1}{2} = \frac{5}{2}$ or $2\frac{1}{2}$.

x	y $y = 5x$	(x,y)
0	0	$(0,0)$
-1	-5	$(-1,-5)$
1	5	$(1,5)$
$\frac{1}{2}$	$\frac{5}{2}$	$\left(\frac{1}{2}, \frac{5}{2}\right)$

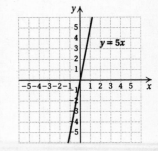

13. Graph: $y = -3x$

We choose any number for x and then determine y. We find several ordered pairs in this manner, plot them, and draw the line.

When $x = 0$, $y = -3 \cdot 0 = 0$.

When $x = -2$, $y = -3(-2) = 6$.

When $x = 1$, $y = -3 \cdot 1 = -3$.

When $x = 2$, $y = -3 \cdot 2 = -6$.

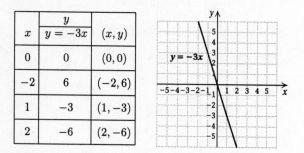

x	$y = -3x$	(x, y)
0	0	$(0, 0)$
-2	6	$(-2, 6)$
1	-3	$(1, -3)$
2	-6	$(2, -6)$

15. Graph: $y = x + 3$

We choose any number for x and then determine y. We find several ordered pairs in this manner, plot them, and draw the line.

When $x = -4$, $y = -4 + 3 = -1$.

When $x = -1$, $y = -1 + 3 = 2$.

When $x = 0$, $y = 0 + 3 = 3$.

When $x = 2$, $y = 2 + 3 = 5$.

x	$y = x + 3$	(x, y)
-4	-1	$(-4, -1)$
-1	2	$(-1, 2)$
0	3	$(0, 3)$
2	5	$(2, 5)$

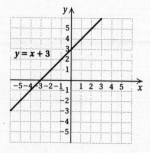

17. Graph: $y = \frac{1}{4}x + 2$

We choose any number for x and then determine y. We find several ordered pairs in this manner, plot them, and

draw the line.

When $x = -4$, $y = \frac{1}{4}(-4) + 2 = -1 + 2 = 1$.

When $x = 0$, $y = \frac{1}{4} \cdot 0 + 2 = 0 + 2 = 2$.

When $x = 2$, $y = \frac{1}{4} \cdot 2 + 2 = \frac{1}{2} + 2 = 2\frac{1}{2}$.

When $x = 4$, $y = \frac{1}{4} \cdot 4 + 2 = 1 + 2 = 3$.

x	$y = \frac{1}{4}x + 2$	(x, y)
-4	1	$(-4, 1)$
0	2	$(0, 2)$
2	$2\frac{1}{2}$	$\left(2, 2\frac{1}{2}\right)$
4	3	$(4, 3)$

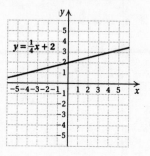

19. Graph: $y = -\frac{1}{5}x + 2$

We choose any number for x and then determine y. For example,

when $x = -5$, $y = -\frac{1}{5}(-5) + 2 = 1 + 2 = 3$;

when $x = 0$, $y = -\frac{1}{5}(0) + 2 = 0 + 2 = 2$.

We compute other pairs, plot them, and draw the line.

x	$y = -\frac{1}{5}x + 2$	(x, y)
-5	3	$(-5, 3)$
0	2	$(0, 2)$
2	$\frac{8}{5}$	$\left(2, \frac{8}{5}\right)$
5	1	$(5, 1)$

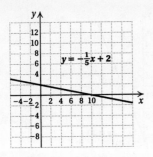

21. Graph: $y = 0.3x - 5$

We choose any number for x and then determine y. For example,

when $x = -1$, $y = 0.3(-1) - 5 = -5.3$;

when $x = 0$, $y = 0.3(0) - 5 = -5$.

x	y $y = 0.3x - 5$	(x, y)
-1	-5.3	$(-1, -5.3)$
0	-5	$(0, -5)$
2	-4.4	$(2, -4.4)$
5	-3.5	$(5, -3.5)$

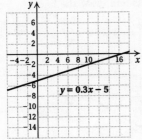

23. a) To find the number of gallons per year that each person drinks in 1990, we substitute 0 for t and calculate.

$$N = 0.733(0) + 8.398 = 8.398 \text{ gal}$$

To find the number of gallons per year that each person drinks in 1998, we substitute 8 for t and calculate.

$$N = 0.733(8) + 8.398 = 14.262 \text{ gal}$$

In 2000, $t = 2000 - 1990$, or 10. To find the number of gallons per year that each person drinks in 2000, we substitute 10 for t and calculate.

$$N = 0.733(10) + 8.398 = 15.728 \text{ gal}$$

b) Using the values computed in part (a) we make a table of values. Note that the number of years t is never negative, since only years from 1990 are considered.

Year	Gallons
1990 ($t = 0$)	8.398
1998 ($t = 8$)	14.262
2000 ($t = 10$)	15.728

Next we plot the points and draw the graph.

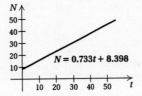

25. $-6x + 2x - 32 < 64$

$-4x - 32 < 64$ Collecting like terms

$-4x < 96$ Adding 32

$x > -24$ Dividing by -4 and reversing the inequality symbol

The solution set is $\{x | x > -24\}$.

27. $128 \div \left(-\dfrac{1}{2}\right) \div (-64) = 128 \cdot (-2) \div (-64)$

$$= -256 \div (-64)$$

$$= 4$$

29. $|3x - 7| < 24$

$-24 < 3x - 7 < 24$

$-17 < 3x < 31$ Adding 7

$-\dfrac{17}{3} < x < \dfrac{31}{3}$ Dividing by 3

The solution set is $\left\{x \left| -\dfrac{17}{3} < x < \dfrac{31}{3}\right.\right\}$.

Exercise Set 3.2

1. $x - 3y = 6$

To find the x-intercept we cover up the y-term and look at the rest of the equation. We have $x = 6$. The x-intercept is $(6, 0)$.

To find the y-intercept we cover up the x-term and look at the rest of the equation. We have $-3y = 6$, or $y = -2$. The y-intercept is $(0, -2)$.

We plot these points and draw the line.

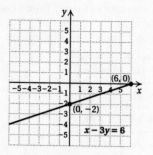

We use a third point as a check. We choose $x = -3$ and solve for y.

$$-3 - 3y = 6$$
$$-3y = 9$$
$$y = -3$$

We plot $(-3, -3)$ and note that it is on the line.

3. $x + 2y = 4$

To find the x-intercept we cover up the y-term and look at the rest of the equation. We have $x = 4$. The x-intercept is $(4, 0)$.

To find the y-intercept we cover up the x-term and look at the rest of the equation. We have $2y = 4$, or $y = 2$. The y-intercept is $(0, 2)$.

We plot these points and draw the line.

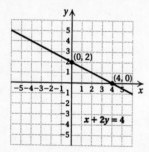

We use a third point as a check. We choose $x = -4$ and solve for y.

$$-4 + 2y = 4$$
$$2y = 8$$
$$y = 4$$

We plot $(-4, 4)$ and note that it is on the line.

5. $5x - 2y = 10$

To find the x-intercept we cover up the y-term and look at the rest of the equation. We have $5x = 10$, or $x = 2$. The x-intercept is $(2, 0)$.

To find the y-intercept we cover up the x-term and look at the rest of the equation. We have $-2y = 10$, or $y = -5$. The y-intercept is $(0, -5)$.

We plot these points and draw the line.

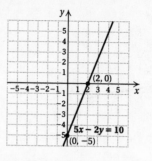

We use a third point as a check. We choose $x = 4$ and solve for y.

$$5(4) - 2y = 10$$
$$20 - 2y = 10$$
$$-2y = -10$$
$$y = 5$$

We plot $(4, 5)$ and note that it is on the line.

7. $5y = -15 + 3x$

To find the x-intercept we let $y = 0$ and solve for x. We have $0 = -15 + 3x$, or $15 = 3x$, or $5 = x$. The x-intercept is $(5, 0)$.

To find the y-intercept we cover up the x-term and look at the rest of the equation. We have $5y = -15$, or $y = -3$. The y-intercept is $(0, -3)$.

We plot these points and draw the line.

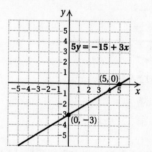

We use a third point as a check. We choose $x = -5$ and solve for y.

$$5y = -15 + 3(-5)$$
$$5y = -15 - 15$$
$$5y = -30$$
$$y = -6$$

We plot $(-5, -6)$ and note that it is on the line.

9. $5x - 10 = 5y$

To find the x-intercept we let $y = 0$ and solve for x. We have $5x - 10 = 0$, or $5x = 10$, or $x = 2$. The x-intercept is $(2, 0)$.

To find the y-intercept we cover up the x-term and look at the rest of the equation. We have $-10 = 5y$, or $-2 = y$. The y-intercept is $(0, -2)$.

We plot these points and draw the line.

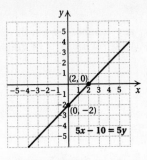

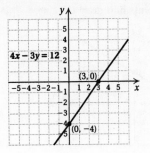

We use a third point as a check. We choose $x = 5$ and solve for y.

$$5(5) - 10 = 5y$$
$$25 - 10 = 5y$$
$$15 = 5y$$
$$3 = y$$

We plot $(5, 3)$ and note that it is on the line.

11. $4x + 5y = 20$

To find the x-intercept we cover up the y-term and look at the rest of the equation. We have $4x = 20$, or $x = 5$. The x-intercept is $(5, 0)$.

To find the y-intercept we cover up the x-term and look at the rest of the equation. We have $5y = 20$, or $y = 4$. The y-intercept is $(0, 4)$.

We plot these points and draw the line.

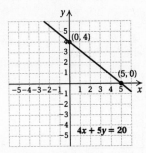

We use a third point as a check. We choose $x = 3$ and solve for y.

$$4(3) + 5y = 20$$
$$12 + 5y = 20$$
$$5y = 8$$
$$y = \frac{8}{5}$$

We plot $\left(3, \frac{8}{5}\right)$ and note that it is on the line.

13. $4x - 3y = 12$

To find the x-intercept we cover up the y-term and look at the rest of the equation. We have $4x = 12$, or $x = 3$. The x-intercept is $(3, 0)$.

To find the y-intercept we cover up the x-term and look at the rest of the equation. We have $-3y = 12$, or $y = -4$. The y-intercept is $(0, -4)$.

We plot these points and draw the line.

We use a third point as a check. We choose $x = 5$ and solve for y.

$$4 \cdot 5 - 3y = 12$$
$$20 - 3y = 12$$
$$-3y = -8$$
$$y = \frac{8}{3}$$

We plot $\left(5, \frac{8}{3}\right)$ and note that it is on the line.

15. $y - 3x = 0$

To find the x-intercept we cover up the y-term and look at the rest of the equation. We have $-3x = 0$, or $x = 0$. The x-intercept is $(0, 0)$. This is also the y-intercept. To find another point we let $x = 2$ and solve for y.

$$y - 3 \cdot 2 = 0$$
$$y - 6 = 0$$
$$y = 6$$

We plot $(0, 0)$ and $(2, 6)$ and draw the line.

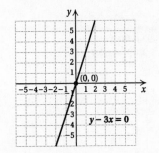

We use a third point as a check. We choose $x = -1$ and solve for y.

$$y - 3(-1) = 0$$
$$y + 3 = 0$$
$$y = -3$$

We plot $(-1, -3)$ and note that it is on the line.

17. $6x - 7 + 3y = 9x - 2y + 8$

$$-3x + 5y = 15 \qquad \text{Collecting like terms}$$

To find the x-intercept we cover up the y-term and look at the rest of the equation. We have $-3x = 15$, or $x = -5$. The x-intercept is $(-5, 0)$.

To find the y-intercept we cover up the x-term and look at the rest of the equation. We have $5y = 15$, or $y = 3$. The y-intercept is $(0, 3)$.

We plot these points and draw the line.

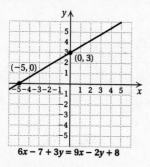

$$6x - 7 + 3y = 9x - 2y + 8$$

We use a third point as a check. We choose $x = 5$ and solve for y.

$$-3 \cdot 5 + 5y = 15$$
$$-15 + 5y = 15$$
$$5y = 30$$
$$y = 6$$

We plot $(5, 6)$ and note that it is on the line.

19. $x = 1$

Since y is missing, any number for y will do. Thus all ordered pairs $(1, y)$ are solutions. The graph is parallel to the y-axis.

x	y
1	-2
1	0	← x-intercept
1	3

x must
be 1.

Choose any
number for y.

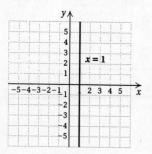

21. $y = -1$

Since x is missing, any number for x will do. Thus all ordered pairs $(x, -1)$ are solutions. The graph is parallel to the x-axis.

x	y
-2	-1
0	-1	← y-intercept
3	-1

Choose
any number
for x.

y must be -1.

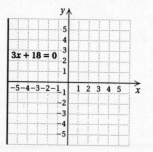

23. $3x + 18 = 0$
$$3x = -18$$
$$x = -6$$

Since y is missing, all ordered pairs $(-6, y)$ are solutions. The graph is parallel to the y-axis.

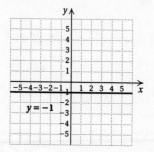

25. $y = 0$

Since x is missing, all ordered pairs $(x, 0)$ are solutions. The graph is the x-axis.

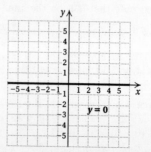

27. $y = -\dfrac{5}{2}$

Since x is missing, all ordered pairs $\left(x, -\dfrac{5}{2}\right)$ are solutions. The graph is parallel to the x-axis.

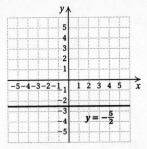

29. *Familiarize.* We will use the formula $I = Prt$.

Translate. We substitute $17.60 for I, $320 for P, and $\dfrac{1}{2}$ for t.

$$I = Prt$$
$$\$17.60 = \$320(r)\left(\dfrac{1}{2}\right)$$

Solve.
$$\$17.60 = \$320(r)\left(\dfrac{1}{2}\right)$$
$$\$17.60 = \$160r$$
$$0.11 = r \qquad \text{Dividing by \$160}$$

Check. $320(0.11)\left(\dfrac{1}{2}\right) = \$17.60$. The number checks.

State. The interest rate would have to be 0.11, or 11%.

31.
$$4x - 2 \le 5x + 7$$
$$-9 \le x \qquad \text{Subtracting } 4x \text{ and } 7$$

The solution set is $\{x| -9 \le x\}$, or $\{x| x \ge -9\}$.

33. $|5x + 7| = 8$
$$5x + 7 = 8 \quad or \quad 5x + 7 = -8$$
$$5x = 1 \quad or \qquad 5x = -15$$
$$x = \dfrac{1}{5} \quad or \qquad x = -3$$

The solution set is $\left\{\dfrac{1}{5}, -3\right\}$.

35. All points on the x-axis are pairs of the form $(x, 0)$. Thus any number for x will do and y must be 0. The equation is $y = 0$.

37. The x-coordinate must be -4, and the y-coordinate must be 5. The point is $(-4, 5)$.

39. The x-coordinate of a point on the line must be 12, and any number for y will do. The equation is $x = 12$.

41. We substitute 4 for x and 0 for y.
$$y = mx + 3$$
$$0 = m(4) + 3$$
$$-3 = 4m$$
$$-\dfrac{3}{4} = m$$

1. Let $(3, 7) = (x_1, y_1)$ and $(7, -5) = (x_2, y_2)$.

Slope $= \dfrac{y_2 - y_1}{x_2 - x_1} = \dfrac{-5 - 7}{7 - 3} = \dfrac{-12}{4} = -3$

3. Let $(16, -12) = (x_1, y_1)$ and $(-8, -15) = (x_2, y_2)$.

Slope $= \dfrac{y_1 - y_2}{x_1 - x_2} = \dfrac{-12 - (-15)}{16 - (-8)} = \dfrac{3}{24} = \dfrac{1}{8}$

5. Let $(-12.4, 9.3) = (x_1, y_1)$ and $(-3.7, 6.4) = (x_2, y_2)$.

Slope $= \dfrac{y_2 - y_1}{x_2 - x_1} = \dfrac{6.4 - 9.3}{-3.7 - (-12.4)} = \dfrac{-2.9}{8.7} = -\dfrac{1}{3}$, or $-0.\overline{3}$

7. Let $(5.1, -14.8) = (x_1, y_1)$ and $(5.1, 3.4) = (x_2, y_2)$.

Slope $= \dfrac{y_1 - y_2}{x_1 - x_2} = \dfrac{-14.8 - 3.4}{5.1 - 5.1} = \dfrac{18.2}{0}$

Since division by 0 is not defined, the slope is undefined.

9.
$$3x = 12 + y$$
$$3x - 12 = y$$
$$y = 3x - 12 \quad (y = mx + b)$$

The slope is 3.

11.
$$5x - 6 = 15$$
$$5x = 21$$
$$x = \dfrac{21}{5}$$

When y is missing, the line is parallel to the y-axis. The line is vertical, and the slope is undefined.

13.
$$5y = 6$$
$$y = \dfrac{6}{5}$$

When x is missing, the line is parallel to the x-axis. The line is horizontal and has slope 0.

15.
$$y - 6 = 14$$
$$y = 20$$

When x is missing, the line is parallel to the x-axis. The line is horizontal and has slope 0.

17.
$$12 - 4x = 9 + x$$
$$3 = 5x$$
$$\dfrac{3}{5} = x$$

When y is missing, the line is parallel to the y-axis. The line is vertical, and the slope is undefined.

19.
$$2y - 4 = 35 + x$$
$$2y = x + 39$$
$$y = \dfrac{1}{2}x + \dfrac{39}{2} \quad (y = mx + b)$$

The slope is $\dfrac{1}{2}$.

21. $3y + x = 3y + 2$

$x = 2$

When y is missing, the line is parallel to the y-axis. The line is vertical, and the slope is undefined.

23. $3y - 2x = 5 + 9y - 2x$

$3y = 5 + 9y$

$-6y = 5$

$y = -\dfrac{5}{6}$

When x is missing, the line is parallel to the x-axis. The line is horizontal and has slope 0.

25. $2y - 7x = 10 - 3x$

$2y = 4x + 10$

$y = 2x + 5 \quad (y = mx + b)$

The slope is 2.

27. $y = \underbrace{-6}_{\downarrow} \; x \; \underbrace{-11}_{\downarrow}$

The slope   The y-intercept
is -6.   is $(0, -11)$.

29. $y = 2.9x$

Think of this as $y = 2.9x + 0$.

The slope is 2.9; the y-intercept is $(0, 0)$.

31. $2x + 3y = 8$

$3y = -2x + 8$

$y = -\dfrac{2}{3}x + \dfrac{8}{3}$

The slope is $-\dfrac{2}{3}$; the y-intercept is $\left(0, \dfrac{8}{3}\right)$.

33. $-8x - 7y = 24$

$-7y = 8x + 24$

$y = -\dfrac{8}{7}x - \dfrac{24}{7}$

The slope is $-\dfrac{8}{7}$; the y-intercept is $\left(0, -\dfrac{24}{7}\right)$.

35. $9x = 3y + 6$

$9x - 6 = 3y$

$3x - 2 = y$

The slope is 3; the y-intercept is $(0, -2)$.

37. $-6x = 4y + 3$

$-6x - 3 = 4y$

$-\dfrac{3}{2}x - \dfrac{3}{4} = y$

The slope is $-\dfrac{3}{2}$; the y-intercept is $\left(0, -\dfrac{3}{4}\right)$.

39.   We use the slope-intercept equation and substitute -8 for m and 4 for b.

$y = mx + b$

$y = -8x + 4$

41.   We use the slope-intercept equation and substitute 2.3 for m and -1 for b.

$y = mx + b$

$y = 2.3x - 1$

43.   We use the slope-intercept equation and substitute $-\dfrac{7}{3}$ for m and -5 for b.

$y = mx + b$

$y = -\dfrac{7}{3}x - 5$

45. $y = \dfrac{5}{2}x + 1$

First we plot the y-intercept $(0, 1)$. Then we consider the slope $\dfrac{5}{2}$. Starting at the y-intercept and using the slope, we find another point by moving 5 units up and 2 units to the right. We get to a new point $(2, 6)$.

We can also think of the slope as $\dfrac{-5}{-2}$. We again start at the y-intercept $(0, 1)$. We move 5 units down and 2 units to the left. We get to another new point $(-2, -4)$. We plot the points and draw the line.

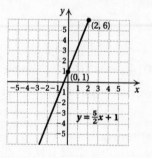

47. $y = -\dfrac{5}{2}x - 4$

First we plot the y-intercept $(0, -4)$. We can think of the slope as $\dfrac{-5}{2}$. Starting at the y-intercept and using the slope, we find another point by moving 5 units down and 2 units to the right. We get to a new point $(2, -9)$.

We can also think of the slope as $\dfrac{5}{-2}$. We again start at the y-intercept $(0, -4)$. We move 5 units up and 2 units to the left. We get to another new point $(-2, 1)$. We plot the points and draw the line.

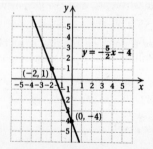

49. $y = 2x - 5$

First we plot the y-intercept $(0, -5)$. We can think of the slope as $\frac{2}{1}$. Starting at the y-intercept and using the slope, we find another point by moving 2 units up and 1 unit to the right. We get to a new point $(1, -3)$.

We can also think of the slope as $\frac{-2}{-1}$. We again start at the y-intercept $(0, -5)$. We move 2 units down and 1 unit to the left. We get to another new point $(-1, -7)$. We plot the points and draw the line.

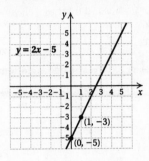

51. $y = \frac{1}{3}x + 6$

First we plot the y-intercept $(0, 6)$. Then we consider the slope $\frac{1}{3}$. Starting at the y-intercept and using the slope, we find another point by moving 1 unit up and 3 units to the right. We get to a new point $(3, 7)$.

We can also think of the slope as $\frac{-1}{-3}$. We again start at the y-intercept $(0, 6)$. We move 1 unit down and 3 units to the left. We get to another new point $(-3, 5)$. We plot the points and draw the line.

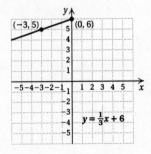

53. $y = -0.25x + 2$, or $y = -\frac{1}{4}x + 2$

First we plot the y-intercept $(0, 2)$. We can think of the slope as $\frac{-1}{4}$. Starting at the y-intercept and using the slope, we move 1 unit down and 4 units to the right. We get to a new point $(4, 1)$.

We can also think of the slope as $\frac{1}{-4}$. We again start at the y-intercept $(0, 2)$. We move 1 unit up and 4 units to the left. We get to another new point $(-4, 3)$. We plot the points and draw the graph.

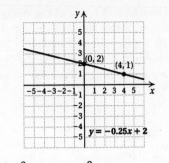

55. $y = -\frac{3}{4}x$, or $y = -\frac{3}{4}x + 0$

First we plot the y-intercept $(0, 0)$. We can think of the slope as $\frac{-3}{4}$. Starting at the y-intercept we move 3 units down and 4 units to the right. We get to a new point $(4, -3)$.

We can also think of the slope as $\frac{3}{-4}$. We again start at the y-intercept $(0, 0)$. We move 3 units up and 4 units to the left. We get to another new point $(-4, 3)$. We plot the points and draw the graph.

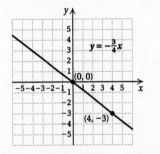

57. Grade $= \dfrac{\text{vertical change}}{\text{horizontal change}} = \dfrac{211.2 \text{ ft}}{5280 \text{ ft}} = 0.04$

The grade of the road is 4%.

59. The grade is 13%, or 0.13. Let v represent the height of the end of the treadmill.

$$\text{Grade} = \frac{\text{vertical change}}{\text{horizontal change}}$$

$$0.13 = \frac{v}{6 \text{ ft}} \qquad \text{Substituting}$$

$$0.78 \text{ ft} = v \qquad \text{Multiplying by 6 ft}$$

61. *Familiarize.* Let t represent the length of a side of the triangle. Then $t - 5$ represents the length of a side of the square.

Translate.

Perimeter of the square	is the same as	perimeter of the triangle
$\downarrow$	$\downarrow$	$\downarrow$
$4(t - 5)$	$=$	$3t$

Solve.

$$4(t - 5) = 3t$$
$$4t - 20 = 3t$$
$$t - 20 = 0$$
$$t = 20$$

Check. If 20 is the length of a side of the triangle, then the length of a side of the square is $20 - 5$, or 15. The perimeter of the square is $4 \cdot 15$, or 60, and the perimeter of the triangle is $3 \cdot 20$, or 60. The numbers check.

State. The square and triangle have sides of length 15 and 20, respectively.

63. $5x - 8 \geq 32$

$\qquad 5x \geq 40 \qquad$ Adding 8

$\qquad x \geq 8 \qquad$ Dividing by 5

The solution set is $\{x | x \geq 8\}$.

65. $|5x - 8| < 32$

$\qquad -32 < 5x - 8 < 32$

$\qquad -24 < 5x < 40 \qquad$ Adding 8

$\qquad -\dfrac{24}{5} < x < 8 \qquad$ Dividing by 5

The solution set is $\left\{ x | -\dfrac{24}{5} < x < 8 \right\}$.

67. $\dfrac{1}{8}y = -x - \dfrac{7}{16}$

$\qquad 8\left(\dfrac{1}{8}y \right) = 8\left(-x - \dfrac{7}{16} \right)$

$\qquad y = -8x - \dfrac{7}{2}$

The slope is -8, and the y-intercept is $\left(0, -\dfrac{7}{2} \right)$.

69.

$\qquad x = -\dfrac{7}{3}y - \dfrac{2}{11}$

$\qquad x + \dfrac{2}{11} = -\dfrac{7}{3}y$

$\qquad -\dfrac{3}{7}\left(x + \dfrac{2}{11} \right) = -\dfrac{3}{7}\left(-\dfrac{7}{3}y \right)$

$\qquad -\dfrac{3}{7}x - \dfrac{6}{77} = y$, or

$\qquad y = -\dfrac{3}{7}x - \dfrac{6}{77}$

The slope is $-\dfrac{3}{7}$, and the y-intercept is $\left(0, -\dfrac{6}{77} \right)$.

71. a) $m = \dfrac{-6c - (-c)}{5b - b} = \dfrac{-6c + c}{4b} = \dfrac{-5c}{4b}$, or $-\dfrac{5c}{4b}$

b) $m = \dfrac{d + e - d}{b - b} = \dfrac{e}{0}$ Undefined

The slope is undefined.

c) $m = \dfrac{a + d - (-a - d)}{c + f - (c - f)} = \dfrac{a + d + a + d}{c + f - c + f} =$

$\qquad \dfrac{2a + 2d}{2f} = \dfrac{2(a + d)}{2f} = \dfrac{a + d}{f}$

Exercise Set 3.4

1. $y - y_1 = m(x - x_1)$ Point-slope equation

$\quad y - 3 = 5(x - 4)$ Substituting 5 for m, 4 for x_1, and 3 for y_1

$\quad y - 3 = 5x - 20$ Simplifying

$\quad y = 5x - 17$

3. $y - y_1 = m(x - x_1)$ Point-slope equation

$\quad y - 6 = -3(x - 9)$ Substituting -3 for m, 9 for x_1, and 6 for y_1

$\quad y - 6 = -3x + 27$

$\quad y = -3x + 33$

5. $y - y_1 = m(x - x_1)$ Point-slope equation

$\quad y - (-7) = 1[x - (-1)]$ Substituting 1 for m, -1 for x_1, and -7 for y_1

$\quad y + 7 = x + 1$

$\quad y = x - 6$

7. $y - y_1 = m(x - x_1)$ Point-slope equation

$\quad y - 0 = -2(x - 8)$ Substituting -2 for m, 8 for x_1, and 0 for y_1

$\quad y = -2x + 16$

9. $y - y_1 = m(x - x_1)$ Point-slope equation

$\quad y - (-7) = 0(x - 0)$ Substituting 0 for m, 0 for x_1, and -7 for y_1

$\quad y + 7 = 0$

$\quad y = -7$

11. $y - y_1 = m(x - x_1)$ Point-slope equation

$\quad y - (-2) = \dfrac{2}{3}(x - 1)$ Substituting $\dfrac{2}{3}$ for m, 1 for x_1, and -2 for y_1

$\quad y + 2 = \dfrac{2}{3}x - \dfrac{2}{3}$

$\quad y = \dfrac{2}{3}x - \dfrac{8}{3}$

13. First find the slope of the line:

$\qquad m = \dfrac{6 - 4}{5 - 1} = \dfrac{2}{4} = \dfrac{1}{2}$

Use the point-slope equation with $m = \dfrac{1}{2}$ and $(1, 4) = (x_1, y_1)$. (We could let $(5, 6) = (x_1, y_1)$ instead and obtain an equivalent equation.)

$\qquad y - 4 = \dfrac{1}{2}(x - 1)$

$\qquad y - 4 = \dfrac{1}{2}x - \dfrac{1}{2}$

$\qquad y = \dfrac{1}{2}x + \dfrac{7}{2}$

15. First find the slope of the line:

$$m = \frac{-3-2}{-3-2} = \frac{-5}{-5} = 1$$

Use the point-slope equation with $m = 1$ and $(2,2) = (x_1, y_1)$. (We could let $(-3,-3) = (x_1, y_1)$ instead and obtain an equivalent equation.)

$$y - 2 = 1(x - 2)$$
$$y - 2 = x - 2$$
$$y = x$$

17. First find the slope of the line:

$$m = \frac{0-7}{-4-0} = \frac{-7}{-4} = \frac{7}{4}$$

Use the point-slope equation with $m = \frac{7}{4}$ and $(-4, 0) = (x_1, y_1)$. (We could let $(0, 7) = (x_1, y_1)$ instead and obtain an equivalent equation.)

$$y - 0 = \frac{7}{4}[x - (-4)]$$
$$y = \frac{7}{4}(x + 4)$$
$$y = \frac{7}{4}x + 7$$

19. First find the slope of the line:

$$m = \frac{-6-(-3)}{-4-(-2)} = \frac{-6+3}{-4+2} = \frac{-3}{-2} = \frac{3}{2}$$

Use the point-slope equation with $m = \frac{3}{2}$ and $(-2, -3) = (x_1, y_1)$.

$$y - (-3) = \frac{3}{2}[x - (-2)]$$
$$y + 3 = \frac{3}{2}(x + 2)$$
$$y + 3 = \frac{3}{2}x + 3$$
$$y = \frac{3}{2}x$$

21. First find the slope of the line:

$$m = \frac{1-0}{6-0} = \frac{1}{6}$$

Use the point-slope equation with $m = \frac{1}{6}$ and $(0, 0) = (x_1, y_1)$.

$$y - 0 = \frac{1}{6}(x - 0)$$
$$y = \frac{1}{6}x$$

23. First find the slope of the line:

$$m = \frac{-\frac{1}{2}-6}{\frac{1}{4}-\frac{3}{4}} = \frac{-\frac{13}{2}}{-\frac{1}{2}} = 13$$

Use the point-slope equation with $m = 13$ and $\left(\frac{3}{4}, 6\right) = (x_1, y_1)$.

$$y - 6 = 13\left(x - \frac{3}{4}\right)$$
$$y - 6 = 13x - \frac{39}{4}$$
$$y = 13x - \frac{15}{4}$$

25. We first solve for y and determine the slope of each line.

$$x + 6 = y$$
$$y = x + 6 \quad \text{Reversing the order}$$

The slope of $y = x + 6$ is 1.

$$y - x = -2$$
$$y = x - 2$$

The slope of $y = x - 2$ is 1.

The slopes are the same, and the y-intercepts are different. The lines are parallel.

27. We first solve for y and determine the slope of each line.

$$y + 3 = 5x$$
$$y = 5x - 3$$

The slope of $y = 5x - 3$ is 5.

$$3x - y = -2$$
$$3x + 2 = y$$
$$y = 3x + 2 \quad \text{Reversing the order}$$

The slope of $y = 3x + 2$ is 3.

The slopes are not the same; the lines are not parallel.

29. We determine the slope of each line.

The slope of $y = 3x + 9$ is 3.

$$2y = 6x - 2$$
$$y = 3x - 1$$

The slope of $y = 3x - 1$ is 3.

The slopes are the same, and the y-intercepts are different. The lines are parallel.

31. We determine the slope of each line.

The slope of $y = 4x - 5$ is 4.

$$4y = 8 - x$$
$$4y = -x + 8$$
$$y = -\frac{1}{4}x + 2$$

The slope of $4y = 8 - x$ is $-\frac{1}{4}$.

The product of their slopes is $4\left(-\frac{1}{4}\right)$, or -1; the lines are perpendicular.

33. We determine the slope of each line.

$$x + 2y = 5$$
$$2y = -x + 5$$
$$y = -\frac{1}{2}x + \frac{5}{2}$$

The slope of $x + 2y = 5$ is $-\frac{1}{2}$.

$$2x + 4y = 8$$
$$4y = -2x + 8$$
$$y = -\frac{1}{2}x + 2$$

The slope of $2x + 4y = 8$ is $-\frac{1}{2}$.

The product of their slopes is $\left(-\frac{1}{2}\right)\left(-\frac{1}{2}\right)$, or $\frac{1}{4}$; the lines are not perpendicular. For the lines to be perpendicular, the product must be -1.

35. We determine the slope of each line.

$$2x - 3y = 7$$
$$-3y = -2x + 7$$
$$y = \frac{2}{3}x - \frac{7}{3}$$

The slope of $2x - 3y = 7$ is $\frac{2}{3}$.

$$2y - 3x = 10$$
$$2y = 3x + 10$$
$$y = \frac{3}{2}x + 5$$

The slope of $2y - 3x = 10$ is $\frac{3}{2}$.

The product of their slopes is $\frac{2}{3} \cdot \frac{3}{2} = 1$; the lines are not perpendicular. For the lines to be perpendicular, the product must be -1.

37. First solve the equation for y and determine the slope of the given line.

$$x + 2y = 6 \qquad \text{Given line}$$
$$2y = -x + 6$$
$$y = -\frac{1}{2}x + 3$$

The slope of the given line is $-\frac{1}{2}$.

The line through $(3, 7)$ must have slope $-\frac{1}{2}$. We find an equation of this new line using the point-slope equation.

$$y - y_1 = m(x - x_1) \quad \text{Point-slope equation}$$
$$y - 7 = -\frac{1}{2}(x - 3) \quad \text{Substituting}$$
$$y - 7 = -\frac{1}{2}x + \frac{3}{2}$$
$$y = -\frac{1}{2}x + \frac{17}{2}$$

39. First solve the equation for y and determine the slope of the given line.

$$5x - 7y = 8 \qquad \text{Given line}$$
$$5x - 8 = 7y$$
$$\frac{5}{7}x - \frac{8}{7} = y$$
$$y = \frac{5}{7}x - \frac{8}{7}$$

The slope of the given line is $\frac{5}{7}$.

The line through $(2, -1)$ must have slope $\frac{5}{7}$. We find an equation of this new line using the point-slope equation.

$$y - y_1 = m(x - x_1) \quad \text{Point-slope equation}$$
$$y - (-1) = \frac{5}{7}(x - 2) \quad \text{Substituting}$$
$$y + 1 = \frac{5}{7}x - \frac{10}{7}$$
$$y = \frac{5}{7}x - \frac{17}{7}$$

41. First solve the equation for y and determine the slope of the given line.

$$3x - 9y = 2 \text{ Given line}$$
$$3x - 2 = 9y$$
$$\frac{1}{3}x - \frac{2}{9} = y$$

The slope of the given line is $\frac{1}{3}$.

The line through $(-6, 2)$ must have slope $\frac{1}{3}$. We find an equation of this new line using the point-slope equation.

$$y - y_1 = m(x - x_1) \qquad \text{Point-slope equation}$$
$$y - 2 = \frac{1}{3}[x - (-6)] \quad \text{Substituting}$$
$$y - 2 = \frac{1}{3}(x + 6)$$
$$y - 2 = \frac{1}{3}x + 2$$
$$y = \frac{1}{3}x + 4$$

43. First solve the equation for y and determine the slope of the given line.

$$2x + y = -3 \qquad \text{Given line}$$
$$y = -2x - 3$$

The slope of the given line is -2.

To find the slope of a perpendicular line, take the reciprocal of -2 and change the sign. The slope is $\frac{1}{2}$.

We find the equation of the line with slope $\frac{1}{2}$ containing the point $(2, 5)$.

$$y - y_1 = m(x - x_1) \quad \text{Point-slope equation}$$

$$y - 5 = \frac{1}{2}(x - 2) \quad \text{Substituting}$$

$$y - 5 = \frac{1}{2}x - 1$$

$$y = \frac{1}{2}x + 4$$

45. First solve the equation for y and determine the slope of the given line.

$$3x + 4y = 5 \qquad \text{Given line}$$

$$4y = -3x + 5$$

$$y = -\frac{3}{4}x + \frac{5}{4}$$

The slope of the given line is $-\frac{3}{4}$.

To find the slope of a perpendicular line, take the reciprocal of $-\frac{3}{4}$ and change the sign. The slope is $\frac{4}{3}$.

We find the equation of the line with slope $\frac{4}{3}$ containing the point $(3, -2)$.

$$y - y_1 = m(x - x_1) \quad \text{Point-slope equation}$$

$$y - (-2) = \frac{4}{3}(x - 3) \quad \text{Substituting}$$

$$y + 2 = \frac{4}{3}x - 4$$

$$y = \frac{4}{3}x - 6$$

47. First solve the equation for y and determine the slope of the given line.

$$2x + 5y = 7 \qquad \text{Given line}$$

$$5y = -2x + 7$$

$$y = -\frac{2}{5}x + \frac{7}{5}$$

The slope of the given line is $-\frac{2}{5}$.

To find the slope of a perpendicular line, take the reciprocal of $-\frac{2}{5}$ and change the sign. The slope is $\frac{5}{2}$.

We find the equation of the line with slope $\frac{5}{2}$ containing the point $(0, 9)$.

$$y - y_1 = m(x - x_1) \quad \text{Point-slope equation}$$

$$y - 9 = \frac{5}{2}(x - 0) \quad \text{Substituting}$$

$$y - 9 = \frac{5}{2}x$$

$$y = \frac{5}{2}x + 9$$

49. $2x + 3 > 51$

$\qquad 2x > 48 \qquad \text{Subtracting 3}$

$\qquad x > 24 \qquad \text{Dividing by 2}$

The solution set is $\{x | x > 24\}$.

51. $2x + 3 \leq 51$

$\qquad 2x \leq 48 \qquad \text{Subtracting 3}$

$\qquad x \leq 24 \qquad \text{Dividing by 2}$

The solution set is $\{x | x \leq 24\}$.

53. $|2x + 3| \leq 13$

$\qquad -13 \leq 2x + 3 \leq 13$

$\qquad -16 \leq 2x \leq 10 \qquad \text{Subtracting 3}$

$\qquad -8 \leq x \leq 5 \qquad \text{Dividing by 2}$

The solution set is $\{x | -8 \leq x \leq 5\}$.

55. Find the slope of the line containing $(-1, 4)$ and $(2, -3)$.

$$m - \frac{4 - (-3)}{-1 - 2} = \frac{7}{-3} = -\frac{7}{3}$$

Use the point-slope equation to find an equation of the line having slope $-\frac{7}{3}$ and containing $(4, -2)$.

$$y - (-2) = -\frac{7}{3}(x - 4) \quad \text{Substituting}$$

$$y + 2 = -\frac{7}{3}x + \frac{28}{3}$$

$$y = -\frac{7}{3}x + \frac{22}{3}$$

57. Find the slope of each line.

$$5y = ax + 5$$

$$y = \frac{a}{5}x + 1$$

The slope of $5y = ax + 5$ is $\frac{a}{5}$.

$$\frac{1}{4}y = \frac{1}{10}x - 1$$

$$4 \cdot \frac{1}{4}y = 4\left(\frac{1}{10}x - 1\right)$$

$$y = \frac{2}{5}x - 4$$

The slope of $\frac{1}{4}y = \frac{1}{10}x - 1$ is $\frac{2}{5}$.

In order for the graphs to be parallel, their slopes must be the same. (Note that the y-intercepts are different.)

$$\frac{a}{5} = \frac{2}{5}$$

$$a = 2 \qquad \text{Multiplying by 5}$$

59. The y-intercept is $\left(0, \frac{2}{5}\right)$, so the equation is of the form $y = mx + \frac{2}{5}$. We substitute -3 for x and 0 for y in this equation to find m.

$$y = mx + \frac{2}{5}$$

$$0 = m(-3) + \frac{2}{5} \qquad \text{Substituting}$$

$$0 = -3m + \frac{2}{5}$$

$$3m = \frac{2}{5} \qquad \text{Adding } 3m$$

$$m = \frac{2}{15} \qquad \text{Multiplying by } \frac{1}{3}$$

The equation is $y = \frac{2}{15}x + \frac{2}{5}$.

(We could also have found the slope as follows:

$$m = \frac{\frac{2}{5} - 0}{0 - (-3)} = \frac{\frac{2}{5}}{3} = \frac{2}{15})$$

Exercise Set 3.5

1. a) When $r = 50$, $M = 146$, so one data point is $(50, 146)$. When $r = 80$, $M = 152$, so another data point is $(80, 152)$.

b) We use the data points to find the slope of the line:
$$m = \frac{152 - 146}{80 - 50} = \frac{6}{30} = \frac{1}{5}$$
Then we substitute in the point-slope equation, choosing $(50, 146)$ for the point:
$$M - 146 = \frac{1}{5}(r - 50)$$
$$M - 146 = \frac{1}{5}r - 10$$
$$M = \frac{1}{5}r + 136$$

c) We find M when $r = 55$:
$$M = \frac{1}{5} \cdot 55 + 136 = 11 + 136 = 147$$
We find M when $r = 60$:
$$M = \frac{1}{5} \cdot 60 + 136 = 12 + 136 = 148$$

3. a) In 1950, $t = 0$. One data point is $(0, 65)$. In 1970, $t = 1970 - 1950$, or 20. Another data point is $(20, 68)$.

b) We use the data points to find the slope of the line:
$$m = \frac{68 - 65}{20 - 0} = \frac{3}{20}, \text{ or } 0.15$$
Then we substitute in the point-slope equation, choosing $(0, 65)$ for the point:
$$E - 65 = 0.15(t - 0)$$
$$E - 65 = 0.15t$$
$$E = 0.15t + 65$$

c) In 1998, $t = 1998 - 1950$, or 48. We find E when $t = 48$:
$$E = 0.15(48) + 65 = 7.2 + 65 = 72.2$$
In 2001, $t = 2001 - 1950$, or 51. We find E when $t = 51$:
$$E = 0.15(51) + 65 = 7.65 + 65 = 72.65$$

5. a) When $t = 0$, $R = 360$, so one data point is $(0, 360)$. When $t = 30$, $R = 0$, so another data point is $(30, 0)$.

b) We use the data points to find the slope of the line:
$$m = \frac{360 - 0}{0 - 30} = \frac{360}{-30} = -12$$
Then we substitute in the point-slope equation, choosing $(0, 360)$ for the point:
$$R - 360 = -12(t - 0)$$
$$R - 360 = -12t$$
$$R = -12t + 360$$

c) We find R when $t = 5$:
$$R = -12 \cdot 5 + 360 = -60 + 360 = \$300$$
We find R when $t = 22$:
$$R = -12 \cdot 22 + 360 = -264 + 360 = \$96$$

7. a) In 1920, $t = 0$. One data point is $(0, 10.43)$. In 1970, $t = 1970 - 1920$, or 50. Another data point is $(50, 9.93)$.

b) We use the data points to find the slope of the line:
$$m = \frac{10.43 - 9.93}{0 - 50} = \frac{0.5}{-50} = -0.01$$
Then we substitute in the point-slope equation, choosing $(0, 10.43)$ for the point:
$$R - 10.43 = -0.01(t - 0)$$
$$R - 10.43 = -0.01t$$
$$R = -0.01t + 10.43$$

c) In 1998, $t = 1998 - 1920$, or 78. We find R when $t = 78$:
$$R = -0.01(78) + 10.43 = -0.78 + 10.43 = 9.65 \text{ sec}$$
In 2004, $t = 2004 - 1920$, or 84. We find R when $t = 84$:
$$R = -0.01(84) + 10.43 = -0.84 + 10.43 = 9.59 \text{ sec}$$

d) We substitute 9.0 for R and solve for t:
$$9.0 = -0.01t + 10.43$$
$$-1.43 = -0.01t$$
$$143 = t$$

The record will be 9.0 sec 143 years after 1920, or in 2063.

9. $5 - 4x < 45$
$$-4x < 40$$
$$x > -10 \quad \text{Dividing by } -4 \text{ and reversing the inequality symbol}$$

The solution set is $\{x | x > -10\}$.

11. $|5 - 4x| = 45$

$\qquad 5 - 4x = 45 \quad or \quad 5 - 4x = -45$

$\qquad -4x = 40 \quad or \qquad -4x = -50$

$\qquad x = -10 \quad or \qquad x = \dfrac{50}{4}, \text{ or } \dfrac{25}{2}$

The solution set is $\left\{ -10, \dfrac{25}{2} \right\}$.

13. $|5 - 4x| < 45$

$\qquad -45 < 5 - 4x < 45$

$\qquad -50 < -4x < 40$

$\qquad \dfrac{50}{4} > x > -10$

$\qquad \dfrac{25}{2} > x > -10, \text{ or}$

$\qquad -10 < x < \dfrac{25}{2}$

The solution set is $\left\{ x \middle| -10 < x < \dfrac{25}{2} \right\}$.

15. The data points (t, V) are $(0, 5200)$ and $(2, 4225)$. We find the slope of the line:

$$m = \frac{4225 - 5200}{2 - 0} = \frac{-975}{2} = -487.5$$

Then we find the equation of the line:

$$V - 5200 = -487.5(t - 0)$$

$$V = -487.5t + 5200$$

Finally we find V when $t = 8$:

$$V = -487.5(8) + 5200 = 1300$$

The value after 8 years is \$1300.

17. The data points (T, L) are $(18, 100)$ and $(20, 100.00356)$. We find the slope of the line:

$$m = \frac{100.00356 - 100}{20 - 18} = \frac{0.00356}{2} = 0.00178$$

Then we find the equation of the line:

$$L - 100 = 0.00178(T - 18)$$

$$L - 100 = 0.00178T - 0.03204$$

$$L = 0.00178T + 99.96796$$

Find L when $T = 40$:

$$L = 0.00178(40) + 99.96796 = 100.03916$$

At 40°C, the length of the wire is 100.03916 cm.

Find L when $T = 0$:

$$L = 0.00178(0) + 99.96796 = 99.96796$$

At 0°C, the length of the wire is 99.96796 cm.

Exercise Set 3.6

1. We use alphabetical order to replace x by -3 and y by 3.

$$\begin{array}{c|c} \multicolumn{2}{c}{3x + y < -5} \\ \hline 3(-3) + 3 & -5 \\ -9 + 3 & \\ -6 & \text{TRUE} \end{array}$$

Since $-6 < -5$ is true, $(-3, 3)$ is a solution.

3. We use alphabetical order to replace x by 5 and y by -2.

$$\begin{array}{c|c} \multicolumn{2}{c}{6y - x > 2} \\ \hline 6(-2) - 5 & 2 \\ -12 - 5 & \\ -17 & \text{FALSE} \end{array}$$

Since $-17 > 2$ is false, $(5, -2)$ is not a solution.

5. Graph: $y > 2x$

We first graph the line $y = 2x$. We draw the line dashed since the inequality symbol is $>$. To determine which half-plane to shade, test a point not on the line. We try $(1, 1)$ and substitute:

$$\begin{array}{c|c} \multicolumn{2}{c}{y > 2x} \\ \hline 1 & 2 \cdot 1 \\ & 2 \quad \text{FALSE} \end{array}$$

Since $1 > 2$ is false, $(1, 1)$ is not a solution, nor are any points in the half-plane containing $(1, 1)$. The points in the opposite half-plane are solutions, so we shade that half-plane and obtain the graph.

7. Graph: $y < x + 1$

First graph the line $y = x + 1$. Draw it dashed since the inequality symbol is $<$. Test the point $(0, 0)$ to determine if it is a solution.

$$\begin{array}{c|c} \multicolumn{2}{c}{y < x + 1} \\ \hline 0 & 0 + 1 \\ & 1 \quad \text{TRUE} \end{array}$$

Since $0 < 1$ is true, we shade the half-plane containing $(0, 0)$ and obtain the graph.

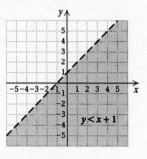

9. Graph: $y > x - 2$

First graph the line $y = x-2$. Draw a dashed line since the inequality symbol is $>$. Test the point $(0,0)$ to determine if it is a solution.

$$\frac{y > x - 2}{0 \mid 0 - 2}$$
$$-2 \quad \text{TRUE}$$

Since $0 > -2$ is true, we shade the half-plane containing $(0,0)$ and obtain the graph.

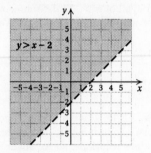

11. Graph: $x + y < 4$

First graph $x + y = 4$. Draw the line dashed since the inequality symbol is $<$. Test the point $(0,0)$ to determine if it is a solution.

$$\frac{x + y < 4}{0 + 0 \mid 4}$$
$$0 \quad \text{TRUE}$$

Since $0 < 4$ is true, we shade the half-plane containing $(0,0)$ and obtain the graph.

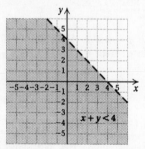

13. Graph: $3x + 4y \le 12$

We first graph $3x + 4y = 12$. Draw the line solid since the inequality symbol is $\le$. Test the point $(0,0)$ to determine if it is a solution.

$$\frac{3x + 4y \le 12}{3 \cdot 0 + 4 \cdot 0 \mid 12}$$
$$0 \quad \text{TRUE}$$

Since $0 \le 12$ is true, we shade the half-plane containing $(0,0)$ and obtain the graph.

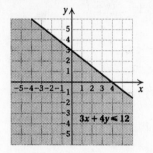

15. Graph: $2y - 3x > 6$

We first graph $2y - 3x = 6$. Draw the line dashed since the inequality symbol is $>$. Test the point $(0,0)$ to determine if it is a solution.

$$\frac{2y - 3x > 6}{2 \cdot 0 - 3 \cdot 0 \mid 6}$$
$$0 \quad \text{FALSE}$$

Since $0 > 6$ is false, we shade the half-plane that does not contain $(0,0)$ and obtain the graph.

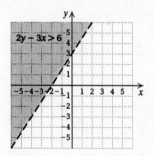

17. Graph: $3x - 2 \le 5x + y$
$$-2 \le 2x + y$$

We first graph $-2 = 2x + y$. Draw the line solid since the inequality symbol is $\le$. Test the point $(0,0)$ to determine if it is a solution.

$$\frac{-2 \le 2x + y}{-2 \mid 2 \cdot 0 + 0}$$
$$0 \quad \text{TRUE}$$

Since $-2 \le 0$ is true, we shade the half-plane containing $(0,0)$ and obtain the graph.

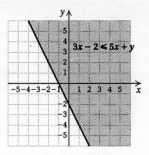

19. Graph: $x < 5$

We first graph $x = 5$. Draw the line dashed since the inequality symbol is $<$. Test the point $(0, 0)$ to determine if it is a solution.

$$\frac{x < 5}{0 \mid 5} \text{ TRUE}$$

Since $0 < 5$ is true, we shade the half-plane containing $(0, 0)$ and obtain the graph.

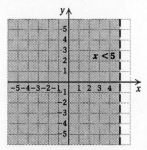

21. Graph: $y > 2$

We first graph $y = 2$. We draw the line dashed since the inequality symbol is $>$. Test the point $(0, 0)$ to determine if it is a solution.

$$\frac{y > 2}{0 \mid 2} \text{ FALSE}$$

Since $0 > 2$ is false, we shade the half-plane that does not contain $(0, 0)$ and obtain the graph.

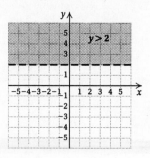

23. Graph: $2x + 3y \le 6$

We first graph $2x + 3y = 6$. We draw the line solid since the inequality symbol is $\le$. Test the point $(0, 0)$ to determine if it is a solution.

$$\frac{2x + 3y \le 6}{2 \cdot 0 + 3 \cdot 0 \mid 6}$$
$$0 \mid \text{TRUE}$$

Since $0 \le 6$ is true, we shade the half-plane containing $(0, 0)$ and obtain the graph.

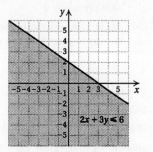

25. **Familiarize**. We make a drawing. We let x represent the length of the shorter piece of rope. Then $x + 5$ represents the length of the longer piece.

$$\underbrace{\overbrace{}^{x} \cdot \overbrace{}^{x+5}}_{78 \text{ ft}}$$

Translate.

$$\underbrace{\text{Shorter length}}_{x} \underbrace{\text{plus}}_{+} \underbrace{\text{longer length}}_{x+5} \underbrace{\text{is}}_{=} \underbrace{78 \text{ ft.}}_{78}$$

Solve.

$$x + x + 5 = 78$$
$$2x + 5 = 78$$
$$2x = 73$$
$$x = 36.5$$

Check. If $x = 36.5$, then $x + 5 = 36.5 + 5$, or 41.5, and $36.5 + 41.5 = 78$. The numbers check.

State. The lengths of the pieces are 36.5 ft and 41.5 ft.

27. $2x - 5 < 10$ or $3x - 4 \ge 12$
$$\quad 2x < 15 \quad or \quad \quad 3x \ge 16$$
$$\quad x < \frac{15}{2} \quad or \quad \quad x \ge \frac{16}{3}$$

The solution set is all real numbers.

29. $|7x - 4| < 20$
$$-20 < 7x - 4 < 20$$
$$-16 < 7x < 24$$
$$-\frac{16}{7} < x < \frac{24}{7}$$

The solution set is $\left\{ x \mid -\frac{16}{7} < x < \frac{24}{7} \right\}$.

31. The length is at most 94 ft, so

$$L \leq 94 \quad \text{and}$$

$$2L \leq 188. \quad \text{Multiplying by 2}$$

The width is at most 50 ft, so

$$W \leq 50 \quad \text{and}$$

$$2W \leq 100. \quad \text{Multiplying by 2}$$

(Of course, L and W would also have to be positive, but we will consider only the upper bound on the dimensions here.)

Then we have

$$2L + 2W \leq 188 + 100, \text{ or}$$

$$2L + 2W \leq 288.$$

(See Exercise 93, Exercise Set 2.4.)

To graph the inequality, we first graph $2L + 2W = 288$ using a solid line since the inequality symbol is $\leq$. (We will let W be the first coordinate and L the second. This is a case where alphabetical order of variables is not used.) Test the point $(0, 0)$ to determine if it is a solution.

$$
\begin{array}{c|c}
\multicolumn{2}{c}{2L + 2W \leq 288} \\
\hline
2 \cdot 0 + 2 \cdot 0 & 288 \\
0 & \text{TRUE}
\end{array}
$$

Since $0 \leq 288$ is true, we shade the half-plane containing $(0, 0)$ and obtain the graph.

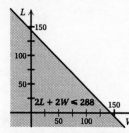

33. The total weight of c children is $35c$ kg, and the total weight of a adults is $75a$ kg. When the total weight is more than 1000 kg the elevator is overloaded, so we have

$$35c + 75a > 1000.$$

(Of course, c and a would also have to be nonnegative, but we will not deal with those constraints here.)

To graph the inequality, we first graph $35c + 75a = 1000$ using a dashed line since the inequality symbol is $>$. Test the point $(0, 0)$ to determine if it is a solution.

$$
\begin{array}{c|c}
\multicolumn{2}{c}{35c + 75a > 1000} \\
\hline
35 \cdot 0 + 75 \cdot 0 & 1000 \\
0 & \text{FALSE}
\end{array}
$$

Since $0 > 1000$ is false, we shade the half-plane that does not contain $(0, 0)$ and obtain the graph.

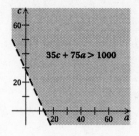

Chapter 4

Systems of Equations and Inequalities

1. We replace x with -1 and y with 2.

$$
\begin{array}{c|c}
\multicolumn{2}{l}{3x - y = -5} \\
\hline
3(-1) - 2 & -5 \\
-3 - 2 & \\
-5 & \text{TRUE}
\end{array}
\qquad
\begin{array}{c|c}
\multicolumn{2}{l}{10x + 7y = 4} \\
\hline
10(-1) + 7 \cdot 2 & 4 \\
-10 + 14 & \\
4 & \text{TRUE}
\end{array}
$$

The pair $(-1, 2)$ makes both equations true, so it is a solution of the system.

3. We replace a with 4 and b with 9.

$$
\begin{array}{c|c}
\multicolumn{2}{l}{b = 2a + 1} \\
\hline
9 & 2 \cdot 4 + 1 \\
 & 8 + 1 \\
 & 9 \quad \text{TRUE}
\end{array}
\qquad
\begin{array}{c|c}
\multicolumn{2}{l}{3a + b = 20} \\
\hline
3 \cdot 4 + 9 & 20 \\
12 + 9 & \\
21 & \text{FALSE}
\end{array}
$$

The pair $(4, 9)$ makes the second equation false, so it is not a solution of the system.

5. We replace x with 2 and y with 6.

$$
\begin{array}{c|c}
\multicolumn{2}{l}{x + y = 8} \\
\hline
2 + 6 & 8 \\
8 & \text{TRUE}
\end{array}
\qquad
\begin{array}{c|c}
\multicolumn{2}{l}{y = 5x - 4} \\
\hline
6 & 5 \cdot 2 - 4 \\
 & 10 - 6 \\
 & 6 \quad \text{TRUE}
\end{array}
$$

The pair $(2, 6)$ makes both equations true, so it is a solution of the system.

7. Graph both lines on the same set of axes.

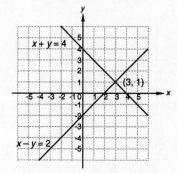

The solution (point of intersection) seems to be the point $(3, 1)$. Check:

$$
\begin{array}{c|c}
\multicolumn{2}{l}{x + y = 4} \\
\hline
3 + 1 & 4 \\
4 & \text{TRUE}
\end{array}
\qquad
\begin{array}{c|c}
\multicolumn{2}{l}{x - y = 2} \\
\hline
3 - 1 & 2 \\
2 & \text{TRUE}
\end{array}
$$

The solution is $(3, 1)$.

Since the system of equations has a solution it is consistent. Since there is exactly one solution, the system is independent.

9. Graph both lines on the same set of axes.

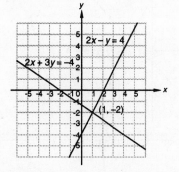

The solution (point of intersection) seems to be the point $(1, -2)$.

Check:

$$
\begin{array}{c|c}
\multicolumn{2}{l}{2x - y = 4} \\
\hline
2 \cdot 1 - (-2) & 4 \\
2 + 2 & \\
4 & \text{TRUE}
\end{array}
\qquad
\begin{array}{c|c}
\multicolumn{2}{l}{2x + 3y = -4} \\
\hline
2 \cdot 1 + 3(-2) & -4 \\
2 - 6 & \\
-4 & \text{TRUE}
\end{array}
$$

The solution is $(1, -2)$.

Since the system of equations has a solution, it is consistent. Since there is exactly one solution, the system is independent.

11. Graph both lines on the same set of axes.

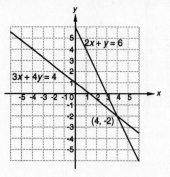

The solution (point of intersection) seems to be the point $(4, -2)$.

Check:

$$
\begin{array}{c|c}
\multicolumn{2}{l}{2x + y = 6} \\
\hline
2 \cdot 4 + (-2) & 6 \\
8 - 2 & \\
6 & \text{TRUE}
\end{array}
\qquad
\begin{array}{c|c}
\multicolumn{2}{l}{3x + 4y = 4} \\
\hline
3 \cdot 4 + 4(-2) & 4 \\
12 - 8 & \\
4 & \text{TRUE}
\end{array}
$$

The solution is $(4, -2)$.

Since the system of equations has a solution, it is consistent. Since there is exactly one solution, the system is independent.

13. Graph both lines on the same set of axes.

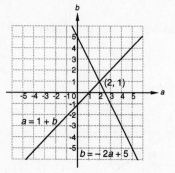

The solution seems to be the point $(2, 1)$.

Check:

$a = 1 + b$		$b = -2a + 5$	
2	$1 + 1$	1	$-2 \cdot 2 + 5$
	2 TRUE		$-4 + 5$
			1 TRUE

The solution is $(2, 1)$.

Since the system of equations has a solution, it is consistent. Since there is exactly one solution, the system is independent.

15. Graph both lines on the same set of axes.

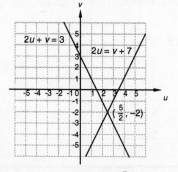

The solution seems to be $\left(\frac{5}{2}, -2\right)$.

Check:

$2u + v = 3$		$2u = v + 7$	
$2 \cdot \frac{5}{2} + (-2)$	3	$2 \cdot \frac{5}{2}$	$-2 + 7$
$5 - 2$		5	5 TRUE
	3 TRUE		

The solution is $\left(\frac{5}{2}, -2\right)$.

Since the system of equations has a solution, it is consistent. Since there is exactly one solution, the system is independent.

17. Graph both lines on the same set of axes.

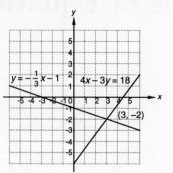

The ordered pair $(3, -2)$ checks in both equations. It is the solution.

Since the system of equations has a solution, it is consistent. Since there is exactly one solution, the system is independent.

19. Graph both lines on the same set of axes.

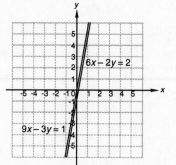

The lines are parallel. There is no solution.

Since the system of equations has no solution, it is inconsistent. Since there is no solution, the system is independent.

21. Graph both lines on the same set of axes.

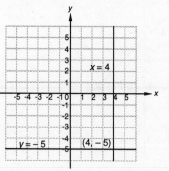

The ordered pair $(4, -5)$ checks in both equations. It is the solution.

Since the system of equations has a solution, it is consistent. Since there is exactly one solution, the system is independent.

23. Graph both lines on the same set of axes.

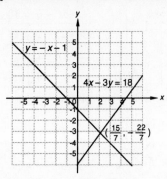

The ordered pair $\left(\dfrac{15}{7}, -\dfrac{22}{7}\right)$ checks in both equations. It is the solution.

Since the system of equations has a solution, it is consistent. Since there is exactly one solution, the system is independent.

25. Graph both lines on the same set of axes.

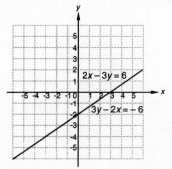

The graphs are the same. Any solution of one of the equations is also a solution of the other. Each equation has an infinite number of solutions. Thus the system of equations has an infinite number of solutions. Since the system of equations has a solution, it is consistent. Since there are infinitely many solutions,the system is dependent.

27. Substitute -5 for x and -1 for y in the first equation.
$$A(-5) - 7(-1) = -3$$
$$-5A + 7 = -3$$
$$-5A = -10$$
$$A = 2$$

Then substitute -5 for x and -1 for y in the second equation.
$$-5 - B(-1) = -1$$
$$-5 + B = -1$$
$$B = 4$$

We have $A = 2$, $B = 4$.

29. There are many correct answers. One can be found by expressing the sum and the difference of the two numbers:
$$x + y = 22,$$
$$x - y = -18$$

31. There are many correct answers. One can be found by writing one equation such as $x + 2y = 5$ and then writing a second equation in which one side is multiplied by a constant while the other side is multiplied by a different constant. For example, we could write
$$x + 2y = 5,$$
$$3x + 6y = 10.$$

Exercise Set 4.2

1. $2x + 5y = 4,$ (1)
   $x = 3 - 3y$ (2)

We substitute $3 - 3y$ for x in Equation (1) and solve for y.

$\quad\ 2x + 5y = 4$ Equation (1)

$2(3 - 3y) + 5y = 4$ Substituting

$\ 6 - 6y + 5y = 4$ Removing parentheses

$\qquad 6 - y = 4$ Collecting like terms

$\qquad\ -y = -2$ Subtracting 6

$\qquad\quad y = 2$ Multiplying by -1

Next we substitute 2 for y in either of the original equations and solve for x.

$x = 3 - 3y$ Equation (2)

$x = 3 - 3 \cdot 2$ Substituting

$x = 3 - 6$

$x = -3$

We check the ordered pair $(-3, 2)$.

$2x + 5y = 4$		$x = 3 - 3y$	
$2(-3) + 5 \cdot 2$	4	-3	$3 - 3 \cdot 2$
$-6 + 10$			$3 - 6$
4	TRUE	-3	TRUE

Since $(-3, 2)$ checks, it is the solution.

3. $9x - 2y = -6,$ (1)
   $7x + 8 = y$ (2)

We substitute $7x + 8$ for y in Equation (1) and solve for x.

$\quad\ 9x - 2y = -6$ Equation (1)

$9x - 2(7x + 8) = -6$ Substituting

$9x - 14x - 16 = -6$

$\quad -5x - 16 = -6$

$\qquad\quad -5x = 10$

$\qquad\qquad x = -2$

Next we substitute -2 for x in either of the original equations and solve for y.

$7x + 8 = y$ Equation (2)

$7(-2) + 8 = y$

$-14 + 8 = y$

$-6 = y$

We check the ordered pair $(-2, -6)$.

$9x - 2y = -6$		$7x + 8 = y$	
$9(-2) - 2(-6)$	-6	$7(-2) + 8$	-6
$-18 + 12$		$-14 + 8$	
-6	TRUE	-6	TRUE

Since $(-2, -6)$ checks, it is the solution.

5. $5m + n = 8$,   (1)

$3m - 4n = 14$ (2)

We solve Equation (1) for n.

$5m + n = 8$ Equation (1)

$n = 8 - 5m$ (3)

We substitute $8 - 5m$ for n in Equation (2) and solve for m.

$3m - 4n = 14$ Equation (2)

$3m - 4(8 - 5m) = 14$ Substituting

$3m - 32 + 20m = 14$

$23m - 32 = 14$

$23m = 46$

$m = 2$

Now we substitute 2 for m in Equation (3) and solve for n.

$n = 8 - 5m$

$n = 8 - 5 \cdot 2 = 8 - 10 = -2$

We check the ordered pair $(2, -2)$.

$5m + n = 8$		$3m - 4n = 14$	
$5 \cdot 2 + (-2)$	8	$3 \cdot 2 - 4(-2)$	14
$10 - 2$		$6 + 8$	
8	TRUE	14	TRUE

Since $(2, -2)$ checks, it is the solution.

7. $4x + 13y = 5$,   (1)

$-6x + y = 13$ (2)

We solve Equation (2) for y.

$-6x + y = 13$ Equation (2)

$y = 6x + 13$ (3)

We substitute $6x + 13$ for y in Equation (1) and solve for x.

$4x + 13y = 5$ Equation (2)

$4x + 13(6x + 13) = 5$ Substituting

$4x + 78x + 169 = 5$

$82x + 169 = 5$

$82x = -164$

$x = -2$

Now we substitute -2 for x in Equation (3) and solve for y.

$y = 6x + 13$

$y = 6(-2) + 13 = -12 + 13 = 1$

We check the ordered pair $(-2, 1)$.

$4x + 13y = 5$		$-6x + y = 13$	
$4(-2) + 13 \cdot 1$	5	$-6(-2) + 1$	13
$-8 + 13$		$12 + 1$	
5	TRUE	13	TRUE

Since $(-2, 1)$ checks, it is the solution.

9. $x + 3y = 7$ (1)

$\underline{-x + 4y = 7}$ (2)

$0 + 7y = 14$ Adding

$7y = 14$

$y = 2$

Substitute 2 for y in one of the original equations and solve for x.

$x + 3y = 7$ Equation (1)

$x + 3 \cdot 2 = 7$ Substituting

$x + 6 = 7$

$x = 1$

Check:

$x + 3y = 7$		$-x + 4y = 7$	
$1 + 3 \cdot 2$	7	$-1 + 4 \cdot 2$	7
$1 + 6$		$-1 + 8$	
7	TRUE	7	TRUE

Since $(1, 2)$ checks, it is the solution.

11. $9x + 5y = 6$ (1)

$\underline{2x - 5y = -17}$ (2)

$11x + 0 = -11$ Adding

$11x = -11$

$x = -1$

Substitute -1 for x in one of the original equations and solve for y.

$9x + 5y = 6$ Equation (1)

$9(-1) + 5y = 6$ Substituting

$-9 + 5y = 6$

$5y = 15$

$y = 3$

We obtain $(-1, 3)$. This checks, so it is the solution.

13. $5x + 3y = 19$,   (1)

$2x - 5y = 11$ (2)

We multiply twice to make two terms become additive inverses.

From (1): $25x + 15y = 95$ Multiplying by 5

From (2): $\underline{6x - 15y = 33}$ Multiplying by 3

$31x + 0 = 128$ Adding

$31x = 128$

$x = \dfrac{128}{31}$

Substitute $\frac{128}{31}$ for x in one of the original equations and solve for y.

$$5x + 3y = 19 \qquad \text{Equation (1)}$$

$$5 \cdot \frac{128}{31} + 3y = 19 \qquad \text{Substituting}$$

$$\frac{640}{31} + 3y = \frac{589}{31}$$

$$3y = -\frac{51}{31}$$

$$\frac{1}{3} \cdot 3y = \frac{1}{3} \cdot \left(-\frac{51}{31}\right)$$

$$y = -\frac{17}{31}$$

We obtain $\left(\frac{128}{31}, -\frac{17}{31}\right)$. This checks, so it is the solution.

15. $5r - 3s = 24$, (1)

$3r + 5s = 28$ (2)

We multiply twice to make two terms become additive inverses.

From (1): $\quad 25r - 15s = 120 \quad$ Multiplying by 5

From (2): $\quad \underline{9r + 15s = 84} \quad$ Multiplying by 3

$$34r + 0 = 204 \quad \text{Adding}$$

$$34r = 204$$

$$r = 6$$

Substitute 6 for r in one of the original equations and solve for s.

$$3r + 5s = 28 \qquad \text{Equation (2)}$$

$$3 \cdot 6 + 5s = 28 \qquad \text{Substituting}$$

$$18 + 5s = 28$$

$$5s = 10$$

$$s = 2$$

We obtain $(6, 2)$. This checks, so it is the solution.

17. $0.3x - 0.2y = 4$,

$0.2x + 0.3y = 1$

We first multiply each equation by 10 to clear decimals.

$3x - 2y = 40$ (1)

$2x + 3y = 10$ (2)

We use the multiplication principle with both equations of the resulting system.

From (1): $\quad 9x - 6y = 120 \quad$ Multiplying by 3

From (2): $\quad \underline{4x + 6y = 20} \quad$ Multiplying by 2

$$13x + 0 = 140 \quad \text{Adding}$$

$$13x = 140$$

$$x = \frac{140}{13}$$

Substitute $\frac{140}{13}$ for x in one of the equations in which the

decimals were cleared and solve for y.

$$2x + 3y = 10 \qquad \text{Equation (2)}$$

$$2 \cdot \frac{140}{13} + 3y = 10 \qquad \text{Substituting}$$

$$\frac{280}{13} + 3y = \frac{130}{13}$$

$$3y = -\frac{150}{13}$$

$$y = -\frac{50}{13}$$

We obtain $\left(\frac{140}{13}, -\frac{50}{13}\right)$. This checks, so it is the solution.

19. $\frac{1}{2}x + \frac{1}{3}y = 4$,

$\frac{1}{4}x + \frac{1}{3}y = 3$

We first multiply each equation by the LCM of the denominators to clear fractions.

$3x + 2y = 24 \quad$ Multiplying by 6

$3x + 4y = 36 \quad$ Multiplying by 12

We multiply by -1 on both sides of the first equation and then add.

$$-3x - 2y = -24 \quad \text{Multiplying by } -1$$

$$\underline{3x + 4y = 36}$$

$$0 + 2y = 12 \quad \text{Adding}$$

$$2y = 12$$

$$y = 6$$

Substitute 6 for y in one of the equations in which the fractions were cleared and solve for x.

$$3x + 2y = 24$$

$$3x + 2 \cdot 6 = 24 \quad \text{Substituting}$$

$$3x + 12 = 24$$

$$3x = 12$$

$$x = 4$$

We obtain $(4, 6)$. This checks, so it is the solution.

21. $\frac{2}{5}x + \frac{1}{2}y = 2$,

$\frac{1}{2}x - \frac{1}{6}y = 3$

We first multiply each equation by the LCM of the denominators to clear fractions.

$4x + 5y = 20 \quad$ Multiplying by 10

$3x - y = 18 \quad$ Multiplying by 6

We multiply by 5 on both sides of the second equation and then add.

$$4x + 5y = 20$$

$$\underline{15x - 5y = 90} \quad \text{Multiplying by 5}$$

$$19x + 0 = 110 \quad \text{Adding}$$

$$19x = 110$$

$$x = \frac{110}{19}$$

Substitute $\dfrac{110}{19}$ for x in one of the equations in which the fractions were cleared and solve for y.

$$3x - y = 18$$

$$3\left(\frac{110}{19}\right) - y = 18 \qquad \text{Substituting}$$

$$\frac{330}{19} - y = \frac{342}{19}$$

$$-y = \frac{12}{19}$$

$$y = -\frac{12}{19}$$

We obtain $\left(\dfrac{110}{19}, -\dfrac{12}{19}\right)$. This checks, so it is the solution.

23. $2x + 3y = 1,$

$4x + 6y = 2$

Multiply the first equation by -2 and then add.

$$-4x - 6y = -2$$
$$\underline{4x + 6y = 2}$$
$$0 = 0 \qquad \text{Adding}$$

We have an equation that is true for all numbers x and y. The system is dependent and has an infinite number of solutions.

25. $2x - 4y = 5,$

$2x - 4y = 6$

Multiply the first equation by -1 and then add.

$$-2x + 4y = -5$$
$$\underline{2x - 4y = 6}$$
$$0 = 1$$

We have a false equation. The system has no solution.

27. $5x - 9y = 7,$

$7y - 3x = -5$

We first write the second equation in the form $Ax + By = C$.

$$5x - 9y = 7 \qquad (1)$$
$$-3x + 7y = -5 \qquad (2)$$

We use the multiplication principle with both equations and then add.

$$15x - 27y = 21 \qquad \text{Multiplying by 3}$$
$$\underline{-15x + 35y = -25} \qquad \text{Multiplying by 5}$$
$$0 + 8y = -4 \qquad \text{Adding}$$
$$8y = -4$$
$$y = -\frac{1}{2}$$

Substitute $-\dfrac{1}{2}$ for y in one of the original equations and solve for x.

$$5x - 9y = 7 \qquad \text{Equation (1)}$$

$$5x - 9\left(-\frac{1}{2}\right) = 7 \qquad \text{Substituting}$$

$$5x + \frac{9}{2} = \frac{14}{2}$$

$$5x = \frac{5}{2}$$

$$x = \frac{1}{2}$$

We obtain $\left(\dfrac{1}{2}, -\dfrac{1}{2}\right)$. This checks, so it is the solution.

29. $3(a - b) = 15,$

$4a = b + 1$

We first write each equation in the form $Ax + By = C$.

$$3a - 3b = 15 \qquad (1)$$
$$4a - b = 1 \qquad (2)$$

We multiply by -3 on both sides of the second equation and then add.

$$3a - 3b = 15$$
$$\underline{-12a + 3b = -3} \qquad \text{Multiplying by } -3$$
$$-9a + 0 = 12$$
$$-9a = 12$$
$$a = -\frac{12}{9}$$
$$a = -\frac{4}{3}$$

Substitute $-\dfrac{4}{3}$ for a in either Equation (1) or Equation (2) and solve for b.

$$4a - b = 1 \qquad \text{Equation (2)}$$

$$4\left(-\frac{4}{3}\right) - b = 1 \qquad \text{Substituting}$$

$$-\frac{16}{3} - b = \frac{3}{3}$$

$$-b = \frac{19}{3}$$

$$b = -\frac{19}{3}$$

We obtain $\left(-\dfrac{4}{3}, -\dfrac{19}{3}\right)$. This checks, so it is the solution.

31. $x - \dfrac{1}{10}y = 100,$

$y - \dfrac{1}{10}x = -100$

We first write the second equation in the form $Ax + By = C$.

$$x - \frac{1}{10}y = 100$$

$$-\frac{1}{10}x + y = -100$$

Next we multiply each equation by 10 to clear fractions.

$$10x - y = 1000 \qquad (1)$$
$$-x + 10y = -1000 \qquad (2)$$

We multiply by 10 on both sides of Equation (1) and then add.

$100x - 10y = 10,000$ Multiplying by 10

$\underline{-x + 10y = -1000}$

$99x + \quad 0 = 9000$

$99x = 9000$

$x = \dfrac{9000}{99}$

$x = \dfrac{1000}{11}$

Substitute $\dfrac{1000}{11}$ for x in one of the equations in which the fractions were cleared and solve for y.

$10x - y = 1000$ Equation (1)

$10\left(\dfrac{1000}{11}\right) - y = 1000$ Substituting

$\dfrac{10,000}{11} - y = \dfrac{11,000}{11}$

$-y = \dfrac{1000}{11}$

$y = -\dfrac{1000}{11}$

We obtain $\left(\dfrac{1000}{11}, -\dfrac{1000}{11}\right)$. This checks, so it is the solution.

33. $0.05x + 0.25y = 22,$

$0.15x + 0.05y = 24$

We first multiply each equation by 100 to clear decimals.

$5x + 25y = 2200$ (1)

$15x + 5y = 2400$ (2)

We multiply by -5 on both sides of the second equation and add.

$5x + \quad 25y = \quad 2200$

$\underline{-75x - \quad 25y = -12,000}$ Multiplying by -5

$-70x + \quad 0 = \quad -9800$ Adding

$-70x = -9800$

$x = \dfrac{-9800}{-70}$

$x = 140$

Substitute 140 for x in one of the equations in which the decimals were cleared and solve for y.

$5x + 25y = 2200$ Equation (1)

$5 \cdot 140 + 25y = 2200$ Substituting

$700 + 25y = 2200$

$25y = 1500$

$y = 60$

We obtain $(140, 60)$. This checks, so it is the solution.

35. $y = 1.3x - 7$

The equation is in slope-intercept form, $y = mx + b$. The slope is 1.3.

37. $A = \dfrac{pq}{7}$

$7A = pq$ Multiplying by 7

$\dfrac{7A}{q} = p$ Dividing by q

39. $-4x + 5(x - 7) = 8x - 6(x + 2)$

$-4x + 5x - 35 = 8x - 6x - 12$ Removing parentheses

$x - 35 = 2x - 12$ Collecting like terms

$-35 = x - 12$ Subtracting x

$-23 = x$ Adding 12

41. $3.5x - \quad 2.1y = 106.2,$

$4.1x + 16.7y = -106.28$

Since this is a calculator exercise, you may choose not to clear the decimals. We will do so here, however.

$35x - \quad 21y = 1062$ Multiplying by 10

$410x + 1670y = -10,628$ Multiplying by 100

Multiply twice to make two terms become additive inverses.

$58,450x - 35,070y = 1,773,540$ Multiplying by 1670

$\underline{8610x + 35,070y = -223,188}$ Multiplying by 21

$67,060x + \quad 0 = 1,550,352$ Adding

$67,060x = 1,550,352$

$x \approx 23.118879$

Substitute 23.118879 for x in one of the equations in which the decimals were cleared and solve for y.

$35x - 21y = 1062$

$35(23.118879) - 21y = 1062$ Substituting

$809.160765 - 21y = 1062$

$-21y = 252.839235$

$y \approx -12.039964$

The numbers check, so the solution is $(23.118879, -12.039964)$.

43. $5x + 2y = a,$

$x - y = b$

We multiply by 2 on both sides of the second equation and then add.

$5x + 2y = a$

$\underline{2x - 2y = 2b}$ Multiplying by 2

$7x + \quad 0 = a + 2b$ Adding

$7x = a + 2b$

$x = \dfrac{a + 2b}{7}$

Next we multiply by -5 on both sides of the second equation and then add.

$$5x + 2y = a$$
$$\underline{-5x + 5y = -5b} \quad \text{Multiplying by } -5$$
$$7y = a - 5b$$
$$y = \frac{a - 5b}{7}$$

We obtain $\left(\dfrac{a + 2b}{7}, \dfrac{a - 5b}{7}\right)$. This checks, so it is the solution.

45. $(0, -3)$ and $\left(-\dfrac{3}{2}, 6\right)$ are two solutions of $px - qy = -1$.

Substitute 0 for x and -3 for y.
$$p \cdot 0 - q \cdot (-3) = -1$$
$$3q = -1$$
$$q = -\frac{1}{3}$$

Substitute $-\dfrac{3}{2}$ for x and 6 for y.
$$p \cdot \left(-\frac{3}{2}\right) - q \cdot 6 = -1$$
$$-\frac{3}{2}p - 6q = -1$$

Substitute $-\dfrac{1}{3}$ for q and solve for p.
$$-\frac{3}{2}p - 6 \cdot \left(-\frac{1}{3}\right) = -1$$
$$-\frac{3}{2}p + 2 = -1$$
$$-\frac{3}{2}p = -3$$
$$-\frac{2}{3} \cdot \left(-\frac{3}{2}p\right) = -\frac{2}{3} \cdot (-3)$$
$$p = 2$$

Thus, $p = 2$ and $q = -\dfrac{1}{3}$.

Exercise Set 4.3

1. Familiarize. Let $x =$ the first number and $y =$ the second number.

Translate.

The sum of two numbers is -42.

Rewording:

The first number	plus	the second number	is	-42.
$\downarrow$	$\downarrow$	$\downarrow$	$\downarrow$	$\downarrow$
x	$+$	y	$=$	-42

The first number	minus	the second number	is	52.
$\downarrow$	$\downarrow$	$\downarrow$	$\downarrow$	$\downarrow$
x	$-$	y	$=$	52

We have a system of equations:
$$x + y = -42, \quad (1)$$
$$x - y = 52 \quad (2)$$

Solve. We solve the system of equations. We use the elimination method.
$$x + y = -42$$
$$\underline{x - y = 52}$$
$$2x = 10 \quad \text{Adding}$$
$$x = 5$$

Substitute 5 for x in one of the equations and solve for y.
$$x + y = -42 \quad \text{Equation (1)}$$
$$5 + y = -42$$
$$y = -47$$

Check. The sum of the numbers is $5 + (-47)$, or -42. The difference is $5 - (-47)$, or 52. The numbers check.

State. The numbers are 5 and -47.

3. Familiarize. Let $x =$ the number of white sweatshirts sold and $y =$ the number of neon sweatshirts sold. List the information in a table.

Kind of sweatshirt	White	Neon	Total	
Number sold	x	y	30	$\to x + y = 30$
Price	\$10.95	\$18.50		
Amount taken in	$10.95x$	$18.50y$	419.10	$\to \begin{array}{l}10.95x+ \\ 18.50y = \\ 419.10\end{array}$

Translate. Using the "Number sold" and "Amount taken in" rows we have a system of equations:
$$x + y = 30,$$
$$10.95x + 18.50y = 419.10$$

After clearing decimals we have
$$x + y = 30, \quad (1)$$
$$1095x + 1850y = 41{,}910. \quad (2)$$

Solve. We solve the system of equations. We use elimination.
$$-1095x - 1095y = -32{,}850 \quad \text{Multiplying (1) by } -1095$$
$$\underline{1095x + 1850y = 41{,}910}$$
$$755y = 9060 \quad \text{Adding}$$
$$y = 12$$

Then we substitute 12 for y in Equation (1) and solve for x.
$$x + y = 30 \quad \text{Equation (1)}$$
$$x + 12 = 30 \quad \text{Substituting}$$
$$x = 18$$

Check. The total number of sweatshirts sold was $18 + 12$, or 30.

Money from white: $\$10.95 \times 18 = \$197.10$

Money from neon: $\underline{\$18.50 \times 12 = \$222.00}$

Total $= \$419.10$

The numbers check.

State. 18 white sweatshirts and 12 neon sweatshirts were sold.

5. *Familiarize*. The basketball court is a rectangle with perimeter 288 ft. Let l = the length and w = width. Recall that for a rectangle with length l and width w, the perimeter P is given by $P = 2l + 2w$.

Translate. The formula for perimeter gives us one equation:

$$2l + 2w = 288$$

The statement relating length and width gives us a second equation:

Length   is   44 ft longer than width

l $=$ $44 + w$

We have a system of equations:

$$2l + 2w = 288, \quad (1)$$
$$l = 44 + w \quad (2)$$

Solve. We solve the system of equations. We use substitution.

$$2(44 + w) + 2w = 288 \quad \text{Substituting } 44 + w$$
$$\text{for } l \text{ in } (1)$$
$$88 + 2w + 2w = 288$$
$$88 + 4w = 288$$
$$4w = 200$$
$$w = 50$$
$$l = 44 + 50 \quad \text{Substituting 50 for } w \text{ in } (2)$$
$$l = 94$$

Check. The perimeter of a 94 ft by 50 ft rectangle is $2 \cdot 94 + 2 \cdot 50 = 188 + 100 = 288$. Also, 94 ft is 44 ft longer than 50 ft. The numbers check.

State. The length is 94 ft, and the width is 50 ft.

7. *Familiarize*. Let x = the measure of the first angle and y = the measure of the second angle. Recall that two angles are complementary if the sum of their measures is $90°$.

Translate. The fact that the angles are complementary gives us one equation.

Rewording: The sum of the measures   is $90°$.

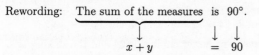

$$x + y \qquad = \quad 90$$

The second statement gives us another equation.

Rewording: The measure of the first angle   plus   the measure of half the second angle   is $64°$.

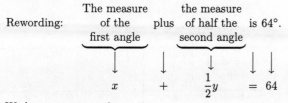

$$x \qquad + \qquad \frac{1}{2}y \qquad = 64$$

We have a system of equations:

$$x + y = 90,$$
$$x + \frac{1}{2}y = 64$$

After clearing the fraction we have

$$x + y = 90, \quad (1)$$
$$2x + y = 128. \quad (2)$$

Solve. We solve the system of equations. We use elimination.

$$-x - y = -90$$
$$\underline{2x + y = 128}$$
$$x = 38 \quad \text{Adding}$$

Now we substitute 38 for x in Equation (1) and solve for y.

$$x + y = 90 \quad \text{Equation (1)}$$
$$38 + y = 90 \quad \text{Substituting}$$
$$y = 52$$

Check. The sum of the measures of the angles is $38° + 52°$, or $90°$. The sum of the measure of the first angle, $38°$, and half the measure of the second angle, $\frac{1}{2} \cdot 52°$, or $26°$, is $64°$. The numbers check.

State. The measure of the first angle is $38°$, and the measure of the second angle is $52°$.

9. *Familiarize*. List the information in a table.

Type of score	Field goal	Free throw	Total
Number scored	x	y	18
Points per score	2	1	
Points scored	$2x$	$1 \cdot y$, or y	30

Translate. The "Number scored" row of the table gives us one equation:

$$x + y = 18$$

The "Points scored" row gives us a second equation:

$$2x + y = 30$$

We have a system of equations:

$$x + y = 18$$
$$2x + y = 30$$

Solve. We solve the system of equations. We use the elimination method.

$$-x - y = -18 \quad \text{Multiplying (1) by } -1$$
$$\underline{2x + y = 30}$$
$$x = 12$$

$$12 + y = 18 \quad \text{Substituting 12 for } x \text{ in (1)}$$
$$y = 6$$

Check. The total number of times the player scored is $12 + 6$, or 18.

Points from field goals: $12 \times 2 = 24$

Points from free throws: $\underline{6 \times 1 = 6}$

              Total   30

The numbers check.

State. The player made 12 field goals and 6 free throws.

11. Familiarize. Let x = number of games won and y = number of games tied. The total points earned in x wins is $2x$; the total points earned in y ties is $1 \cdot y$, or y.

Translate.

$\underbrace{\text{Points from wins}}_{2x}$ plus $\underbrace{\text{points from ties}}_{y}$ is 60.

$$2x + y = 60$$

$\underbrace{\text{Number of wins}}_{x}$ is $\underbrace{\text{9 more than the number of ties.}}_{9 + y}$

$$x = 9 + y$$

We have a system of equations:

$$2x + y = 60,$$
$$x = 9 + y$$

Solve. We solve the system of equations. We use substitution.

$$2(9 + y) + y = 60 \quad \text{Substituting } 9+y \text{ for } x \text{ in (1)}$$
$$18 + 2y + y = 60$$
$$18 + 3y = 60$$
$$3y = 42$$
$$y = 14$$

$$x = 9 + 14 \quad \text{Substituting 14 for } y \text{ in (2)}$$
$$x = 23$$

Check. The number of wins, 23, is 9 more than the number of ties, 14.

Points from wins: $23 \times 2 = 46$

Points from ties: $14 \times 1 = \underline{14}$

Total   60

The numbers check.

State. The team had 23 wins and 14 ties.

13. Familiarize. Let x = number of 30-sec commercials and y = number of 60-sec commercials. The total time used by x 30-sec commercials is $30x$; the total time used by y 60-sec commercials is $60y$. Also note that 10 min = 10×60, or 600 sec.

Translate.

$\underbrace{\text{Total number of commercials}}_{x + y}$ is 12.

$$x + y = 12$$

$\underbrace{\text{Total commercial time}}_{30x + 60y}$ is $\underbrace{\text{10 min, or 600 sec.}}_{600}$

$$30x + 60y = 600$$

We have a system of equations:

$$x + y = 12,$$
$$30x + 60y = 600$$

Solve. We solve the system of equations. We use the elimination method.

$$-30x - 30y = -360 \quad \text{Multiplying (1) by } -30$$
$$\underline{30x + 60y = 600}$$
$$30y = 240 \quad \text{Adding}$$
$$y = 8$$

$$x + 8 = 12 \quad \text{Substituting 8 for } y \text{ in (1)}$$
$$x = 4$$

Check. The total number of commercials is $4 + 8$, or 12.

Time for 30-sec commercials: $30 \times 4 = 120$ sec

Time for 60-sec commercials: $\underline{60 \times 8 = 480 \text{ sec}}$

600 sec,
or 10 min

The numbers check.

State. There were 4 30-sec commercials and 8 60-sec commercials.

15. Familiarize. Let x = the larger number and y = the smaller number.

Translate.

$\underbrace{\text{The difference of the numbers}}_{x - y}$ is 16.

$$x - y = 16$$

$\underbrace{\text{Three times the larger}}_{3x}$ is $\underbrace{\text{nine times the smaller.}}_{9y}$

$$3x = 9y$$

We have a system of equations:

$$x - y = 16, \quad (1)$$
$$3x = 9y \quad (2)$$

Solve. Solve the system of equations. We use the substitution method.

$$x = 3y \quad \text{Solving (2) for } x$$
$$3y - y = 16 \quad \text{Substituting } 3y \text{ for } x \text{ in (1)}$$
$$2y = 16$$
$$y = 8$$

$$x - 8 = 16 \quad \text{Substituting 8 for } y \text{ in (1)}$$
$$x = 24$$

Check. The difference of the numbers is $24 - 8$, or 16. Also, $3 \cdot 24 = 72 = 9 \cdot 8$ The numbers check.

State. The larger number is 24 and the smaller is 8.

17. Familiarize. We organize the information in a table.

Let x = the number of pounds of soybean meal and y = the number of pounds of corn meal.

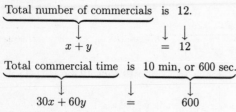

Type of meal	Pounds of meal	Percent of protein	Pounds of protein in meal
Soybean	x	16%	$0.16x$
Corn	y	9%	$0.09y$
Mixture	350	12%	0.12×350 or 42

Translate. The "Pounds of meal" column gives us one equation: $x + y = 350$

The last column gives us a second equation:
$0.16x + 0.09y = 42$

After clearing decimals, we have this system:
$$x + y = 350, \quad (1)$$
$$16x + 9y = 4200 \quad (2)$$

Solve. Solve the system of equations.
$$-9x - 9y = -3150 \quad \text{Multiplying (1) by } -9$$
$$\underline{16x + 9y = 4200}$$
$$7x = 1050$$
$$x = 150$$
$$150 + y = 350 \quad \text{Substituting 150 for } x \text{ in (1)}$$
$$y = 200$$

Check. The total number of pounds is $150 + 200$, or 350. Also, 16% of 150 is 24, and 9% of 200 is 18. Their total is 42. The numbers check.

State. 150 lb of soybean meal and 200 lb of corn meal should be mixed.

19. *Familiarize.* We can organize the information in a table. Let $x =$ the number of liters of the drink containing 15% orange juice and $y =$ the number of liters of the drink containing 5% orange juice.

Type of canned juice drink	Amount of drink	Percent of orange juice	Amount of orange juice in drink
15% juice	x	15%	$0.15x$
5% juice	y	5%	$0.05y$
Mixture	10	10%	0.1×10 or 1

Translate. The "Amount of drink" column gives us one equation: $x + y = 10$

The last column gives us a second equation:
$0.15x + 0.05y = 1$

After clearing decimals, we have this system:
$$x + y = 10, \quad (1)$$
$$15x + 5y = 100 \quad (2)$$

Solve. Solve the system of equations.
$$-5x - 5y = -50 \quad \text{Multiplying (1) by } -5$$
$$\underline{15x + 5y = 100}$$
$$10x = 50 \quad \text{Adding}$$
$$x = 5$$
$$5 + y = 10 \quad \text{Substituting 5 for } x \text{ in (1)}$$
$$y = 5$$

Check. The total number of liters is $5 + 5$, or 10. Also, 15% of 5 is 0.75, and 5% of 5 is 0.25. Their sum is 1. The numbers check.

State. 5 L of each drink should be used.

21. *Familiarize.* Let $x =$ one investment and $y =$ the other investment. We list the information in a table.

	Principal	Rate	Time	Interest ($I = Prt$)
First investment	x	7%	1 yr	$0.07x$
Second investment	y	5%	1 yr	$0.05y$
Total	$8800			$524

Translate. The first column gives us one equation: $x + y = 8800$

The last column gives us a second equation:
$0.07x + 0.05y = 524$

After clearing decimals we have this system:
$$x + y = 8800, \quad (1)$$
$$7x + 5y = 52,400 \quad (2)$$

Solve. We solve the system of equations using elimination.
$$-5x - 5y = -44,000 \quad \text{Multiplying (1) by } -5$$
$$\underline{7x + 5y = 52,400}$$
$$2x = 8400 \quad \text{Adding}$$
$$x = 4200$$
$$4200 + y = 8800 \quad \text{Substituting 4200 for } x \text{ in (1)}$$
$$y = 4600$$

Check. The sum of the investments is $4200 + 4600, or $8800. The interest earned is 7% of $4200, or $294, and 5% of $4600, or $230. The total interest is $294 + $230, or $524. The values check.

State. $4200 is invested at 7%, and $4600 is invested at 5%.

23. *Familiarize.* Let $x =$ one investment and $y =$ the other investment. List the information in a table. (We assume the investment is made for 1 year.)

	Principal	Rate	Time	Interest ($I = Prt$)
First investment	x	6%	1 yr	$0.06x$
Second investment	y	8.5%	1 yr	$0.085y$
Total	$3270			$225.45

Translate. The first column gives us one equation: $x + y = 3270$

The last column gives us a second equation:
$0.06x + 0.085y = 225.45$

After clearing decimals we have this system:
$$x + y = 3270, \quad (1)$$
$$60x + 85y = 225,450 \quad (2)$$

Solve. We solve the system of equations using elimination.

$$-60x - 60y = -196,200 \quad \text{Multiplying (1) by } -60$$
$$\underline{60x + 85y = 225,450}$$
$$25y = 29,250 \quad \text{Adding}$$
$$y = 1170$$

$$x + 1170 = 3270 \quad \text{Substituting 1170 for } y \text{ in (1)}$$
$$x = 2100$$

Check. The sum of the investments is $1170 + $2100, or $3270. The interest earned is 6% of $2100, or $126, and 8.5% of $1170, or $99.45. The total interest is $225.45. The values check.

State. $2100 is invested at 6%, and $1170 is invested at 8.5%.

25. *Familiarize*. Let x = the cost of one hot dog and y = the cost of one hamburger.

Translate. The first statement gives us one equation.

$$\underbrace{\text{Three}}_{3x} \underbrace{\text{hot-dogs}} \underbrace{\text{and}}_{+} \underbrace{\text{five}}_{5y} \underbrace{\text{hamburgers}} \underbrace{\text{cost}}_{=} \underbrace{\$18.50.}_{18.50}$$

The second statement gives us another equation.

$$\underbrace{\text{Five}}_{5x} \underbrace{\text{hot-dogs}} \underbrace{\text{and}}_{+} \underbrace{\text{three}}_{3y} \underbrace{\text{hamburgers}} \underbrace{\text{cost}}_{=} \underbrace{\$16.70.}_{16.70}$$

After clearing decimals, we have this system:

$$30x + 50y = 185, \quad (1)$$
$$50x + 30y = 167 \quad (2)$$

Solve. We solve the system of equations.

$$900x + 1500y = 5550 \quad \text{Multiplying (1) by 30}$$
$$\underline{-2500x - 1500y = -8350} \quad \text{Multiplying (2) by } -50$$
$$-1600x = -2800 \quad \text{Adding}$$
$$x = 1.75$$

$$30(1.75) + 50y = 185 \quad \text{Substituting 1.75 for } x \text{ in (1)}$$
$$52.5 + 50y = 185$$
$$50y = 132.5$$
$$y = 2.65$$

Check. If one hot dog costs $1.75 and one hamburger costs $2.65, then 3 hot dogs and 5 hamburgers cost $3(\$1.75) + 5(\$2.65) = \$5.25 + \$13.25 = \$18.50$. Also, 5 hot dogs and 3 hamburgers cost $5(\$1.75) + 3(\$2.65) = \$8.75 + \$7.95$, or $16.70. The numbers check.

State. One hot dog costs $1.75 and one hamburger costs $2.65.

27. *Familiarize*. Let x = Hannah's age now and y = Irving's age now. Four years from now, Hannah's age will be $x + 4$ and Irving's age will be $y + 4$.

Translate.

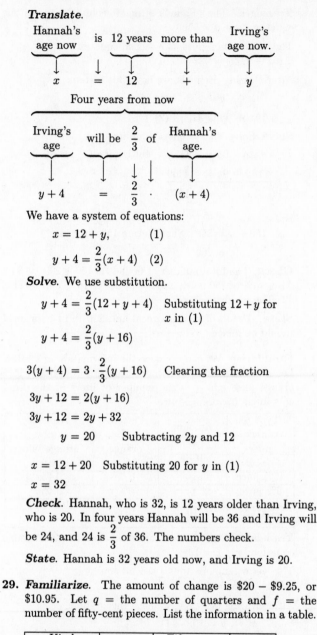

We have a system of equations:

$$x = 12 + y, \quad (1)$$
$$y + 4 = \frac{2}{3}(x + 4) \quad (2)$$

Solve. We use substitution.

$$y + 4 = \frac{2}{3}(12 + y + 4) \quad \text{Substituting } 12 + y \text{ for } x \text{ in (1)}$$
$$y + 4 = \frac{2}{3}(y + 16)$$
$$3(y + 4) = 3 \cdot \frac{2}{3}(y + 16) \quad \text{Clearing the fraction}$$
$$3y + 12 = 2(y + 16)$$
$$3y + 12 = 2y + 32$$
$$y = 20 \quad \text{Subtracting } 2y \text{ and } 12$$

$$x = 12 + 20 \quad \text{Substituting 20 for } y \text{ in (1)}$$
$$x = 32$$

Check. Hannah, who is 32, is 12 years older than Irving, who is 20. In four years Hannah will be 36 and Irving will be 24, and 24 is $\frac{2}{3}$ of 36. The numbers check.

State. Hannah is 32 years old now, and Irving is 20.

29. *Familiarize*. The amount of change is $20 − $9.25, or $10.95. Let q = the number of quarters and f = the number of fifty-cent pieces. List the information in a table.

Kind of coin	Quarter	Fifty-cent piece	Total
Number	q	f	30
Amount of change	$0.25q$	$0.50f$	$10.75

Translate. The rows of the table give us two equations:

$$q + f = 30,$$
$$0.25q + 0.50f = 10.75$$

After clearing decimals, we have

$$q + f = 30 \quad (1)$$
$$25q + 50f = 1075. \quad (2)$$

Solve. Solve the system of equations.

$$-25q - 25f = -750 \quad \text{Multiplying (1) by } -25$$
$$\underline{25q + 50f = 1075}$$
$$25f = 325 \quad \text{Adding}$$
$$f = 13$$

$$q + 13 = 30 \quad \text{Substituting 13 for } f \text{ in (1)}$$
$$q = 17$$

Check. The total number of coins is $17 + 13$, or 30. The value of 17 quarters is \$4.25, and the value of 13 fifty-cent pieces is \$6.50, so the total value of the coins is \$4.25 + \$6.50, or \$10.75. The numbers check.

State. There are 17 quarters and 13 fifty-cent pieces.

31. *Familiarize.* We first make a drawing.

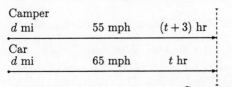

Camper
d mi   55 mph $(t+3)$ hr

Car
d mi   65 mph t hr

Car overtakes
camper here.

From the drawing we see that the distances are the same. Now we complete the table in the text.

$$d = r \cdot t$$

	Distance	Rate	Time	
Camper	d	55	$t+3$	$\rightarrow d = 55(t+3)$
Car	d	65	t	$\rightarrow d = 65t$

Translate. Using $d = rt$ in each row of the table, we get a system of equations:

$$d = 55(t+3), \quad (1)$$
$$d = 65(t) \quad\quad\; (2)$$

Solve. We solve the system.

$$65t = 55(t+3) \quad \text{Substituting } 65t \text{ for } d \text{ in (1)}$$
$$65t = 55t + 165 \quad \text{Removing parentheses}$$
$$10t = 165 \quad \text{Subtracting } 55t$$
$$t = 16.5 \quad \text{Dividing by 10}$$

The time for the car is 16.5 hr, and the time for the camper is $16.5 + 3$, or 19.5 hr.

Check. At 55 mph, in 19.5 hr the camper will travel $55(19.5)$, or 1072.5 mi. At 65 mph, in 16.5 hr the car will travel $65(16.5)$, or 1072.5 mi. Since the distances are the same, the numbers check.

State. The car will overtake the camper after 1072.5 mi.

33. *Familiarize.* We first make a drawing. Let $d =$ the distance and $r =$ the speed of the boat in still water. Then when the boat travels downstream its speed is $r + 6$, and its speed upstream is $r - 6$.

Downstream, 6 mph current
d mi, $r + 6$, 3 hr

Upstream, 6 mph current
d mi, $r - 6$, 5 hr

From the drawing we see that the distances are the same. List the information in a table.

$$d = r \cdot t$$

	Distance	Rate	Time	
Down-stream	d	$r+6$	3	$\rightarrow d = (r+6)3$
Up-stream	d	$r-6$	5	$\rightarrow d = (r-6)5$

Translate. Using $d = rt$ in each row of the table, we get a system of equations:

$$d = 3r + 18,$$
$$d = 5r - 30$$

Solve. Solve the system of equations.

$$3r + 18 = 5r - 30 \quad \text{Using substitution}$$
$$18 = 2r - 30$$
$$48 = 2r$$
$$24 = r$$

Check. When $r = 24$, $r+6 = 30$, and the distance traveled in 3 hr is $30 \cdot 3$, or 90 mi. Also, $r - 6 = 18$, and the distance traveled in 5 hr is $18 \cdot 5$, or 90 mi. The value checks.

State. The speed of the boat in still water is 24 mph.

35. *Familiarize.* We first make a drawing. Let $d =$ the distance and $t =$ the time at 32 mph. At 4 mph faster, the speed is 36 mph.

32 mph t hr d mi

36 mph $\left(t - \dfrac{1}{2}\right)$ hr d mi

From the drawing, we see that the distances are the same. List the information in a table.

$$d = r \cdot t$$

	Distance	Rate	Time	
Slower trip	d	32	t	$\rightarrow d = 32t$
Faster trip	d	36	$t - \dfrac{1}{2}$	$\rightarrow d = 36\left(t - \dfrac{1}{2}\right)$

Translate. Using $d = rt$ in each row of the table, we get a system of equations:

$$d = 32t, \quad\quad\quad (1)$$
$$d = 36\left(t - \frac{1}{2}\right) \quad (2)$$

Solve. We solve the system of equations.

$$32t = 36\left(t - \frac{1}{2}\right) \quad \text{Substituting } 32t \text{ for } d \text{ in } (2)$$
$$32t = 36t - 18$$
$$-4t = -18$$
$$t = \frac{18}{4}, \text{ or } \frac{9}{2}$$

The time at 32 mph is $\frac{9}{2}$ hr, and the time at 36 mph is $\frac{9}{2} - \frac{1}{2}$, or 4 hr.

Check. At 32 mph, in $\frac{9}{2}$ hr the salesperson will travel $32 \cdot \frac{9}{2}$, or 144 mi. At 36 mph, in 4 hr she will travel $36 \cdot 4$, or 144 mi. Since the distances are the same, the numbers check.

State. The towns are 144 mi apart.

37. *Familiarize*. We first make a drawing. Let $t =$ the time, $d =$ the distance traveled at 190 km/h, and $780 - d =$ the distance traveled at 200 km/h.

190 km/h t hr t hr   200 km/h

|————————— 780 km —————————|

We list the information in a table.

	Distance	Rate	Time	
Slower plane	d	190	t	$\rightarrow d = 190t$
Faster plane	$780 - d$	200	t	$\rightarrow 780 - d = 200t$

$d = r \cdot t$

Translate. Using $d = rt$ in each row of the table, we get a system of equations:

$$d = 190t, \quad (1)$$
$$780 - d = 200t \quad (2)$$

Solve. We solve the system of equations.

$$780 - 190t = 200t \quad \text{Substituting } 190t \text{ for } d \text{ in } (2)$$
$$780 = 390t$$
$$2 = t$$

Check. In 2 hr the slower plane will travel $190 \cdot 2$, or 380 km, and the faster plane will travel $200 \cdot 2$, or 400 km. The sum of the distances is $380 + 400$, or 780 km. The value checks.

State. The planes will meet in 2 hr.

39. *Familiarize*. We first make a drawing. Let $d =$ the distance traveled at 420 km/h and $t =$ the time traveled. Then $1000 - d =$ the distance traveled at 330 km/h.

d km, 420 km/h, t hr $1000 - d$ km, 330 km/h, t hr

|————————— 1000 km —————————|

We list the information in a table.

	Distance	Rate	Time	
Faster airplane	d	420	t	$\rightarrow d = 420t$
Slower airplane	$1000 - d$	330	t	$\rightarrow 1000 - d = 330t$

$d = r \cdot t$

Translate. Using $d = rt$ in each row of the table, we get a system of equations:

$$d = 420t, \quad (1)$$
$$1000 - d = 330t \quad (2)$$

Solve. We use substitution.

$$1000 - 420t = 330t \quad \text{Substituting } 420t \text{ for } d \text{ in } (2)$$
$$1000 = 750t$$
$$\frac{4}{3} = t$$

Check. If $t = \frac{4}{3}$, then $420 \cdot \frac{4}{3} = 560$, the distance traveled by the faster airplane. Also, $330 \cdot \frac{4}{3} = 440$, the distance traveled by the slower plane. The sum of the distances is $560 + 440$, or 1000 km. The values check.

State. The airplanes will meet after $\frac{4}{3}$ hr, or $1\frac{1}{3}$ hr.

41. *Familiarize*. We first make a drawing. Let $l =$ the length and $w =$ the width of the original piece of posterboard. Then $w - 6 =$ the width after cutting off 6 in.

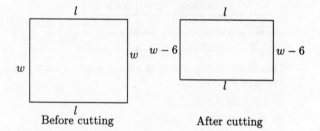

Before cutting   After cutting

Translate. The first statement gives us one equation.

$$\underbrace{\text{The perimeter of the original piece of posterboard}}_{2l + 2w} \underbrace{\text{is}}_{=} \underbrace{156 \text{ in.}}_{156}$$

Rewording the second statement, we get another equation.

$$\underbrace{\text{The length}}_{l} \underbrace{\text{is}}_{=} \underbrace{4}_{4} \underbrace{\text{times}}_{\cdot} \underbrace{\text{the width after cutting off 6 in.}}_{(w - 6)}$$

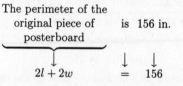

We have a system of equations:

$$2l + 2w = 156, \quad (1)$$
$$l = 4(w - 6) \quad (2)$$

Solve. We solve the system of equations.

$2 \cdot 4(w - 6) + 2w = 156$ Substituting $4(w-6)$
                                  for l in (1)

$8w - 48 + 2w = 156$

$10w - 48 = 156$

$10w = 204$

$w = \dfrac{204}{10}, \text{ or } \dfrac{102}{5}$

$l = 4\left(\dfrac{102}{5} - 6\right)$ Substituting $\dfrac{102}{5}$ for w in (2)

$l = 4\left(\dfrac{102}{5} - \dfrac{30}{5}\right)$

$l = 4\left(\dfrac{72}{5}\right)$

$l = \dfrac{288}{5}$

Check. The perimeter of a rectangle with width $\dfrac{102}{5}$ in. and length $\dfrac{288}{5}$ in. is $2\left(\dfrac{288}{5}\right) + 2\left(\dfrac{102}{5}\right) = \dfrac{576}{5} + \dfrac{204}{5} = \dfrac{780}{5} = 156$ in. If 6 in. is cut off the width, the new width is $\dfrac{102}{5} - 6 = \dfrac{102}{5} - \dfrac{30}{5} = \dfrac{72}{5}$. The length, $\dfrac{288}{5}$, is $4\left(\dfrac{72}{5}\right)$. The numbers check.

State. The original piece of posterboard has width $\dfrac{102}{5}$ in. or $20\dfrac{2}{5}$ in. and length $\dfrac{288}{5}$ in., or $57\dfrac{3}{5}$ in.

43. *Familiarize*. Let x and y represent the number of members who ordered one and two books, respectively. Note that the y members ordered $2y$ books.

Translate.

The number of books sold was 880.

$$x + 2y = 880$$

The amount taken in was \$9840.

$$12x + 20y = 9840$$

We have a system of equations:

$x + 2y = 880,$ (1)

$12x + 20y = 9840$ (2)

Solve. We use the elimination method.

$-12x - 24y = -10,560$ Multiplying (1) by -12

$\underline{12x + 20y = \quad\ \ 9840}$

$-4y = \quad -720$ Adding

$y = 180$

Check. If 180 members each buy 2 books, they buy a total of 360 books. Then $880 - 360$, or 520 members, each buy one book. The amount taken in from 520 single book orders and 180 orders of 2 books is $520(\$12) + 180(\$20)$, or \$9840. The result checks.

State. 180 members ordered two books.

45. *Familiarize*. Let x and y represent the number of miles of city driving and highway driving, respectively.

Translate. We write two equations.

The number of miles driven was 465.

$$x + y = 465$$

The gallons used in city driving plus the gallons used in highway driving were 23 gal.

$$\dfrac{x}{18} + \dfrac{y}{24} = 23$$

After clearing fractions, we have this system of equations:

$x + y = 465,$ (1)

$4x + 3y = 1656$ (2)

Solve. We use the elimination method.

$-3x - 3y = -1395$ Multiplying (1) by -3

$\underline{4x + 3y = \quad 1656}$

$x = \quad 261$ Adding

$261 + y = 465$ Substituting 261 for x in (1)

$y = 204$

Check. The total numbers of miles driven is $261 + 204$, or 465. In city driving, 261/18, or 14.5 gal of gasoline were used. In highway driving, 204/24, or 8.5 gal of gasoline were used. The total amount of gasoline used was $14.5 + 8.5$, or 23 gal. The numbers check.

State. 261 mi were driven in the city, and 204 mi were driven on the highway.

47. *Familiarize*. We first make a drawing. We let $d =$ the distance traveled in each direction. Since 1/2 hr of fuel must be reserved, the plane can travel for a total of $2\dfrac{1}{2}$ hr. We let $t =$ the time the plane travels against the 20 mph head wind. Then $2\dfrac{1}{2} - t$, or $2.5 - t =$ the time the plane travels with the head wind. The plane's speed against the head wind is $120 - 20$, or 100 km/h, and the speed with the head wind is $120 + 20$, or 140 km/h.

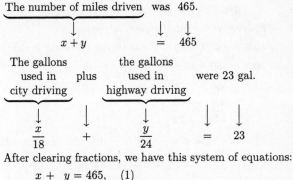

Against head wind
100 mph t hr d mi

With head wind
140 mph $(2.5 - t)$ hr d mi

From the drawing we see that the distances are the same. We list the information in a table.

	d	$=$	r	$\cdot$	t	
	Distance		Rate		Time	
Against head wind	d		100		t	$\rightarrow d = 100t$
With head wind	d		140		$2.5 - t$	$\rightarrow d = 140(2.5 - t)$

Translate. Using $d = rt$ in each row of the table, we get a system of equations:

$$d = 100t, \qquad (1)$$
$$d = 140(2.5 - t) \quad (2)$$

Solve. We solve the system of equations.

$100t = 140(2.5 - t)$ Substituting $100t$ for d in (2)

$100t = 350 - 140t$

$240t = 350$

$$t = \frac{350}{240}, \text{ or } \frac{35}{24}$$

The time against the wind is $\frac{35}{24}$ hr, and the time with the wind is $2.5 - \frac{35}{24}$, or $\frac{5}{2} - \frac{35}{24}$, or $\frac{25}{24}$ hr.

Check. Against the wind, the plane will travel $100 \cdot \frac{35}{24}$, or $145\frac{5}{6}$ mi. With the wind, it will travel $140 \cdot \frac{25}{24}$, or $145\frac{5}{6}$. Since the distances are the same, the numbers check.

State. The instructor should allow the student to fly $145\frac{5}{6}$ mi before returning.

Exercise Set 4.4

1. Substitute $(1, -2, 3)$ into the three equations, using alphabetical order.

$x + y + z = 2$	
$1 + (-2) + 3$	2
2	TRUE

$x - 2y - z = 2$	
$1 - 2(-2) - 3$	2
$1 + 4 - 3$	
2	TRUE

$3x + 2y + z = 2$	
$3 \cdot 1 + 2(-2) + 3$	2
$3 - 4 + 3$	
2	TRUE

The triple $(1, -2, 3)$ makes all three equations true, so it is a solution.

3.
$$x + y + z = 2, \quad (1)$$
$$2x - y + 5z = -5, \quad (2)$$
$$-x + 2y + 2z = 1 \quad (3)$$

Add Equations (1) and (2) to eliminate y:

$$\begin{array}{l} x + y + z = 2 \quad (1) \\ \underline{2x - y + 5z = -5} \quad (2) \\ 3x \qquad + 6z = -3 \quad (4) \quad \text{Adding} \end{array}$$

Use a different pair of equations and eliminate y:

$$\begin{array}{l} 4x - 2y + 10z = -10 \quad \text{Multiplying (2) by 2} \\ \underline{-x + 2y + 2z = 1} \quad (3) \\ 3x \qquad + 12z = -9 \quad (5) \quad \text{Adding} \end{array}$$

Now solve the system of Equations (4) and (5).

$$3x + 6z = -3 \quad (4)$$
$$3x + 12z = -9 \quad (5)$$

$$\begin{array}{l} -3x - 6z = 3 \quad \text{Multiplying (4) by } -1 \\ \underline{3x + 12z = -9} \quad (5) \\ 6z = -6 \quad \text{Adding} \\ z = -1 \end{array}$$

$3x + 6(-1) = -3$ Substituting -1 for z in (4)

$3x - 6 = -3$

$3x = 3$

$x = 1$

$1 + y + (-1) = 2$ Substituting 1 for x and -1 for z in (1)

$y = 2$ Simplifying

We obtain $(1, 2, -1)$. This checks, so it is the solution.

5.
$$2x - y + z = 5, \quad (1)$$
$$6x + 3y - 2z = 10, \quad (2)$$
$$x - 2y + 3z = 5 \quad (3)$$

We start by eliminating z from two different pairs of equations.

$$\begin{array}{l} 4x - 2y + 2z = 10 \quad \text{Multiplying (1) by 2} \\ \underline{6x + 3y - 2z = 10} \quad (2) \\ 10x + y \qquad = 20 \quad (4) \quad \text{Adding} \end{array}$$

$$\begin{array}{l} -6x + 3y - 3z = -15 \quad \text{Multiplying (1) by } -3 \\ \underline{x - 2y + 3z = 5} \quad (3) \\ -5x + y \qquad = -10 \quad (5) \quad \text{Adding} \end{array}$$

Now solve the system of Equations (4) and (5).

$$\begin{array}{l} 10x + y = 20 \quad (4) \\ \underline{5x - y = 10} \quad \text{Multiplying (5) by } -1 \\ 15x \qquad = 30 \quad \text{Adding} \\ x = 2 \end{array}$$

$10 \cdot 2 + y = 20$ Substituting 2 for x in (4)

$20 + y = 20$

$y = 0$

$2 \cdot 2 - 0 + z = 5$ Substituting 2 for x and 0 for y in (1)

$4 + z = 5$

$z = 1$

We obtain $(2, 0, 1)$. This checks, so it is the solution.

7.
$$2x - 3y + z = 5, \quad (1)$$
$$x + 3y + 8z = 22, \quad (2)$$
$$3x - y + 2z = 12 \quad (3)$$

We start by eliminating y from two different pairs of equations.

$$2x - 3y + z = 5 \quad (1)$$
$$\underline{x + 3y + 8z = 22} \quad (2)$$
$$3x \qquad + 9z = 27 \quad (4) \quad \text{Adding}$$

$$x + 3y + 8z = 22 \quad (2)$$
$$\underline{9x - 3y + 6z = 36} \quad \text{Multiplying (3) by 3}$$
$$10x \qquad + 14z = 58 \quad (5) \quad \text{Adding}$$

Solve the system of Equations (4) and (5).

$$3x + 9z = 27 \quad (4)$$
$$10x + 14z = 58 \quad (5)$$

$$30x + 90z = 270 \quad \text{Multiplying (4) by 10}$$
$$\underline{-30x - 42z = -174} \quad \text{Multiplying (5) by } -3$$
$$48z = 96 \quad \text{Adding}$$
$$z = 2$$

$$3x + 9 \cdot 2 = 27 \quad \text{Substituting 2 for } z \text{ in (4)}$$
$$3x + 18 = 27$$
$$3x = 9$$
$$x = 3$$

$$2 \cdot 3 - 3y + 2 = 5 \qquad \text{Substituting 3 for } x \text{ and 2}$$
$$\text{for } z \text{ in (1)}$$
$$-3y + 8 = 5$$
$$-3y = -3$$
$$y = 1$$

We obtain $(3, 1, 2)$. This checks, so it is the solution.

9.
$$3a - 2b + 7c = 13, \quad (1)$$
$$a + 8b - 6c = -47, \quad (2)$$
$$7a - 9b - 9c = -3 \quad (3)$$

We start by eliminating a from two different pairs of equations.

$$3a - 2b + 7c = 13 \quad (1)$$
$$\underline{-3a - 24b + 18c = 141} \quad \text{Multiplying (2) by } -3$$
$$- 26b + 25c = 154 \quad (4) \quad \text{Adding}$$

$$-7a - 56b + 42c = 329 \quad \text{Multiplying (2) by } -7$$
$$\underline{7a - 9b - 9c = -3} \quad (3)$$
$$- 65b + 33c = 326 \quad (5) \quad \text{Adding}$$

Now solve the system of Equations (4) and (5).

$$-26b + 25c = 154 \quad (4)$$
$$-65b + 33c = 326 \quad (5)$$

$$-130b + 125c = 770 \quad \text{Multiplying (4) by 5}$$
$$\underline{130b - 66c = -652} \quad \text{Multiplying (5) by } -2$$
$$59c = 118$$
$$c = 2$$

$$-26b + 25 \cdot 2 = 154 \quad \text{Substituting 2 for } c \text{ in (4)}$$
$$-26b + 50 = 154$$
$$-26b = 104$$
$$b = -4$$

$$a + 8(-4) - 6(2) = -47 \quad \text{Substituting } -4 \text{ for } b \text{ and}$$
$$2 \text{ for } c \text{ in (2)}$$
$$a - 32 - 12 = -47$$
$$a - 44 = -47$$
$$a = -3$$

We obtain $(-3, -4, 2)$. This checks, so it is the solution.

11.
$$2x + 3y + z = 17, \quad (1)$$
$$x - 3y + 2z = -8, \quad (2)$$
$$5x - 2y + 3z = 5 \quad (3)$$

We start by eliminating y from two different pairs of equations.

$$2x + 3y + z = 17 \quad (1)$$
$$\underline{x - 3y + 2z = -8} \quad (2)$$
$$3x \qquad + 3z = 9 \quad (4) \quad \text{Adding}$$

$$4x + 6y + 2z = 34 \quad \text{Multiplying (1) by 2}$$
$$\underline{15x - 6y + 9z = 15} \quad \text{Multiplying (3) by 3}$$
$$19x \qquad + 11z = 49 \quad (5) \quad \text{Adding}$$

Now solve the system of Equations (4) and (5).

$$3x + 3z = 9 \quad (4)$$
$$19x + 11z = 49 \quad (5)$$

$$33x + 33z = 99 \quad \text{Multiplying (4) by 11}$$
$$\underline{-57x - 33z = -147} \quad \text{Multiplying (5) by } -3$$
$$-24x = -48$$
$$x = 2$$

$$3 \cdot 2 + 3z = 9 \quad \text{Substituting 2 for } x \text{ in (4)}$$
$$6 + 3z = 9$$
$$3z = 3$$
$$z = 1$$

$$2 \cdot 2 + 3y + 1 = 17 \quad \text{Substituting 2 for } x \text{ and 1 for}$$
$$z \text{ in (1)}$$
$$3y + 5 = 17$$
$$3y = 12$$
$$y = 4$$

We obtain $(2, 4, 1)$. This checks, so it is the solution.

13.
$$2x + y + z = -2, \quad (1)$$
$$2x - y + 3z = 6, \quad (2)$$
$$3x - 5y + 4z = 7 \quad (3)$$

We start by eliminating y from two different pairs of equations.

$$2x + y + z = -2 \quad (1)$$
$$\underline{2x - y + 3z = 6} \quad (2)$$
$$4x \qquad + 4z = 4 \quad (4) \quad \text{Adding}$$

$$10x + 5y + 5z = -10 \quad \text{Multiplying (1) by 5}$$
$$\underline{3x - 5y + 4z = 7} \quad (3)$$
$$13x \qquad + 9z = -3 \quad (5) \quad \text{Adding}$$

Now solve the system of Equations (4) and (5).

$$4x + 4z = 4 \quad (4)$$
$$13x + 9z = -3 \quad (5)$$

$$36x + 36z = 36 \quad \text{Multiplying (4) by 9}$$
$$\underline{-52x - 36z = 12} \quad \text{Multiplying (5) by } -4$$
$$-16x \qquad = 48 \quad \text{Adding}$$
$$x = -3$$

$$4(-3) + 4z = 4 \quad \text{Substituting } -3 \text{ for } x \text{ in (4)}$$
$$-12 + 4z = 4$$
$$4z = 16$$
$$z = 4$$

$$2(-3) + y + 4 = -2 \quad \text{Substituting } -3 \text{ for } x \text{ and } 4$$
$$\text{for } z \text{ in (1)}$$
$$y - 2 = -2$$
$$y = 0$$

We obtain $(-3, 0, 4)$. This checks, so it is the solution.

15.
$$x - y + z = 4, \quad (1)$$
$$5x + 2y - 3z = 2, \quad (2)$$
$$3x - 7y + 4z = 8 \quad (3)$$

We start by eliminating z from two different pairs of equations.

$$3x - 3y + 3z = 12 \quad \text{Multiplying (1) by 3}$$
$$\underline{5x + 2y - 3z = 2} \quad (2)$$
$$8x - y \qquad = 14 \quad (4) \quad \text{Adding}$$

$$-4x + 4y - 4z = -16 \quad \text{Multiplying (1) by } -4$$
$$\underline{3x - 7y + 4z = 8} \quad (3)$$
$$-x - 3y \qquad = -8 \quad (5) \quad \text{Adding}$$

Now solve the system of Equations (4) and (5).

$$8x - y = 14 \quad (4)$$
$$-x - 3y = -8 \quad (5)$$

$$8x - y = 14 \quad (4)$$
$$\underline{-8x - 24y = -64} \quad \text{Multiplying (5) by 8}$$
$$-25y = -50$$
$$y = 2$$

$$8x - 2 = 14 \quad \text{Substituting 2 for } y \text{ in (4)}$$
$$8x = 16$$
$$x = 2$$

$$2 - 2 + z = 4 \quad \text{Substituting 2 for } x \text{ and 2 for}$$
$$y \text{ in (1)}$$
$$z = 4$$

We obtain $(2, 2, 4)$. This checks, so it is the solution.

17.
$$4x - y - z = 4, \quad (1)$$
$$2x + y + z = -1, \quad (2)$$
$$6x - 3y - 2z = 3 \quad (3)$$

We start by eliminating y from two different pairs of equations.

$$4x - y - z = 4 \quad (1)$$
$$\underline{2x + y + z = -1} \quad (2)$$
$$6x \qquad = 3 \quad (4) \quad \text{Adding}$$

At this point we can either continue by eliminating y from a second pair of equations or we can solve (4) for x and substitute that value in a different pair of the original equations to obtain a system of two equations in two variables. We take the second option.

$$6x = 3 \quad (4)$$
$$x = \frac{1}{2}$$

Substitute $\frac{1}{2}$ for x in (1):

$$4\left(\frac{1}{2}\right) - y - z = 4$$
$$2 - y - z = 4$$
$$-y - z = 2 \quad (5)$$

Substitute $\frac{1}{2}$ for x in (3):

$$6\left(\frac{1}{2}\right) - 3y - 2z = 3$$
$$3 - 3y - 2z = 3$$
$$-3y - 2z = 0 \quad (6)$$

Solve the system of Equations (5) and (6).

$$2y + 2z = -4 \quad \text{Multiplying (5) by } -2$$
$$\underline{-3y - 2z = 0} \quad (6)$$
$$-y \qquad = -4$$
$$y = 4$$

$$-4 - z = 2 \quad \text{Substituting 4 for } y \text{ in (5)}$$
$$-z = 6$$
$$z = -6$$

We obtain $\left(\frac{1}{2}, 4, -6\right)$. This checks, so it is the solution.

19.
$$2r + 3s + 12t = 4, \quad (1)$$
$$4r - 6s + 6t = 1, \quad (2)$$
$$r + s + t = 1 \quad (3)$$

We start by eliminating s from two different pairs of equations.

$$4r + 6s + 24t = 8 \quad \text{Multiplying (1) by 2}$$
$$\underline{4r - 6s + 6t = 1} \quad (2)$$
$$8r \qquad + 30t = 9 \quad (4) \quad \text{Adding}$$

$$4r - 6s + 6t = 1 \quad (2)$$
$$\underline{6r + 6s + 6t = 6} \quad \text{Multiplying (3) by 6}$$
$$10r \qquad + 12t = 7 \quad (5) \quad \text{Adding}$$

Solve the system of Equations (4) and (5).

$$40r + 150t = 45 \quad \text{Multiplying (4) by 5}$$
$$\underline{-40r - 48t = -28} \quad \text{Multiplying (5) by } -4$$
$$102t = 17$$
$$t = \frac{17}{102}$$
$$t = \frac{1}{6}$$

$$8r + 30\left(\frac{1}{6}\right) = 9 \quad \text{Substituting } \frac{1}{6} \text{ for } t \text{ in (4)}$$
$$8r + 5 = 9$$
$$8r = 4$$
$$r = \frac{1}{2}$$

$$\frac{1}{2} + s + \frac{1}{6} = 1 \quad \text{Substituting } \frac{1}{2} \text{ for } r \text{ and}$$
$$\frac{1}{6} \text{ for } t \text{ in (3)}$$
$$s + \frac{2}{3} = 1$$
$$s = \frac{1}{3}$$

We obtain $\left(\frac{1}{2}, \frac{1}{3}, \frac{1}{6}\right)$. This checks, so it is the solution.

21.
$$4a + 9b \qquad = 8, \quad (1)$$
$$8a \qquad + 6c = -1, \quad (2)$$
$$6b + 6c = -1 \quad (3)$$

We will use the elimination method. Note that there is no c in Equation (1). We will use equations (2) and (3) to obtain another equation with no c terms.

$$8a \qquad + 6c = -1 \quad (2)$$
$$\underline{\qquad - 6b - 6c = 1} \quad \text{Multiplying (3) by } -1$$
$$8a - 6b \qquad = 0 \quad (4) \quad \text{Adding}$$

Now solve the system of Equations (1) and (4).

$$-8a - 18b = -16 \quad \text{Multiplying (1) by } -2$$
$$\underline{8a - 6b = 0}$$
$$-24b = -16$$
$$b = \frac{2}{3}$$

$$8a - 6\left(\frac{2}{3}\right) = 0 \quad \text{Substituting } \frac{2}{3} \text{ for } b \text{ in (4)}$$
$$8a - 4 = 0$$
$$8a = 4$$
$$a = \frac{1}{2}$$

$$8\left(\frac{1}{2}\right) + 6c = -1 \quad \text{Substituting } \frac{1}{2} \text{ for } a \text{ in (2)}$$
$$4 + 6c = -1$$
$$6c = -5$$
$$c = -\frac{5}{6}$$

We obtain $\left(\frac{1}{2}, \frac{2}{3}, -\frac{5}{6}\right)$. This checks, so it is the solution.

23.
$$x + y + z = 57, \quad (1)$$
$$-2x + y \qquad = 3, \quad (2)$$
$$x \qquad - z = 6 \quad (3)$$

We will use the substitution method. Solve Equations (2) and (3) for y and z, respectively. Then substitute in Equation (1) to solve for x.

$$-2x + y = 3 \qquad \text{Solving (2) for } y$$
$$y = 2x + 3$$
$$x - z = 6 \qquad \text{Solving (3) for } z$$
$$-z = -x + 6$$
$$z = x - 6$$

$$x + (2x + 3) + (x - 6) = 57 \quad \text{Substituting in (1)}$$
$$4x - 3 = 57$$
$$4x = 60$$
$$x = 15$$

To find y, substitute 15 for x in $y = 2x + 3$:
$$y = 2 \cdot 15 + 3 = 33$$

To find z, substitute 15 for x in $z = x - 6$:
$$z = 15 - 6 = 9$$

We obtain $(15, 33, 9)$. This checks, so it is the solution.

25.
$$r + s \qquad = 5, \quad (1)$$
$$3s + 2t = -1, \quad (2)$$
$$4r \qquad + t = 14 \quad (3)$$

We will use the elimination method. Note that there is no t in Equation (1). We will use Equations (2) and (3) to obtain another equation with no t terms.

$$3s + 2t = -1 \quad (2)$$
$$\underline{-8r \qquad - 2t = -28} \quad \text{Multiplying (3) by } -2$$
$$-8r + 3s \qquad = -29 \quad (4) \quad \text{Adding}$$

Now solve the system of Equations (1) and (4).

$$r + s = 5 \quad (1)$$
$$-8r + 3s = -29 \quad (4)$$

$$8r + 8s = 40 \quad \text{Multiplying (1) by 8}$$
$$\underline{-8r + 3s = -29} \quad (4)$$
$$11s = 11 \quad \text{Adding}$$
$$s = 1$$

$$r + 1 = 5 \quad \text{Substituting 1 for } s \text{ in (1)}$$
$$r = 4$$

$$4 \cdot 4 + t = 14 \quad \text{Substituting 4 for } r \text{ in (3)}$$
$$16 + t = 14$$
$$t = -2$$

We obtain $(4, 1, -2)$. This checks, so it is the solution.

27.
$$F = 3ab$$
$$\frac{F}{3b} = a \quad \text{Dividing by } 3b$$

29.
$$F = \frac{1}{2}t(c - d)$$

$2F = t(c - d)$ Multiplying by 2

$2F = tc - td$ Removing parentheses

$2F + td = tc$ Adding td

$\dfrac{2F + td}{t} = c$, or   Dividing by t

$\dfrac{2F}{t} + d = c$

31. $y = 5 - 4x$

$y = -4x + 5$ Rewriting

The equation is now in the form $y = mx + b$. The slope is -4.

33.
$$w + x + y + z = 2, \quad (1)$$
$$w + 2x + 2y + 4z = 1, \quad (3)$$
$$w - x + y + z = 6, \quad (3)$$
$$w - 3x - y + z = 2 \quad (4)$$

Start by eliminating w from three different pairs of equations.

$\begin{aligned} w + x + y + z &= 2 \quad (1) \\ \underline{-w - 2x - 2y - 4z} &= \underline{-1} \quad \text{Multiplying (2) by } -1 \\ -x - y - 3z &= 1 \quad (5) \quad \text{Adding} \end{aligned}$

$\begin{aligned} w + x + y + z &= 2 \quad (1) \\ \underline{-w + x - y - z} &= \underline{-6} \quad \text{Multiplying (3) by } -1 \\ 2x \qquad\qquad &= -4 \quad (6) \quad \text{Adding} \end{aligned}$

$\begin{aligned} w + x + y + z &= 2 \quad (1) \\ \underline{-w + 3x + y - z} &= \underline{-2} \quad \text{Multiplying (4) by } -1 \\ 4x + 2y \qquad &= 0 \quad (7) \quad \text{Adding} \end{aligned}$

We can solve (6) for x:

$2x = -4$

$x = -2$

Substitute -2 for x in (7):

$4(-2) + 2y = 0$

$-8 + 2y = 0$

$2y = 8$

$y = 4$

Substitute -2 for x and 4 for y in (5):

$-(-2) - 4 - 3z = 1$

$-2 - 3z = 1$

$-3z = 3$

$z = -1$

Substitute -2 for x, 4 for y, and -1 for z in (1):

$w - 2 + 4 - 1 = 2$

$w + 1 = 2$

$w = 1$

We obtain $(1, -2, 4, -1)$. This checks, so it is the solution.

Exercise Set 4.5

1. ***Familiarize***. Let $x =$ the first number, $y =$ the second number, and $z =$ the third number.

Translate.

$\underbrace{\text{The sum of the three numbers}}$ is   5.

$\qquad\qquad x + y + z \qquad\qquad = 5$

$\underbrace{\text{The first number}}$ minus $\underbrace{\text{the second}}$ plus $\underbrace{\text{the third}}$ is 1.

$\qquad x \qquad\quad - \qquad\quad y \qquad + \qquad z \qquad = 1$

$\underbrace{\text{The first number}}$ minus $\underbrace{\text{the third}}$ is $\underbrace{\text{3 more than the second.}}$

$\qquad x \qquad\quad - \qquad\quad z \qquad = \qquad y + 3$

We now have a system of equations.

$\begin{aligned} x + y + z &= 5, & \text{or} && x + y + z &= 5, \\ x - y + z &= 1, &&& x - y + z &= 1, \\ x - z &= y + 3 &&& x - y - z &= 3 \end{aligned}$

Solve. Solving the system we get $(4, 2, -1)$.

Check. The sum of the numbers is 5. The first minus the second plus the third is $4 - 2 + (-1)$, or 1. The first minus the third is 5, which is three more than the second. The numbers check.

State. The numbers are 4, 2, and -1.

3. ***Familiarize***. We first make a drawing.

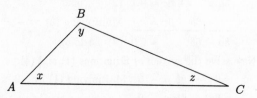

We let x, y, and z represent the measures of angles A, B, and C, respectively. The measures of the angles of a triangle add up to $180°$.

Translate.

$\underbrace{\text{The sum of the measures}}$ is $180°$.

$\qquad\quad x + y + z \qquad\quad = 180$

$\underbrace{\text{The measure of angle B}}$ is $\underbrace{\text{2° more than three times the measure of angle A.}}$

$\qquad\quad y \qquad\qquad = \qquad\qquad 3x + 2$

$\underbrace{\text{The measure of angle C}}$ is $\underbrace{\text{8° more than the measure of angle A.}}$

$\qquad\quad z \qquad\qquad = \qquad\qquad x + 8$

We now have a system of equations.

$$x + y + z = 180,$$
$$y = 3x + 2,$$
$$z = x + 8$$

Solve. Solving the system we get $(34, 104, 42)$.

Check. The sum of the numbers is 180, so that checks. Three times the measure of angle A is $3 \cdot 34°$, or $102°$, and $2°$ added to $102°$ is $104°$, the measure of angle B. The measure of angle C, $42°$, is $8°$ more than $34°$, the measure of angle A. These values check.

State. Angles A, B, and C measure $34°$, $104°$, and $42°$, respectively.

5. **Familiarize**. Let x, y, and z represent the amount spent on newspaper, television and radio ads, respectively, in billions of dollars.

Translate.

The total expenditure was \$84.8 billion.

$$x + y + z \qquad = \qquad 84.8$$

The total amount spent on television and radio ads was \$2.6 billion more than the amount spent on newspaper ads.

$$y + z \qquad = \qquad 2.6 \qquad + \qquad x$$

The amount spent on newspaper ads was \$5.1 billion more than the amount spent on television ads.

$$x \qquad = \qquad 5.1 \qquad + \qquad y$$

We now have a system of equations.

$$x + y + z = 84.8,$$
$$y + z = 2.6 + x,$$
$$x = 5.1 + y$$

Solve. Solving the system we get $(41.1., 36, 7.7)$.

Check. The sum of the numbers is 84.8, so that checks. Also, $36 + 7.7 = 43.7$ which is $2.6 + 41.1$, and $41.1 = 5.1 + 36$. These values check.

State. \$41.1 billion was spent on newspaper ads, \$36 billion was spent on television ads, and \$7.7 billion was spent on radio ads.

7. **Familiarize**. We first make a drawing.

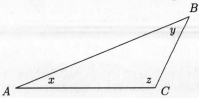

We let x, y, and z represent the measures of angles, A, B, and C, respectively. The measures of the angles of a triangle add up to $180°$.

Translate.

The sum of the measures is $180°$.

$$x + y + z \qquad = \qquad 180$$

The measure of angle B is twice the measure of angle A.

$$y \qquad = \qquad 2 \cdot \qquad x$$

The measure of angle C is $80°$ more than the measure of angle A.

$$z \qquad = \qquad x + 80$$

We now have a system of equations.

$$x + y + z = 180,$$
$$y = 2x,$$
$$z = x + 80$$

Solve. Solving the system we get $(25, 50, 105)$.

Check. The sum of the numbers is 180, so that checks. The measure of angle B, $50°$, is twice $25°$, the measure of angle A. The measure of angle C, $105°$, is $80°$ more than $25°$, the measure of angle A. The values check.

State. Angles A, B, and C measure $25°$, $50°$, and $105°$, respectively.

9. Let x, y, and z represent the amount of cholesterol in one egg, one cupcake, and one slice of pizza, respectively, in mg.

Translate.

The cholesterol in 1 egg and 1 cupcake and 1 slice of pizza is 302 mg.

$$x \quad + \quad y \quad + \quad z \quad = \quad 302$$

The cholesterol in 2 cupcakes and 3 slices of pizza is 65 mg.

$$2y \quad + \quad 3z \quad = \quad 65$$

The cholesterol in 2 eggs and 1 cupcake is 567 mg.

$$2x \quad + \quad y \quad = \quad 567$$

We have a system of equations.

$$x + y + z = 302,$$
$$2y + 3z = 65,$$
$$2x + y = 567$$

Solve. Solving the system we get $(274, 19, 9)$.

Check. The cholesterol in 1 egg, 1 cupcake, and 1 slice of pizza would be $274 + 19 + 9$, or 302 mg. The cholesterol in 2 cupcakes and 3 slices of pizza would be $2 \cdot 19 + 3 \cdot 9$, or 65 mg. The cholesterol in 2 eggs and 1 cupcake would be $2 \cdot 274 + 19$, or 567 mg. These values check.

State. One egg contains 274 mg of cholesterol, 1 cupcake contains 19 mg, and 1 slice of pizza contains 9 mg.

11. *Familiarize*. Let s = the number of servings of steak, p = the number of baked potatoes, and a = the number of servings of asparagus. Then s servings of steak contain $300s$ calories, $20s$ g of protein, and no vitamin C. In p baked potatoes there are $100p$ calories, $5p$ g of protein, and $20p$ mg of vitamin C. And a servings of asparagus contain $50a$ calories, $5a$ g of protein, and $40a$ mg of vitamin C. The patient requires 800 calories, 55 g of protein, and 220 mg of vitamin C.

Translate. Write equations for the total number of calories, the total amount of protein, and the total amount of vitamin C.

$$300s + 100p + 50a = 800 \quad \text{(calories)}$$
$$20s + 5p + 5a = 55 \quad \text{(protein)}$$
$$20p + 40a = 220 \quad \text{(vitamin C)}$$

We now have a system of equations.

Solve. Solving the system we get $s = 1$, $p = 3$, and $a = 4$.

Check. One serving of steak provides 300 calories, 20 g of protein, and no vitamin C. Three baked potatoes provide 300 calories, 15 g of protein, and 60 mg of vitamin C. Four servings of asparagus provide 200 calories, 20 g of protein, and 160 mg of vitamin C. Together, then, they provide 800 calories, 55 g of protein, and 220 mg of vitamin C. The values check.

State. One serving of steak, 3 baked potatoes, and 4 servings of asparagus are required.

13. *Familiarize*. Let x, y, and z represent the number of fraternal twin births for Orientals, African-Americans, and Caucasians in the U.S., respectively, out of every 15,400 births.

Translate. Out of every 15,400 births, we have the following statistics:

The total number of fraternal twin births is 739.

$$x + y + z = 739$$

The number of fraternal twin births for Orientals is 185 more than the number for African-Americans.

$$x = 185 + y$$

The number of fraternal twin births for Orientals is 231 more than the number for Caucasians.

$$x = 231 + z$$

We have a system of equations.

$$x + y + z = 739,$$
$$x = 185 + y,$$
$$x = 231 + y$$

Solve. Solving the system we get $(385, 200, 154)$.

Check. The total of the numbers is 739. Also 385 is 185 more than 200, and it is 231 more than 154.

State. Out of every 15,400 births, there are 385 births of fraternal twins for Orientals, 200 for African-Americans, and 154 for Caucasians.

15. *Familiarize*. It helps to organize the information in a table.

Pumps Working	A	B	C	A + B	A + C	A, B, & C
Gallons Per hour	x	y	z	2200	2400	3700

Wet let x, y, and z represent the number of gallons per hour which can be pumped by pumps A, B, and C, respectively.

Translate. From the table, we obtain three equations.

$$x + y + z = 3700 \quad \text{(All three pumps working)}$$
$$x + y = 2200 \quad \text{(A and B working)}$$
$$x + z = 2400 \quad \text{(A and C working)}$$

Solve. Solving the system we get $(900, 1300, 1500)$.

Check. The sum of the gallons per hour pumped when all three are pumping is $900 + 1300 + 1500$, or 3700. The sum of the gallons per hour pumped when only pump A and pump B are pumping is $900 + 1300$, or 2200. The sum of the gallons per hour pumped when only pump A and pump C are pumping is $900 + 1500$, or 2400. The numbers check.

State. The pumping capacities of pumps A, B, and C are respectively 900, 1300, and 1500 gallons per hour.

17. Let x, y, and z represent the amount invested at 8%, 6%, and 9%, respectively. The interest earned is $0.08x$, $0.06y$, and $0.09z$.

Translate.

The total invested was $80,000.

$$x + y + z = 80,000$$

The total interest was $6300.

$$0.08x + 0.06y + 0.09z = 6300$$

The interest at 8% was 4 times the interest at 6%.

$$0.08x = 4 \cdot 0.06y$$

We have a system of equations.

$$x + y + z = 80,000,$$
$$0.08x + 0.06y + 0.09z = 6300,$$
$$0.08x = 4(0.06y)$$

Solve. Solving the system we get $(45,000, 15,000, 20,000)$.

Check. The numbers add up to 80,000. Also, $0.08(45,000) + 0.06(15,000) + 0.09(20,000) = 6300$. In addition, $0.08(45,000) = 3600$ which is $4[0.06(15,000)]$. The values check.

State. $45,000 was invested at 8%, $15,000 at 6%, and $20,000 at 9%.

19. *Familiarize.* Let x, y, and z represent the number of par-3, par-4, and par-5 holes, respectively. Then a par golfer shoots $3x$ on the par-3 holes, $4x$ on the par-4 holes, and $5x$ on the par-5 holes.

Translate.

The total number of holes is 18.

$$x + y + z = 18$$

A par golfer's score is 70.

$$3x + 4y + 5z = 70$$

The number of par-4 holes is 2 times the number of par-5 holes.

$$y = 2 \cdot z$$

We have a system of equations.

$$x + y + z = 18,$$
$$3x + 4y + 5z = 70,$$
$$y = 2z$$

Solve. Solving the system we get $(6, 8, 4)$.

Check. The numbers add up to 18. A par golfer would shoot $3 \cdot 6 + 4 \cdot 8 + 5 \cdot 4$, or 70. The number of par-4 holes, 8, is twice the number of par-5 holes, 4. The numbers check.

State. There are 6 par-3 holes, 8 par-4 holes, and 4 par-5 holes.

21. $-4x + 5 - 2(x - 3) = 7$

$$\begin{array}{ll} -4x + 5 - 2x + 6 = 7 & \text{Removing parentheses} \\ -6x + 11 = 7 & \text{Collecting like terms} \\ -6x = -4 & \text{Subtracting 11} \\ x = \dfrac{4}{6} & \text{Dividing by } -6 \\ x = \dfrac{2}{3} & \text{Simplifying} \end{array}$$

The solution is $\dfrac{2}{3}$.

23. $10(3x - 4) = 8(3x - 8)$

$$\begin{array}{ll} 30x - 40 = 24x - 64 & \text{Removing parentheses} \\ 6x - 40 = -64 & \text{Subtracting } 24x \\ 6x = -24 & \text{Adding 40} \\ x = -4 & \text{Dividing by 6} \end{array}$$

The solution is -4.

25. We let $(-8, -2) = (x_1, y_1)$ and $(5, 6) = (x_2, y_2)$.

$$m = \frac{y_1 - y_2}{x_1 - x_2} = \frac{-2 - 6}{-8 - 5} = \frac{-8}{-13} = \frac{8}{13}$$

27. *Familiarize.* Let $x =$ the amount of the car loan, $y =$ the amount of the house loan, and $z =$ the amount due on credit cards. Then the interest on the car loan was $1\%x$, or $\dfrac{1}{100}x$. The interest on the house loan was $\dfrac{2}{3}\%y =$

$\dfrac{2}{3} \times 0.01 \times y = \dfrac{0.02}{3}y = \dfrac{2}{300}y = \dfrac{1}{150}y$. And the interest on the credit cards was $1.5\%z = 0.015z = \dfrac{15}{1000}z = \dfrac{3}{200}z$.

Translate.

The total of the loans is \$75,300.

$$x + y + z = 75{,}300$$

The total interest was \$742.

$$\frac{1}{100}x + \frac{1}{150}y + \frac{3}{200}z = 742$$

Credit card interest was \$2 more than car loan interest.

$$\frac{3}{200}z = 2 + \frac{1}{100}x$$

We have a system of equations.

$$x + y + z = 75{,}300,$$
$$\frac{1}{100}x + \frac{1}{150}y + \frac{3}{200}z = 742,$$
$$\frac{3}{200}z = 2 + \frac{1}{100}x$$

Solve. Solving the system we get $(26{,}875, \ 30{,}375, \ 18{,}050)$.

Check. The total of the loans is $\$26{,}875 + \$30{,}375 + \$18{,}050$, or \$75,300. The interest on the car loan was $\dfrac{1}{100}(\$26{,}875)$, or \$268.75; the interest on the house loan was $\dfrac{1}{150}(\$30{,}375)$, or \$202.50; and the interest on the credit cards was $\dfrac{3}{200}(\$18{,}050)$, or \$270.75. Then the total interest was $\$268.75 + \$202.50 + \$270.75$, or \$742. Also, the credit card interest, \$270.75, was \$2 more than the interest on the car loan, \$268.75.

State. The car loan is \$26,875; the house loan is \$30,375; and the amount due on the credit cards is \$18,050.

29. *Familiarize.* Let $x =$ the number of men in the audience, $y =$ the number of women, and $z =$ the number of children. Then $10x$ is taken in from ticket sales to men, $3y$ from ticket sales to women, and $0.5z$ from ticket sales to children.

Translate.

Total attendance is 100.

$$x + y + z = 100$$

Amount taken in was \$100.

$$10x + 3y + 0.5z = 100$$

There is no other information that can be translated to an equation. We have a system of two equations in three variables:

$$x + y + z = 100,$$
$$10x + 3y + 0.5z = 100$$

Clearing decimals, we have

$$x + \quad y + \quad z = 100, \quad (1)$$
$$100x + 30y + 5z = 1000 \quad (2)$$

Solve. We start by eliminating z.

$$\begin{array}{ll} -5x - \quad 5y - 5z = -500 & \text{Multiplying (1) by } -5 \\ \underline{100x + 30y + 5z = \quad 1000} & (2) \\ 95x + 25y \qquad\quad = \quad 500 & (4) \quad \text{Adding} \\ 19x + 5y = 100 & (3) \quad \text{Multiplying by } \frac{1}{5} \end{array}$$

Since we have only two equations, it is not possible to eliminate z from another pair of equations. However, in Equation (3), note that 5 is a factor of both $5y$ and 100. Therefore, 5 must also be a factor of $19x$, and hence of x, since 5 is not a factor of 19. Then for some positive integer n, $x = 5n$. (We require n to be positive, since the number of men clearly cannot be negative and must also be nonzero since the exercise states that the audience consists of <u>men</u>, women and children.) We have:

$$19 \cdot 5n + 5y = 100$$
$$19n + \quad y = 20 \quad \text{Dividing by 5}$$

Since n and y must both be positive, $n = 1$. (If $n > 1$, then $19n + y > 20$.) Then $x = 5 \cdot 1$, or 5.

$$19 \cdot 5 + 5y = 100 \quad \text{Substituting in (3)}$$
$$y = 1$$
$$5 + 1 + x = 100 \quad \text{Substituting in (1)}$$
$$z = 94$$

Check. The total attendance is $5 + 1 + 94$, or 100. The amount taken in is $10 \cdot 5 + 3 \cdot 1 + 0.5(94) = 50 + 3 + 47$, or \$100. The numbers check.

State. There were 5 men, 1 woman, and 94 children.

Exercise Set 4.6

1. $4x + 2y = 11,$

$3x - \quad y = \quad 2$

Write a matrix using only the constants.

$$\begin{bmatrix} 4 & 2 & \vdots & 11 \\ 3 & -1 & \vdots & 2 \end{bmatrix}$$

Multiply row 2 by 4 to make the first number in row 2 a multiple of 4.

$$\begin{bmatrix} 4 & 2 & \vdots & 11 \\ 12 & -4 & \vdots & 8 \end{bmatrix}$$

Multiply row 1 by -3 and add it to row 2.

$$\begin{bmatrix} 4 & 2 & \vdots & 11 \\ 0 & -10 & \vdots & -25 \end{bmatrix}$$

Reinserting the variables, we have

$$4x + 2y = 11, \quad (1)$$
$$-10y = -25. \quad (2)$$

Solve Equation (2) for y.

$$-10y = -25$$
$$y = \frac{5}{2}$$

Back-substitute $\frac{5}{2}$ for y in Equation (1) and solve for x.

$$4x + 2y = 11$$
$$4x + 2 \cdot \frac{5}{2} = 11$$
$$4x + 5 = 11$$
$$4x = 6$$
$$x = \frac{3}{2}$$

The solution is $\left(\frac{3}{2}, \frac{5}{2}\right)$.

3. $x + 4y = 8,$

$3x + 5y = 3$

We first write a matrix using only the constants.

$$\begin{bmatrix} 1 & 4 & \vdots & 8 \\ 3 & 5 & \vdots & 3 \end{bmatrix}$$

Multiply the first row by -3 and add it to the second row.

$$\begin{bmatrix} 1 & 4 & \vdots & 8 \\ 0 & -7 & \vdots & -21 \end{bmatrix}$$

Reinserting the variables, we have

$$x + 4y = 8, \quad (1)$$
$$-7y = -21. \quad (2)$$

Solve Equation (2) for y.

$$-7y = -21$$
$$y = 3$$

Back-substitute 3 for y in Equation (1) and solve for x.

$$x + 4 \cdot 3 = 8$$
$$x + 12 = 8$$
$$x = -4$$

The solution is $(-4, 3)$.

5. $5x - 3y = -2,$

$4x + 2y = 5$

Write a matrix using only the constants.

$$\begin{bmatrix} 5 & -3 & \vdots & -2 \\ 4 & 2 & \vdots & 5 \end{bmatrix}$$

Multiply the second row by 5 to make the first number in row 2 a multiple of 5.

$$\begin{bmatrix} 5 & -3 & \vdots & -2 \\ 20 & 10 & \vdots & 25 \end{bmatrix}$$

Now multiply the first row by -4 and add it to the second row.

$$\begin{bmatrix} 5 & -3 & \vdots & -2 \\ 0 & 22 & \vdots & 33 \end{bmatrix}$$

Reinserting the variables, we have

$$5x - 3y = -2, \quad (1)$$
$$22y = 33. \quad (2)$$

Solve Equation (2) for y.

$$22y = 33$$
$$y = \frac{3}{2}$$

Back-substitute $\frac{3}{2}$ for y in Equation (1) and solve for x.

$$5x - 3y = -2$$
$$5x - 3 \cdot \frac{3}{2} = -2$$
$$5x - \frac{9}{2} = -\frac{4}{2}$$
$$5x = \frac{5}{2}$$
$$x = \frac{1}{2}$$

The solution is $\left(\frac{1}{2}, \frac{3}{2}\right)$.

7. $2x - 3y = 50,$
$\quad 5x + y = 40$

Write a matrix using only the constants.

$$\begin{bmatrix} 2 & -3 & | & 50 \\ 5 & 1 & | & 40 \end{bmatrix}$$

Multiply the second row by 2 to make the first number in the second row a multiple of 2.

$$\begin{bmatrix} 2 & -3 & | & 50 \\ 10 & 2 & | & 80 \end{bmatrix}$$

Now multiply the first row by -5 and add it to the second row.

$$\begin{bmatrix} 2 & -3 & | & 50 \\ 0 & 17 & | & -170 \end{bmatrix}$$

Reinserting the variables, we have

$$2x - 3y = 50, \quad (1)$$
$$17y = -170. \quad (2)$$

Solve Equation (2) for y.

$$17y = -170$$
$$y = -10$$

Back-substitute -10 for y in Equation (1) and solve for x.

$$2x - 3(-10) = 50$$
$$2x + 30 = 50$$
$$2x = 20$$
$$x = 10$$

The solution is $(10, -10)$.

9. $4x - y - 3z = 1,$
$\quad 8x + y - z = 5,$
$\quad 2x + y + 2z = 5$

Write a matrix using only the constants.

$$\begin{bmatrix} 4 & -1 & -3 & | & 1 \\ 8 & 1 & -1 & | & 5 \\ 2 & 1 & 2 & | & 5 \end{bmatrix} \begin{array}{l} (P1) \\ (P2) \\ (P3) \end{array}$$

First interchange rows 1 and 3 so that each number below the first number in the first row is a multiple of that number.

$$\begin{bmatrix} 2 & 1 & 2 & | & 5 \\ 8 & 1 & -1 & | & 5 \\ 4 & -1 & -3 & | & 1 \end{bmatrix}$$

Multiply row 1 by -4 and add it to row 2.

Multiply row 1 by -2 and add it to row 3.

$$\begin{bmatrix} 2 & 1 & 2 & | & 5 \\ 0 & -3 & -9 & | & -15 \\ 0 & -3 & -7 & | & -9 \end{bmatrix}$$

Multiply row 2 by -1 and add it to row 3.

$$\begin{bmatrix} 2 & 1 & 2 & | & 5 \\ 0 & -3 & -9 & | & -15 \\ 0 & 0 & 2 & | & 6 \end{bmatrix}$$

Reinserting the variables, we have

$$2x + y + 2z = 5, \quad (P1)$$
$$-3y - 9z = -15, \quad (P2)$$
$$2z = 6. \quad (P3)$$

Solve (P3) for z.

$$2z = 6$$
$$z = 3$$

Back-substitute 3 for z in (P2) and solve for y.

$$-3y - 9z = -15$$
$$-3y - 9(3) = -15$$
$$-3y - 27 = -15$$
$$-3y = 12$$
$$y = -4$$

Back-substitute 3 for z and -4 for y in (P1) and solve for x.

$$2x + y + 2z = 5$$
$$2x + (-4) + 2(3) = 5$$
$$2x - 4 + 6 = 5$$
$$2x = 3$$
$$x = \frac{3}{2}$$

The solution is $\left(\frac{3}{2}, -4, 3\right)$.

11. $p + q + r = 1,$
$\quad p - 2q - 3r = 3,$
$\quad 4p + 5q + 6r = 4$

We first write a matrix using only the constants.

$$\begin{bmatrix} 1 & 1 & 1 & | & 1 \\ 1 & -2 & -3 & | & 3 \\ 4 & 5 & 6 & | & 4 \end{bmatrix} \begin{array}{l} (P1) \\ (P2) \\ (P3) \end{array}$$

Multiply row 1 by -1 and add it to row 2.

Multiply row 1 by -4 and add it to row 3.

$$\begin{bmatrix} 1 & 1 & 1 & | & 1 \\ 0 & -3 & -4 & | & 2 \\ 0 & 1 & 2 & | & 0 \end{bmatrix}$$

Multiply row 3 by 3.

$$\begin{bmatrix} 1 & 1 & 1 & | & 1 \\ 0 & -3 & -4 & | & 2 \\ 0 & 3 & 6 & | & 0 \end{bmatrix}$$

Add row 2 to row 3.

$$\begin{bmatrix} 1 & 1 & 1 & \vdots & 1 \\ 0 & -3 & -4 & \vdots & 2 \\ 0 & 0 & 2 & \vdots & 2 \end{bmatrix}$$

Reinserting the variables, we have

$$\begin{aligned} p + \quad q + \quad r &= 1, \quad (P1) \\ -3q - 4r &= 2, \quad (P2) \\ 2r &= 2. \quad (P3) \end{aligned}$$

Solve (P3) for r.

$$2r = 2$$
$$r = 1$$

Back-substitute 1 for r in (P2) and solve for q.

$$-3q - 4 \cdot 1 = 2$$
$$-3q - 4 = 2$$
$$-3q = 6$$
$$q = -2$$

Back-substitute -2 for q and 1 for r in (P1) and solve for p.

$$p + (-2) + 1 = 1$$
$$p - 1 = 1$$
$$p = 2$$

The solution is $(2, -2, 1)$.

13.
$$\begin{aligned} x - \quad y + 2z &= 0, \\ x - 2y + 3z &= -1, \\ 2x - 2y + \quad z &= -3 \end{aligned}$$

We first write a matrix using only the constants.

$$\begin{bmatrix} 1 & -1 & 2 & \vdots & 0 \\ 1 & -2 & 3 & \vdots & -1 \\ 2 & -2 & 1 & \vdots & -3 \end{bmatrix} \begin{matrix} (P1) \\ (P2) \\ (P3) \end{matrix}$$

Multiply row 1 by -1 and add it to row 2.

Multiply row 1 by -2 and add it to row 3.

$$\begin{bmatrix} 1 & -1 & 2 & \vdots & 0 \\ 0 & -1 & 1 & \vdots & -1 \\ 0 & 0 & -3 & \vdots & -3 \end{bmatrix}$$

Reinserting the variables, we have

$$\begin{aligned} x - y + 2z &= 0, \quad (P1) \\ - y + \quad z &= -1, \quad (P2) \\ - 3z &= -3. \quad (P3) \end{aligned}$$

Solve (P3) for z.

$$-3z = -3$$
$$z = 1$$

Back-substitute 1 for z in (P2) and solve for y.

$$-y + 1 = -1$$
$$-y = -2$$
$$y = 2$$

Back-substitute 2 for y and 1 for z in (P1) and solve for x.

$$x - 2 + 2 \cdot 1 = 0$$
$$x - 2 + 2 = 0$$
$$x = 0$$

The solution is $(0, 2, 1)$.

15.
$$\begin{aligned} 3p \quad\quad + 2r &= 11, \\ q - 7r &= 4, \\ p - 6q \quad\quad &= 1 \end{aligned}$$

We first write a matrix using only the constants.

$$\begin{bmatrix} 3 & 0 & 2 & \vdots & 11 \\ 0 & 1 & -7 & \vdots & 4 \\ 1 & -6 & 0 & \vdots & 1 \end{bmatrix} \begin{matrix} (P1) \\ (P2) \\ (P3) \end{matrix}$$

Interchange the first and third rows.

$$\begin{bmatrix} 1 & -6 & 0 & \vdots & 1 \\ 0 & 1 & -7 & \vdots & 4 \\ 3 & 0 & 2 & \vdots & 11 \end{bmatrix}$$

Multiply row 1 by -3 and add it to row 3.

$$\begin{bmatrix} 1 & -6 & 0 & \vdots & 1 \\ 0 & 1 & -7 & \vdots & 4 \\ 0 & 18 & 2 & \vdots & 8 \end{bmatrix}$$

Multiply row 2 by -18 and add it to row 3.

$$\begin{bmatrix} 1 & -6 & 0 & \vdots & 1 \\ 0 & 1 & -7 & \vdots & 4 \\ 0 & 0 & 128 & \vdots & -64 \end{bmatrix}$$

Reinserting the variables, we have

$$\begin{aligned} p - 6q \quad\quad &= 1, \quad (P1) \\ q - \quad 7r &= 4, \quad (P2) \\ 128r &= -64. \quad (P3) \end{aligned}$$

Solve (P3) for r.

$$128r = -64$$
$$r = -\frac{1}{2}$$

Back-substitute $-\frac{1}{2}$ for r in (P2) and solve for q.

$$q - 7r = 4$$
$$q - 7\left(-\frac{1}{2}\right) = 4$$
$$q + \frac{7}{2} = 4$$
$$q = \frac{1}{2}$$

Back-substitute $\frac{1}{2}$ for q in (P1) and solve for p.

$$p - 6 \cdot \frac{1}{2} = 1$$
$$p - 3 = 1$$
$$p = 4$$

The solution is $\left(4, \frac{1}{2}, -\frac{1}{2}\right)$.

17. We will rewrite the equations with the variables in alphabetical order:

$$-2w + 2x + 2y - 2z = -10,$$
$$w + x + y + z = -5,$$
$$3w + x - y + 4z = -2,$$
$$w + 3x - 2y + 2z = -6$$

Write a matrix using only the constants.

$$\begin{bmatrix} -2 & 2 & 2 & -2 & | & -10 \\ 1 & 1 & 1 & 1 & | & -5 \\ 3 & 1 & -1 & 4 & | & -2 \\ 1 & 3 & -2 & 2 & | & -6 \end{bmatrix} \begin{array}{c} (P1) \\ (P2) \\ (P3) \\ (P4) \end{array}$$

Multiply row 1 by $\frac{1}{2}$ so that each number in the first column below the first number is a multiple of that number.

$$\begin{bmatrix} -1 & 1 & 1 & -1 & | & -5 \\ 1 & 1 & 1 & 1 & | & -5 \\ 3 & 1 & -1 & 4 & | & -2 \\ 1 & 3 & -2 & 2 & | & -6 \end{bmatrix}$$

Add row 1 to row 2.

Multiply row 1 by 3 and add it to row 3.

Add row 1 to row 4.

$$\begin{bmatrix} -1 & 1 & 1 & -1 & | & -5 \\ 0 & 2 & 2 & 0 & | & -10 \\ 0 & 4 & 2 & 1 & | & -17 \\ 0 & 4 & -1 & 1 & | & -11 \end{bmatrix}$$

Multiply row 2 by -2 and add it to row 3 and to row 4.

$$\begin{bmatrix} -1 & 1 & 1 & -1 & | & -5 \\ 0 & 2 & 2 & 0 & | & -10 \\ 0 & 0 & -2 & 1 & | & 3 \\ 0 & 0 & -5 & 1 & | & 9 \end{bmatrix}$$

Multiply row 4 by 2.

$$\begin{bmatrix} -1 & 1 & 1 & -1 & | & -5 \\ 0 & 2 & 2 & 0 & | & -10 \\ 0 & 0 & -2 & 1 & | & 3 \\ 0 & 0 & -10 & 2 & | & 18 \end{bmatrix}$$

Multiply row 3 by -5 and add it to row 4.

$$\begin{bmatrix} -1 & 1 & 1 & -1 & | & -5 \\ 0 & 2 & 2 & 0 & | & -10 \\ 0 & 0 & -2 & 1 & | & 3 \\ 0 & 0 & 0 & -3 & | & 3 \end{bmatrix}$$

Reinserting the variables, we have

$$-w + x + y - z = -5, \quad (P1)$$
$$2x + 2y = -10, \quad (P2)$$
$$-2y + z = 3, \quad (P3)$$
$$-3z = 3. \quad (P4)$$

Solve (P4) for z.

$$-3z = 3$$
$$z = -1$$

Back-substitute -1 for z in (P3) and solve for y.

$$-2y + (-1) = 3$$
$$-2y = 4$$
$$y = -2$$

Back-substitute -2 for y in (P2) and solve for x.

$$2x + 2(-2) = -10$$
$$2x - 4 = -10$$
$$2x = -6$$
$$x = -3$$

Back-substitute -3 for x, -2 for y, and -1 for z in (P1) and solve for w.

$$-w + (-3) + (-2) - (-1) = -5$$
$$-w - 3 - 2 + 1 = -5$$
$$-w - 4 = -5$$
$$-w = -1$$
$$w = 1$$

The solution is $(1, -3, -2, -1)$.

19. $8a - [10 - 4(6a - 3)] = 8a - [10 - 24a + 12]$
$$= 8a - [22 - 24a]$$
$$= 8a - 22 + 24a$$
$$= 32a - 22$$

21. $[10(x + 5) - 6] + [3(x - 2) + 8]$
$$= [10x + 50 - 6] + [3x - 6 + 8]$$
$$= [10x + 44] + [3x + 2]$$
$$= 10x + 44 + 3x + 2$$
$$= 13x + 46$$

23. $6\{[4(x - 3) + 2^2] - 2[3(x + 6) - 8^2]\}$
$$= 6\{[4(x - 3) + 4] - 2[3(x + 6) - 64]\}$$
$$= 6\{[4x - 12 + 4] - 2[3x + 18 - 64]\}$$
$$= 6\{[4x - 8] - 2[3x - 46]\}$$
$$= 6\{4x - 8 - 6x + 92\}$$
$$= 6\{-2x + 84\}$$
$$= -12x + 504$$

Exercise Set 4.7

1. $\begin{vmatrix} 3 & 7 \\ 2 & 8 \end{vmatrix} = 3 \cdot 8 - 2 \cdot 7 = 24 - 14 = 10$

3. $\begin{vmatrix} -3 & -6 \\ -5 & -10 \end{vmatrix} = -3(-10) - (-5)(-6) = 30 - 30 = 0$

5. $\begin{vmatrix} 8 & 2 \\ 12 & -3 \end{vmatrix} = 8(-3) - 12 \cdot 2 = -24 - 24 = -48$

7. $\begin{vmatrix} 2 & -7 \\ 0 & 0 \end{vmatrix} = 2 \cdot 0 - 0(-7) = 0 - 0 = 0$

9.
$$\begin{vmatrix} 0 & 2 & 0 \\ 3 & -1 & 1 \\ 1 & -2 & 2 \end{vmatrix}$$

$$= 0\begin{vmatrix} -1 & 1 \\ -2 & 2 \end{vmatrix} - 3\begin{vmatrix} 2 & 0 \\ -2 & 2 \end{vmatrix} + 1\begin{vmatrix} 2 & 0 \\ -1 & 1 \end{vmatrix}$$

$$= 0 - 3[2 \cdot 2 - (-2) \cdot 0] + 1[2 \cdot 1 - (-1) \cdot 0]$$

$$= 0 - 3 \cdot 4 + 1 \cdot 2$$

$$= 0 - 12 + 2$$

$$= -10$$

11.
$$\begin{vmatrix} -1 & -2 & -3 \\ 3 & 4 & 2 \\ 0 & 1 & 2 \end{vmatrix}$$

$$= -1\begin{vmatrix} 4 & 2 \\ 1 & 2 \end{vmatrix} - 3\begin{vmatrix} -2 & -3 \\ 1 & 2 \end{vmatrix} + 0\begin{vmatrix} -2 & -3 \\ 4 & 2 \end{vmatrix}$$

$$= -1[4 \cdot 2 - 1 \cdot 2] - 3[-2 \cdot 2 - 1(-3)] + 0$$

$$= -1 \cdot 6 - 3 \cdot (-1) + 0$$

$$= -6 + 3 + 0$$

$$= -3$$

13.
$$\begin{vmatrix} 3 & 2 & -2 \\ -2 & 1 & 4 \\ -4 & -3 & 3 \end{vmatrix}$$

$$= 3\begin{vmatrix} 1 & 4 \\ -3 & 3 \end{vmatrix} - (-2)\begin{vmatrix} 2 & -2 \\ -3 & 3 \end{vmatrix} + (-4)\begin{vmatrix} 2 & -2 \\ 1 & 4 \end{vmatrix}$$

$$= 3[1 \cdot 3 - (-3) \cdot 4] + 2[2 \cdot 3 - (-3)(-2)] - 4[2 \cdot 4 - 1(-2)]$$

$$= 3 \cdot 15 + 2 \cdot 0 - 4 \cdot 10$$

$$= 45 + 0 - 40$$

$$= 5$$

15.
$$\begin{vmatrix} 3 & 2 & 4 \\ 1 & 1 & 1 \\ 1 & 1 & 1 \end{vmatrix}$$

$$= 3\begin{vmatrix} 1 & 1 \\ 1 & 1 \end{vmatrix} - 1\begin{vmatrix} 2 & 4 \\ 1 & 1 \end{vmatrix} + 1\begin{vmatrix} 2 & 4 \\ 1 & 1 \end{vmatrix}$$

$$= 3(1 \cdot 1 - 1 \cdot 1) - (2 \cdot 1 - 1 \cdot 4) + (2 \cdot 1 - 1 \cdot 4)$$

$$= 3 \cdot 0 - (-2) + (-2)$$

$$= 0 + 2 - 2$$

$$= 0$$

17. $3x - 4y = 6,$

$\qquad 5x + 9y = 10$

We compute D, D_x, and D_y.

$$D = \begin{vmatrix} 3 & -4 \\ 5 & 9 \end{vmatrix} = 27 - (-20) = 47$$

$$D_x = \begin{vmatrix} 6 & -4 \\ 10 & 9 \end{vmatrix} = 54 - (-40) = 94$$

$$D_y = \begin{vmatrix} 3 & 6 \\ 5 & 10 \end{vmatrix} = 30 - 30 = 0$$

Then,

$$x = \frac{D_x}{D} = \frac{94}{47} = 2$$

and

$$y = \frac{D_y}{D} = \frac{0}{47} = 0.$$

The solution is $(2, 0)$.

19. $-2x + 4y = 3,$

$\qquad 3x - 7y = 1$

We compute D, D_x, and D_y.

$$D = \begin{vmatrix} -2 & 4 \\ 3 & -7 \end{vmatrix} = 14 - 12 = 2$$

$$D_x = \begin{vmatrix} 3 & 4 \\ 1 & -7 \end{vmatrix} = -21 - 4 = -25$$

$$D_y = \begin{vmatrix} -2 & 3 \\ 3 & 1 \end{vmatrix} = -2 - 9 = -11$$

Then,

$$x = \frac{D_x}{D} = \frac{-25}{2} = -\frac{25}{2}$$

and

$$y = \frac{D_y}{D} = \frac{-11}{2} = -\frac{11}{2}.$$

The solution is $\left(-\dfrac{25}{2}, -\dfrac{11}{2}\right)$.

21. $4x + 2y = 11,$

$\qquad 3x - y = 2$

We compute D, D_x, and D_y.

$$D = \begin{vmatrix} 4 & 2 \\ 3 & -1 \end{vmatrix} = 4(-1) - 3 \cdot 2 = -10$$

$$D_x = \begin{vmatrix} 11 & 2 \\ 2 & -1 \end{vmatrix} = 11(-1) - 2 \cdot 2 = -15$$

$$D_y = \begin{vmatrix} 4 & 11 \\ 3 & 2 \end{vmatrix} = 4 \cdot 2 - 3 \cdot 11 = -25$$

Then,

$$x = \frac{D_x}{D} = \frac{-15}{-10} = \frac{3}{2}$$

and

$$y = \frac{D_y}{D} = \frac{-25}{-10} = \frac{5}{2}.$$

The solution is $\left(\dfrac{3}{2}, \dfrac{5}{2}\right)$.

23. $x + 4y = 8,$

$\quad 3x + 5y = 3$

We compute D, D_x, and D_y.

$$D = \begin{vmatrix} 1 & 4 \\ 3 & 5 \end{vmatrix} = 1 \cdot 5 - 3 \cdot 4 = -7$$

$$D_x = \begin{vmatrix} 8 & 4 \\ 3 & 5 \end{vmatrix} = 8 \cdot 5 - 3 \cdot 4 = 28$$

$$D_y = \begin{vmatrix} 1 & 8 \\ 3 & 3 \end{vmatrix} = 1 \cdot 3 - 3 \cdot 8 = -21$$

Then,

$$x = \frac{D_x}{D} = \frac{28}{-7} = -4$$

and

$$y = \frac{D_y}{D} = \frac{-21}{-7} = 3.$$

The solution is $(-4, 3)$.

25. $2x - 3y + 5z = 27,$

$\quad x + 2y - z = -4,$

$\quad 5x - y + 4z = 27$

We compute D, D_x, D_y, and D_z.

$$D = \begin{vmatrix} 2 & -3 & 5 \\ 1 & 2 & -1 \\ 5 & -1 & 4 \end{vmatrix}$$

$$= 2 \begin{vmatrix} 2 & -1 \\ -1 & 4 \end{vmatrix} - 1 \begin{vmatrix} -3 & 5 \\ -1 & 4 \end{vmatrix} + 5 \begin{vmatrix} -3 & 5 \\ 2 & -1 \end{vmatrix}$$

$$= 2(7) - 1(-7) + 5(-7)$$

$$= 14 + 7 - 35$$

$$= -14$$

$$D_x = \begin{vmatrix} 27 & -3 & 5 \\ -4 & 2 & -1 \\ 27 & -1 & 4 \end{vmatrix}$$

$$= 27 \begin{vmatrix} 2 & -1 \\ -1 & 4 \end{vmatrix} - (-4) \begin{vmatrix} -3 & 5 \\ -1 & 4 \end{vmatrix} + 27 \begin{vmatrix} -3 & 5 \\ 2 & -1 \end{vmatrix}$$

$$= 27(7) + 4(-7) + 27(-7)$$

$$= 189 - 28 - 189$$

$$= -28$$

$$D_y = \begin{vmatrix} 2 & 27 & 5 \\ 1 & -4 & -1 \\ 5 & 27 & 4 \end{vmatrix}$$

$$= 2 \begin{vmatrix} -4 & -1 \\ 27 & 4 \end{vmatrix} - 1 \begin{vmatrix} 27 & 5 \\ 27 & 4 \end{vmatrix} + 5 \begin{vmatrix} 27 & 5 \\ -4 & -1 \end{vmatrix}$$

$$= 2(11) - 1(-27) + 5(-7)$$

$$= 22 + 27 - 35$$

$$= 14$$

$$D_z = \begin{vmatrix} 2 & -3 & 27 \\ 1 & 2 & -4 \\ 5 & -1 & 27 \end{vmatrix}$$

$$= 2 \begin{vmatrix} 2 & -4 \\ -1 & 27 \end{vmatrix} - 1 \begin{vmatrix} -3 & 27 \\ -1 & 27 \end{vmatrix} + 5 \begin{vmatrix} -3 & 27 \\ 2 & -4 \end{vmatrix}$$

$$= 2(50) - 1(-54) + 5(-42)$$

$$= 100 + 54 - 210$$

$$= -56$$

Then,

$$x = \frac{D_x}{D} = \frac{-28}{-14} = 2,$$

$$y = \frac{D_y}{D} = \frac{14}{-14} = -1,$$

and

$$z = \frac{D_z}{D} = \frac{-56}{-14} = 4.$$

(Note that we could have found z by substitution once we had found x and y.)

The solution is $(2, -1, 4)$.

27. $r - 2s + 3t = 6,$

$\quad 2r - s - t = -3,$

$\quad r + s + t = 6$

We compute D, D_r, D_s, and D_t.

$$D = \begin{vmatrix} 1 & -2 & 3 \\ 2 & -1 & -1 \\ 1 & 1 & 1 \end{vmatrix}$$

$$= 1 \begin{vmatrix} -1 & -1 \\ 1 & 1 \end{vmatrix} - 2 \begin{vmatrix} -2 & 3 \\ 1 & 1 \end{vmatrix} + 1 \begin{vmatrix} -2 & 3 \\ -1 & -1 \end{vmatrix}$$

$$= 1(0) - 2(-5) + 1(5)$$

$$= 0 + 10 + 5$$

$$= 15$$

$$D_r = \begin{vmatrix} 6 & -2 & 3 \\ -3 & -1 & -1 \\ 6 & 1 & 1 \end{vmatrix}$$

$$= 6 \begin{vmatrix} -1 & -1 \\ 1 & 1 \end{vmatrix} - (-3) \begin{vmatrix} -2 & 3 \\ 1 & 1 \end{vmatrix} + 6 \begin{vmatrix} -2 & 3 \\ -1 & -1 \end{vmatrix}$$

$$= 6(0) + 3(-5) + 6(5)$$

$$= 0 - 15 + 30$$

$$= 15$$

$$D_s = \begin{vmatrix} 1 & 6 & 3 \\ 2 & -3 & -1 \\ 1 & 6 & 1 \end{vmatrix}$$

$$= 1\begin{vmatrix} -3 & -1 \\ 6 & 1 \end{vmatrix} - 2\begin{vmatrix} 6 & 3 \\ 6 & 1 \end{vmatrix} + 1\begin{vmatrix} 6 & 3 \\ -3 & -1 \end{vmatrix}$$

$$= 1(3) - 2(-12) + 1(3)$$

$$= 3 + 24 + 3$$

$$= 30$$

$$D_t = \begin{vmatrix} 1 & -2 & 6 \\ 2 & -1 & -3 \\ 1 & 1 & 6 \end{vmatrix}$$

$$= 1\begin{vmatrix} -1 & -3 \\ 1 & 6 \end{vmatrix} - 2\begin{vmatrix} -2 & 6 \\ 1 & 6 \end{vmatrix} + 1\begin{vmatrix} -2 & 6 \\ -1 & -3 \end{vmatrix}$$

$$= 1(-3) - 2(-18) + 1(12)$$

$$= -3 + 36 + 12$$

$$= 45$$

Then,

$$r = \frac{D_r}{D} = \frac{15}{15} = 1,$$

$$s = \frac{D_s}{D} = \frac{30}{15} = 2,$$

and

$$t = \frac{D_t}{D} = \frac{45}{15} = 3.$$

(Note that we could have found t by substitution once we had found r and s.)

The solution is $(1, 2, 3)$.

29. $4x - y - 3z = 1,$
    $8x + y - z = 5,$
    $2x + y + 2z = 5$

We compute D, D_x, D_y, and D_z.

$$D = \begin{vmatrix} 4 & -1 & -3 \\ 8 & 1 & -1 \\ 2 & 1 & 2 \end{vmatrix}$$

$$= 4\begin{vmatrix} 1 & -1 \\ 1 & 2 \end{vmatrix} - 8\begin{vmatrix} -1 & -3 \\ 1 & 2 \end{vmatrix} + 2\begin{vmatrix} -1 & -3 \\ 1 & -1 \end{vmatrix}$$

$$= 4(3) - 8(1) + 2(4)$$

$$= 12 - 8 + 8$$

$$= 12$$

$$D_x = \begin{vmatrix} 1 & -1 & -3 \\ 5 & 1 & -1 \\ 5 & 1 & 2 \end{vmatrix}$$

$$= 1\begin{vmatrix} 1 & -1 \\ 1 & 2 \end{vmatrix} - 5\begin{vmatrix} -1 & -3 \\ 1 & 2 \end{vmatrix} + 5\begin{vmatrix} -1 & -3 \\ 1 & -1 \end{vmatrix}$$

$$= 1(3) - 5(1) + 5(4)$$

$$= 3 - 5 + 20$$

$$= 18$$

$$D_y = \begin{vmatrix} 4 & 1 & -3 \\ 8 & 5 & -1 \\ 2 & 5 & 2 \end{vmatrix}$$

$$= 4\begin{vmatrix} 5 & -1 \\ 5 & 2 \end{vmatrix} - 8\begin{vmatrix} 1 & -3 \\ 5 & 2 \end{vmatrix} + 2\begin{vmatrix} 1 & -3 \\ 5 & -1 \end{vmatrix}$$

$$= 4(15) - 8(17) + 2(14)$$

$$= 60 - 136 + 28$$

$$= -48$$

$$D_z = \begin{vmatrix} 4 & -1 & 1 \\ 8 & 1 & 5 \\ 2 & 1 & 5 \end{vmatrix}$$

$$= 4\begin{vmatrix} 1 & 5 \\ 1 & 5 \end{vmatrix} - 8\begin{vmatrix} -1 & 1 \\ 1 & 5 \end{vmatrix} + 2\begin{vmatrix} -1 & 1 \\ 1 & 5 \end{vmatrix}$$

$$= 4(0) - 8(-6) + 2(-6)$$

$$= 0 + 48 - 12$$

$$= 36$$

Then,

$$x = \frac{D_x}{D} = \frac{18}{12} = \frac{3}{2},$$

$$y = \frac{D_y}{D} = \frac{-48}{12} = -4,$$

and

$$z = \frac{D_z}{D} = \frac{36}{12} = 3.$$

(Note that we could have found z by substitution once we had found x and y.)

The solution is $\left(\dfrac{3}{2}, -4, 3\right)$.

31. $p + q + r = 1,$
    $p - 2q - 3r = 3,$
    $4p + 5q + 6r = 4$

We compute D, D_p, D_q, and D_r.

$$D = \begin{vmatrix} 1 & 1 & 1 \\ 1 & -2 & -3 \\ 4 & 5 & 6 \end{vmatrix}$$

$$= 1 \begin{vmatrix} -2 & -3 \\ 5 & 6 \end{vmatrix} - 1 \begin{vmatrix} 1 & 1 \\ 5 & 6 \end{vmatrix} + 4 \begin{vmatrix} 1 & 1 \\ -2 & -3 \end{vmatrix}$$

$$= 1(3) - 1(1) + 4(-1)$$

$$= 3 - 1 - 4$$

$$= -2$$

$$D_p = \begin{vmatrix} 1 & 1 & 1 \\ 3 & -2 & -3 \\ 4 & 5 & 6 \end{vmatrix}$$

$$= 1 \begin{vmatrix} -2 & -3 \\ 5 & 6 \end{vmatrix} - 3 \begin{vmatrix} 1 & 1 \\ 5 & 6 \end{vmatrix} + 4 \begin{vmatrix} 1 & 1 \\ -2 & -3 \end{vmatrix}$$

$$= 1(3) - 3(1) + 4(-1)$$

$$= 3 - 3 - 4$$

$$= -4$$

$$D_q = \begin{vmatrix} 1 & 1 & 1 \\ 1 & 3 & -3 \\ 4 & 4 & 6 \end{vmatrix}$$

$$= 1 \begin{vmatrix} 3 & -3 \\ 4 & 6 \end{vmatrix} - 1 \begin{vmatrix} 1 & 1 \\ 4 & 6 \end{vmatrix} + 4 \begin{vmatrix} 1 & 1 \\ 3 & -3 \end{vmatrix}$$

$$= 1(30) - 1(2) + 4(-6)$$

$$= 30 - 2 - 24$$

$$= 4$$

$$D_r = \begin{vmatrix} 1 & 1 & 1 \\ 1 & -2 & 3 \\ 4 & 5 & 4 \end{vmatrix}$$

$$= 1 \begin{vmatrix} -2 & 3 \\ 5 & 4 \end{vmatrix} - 1 \begin{vmatrix} 1 & 1 \\ 5 & 4 \end{vmatrix} + 4 \begin{vmatrix} 1 & 1 \\ -2 & 3 \end{vmatrix}$$

$$= 1(-23) - 1(-1) + 4(5)$$

$$= -23 + 1 + 20$$

$$= -2$$

Then,

$$p = \frac{D_p}{D} = \frac{-4}{-2} = 2,$$

$$q = \frac{D_q}{D} = \frac{4}{-2} = -2,$$

and

$$r = \frac{D_r}{D} = \frac{-2}{-2} = 1.$$

(Note that we could have found r by substitution once we had found p and q.)

The solution is $(2, -2, 1)$.

33.
$$0.5x - 2.34 + 2.4x = 7.8x - 9$$
$$2.9x - 2.34 = 7.8x - 9$$
$$6.66 = 4.9x$$
$$\frac{6.66}{4.9} = x$$
$$\frac{666}{490} = x$$
$$\frac{333}{245} = x$$

35. We solve for y to obtain a slope-intercept equation.
$$5y - 3x = 8$$
$$5y = 3x + 8$$
$$y = \frac{3}{5}x + \frac{8}{5} \qquad (y = mx + b)$$

The slope is $\frac{3}{5}$.

37.
$$-4(x + 8) - 2(3x - 4) = -4x - 32 - 6x + 8$$
$$= -10x - 24$$

39.
$$\begin{vmatrix} y & -2 \\ 4 & 3 \end{vmatrix} = 44$$
$$y \cdot 3 - 4(-2) = 44 \quad \text{Evaluating the determinant}$$
$$3y + 8 = 44$$
$$3y = 36$$
$$y = 12$$

41.
$$\begin{vmatrix} m + 1 & -2 \\ m - 2 & 1 \end{vmatrix} = 27$$
$$(m + 1)(1) - (m - 2)(-2) = 27 \quad \text{Evaluating the determinant}$$
$$m + 1 + 2m - 4 = 27$$
$$3m = 30$$
$$m = 10$$

Exercise Set 4.8

1. Graph: $y \geq x$,

$\qquad\qquad y \leq -x + 2$

We graph the lines $y = x$ and $y = -x + 2$, using solid lines. We indicate the region for each inequality by the arrows at the ends of the lines. Note where the regions overlap, and shade the region of solutions.

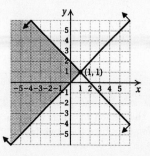

To find the vertex we solve the system of related equations:

$$y = x,$$
$$y = -x + 2$$

Solving, we obtain the vertex $(1, 1)$.

3. Graph: $y > x$,

$\qquad y < -x + 1$

We graph the lines $y = x$ and $y = -x + 1$, using dashed lines. We indicate the region for each inequality by arrows at the ends of the lines. Note where the regions overlap, and shade the region of solutions.

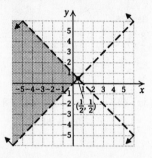

To find the vertex we solve the system of related equations:

$$y = x,$$
$$y = -x + 1$$

Solving, we obtain the vertex $\left(\dfrac{1}{2}, \dfrac{1}{2} \right)$.

5. Graph: $y \geq -2$,

$\qquad x \geq 1$

We graph the lines $y = -2$ and $x = 1$, using solid lines. We indicate the region for each inequality by arrows. Shade the region where they overlap.

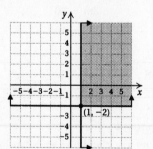

To find the vertex, we solve the system of related equations:

$$y = -2,$$
$$x = 1$$

Solving, we obtain the vertex $(1, -2)$.

7. Graph: $x \leq 3$,

$\qquad y \geq -3x + 2$

Graph the lines $x = 3$ and $y = -3x + 2$, using solid lines. Indicate the region for each inequality by arrows, and shade the region where they overlap.

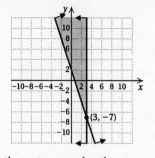

To find the vertex we solve the system of related equations:

$$x = 3,$$
$$y = -3x + 2$$

Solving, we obtain the vertex $(3, -7)$.

9. Graph: $y \geq -2$,

$\qquad y \geq x + 3$

Graph the lines $y = -2$ and $y = x+3$, using solid lines. Indicate the region for each inequality by arrows, and shade the region where they overlap.

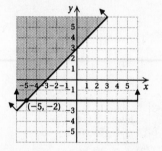

To find the vertex we solve the system of related equations:

$$y = -2,$$
$$y = x + 3$$

The vertex is $(-5, -2)$.

11. Graph: $x + y \leq 1$,

$\qquad x - y \leq 2$

Graph the lines $x + y = 1$ and $x - y = 2$, using solid lines. Indicate the region for each inequality by arrows, and shade the region where they overlap.

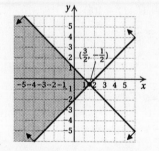

To find the vertex we solve the system of related equations:

$$x + y = 1,$$
$$x - y = 2$$

The vertex is $\left(\dfrac{3}{2}, -\dfrac{1}{2} \right)$.

13. Graph: $y - 2x \geq 1,$

$y - 2x \leq 3$

Graph the lines $y - 2x = 1$ and $y - 2x = 3$, using solid lines. Indicate the region for each inequality by arrows, and shade the region where they overlap.

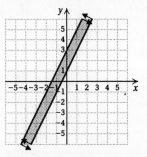

We can see from the graph that the lines are parallel. Hence there are no vertices.

15. Graph: $y \leq 2x + 1,$ (1)

$y \geq -2x + 1,$ (2)

$x \leq 2$ (3)

Shade the intersection of the graphs of $y \leq 2x + 1$, $y \geq -2x + 1$, and $x \leq 2$.

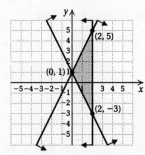

To find the vertices we solve three different systems of equations. From (1) and (2) we obtain the vertex $(0, 1)$. From (1) and (3) we obtain the vertex $(2, 5)$. From (2) and (3) we obtain the vertex $(2, -3)$.

17. Graph: $x + 2y \leq 12,$ (1)

$2x + y \leq 12,$ (2)

$x \geq 0,$ (3)

$y \geq 0$ (4)

Shade the intersection of the graphs of the four inequalities above.

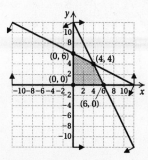

To find the vertices we solve four different systems of equations, as follows:

System of equations	Vertex
From (1) and (2)	$(4, 4)$
From (1) and (3)	$(0, 6)$
From (2) and (4)	$(6, 0)$
From (3) and (4)	$(0, 0)$

19. Graph: $8x + 5y \leq 40,$ (1)

$x + 2y \leq 8,$ (2)

$x \geq 0,$ (3)

$y \geq 0$ (4)

Shade the intersection of the graphs of the four inequalities above.

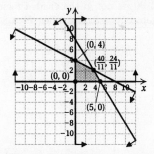

To find the vertices we solve four different systems of equations, as follows:

System of equations	Vertex
From (1) and (2)	$\left(\dfrac{40}{11}, \dfrac{24}{11}\right)$
From (1) and (4)	$(5, 0)$
From (2) and (3)	$(0, 4)$
From (3) and (4)	$(0, 0)$

21. $\dfrac{4^3 + 5 \cdot 6 - 7 \cdot 8}{|3 - 5|^5 + 4(9 - 2)} = \dfrac{64 + 5 \cdot 6 - 7 \cdot 8}{|-2|^5 + 4 \cdot 7}$

$= \dfrac{64 + 30 - 56}{32 + 28}$

$= \dfrac{38}{60} = \dfrac{19}{30}$

23. $5(3x - 4) = -2(x + 5)$

$15x - 20 = -2x - 10$

$17x - 20 = -10$

$17x = 10$

$x = \dfrac{10}{17}$

The solution is $\dfrac{10}{17}$.

25. $2(x - 1) + 3(x - 2) - 4(x - 5) = 10$

$2x - 2 + 3x - 6 - 4x + 20 = 10$

$x + 12 = 10$

$x = -2$

The solution is -2.

27. $4H - 3F < 70,$

$\qquad F + H > 160,$

$\qquad 2F + 3H > 390$

a) Shade the intersection of the graphs of the three
   inequalities above.

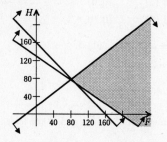

b) The point $(80, 80)$ is not in the shaded region, so
   it is not dangerous to exercise when $F = 80°$ and
   $H = 80\%$.

Chapter 5

Polynomials

1. $-9x^4 - x^3 + 7x^2 + 6x - 8$

Term	$-9x^4$	$-x^3$	$7x^2$	$6x$	-8
Degree	4	3	2	1	0
Degree of polynomial	4				
Leading term	$-9x^4$				
Leading coefficient	-9				

3. $t^3 + 4t^7 + s^2t^4 - 2$

Term	t^3	$4t^7$	s^2t^4	-2
Degree	3	7	6	0
Degree of polynomial	7			
Leading term	$4t^7$			
Leading coefficient	4			

5. $u^7 + 8u^2v^6 + 3uv + 4u - 1$

Term	u^7	$8u^2v^6$	$3uv$	$4u$	-1
Degree	7	8	2	1	0
Degree of polynomial	8				
Leading term	$8u^2v^6$				
Leading coefficient	8				

7. $-4y^3 - 6y^2 + 7y + 23$

9. $-xy^3 + x^2y^2 + x^3y + 1$

11. $-9b^5y^5 - 8b^2y^3 + 2by$

13. $5 + 12x - 4x^3 + 8x^5$

15. $3xy^3 + x^2y^2 - 9x^3y + 2x^4$

17. $-7ab + 4ax - 7ax^2 + 4x^6$

19. $P = 4x^2 - 3x + 2$

When $x = 4$, $P = 4 \cdot 4^2 - 3 \cdot 4 + 2$

$\qquad = 64 - 12 + 2$

$\qquad = 54$

When $x = 0$, $P = 4 \cdot 0^2 - 3 \cdot 0 + 2$

$\qquad = 0 - 0 + 2$

$\qquad = 2$

21. $Q = 6y^3 - 11y - 4$

When $y = -2$, $Q = 6(-2)^3 - 11(-2) - 4$

$\qquad = -48 + 22 - 4$

$\qquad = -30$

When $y = \frac{1}{3}$, $Q = 6\left(\frac{1}{3}\right)^3 - 11\left(\frac{1}{3}\right) - 4$

$\qquad = 6 \cdot \frac{1}{27} - \frac{11}{3} - 4$

$\qquad = \frac{2}{9} - \frac{11}{3} - 4$

$\qquad = \frac{2}{9} - \frac{33}{9} - \frac{36}{9}$

$\qquad = -\frac{67}{9}$

23. To find the number of games played when there are 8 teams in the league, we substitute 8 for n in the polynomial and calculate:

$N = \frac{1}{2}n^2 - \frac{1}{2}n$

$\quad = \frac{1}{2}(8)^2 - \frac{1}{2}(8)$

$\quad = \frac{1}{2} \cdot 64 - 4$

$\quad = 32 - 4$

$\quad = 28$

There will be 28 games.

For 20 teams, we substitute 20 for n and calculate:

$N = \frac{1}{2}n^2 - \frac{1}{2}n$

$\quad = \frac{1}{2}(20)^2 - \frac{1}{2}(20)$

$\quad = \frac{1}{2} \cdot 400 - 10$

$\quad = 200 - 10$

$\quad = 190$

There will be 190 games.

25. We substitute 4.7 for h, 1.2 for r, and 3.14 for π:

$$2\pi rh + 2\pi r^2 = 2(3.14)(1.2)(4.7) + 2(3.14)(1.2)^2$$
$$= 2(3.14)(1.2)(4.7) + 2(3.14)(1.44)$$
$$= 35.4192 + 9.0432$$
$$= 44.4624$$

The surface area is about 44.5 in^2.

27. Substitute 200 for x:

$$R = 280x - 0.4x^2$$
$$= 280(200) - 0.4(200)^2$$
$$= 56,000 - 0.4(40,000)$$
$$= 56,000 - 16,000$$
$$= 40,000$$

The total revenue is \$40,000.

29. Substitute 200 for x:

$$C = 7000 + 0.6x^2$$
$$= 7000 + 0.6(200)^2$$
$$= 7000 + 0.6(40,000)$$
$$= 7000 + 24,000$$
$$= 31,000$$

The total revenue is \$31,000.

31. We subtract:

$$P = R - C$$
$$= 280x - 0.4x^2 - (7000 + 0.6x^2)$$
$$= 280x - 0.4x^2 + (-7000 - 0.6x^2) \quad \text{Adding the opposite}$$
$$= -x^2 + 280x - 7000$$

The total profit is given by $P = -x^2 + 280x - 7000$.

33. $6x^2 - 7x^2 + 3x^2 = (6 - 7 + 3)x^2 = 2x^2$

35.
$$7x - 2y - 4x + 6y$$
$$= (7 - 4)x + (-2 + 6)y$$
$$= 3x + 4y$$

37.
$$3a + 9 - 2 + 8a - 4a + 7$$
$$= (3 + 8 - 4)a + (9 - 2 + 7)$$
$$= 7a + 14$$

39.
$$3a^2b + 4b^2 - 9a^2b - 6b^2$$
$$= (3 - 9)a^2b + (4 - 6)b^2$$
$$= -6a^2b - 2b^2$$

41.
$$8x^2 - 3xy + 12y^2 + x^2 - y^2 + 5xy + 4y^2$$
$$= (8 + 1)x^2 + (-3 + 5)xy + (12 - 1 + 4)y^2$$
$$= 9x^2 + 2xy + 15y^2$$

43.
$$4x^2y - 3y + 2xy^2 - 5x^2y + 7y + 7xy^2$$
$$= (4 - 5)x^2y + (-3 + 7)y + (2 + 7)xy^2$$
$$= -x^2y + 4y + 9xy^2$$

45.
$$(3x^2 + 5y^2 + 6) + (2x^2 - 3y^2 - 1)$$
$$= (3 + 2)x^2 + (5 - 3)y^2 + (6 - 1)$$
$$= 5x^2 + 2y^2 + 5$$

47.
$$(2a + 3b - c) + (4a - 2b + 2c)$$
$$= (2 + 4)a + (3 - 2)b + (-1 + 2)c$$
$$= 6a + b + c$$

49.
$$(a^2 - 3b^2 + 4c^2) + (-5a^2 + 2b^2 - c^2)$$
$$= (1 - 5)a^2 + (-3 + 2)b^2 + (4 - 1)c^2$$
$$= -4a^2 - b^2 + 3c^2$$

51.
$$(x^2 + 3x - 2xy - 3) + (-4x^2 - x + 3xy + 2)$$
$$= (1 - 4)x^2 + (3 - 1)x + (-2 + 3)xy + (-3 + 2)$$
$$= -3x^2 + 2x + xy - 1$$

53.
$$(7x^2y - 3xy^2 + 4xy) + (-2x^2y - xy^2 + xy)$$
$$= (7 - 2)x^2y + (-3 - 1)xy^2 + (4 + 1)xy$$
$$= 5x^2y - 4xy^2 + 5xy$$

55.
$$(2r^2 + 12r - 11) + (6r^2 - 2r + 4) + (r^2 - r - 2)$$
$$= (2 + 6 + 1)r^2 + (12 - 2 - 1)r + (-11 + 4 - 2)$$
$$= 9r^2 + 9r - 9$$

57.
$$\left(\frac{2}{3}xy + \frac{5}{6}xy^2 + 5.1x^2y\right) + \left(-\frac{4}{5}xy + \frac{3}{4}xy^2 - 3.4x^2y\right)$$
$$= \left(\frac{2}{3} - \frac{4}{5}\right)xy + \left(\frac{5}{6} + \frac{3}{4}\right)xy^2 + (5.1 - 3.4)x^2y$$
$$= \left(\frac{10}{15} - \frac{12}{15}\right)xy + \left(\frac{10}{12} + \frac{9}{12}\right)xy^2 + 1.7x^2y$$
$$= -\frac{2}{15}xy + \frac{19}{12}xy^2 + 1.7x^2y$$

59. $5x^3 - 7x^2 + 3x - 6$

   a) $-(5x^3 - 7x^2 + 3x - 6)$ Writing an inverse sign in front

   b) $-5x^3 + 7x^2 - 3x + 6$ Writing the opposite of each term

61. $-13y^2 + 6ay^4 - 5by^2$

   a) $-(-13y + 6ay^4 - 5by^2)$

   b) $13y^2 - 6ay^4 + 5by^2$

63.
$$(7x - 2) - (-4x + 5)$$
$$= (7x - 2) + (4x - 5) \quad \text{Adding the opposite}$$
$$= 11x - 7$$

65.
$$(-3x^2 + 2x + 9) - (x^2 + 5x - 4)$$
$$= (-3x^2 + 2x + 9) + (-x^2 - 5x + 4) \quad \text{Adding the opposite}$$
$$= -4x^2 - 3x + 13$$

67.
$$(5a - 2b + c) - (3a + 2b - 2c)$$
$$= (5a - 2b + c) + (-3a - 2b + 2c)$$
$$= 2a - 4b + 3c$$

69. $\quad (3x^2 - 2x - x^3) - (5x^2 - 8x - x^3)$

$\quad = (3x^2 - 2x - x^3) + (-5x^2 + 8x + x^3)$

$\quad = -2x^2 + 6x$

71. $\quad (5a^2 + 4ab - 3b^2) - (9a^2 - 4ab + 2b^2)$

$\quad = (5a^2 + 4ab - 3b^2) + (-9a^2 + 4ab - 2b^2)$

$\quad = -4a^2 + 8ab - 5b^2$

73. $\quad (6ab - 4a^2b + 6ab^2) - (3ab^2 - 10ab - 12a^2b)$

$\quad = (6ab - 4a^2b + 6ab^2) + (-3ab^2 + 10ab + 12a^2b)$

$\quad = 16ab + 8a^2b + 3ab^2$

75. $\quad (0.09y^4 - 0.052y^3 + 0.93) -$
$\qquad (0.03y^4 - 0.084y^3 + 0.94y^2)$

$\quad = (0.09y^4 - 0.052y^3 + 0.93) +$
$\qquad (-0.03y^4 + 0.084y^3 - 0.94y^2)$

$\quad = 0.06y^4 + 0.032y^3 - 0.94y^2 + 0.93$

77. $\quad \left(\dfrac{5}{8}x^4 - \dfrac{1}{4}x^2 - \dfrac{1}{2} \right) - \left(-\dfrac{3}{8}x^4 + \dfrac{3}{4}x^2 + \dfrac{1}{2} \right)$

$\quad = \left(\dfrac{5}{8}x^4 - \dfrac{1}{4}x^2 - \dfrac{1}{2} \right) + \left(\dfrac{3}{8}x^4 - \dfrac{3}{4}x^2 - \dfrac{1}{2} \right)$

$\quad = x^4 - x^2 - 1$

79. $\quad |4 - 2x| < 18$

$\qquad -18 < 4 - 2x < 18$

$\qquad -22 < -2x < 14 \qquad$ Subtracting 4

$\qquad 11 > x > -7 \qquad$ Dividing by -2

The solution set is $\{x | 11 > x > -7\}$, or $\{x | -7 < x < 11\}$.

81. $\quad |4 - 2x| = 18$

$\quad 4 - 2x = 18 \quad or \quad 4 - 2x = -18$

$\qquad -2x = 14 \quad or \qquad -2x = -22$

$\qquad\quad x = -7 \quad or \qquad\quad x = 11$

The solution set is $\{-7, 11\}$.

83. Graph: $y = -0.4x + 1$

We find several ordered pairs, plot them, and draw the line.

x	$\begin{array}{c} y \\ y = -0.4x + 1 \end{array}$	(x, y)
0	1	$(0, 1)$
-5	3	$(-5, 3)$
5	-1	$(5, -1)$

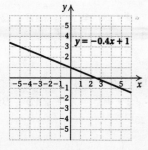

85. The area of the base is $x \cdot x$, or x^2.

The area of each side is $x \cdot (x - 2)$.

The total area of all four sides is $4x(x - 2)$.

The surface area of this box can be expressed as a polynomial.

$x^2 + 4x(x - 2) = x^2 + 4x^2 - 8x = 5x^2 - 8x$

87. $\quad (3x^{6a} - 5x^{5a} + 4x^{3a} + 8) -$
$\qquad\qquad (2x^{6a} + 4x^{4a} + 3x^{3a} + 2x^{2a})$

$\quad = (3x^{6a} - 5x^{5a} + 4x^{3a} + 8) +$
$\qquad\qquad (-2x^{6a} - 4x^{4a} - 3x^{3a} - 2x^{2a})$

$\quad = (3 - 2)x^{6a} - 5x^{5a} - 4x^{4a} + (4 - 3)x^{3a} - 2x^{2a} + 8$

$\quad = x^{6a} - 5x^{5a} - 4x^{4a} + x^{3a} - 2x^{2a} + 8$

Exercise Set 5.2

1. $8y^2 \cdot 3y = (8 \cdot 3)(y^2 \cdot y) = 24y^3$

3. $2x(-10x^2y) = [2(-10)](x \cdot x^2)(y) = -20x^3y$

5. $(5x^5y^4)(-2xy^3) = [5(-2)](x^5 \cdot x)(y^4 \cdot y^3) = -10x^6y^7$

7. $\quad 2z(7 - x)$

$\quad = 2z \cdot 7 - 2z \cdot x \quad$ Using a distributive law

$\quad = 14z - 2zx \qquad$ Multiplying monomials

9. $\quad 6ab(a + b)$

$\quad = 6ab \cdot a + 6ab \cdot b \quad$ Using a distributive law

$\quad = 6a^2b + 6ab^2 \qquad$ Multiplying monomials

11. $\quad 5cd(3c^2d - 5cd^2)$

$\quad = 5cd \cdot 3c^2d - 5cd \cdot 5cd^2$

$\quad = 15c^3d^2 - 25c^2d^3$

13. $\quad (5x + 2)(3x - 1)$

$\quad = 15x^2 - 5x + 6x - 2 \qquad$ FOIL

$\quad = 15x^2 + x - 2$

15. $\quad (s + 3t)(s - 3t)$

$\quad = s^2 - (3t)^2 \qquad (A + B)(A - B) = A^2 - B^2$

$\quad = s^2 - 9t^2$

17. $\quad (x - y)(x - y)$

$\quad = x^2 - 2xy + y^2 \quad (A - B)^2 = A^2 - 2AB + B^2$

19. $\quad (x^3 + 8)(x^3 - 5)$

$\quad = x^6 - 5x^3 + 8x^3 - 40 \qquad$ FOIL

$\quad = x^6 + 3x^3 - 40$

21. $\quad (a^2 - 2b^2)(a^2 - 3b^2)$

$\quad = a^4 - 3a^2b^2 - 2a^2b^2 + 6b^4 \qquad$ FOIL

$\quad = a^4 - 5a^2b^2 + 6b^4$

23. $(x-4)(x^2+4x+16)$

$= (x-4)(x^2) + (x-4)(4x) + (x-4)(16)$
$\qquad\qquad\qquad\qquad$ Using a distributive law
$= x(x^2) - 4(x^2) + x(4x) - 4(4x) + x(16) - 4(16)$
$\qquad\qquad\qquad\qquad$ Using a distributive law
$= x^3 - 4x^2 + 4x^2 - 16x + 16x - 64$
$\qquad\qquad\qquad\qquad$ Multiplying monomials
$= x^3 - 64 \quad$ Collecting like terms

25. $(x+y)(x^2-xy+y^2)$

$= (x+y)x^2 + (x+y)(-xy) + (x+y)(y^2)$
$= x(x^2) + y(x^2) + x(-xy) + y(-xy) + x(y^2) + y(y^2)$
$= x^3 + x^2y - x^2y - xy^2 + xy^2 + y^3$
$= x^3 + y^3$

27.
$$
\begin{array}{r}
a^2 + a - 1 \\
a^2 + 4a - 5 \\
\hline
-5a^2 - 5a + 5 \quad \text{Multiplying by } -5 \\
4a^3 + 4a^2 - 4a \quad\;\; \text{Multiplying by } 4a \\
a^4 + a^3 - a^2 \qquad\qquad \text{Multiplying by } a^2 \\
\hline
a^4 + 5a^3 - 2a^2 - 9a + 5 \quad \text{Adding}
\end{array}
$$

29.
$$
\begin{array}{ll}
4a^2b - 2ab + 3b^2 & \\
ab - 2b + a & \\
\hline
4a^3b - 2a^2b + 3ab^2 & (1) \\
-6b^3 \qquad\quad +4ab^2 - 8a^2b^2 & (2) \\
3ab^3 \qquad\qquad\qquad -2a^2b^2 + 4a^3b^2 & (3) \\
\hline
3ab^3 - 6b^3 + 4a^3b - 2a^2b + 7ab^2 - 10a^2b^2 + 4a^3b^2 & (4)
\end{array}
$$

(1) Multiplying by a

(2) Multiplying by $-2b$

(3) Multiplying by ab

(4) Adding

31. $\left(x+\dfrac{1}{4}\right)\left(x+\dfrac{1}{4}\right)$

$= x^2 + \dfrac{1}{4}x + \dfrac{1}{4}x + \dfrac{1}{16} \qquad$ FOIL

$= x^2 + \dfrac{1}{2}x + \dfrac{1}{16}$

33. $(1.3x-4y)(2.5x+7y)$

$= 3.25x^2 + 9.1xy - 10xy - 28y^2 \qquad$ FOIL

$= 3.25x^2 - 0.9xy - 28y^2$

35. $(a+8)(a+5)$

$= a^2 + 5a + 8a + 40 \qquad$ FOIL

$= a^2 + 13a + 40$

37. $(y+7)(y-4)$

$= y^2 - 4y + 7y - 28 \qquad$ FOIL

$= y^2 + 3y - 28$

39. $\left(3a+\dfrac{1}{2}\right)^2$

$= (3a)^2 + 2(3a)\left(\dfrac{1}{2}\right) + \left(\dfrac{1}{2}\right)^2$
$\qquad\qquad (A+B)^2 = A^2 + 2AB + B^2$
$= 9a^2 + 3a + \dfrac{1}{4}$

41. $(x-2y)^2$

$= x^2 - 2(x)(2y) + (2y)^2$
$\qquad\qquad (A-B)^2 = A^2 - 2AB + B^2$
$= x^2 - 4xy + 4y^2$

43. $\left(b-\dfrac{1}{3}\right)\left(b-\dfrac{1}{2}\right)$

$= b^2 - \dfrac{1}{2}b - \dfrac{1}{3}b + \dfrac{1}{6} \qquad$ FOIL

$= b^2 - \dfrac{3}{6}b - \dfrac{2}{6}b + \dfrac{1}{6}$

$= b^2 - \dfrac{5}{6}b + \dfrac{1}{6}$

45. $(2x+9)(x+2)$

$= 2x^2 + 4x + 9x + 18 \qquad$ FOIL

$= 2x^2 + 13x + 18$

47. $(20a-0.16b)^2$

$= (20a)^2 - 2(20a)(0.16b) + (0.16b)^2$
$\qquad\qquad (A-B)^2 = A^2 - 2AB + B^2$
$= 400a^2 - 6.4ab + 0.0256b^2$

49. $(2x-3y)(2x+y)$

$= 4x^2 + 2xy - 6xy - 3y^2 \qquad$ FOIL

$= 4x^2 - 4xy - 3y^2$

51. $(x^3+2)^2$

$= (x^3)^2 + 2 \cdot x^3 \cdot 2 + 2^2 \quad (A+B)^2 = A^2 + 2AB + B^2$
$= x^6 + 4x^3 + 4$

53. $(2x^2-3y^2)^2$

$= (2x^2)^2 - 2(2x^2)(3y^2) + (3y^2)^2$
$\qquad\qquad (A-B)^2 = A^2 - 2AB + B^2$
$= 4x^4 - 12x^2y^2 + 9y^4$

55. $(a^3b^2+1)^2$

$= (a^3b^2)^2 + 2 \cdot a^3b^2 \cdot 1 + 1^2$
$\qquad\qquad (A+B)^2 = A^2 + 2AB + B^2$
$= a^6b^4 + 2a^3b^2 + 1$

57. $(0.1a^2-5b)^2$

$= (0.1a^2)^2 - 2(0.1a^2)(5b) + (5b)^2$
$\qquad\qquad (A-B)^2 = A^2 - 2AB + B^2$
$= 0.01a^4 - a^2b + 25b^2$

59. $A = P(1+i)^2$

$\quad A = P(1 + 2i + i^2)$ FOIL

$\quad A = P + 2Pi + Pi^2$ Multiplying by P

61. $\quad (d+8)(d-8)$

$\quad = d^2 - 8^2$ $(A+B)(A-B) = A^2 - B^2$

$\quad = d^2 - 64$

63. $\quad (2c+3)(2c-3)$

$\quad = (2c)^2 - 3^2$ $(A+B)(A-B) = A^2 - B^2$

$\quad = 4c^2 - 9$

65. $\quad (6m-5n)(6m+5n)$

$\quad = (6m)^2 - (5n)^2$ $(A+B)(A-B) = A^2 - B^2$

$\quad = 36m^2 - 25n^2$

67. $\quad (x^2 + yz)(x^2 - yz)$

$\quad = (x^2)^2 - (yz)^2$ $(A+B)(A-B) = A^2 - B^2$

$\quad = x^4 - y^2 z^2$

69. $\quad (-mn + m^2)(mn + m^2)$

$\quad = (m^2 - mn)(m^2 + mn)$

$\quad = (m^2)^2 - (mn)^2$ $(A+B)(A-B) = A^2 - B^2$

$\quad = m^4 - m^2 n^2$

71. $\quad \left(\dfrac{1}{2}p - \dfrac{2}{3}q\right)\left(\dfrac{1}{2}p + \dfrac{2}{3}q\right)$

$\quad = \left(\dfrac{1}{2}p\right)^2 - \left(\dfrac{2}{3}q\right)^2$

$\quad = \dfrac{1}{4}p^2 - \dfrac{4}{9}q^2$

73. $\quad (x+1)(x-1)(x^2+1)$

$\quad = (x^2 - 1^2)(x^2 + 1)$

$\quad = (x^2 - 1)(x^2 + 1)$

$\quad = (x^2)^2 - 1^2$

$\quad = x^4 - 1$

75. $\quad (a-b)(a+b)(a^2 - b^2)$

$\quad = (a^2 - b^2)(a^2 - b^2)$

$\quad = (a^2 - b^2)^2$

$\quad = (a^2)^2 - 2(a^2)(b^2) + (b^2)^2$

$\quad = a^4 - 2a^2 b^2 + b^4$

77. $\quad (a+b+1)(a+b-1)$

$\quad = [(a+b)+1][(a+b)-1]$

$\quad = (a+b)^2 - 1^2$

$\quad = a^2 + 2ab + b^2 - 1$

79. $\quad (2x + 3y + 4)(2x + 3y - 4)$

$\quad = [(2x+3y)+4][(2x+3y)-4]$

$\quad = (2x+3y)^2 - 4^2$

$\quad = 4x^2 + 12xy + 9y^2 - 16$

81. *Familiarize.* Let a, b, and c represent the daily production of machines, A, B, and C, respectively.

Translate. Rewording, we have:

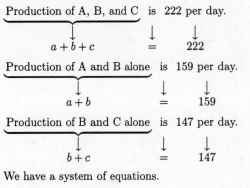

We have a system of equations.

$\quad a + b + c = 222,$

$\quad a + b \quad\quad = 159,$

$\quad\quad\quad b + c = 147$

Solve. Solving the system we get $(75, 84, 63)$.

Check. The daily production of the three machines together is $75 + 84 + 63$, or 222. The daily production of A and B alone is $75 + 84$, or 159. The daily production of B and C alone is $84 + 63$, or 147. The numbers check.

State. The daily production of suitcases by machines A, B, and C is 75, 84, and 63, respectively.

83. $\quad |3x - 7| \le 25$

$\quad\quad -25 \le 3x - 7 \le 25$

$\quad\quad -18 \le 3x \le 32$ Adding 7

$\quad\quad -6 \le x \le \dfrac{32}{3}$ Dividing by 3

The solution set is $\left\{ x \,\middle|\, -6 \le x \le \dfrac{32}{3} \right\}$.

85. $|8x + 3| > 37$

$\quad 8x + 3 > 37$ *or* $8x + 3 < -37$

$\quad 8x > 34$ *or* $8x < -40$

$\quad x > \dfrac{34}{8}$ *or* $x < -5$

$\quad x > \dfrac{17}{4}$ *or* $x < -5$

The solution set is $\left\{ x \,\middle|\, x < -5 \text{ or } x > \dfrac{17}{4} \right\}$.

87. $[(a^{2n})^{2n}]^4 = (a^{2n})^{2n \cdot 4}$

$\quad\quad = (a^{2n})^{8n}$

$\quad\quad = a^{2n \cdot 8n}$

$\quad\quad = a^{16n^2}$

89. $\quad (a^x b^{2y})\left(\dfrac{1}{2} a^{3x} b\right)^2$

$\quad = (a^x b^{2y})\left(\dfrac{1}{4} a^{6x} b^2\right)$

$\quad = \dfrac{1}{4} a^{7x} b^{2y+2}$

91. $\quad y^3 z^n (y^{3n} z^3 - 4yz^{2n})$

$= y^3 z^n (y^{3n} z^3) - y^3 z^n (4yz^{2n})$

$= y^{3+3n} z^{n+3} - 4y^4 z^{3n}$

93. $\quad (y-1)^6 (y+1)^6 = [(y-1)(y+1)]^6$

$= (y^2 - 1)^6 = [(y^2 - 1)^2]^3 = (y^4 - 2y^2 + 1)^3$

$= [(y^4 - 2y^2) + 1]^2 (y^4 - 2y^2 + 1)$

$= (y^8 - 4y^6 + 4y^4 + 2y^4 - 4y^2 + 1)(y^4 - 2y^2 + 1)$

$= (y^8 - 4y^6 + 6y^4 - 4y^2 + 1)(y^4 - 2y^2 + 1)$

$= y^{12} - 4y^{10} + 6y^8 - 4y^6 + y^4 - 2y^{10} + 8y^8 -$

$\quad 12y^6 + 8y^4 - 2y^2 + y^8 - 4y^6 + 6y^4 - 4y^2 + 1)$

$= y^{12} - 6y^{10} + 15y^8 - 20y^6 + 15y^4 - 6y^2 + 1$

95. $\quad \left(3x^5 - \dfrac{5}{11}\right)^2$

$= (3x^5)^2 - 2(3x^5)\left(\dfrac{5}{11}\right) + \left(\dfrac{5}{11}\right)^2$

$= 9x^{10} - \dfrac{30}{11}x^5 + \dfrac{25}{121}$

97. $\quad \left(x - \dfrac{1}{7}\right)\left(x^2 + \dfrac{1}{7}x + \dfrac{1}{49}\right)$

$= x^3 + \dfrac{1}{7}x^2 + \dfrac{1}{49}x - \dfrac{1}{7}x^2 - \dfrac{1}{49}x - \dfrac{1}{343}$

$= x^3 - \dfrac{1}{343}$

99. $\quad [a - (b-1)][(b-1)^2 + a(b-1) + a^2]$

$= (a - b + 1)(b^2 - 2b + 1 + ab - a + a^2)$

$= ab^2 - 2ab + a + a^2 b - a^2 + a^3 - b^3 + 2b^2 -$

$\quad b - ab^2 + ab - a^2 b + b^2 - 2b + 1 + ab - a + a^2$

$= a^3 - b^3 + 3b^2 - 3b + 1$

101. $\quad (x^{a-b})^{a+b} = x^{(a-b)(a+b)} = x^{a^2 - b^2}$

Exercise Set 5.3

1. $\quad 6a^2 + 3a$

$= 3a \cdot 2a + 3a \cdot 1$

$= 3a(2a + 1)$

3. $\quad x^3 + 9x^2$

$= x^2 \cdot x + x^2 \cdot 9$

$= x^2(x + 9)$

5. $\quad 8x^2 - 4x^4$

$= 4x^2 \cdot 2 - 4x^2 \cdot x^2$

$= 4x^2(2 - x^2)$

7. $\quad 4x^2 y - 12xy^2$

$= 4xy \cdot x - 4xy \cdot 3y$

$= 4xy(x - 3y)$

9. $\quad 3y^2 - 3y - 9$

$= 3 \cdot y^2 - 3 \cdot y - 3 \cdot 3$

$= 3(y^2 - y - 3)$

11. $\quad 4ab - 6ac + 12ad$

$= 2a \cdot 2b - 2a \cdot 3c + 2a \cdot 6d$

$= 2a(2b - 3c + 6d)$

13. $\quad 10a^4 + 15a^2 - 25a - 30$

$= 5 \cdot 2a^4 + 5 \cdot 3a^2 - 5 \cdot 5a - 5 \cdot 6$

$= 5(2a^4 + 3a^2 - 5a - 6)$

15. $-5x - 45 = -5(x + 9)$

17. $-6a - 84 = -6(a + 14)$

19. $-2x^2 + 2x - 24 = -2(x^2 - x + 12)$

21. $-3y^2 + 24y = -3y(y - 8)$

23. $-3y^3 + 12y^2 - 15y = -3y(y^2 - 4y + 5)$

25. $-x^2 + 3x - 7 = -1(x^2 - 3x + 7)$

27. $-a^4 + 2a^3 - 13a = -a(a^3 - 2a^2 + 13)$

29. $a(b - 2) + c(b - 2) = (a + c)(b - 2)$

31. $\quad (x - 2)(x + 5) + (x - 2)(x + 8)$

$= (x - 2)[(x + 5) + (x + 8)]$

$= (x - 2)(2x + 13)$

33. $a^2(x - y) + a^2(x - y) = 2a^2(x - y)$

35. $\quad ac + ad + bc + bd$

$= a(c + d) + b(c + d)$

$= (a + b)(c + d)$

37. $\quad b^3 - b^2 + 2b - 2$

$= b^2(b - 1) + 2(b - 1)$

$= (b^2 + 2)(b - 1)$

39. $\quad y^3 - 8y^2 + y - 8$

$= y^2(y - 8) + 1(y - 8)$

$= (y^2 + 1)(y - 8)$

41. $\quad 24x^3 - 36x^2 + 72x - 108$

$= 12(2x^3 - 3x^2 + 6x - 9)$

$= 12[x^2(2x - 3) + 3(2x - 3)]$

$= 12(x^2 + 3)(2x - 3)$

43. $a^4 - a^3 + a^2 + a = a(a^3 - a^2 + a + 1)$

45. $\quad 2y^4 + 6y^2 + 5y^2 + 15$

$= 2y^2(y^2 + 3) + 5(y^2 + 3)$

$= (2y^2 + 5)(y^2 + 3)$

47. $2x + 9y + 6z = 5,$ (1)

$x - y + z = 4,$ (2)

$3x + 2y + 3z = 7$ (3)

We will use the elimination method. First we use Equations (1) and (2) to eliminate y.

$2x + 9y + 6z = 5$ (1)

$\underline{9x - 9y + 9z = 36}$ Multiplying (2) by 9

$11x + 15z = 41$ (4)   Adding

Use a different pair of equations and eliminate y.

$2x - 2y + 2z = 8$ Multiplying (2) by 2

$\underline{3x + 2y + 3z = 7}$ (3)

$5x + 5z = 15$ (5)   Adding

Now solve the system of Equations (4) and (5).

$11x + 15z = 41$ (4)

$5x + 5z = 15$ (5)

$11x + 15z = 41$ (4)

$\underline{-15x - 15z = -45}$ Multiplying (5) by -3

$-4x = -4$ Adding

$x = 1$

We substitute 1 for x in Equation (5).

$5 \cdot 1 + 5z = 15$

$5 + 5z = 15$

$5z = 10$

$z = 2$

Now we substitute 1 for x and 2 for z in Equation (2).

$1 - y + 2 = 4$

$3 - y = 4$

$-y = 1$

$y = -1$

We obtain $(1, -1, 2)$. This checks, so it is the solution.

49. Graph: $y = -\dfrac{3}{2}x$

We find several order pairs, plot them, and draw the line.

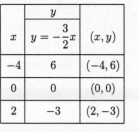

x	$y = -\dfrac{3}{2}x$	(x, y)
-4	6	$(-4, 6)$
0	0	$(0, 0)$
2	-3	$(2, -3)$

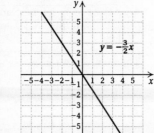

51. *Familiarize*. Let x, y, and z represent the amounts invested at $5\frac{1}{2}\%$, 7% and 8%, respectively. Then the interest is $5\frac{1}{2}\%x$, $7\%y$, and $8\%z$, or $0.055x$, $0.07y$, and $0.08z$.

Translate.

The total interest   is \$244.

$0.055x + 0.07y + 0.08z = 244$

The total investment   is \$3500.

$x + y + z = 3500$

The amount invested at 8%   is \$1100 more than   the amount invested at 7%.

$z = 1100 + y$

We have a system of equations:

$0.055x + 0.07y + 0.08z = 244,$

$x + y + z = 3500,$

$z = 1100 + y$

Solve. Solving the system we get $(1200, 600, 1700)$.

Check. The sum of the numbers is 3500. Also $0.055(\$1200) + 0.07(\$600) + 0.08(\$1700) = \$66 + \$42 + \$136 = \$244$, and \$1700 is \$1100 more than \$600. The numbers check.

State. \$1200 is invested at $5\frac{1}{2}\%$, \$600 is invested at 7%, and \$1700 is invested at 8%.

53. $4y^{4a} + 12y^{2a} + 10y^{2a} + 30$

$= 2(2y^{4a} + 6y^{2a} + 5y^{2a} + 15)$

$= 2[2y^{2a}(y^{2a} + 3) + 5(y^{2a} + 3)]$

$= 2(2y^{2a} + 5)(y^{2a} + 3)$

55. $D = \dfrac{1}{2}(8)^2 - \dfrac{3}{2}(8) = 32 - 12 = 20$

Exercise Set 5.4

1. $x^2 + 13x + 36$

We look for two numbers whose product is 36 and whose sum is 13. Since both 36 and 13 are positive, we need consider only positive factors.

Pairs of Factors	Sums of Factors
1, 36	37
2, 18	20
3, 12	15
4, 9	13
6, 6	12

The numbers we need are 4 and 9. The factorization is $(x + 4)(x + 9)$.

3. $t^2 - 8t + 15$

Since the constant term, 15, is positive and the coefficient of the middle term, -8, is negative, we look for a factorization of 15 in which both factors are negative. Their sum must be -8.

Pairs of Factors	Sums of Factors
$-1, -15$	-16
$-3, -5$	-8

The numbers we need are -3 and -5. The factorization is $(t-3)(t-5)$.

5. $x^2 - 8x - 33$

Since the constant term, -33, is negative, we look for a factorization of -33 in which one factor is positive and one factor is negative. The sum of the factors must be -8, so the negative factor must have the larger absolute value. Thus we consider only pairs of factors in which the negative factor has the larger absolute value.

Pairs of Factors	Sums of Factors
$1, -33$	-32
$3, -11$	-8

The numbers we want are 3 and -11. The factorization is $(x+3)(x-11)$.

7. $2y^2 - 16y + 32$

   $= 2(y^2 - 8y + 16)$ Removing the common factor

We now factor $y^2 - 8y + 16$. We look for two numbers whose product is 16 and whose sum is -8. Since the constant term is positive and the coefficient of the middle term is negative, we look for a factorization of 16 in which both factors are negative.

Pairs of Factors	Sums of Factors
$-1, -16$	-17
$-2, -8$	-10
$-4, -4$	-8

The numbers we need are -4 and -4.

$$y^2 - 8y + 16 = (y-4)(y-4)$$

We must not forget to include the common factor 2.

$$2y^2 - 16y + 32 = 2(y-4)(y-4).$$

9. $p^2 + 3p - 54$

Since the constant term is negative, we look for a factorization of -54 in which one factor is positive and one factor is negative. We consider only pairs of factors in which the positive factor has the larger absolute value, since the sum of the factors, 3, is positive.

Pairs of Factors	Sums of Factors
$54, -1$	53
$27, -2$	25
$18, -3$	15
$9, -6$	3

The numbers we need are 9 and -6. The factorization is $(p+9)(p-6)$.

11. $12x + x^2 + 27 = x^2 + 12x + 27$

We look for two numbers whose product is 27 and whose sum is 12. Since both 27 and 12 are positive, we need consider only positive factors.

Pairs of Factors	Sums of Factors
$1, \ 27$	28
$3, \ \ 9$	12

The numbers we want are 3 and 9. The factorization is $(x+3)(x+9)$.

13. $y^2 - \dfrac{2}{3}y + \dfrac{1}{9}$

Since the constant term, $\dfrac{1}{9}$, is positive and the coefficient of the middle term, $-\dfrac{2}{3}$, is negative, we look for a factorization of $\dfrac{1}{9}$ in which both factors are negative. Their sum must be $-\dfrac{2}{3}$.

Pairs of Factors	Sums of Factors
$-1, -\dfrac{1}{9}$	$-\dfrac{10}{9}$
$-\dfrac{1}{3}, -\dfrac{1}{3}$	$-\dfrac{2}{3}$

The numbers we need are $-\dfrac{1}{3}$ and $-\dfrac{1}{3}$. The factorization is $\left(y - \dfrac{1}{3}\right)\left(y - \dfrac{1}{3}\right)$.

15. $t^2 - 4t + 3$

Since the constant term, 3, is positive and the coefficient of the middle term, -4, is negative, we look for a factorization of 3 in which both factors are negative. Their sum must be -4. The only possibility is $-1, -3$. These are the numbers we need. The factorization is $(t-1)(t-3)$.

17. $5x + x^2 - 14 = x^2 + 5x - 14$

Since the constant term, -14, is negative, we look for a factorization of -14 in which one factor is positive and one factor is negative. Their sum must be 5, so the positive factor must have the larger absolute value. We consider only pairs of factors in which the positive factor has the larger absolute value.

Pairs of Factors	Sums of Factors
$-1, 14$	13
$-2, \ 7$	5

The numbers we need are -2 and 7. The factorization is $(x-2)(x+7)$.

19. $x^2 + 5x + 6$

We look for two numbers whose product is 6 and whose sum is 5. Since 6 and 5 are both positive, we need consider only positive factors.

Pairs of Factors	Sums of Factors
1, 6	7
2, 3	5

The numbers we need are 2 and 3. The factorization is $(x + 2)(x + 3)$.

21. $56 + x - x^2 = -x^2 + x + 56 = -(x^2 - x - 56)$

We now factor $x^2 - x - 56$. Since the constant term, -56, is negative, we look for a factorization of -56 in which one factor is positive and one factor is negative. We consider only pairs of factors in which the negative factor has the larger absolute value, since the sum of the factors, -1, is negative.

Pairs of Factors	Sums of Factors
-56, 1	-55
-28, 2	-26
-14, 4	-10
-8, 7	-1

The numbers we need are -8 and 7. Thus, $x^2 - x - 56 = (x - 8)(x + 7)$. We must not forget to include the factor that was factored out earlier:

$56 + x - x^2 = -(x - 8)(x + 7)$, or

$(-x + 8)(x + 7)$, or $(8 - x)(7 + x)$

23. $32y + 4y^2 - y^3$

There is a common factor, y. We also factor out -1 in order to make the leading coefficient positive.

$$32y + 4y^2 - y^3 = -y(-32 - 4y + y^2)$$
$$= -y(y^2 - 4y - 32)$$

Now we factor $y^2 - 4y - 32$. Since the constant term, -32, is negative, we look for a factorization of -32 in which one factor is positive and one factor is negative. We consider only pairs of factors in which the negative factor has the larger absolute value, since the sum of the factors, -4, is negative.

Pairs of Factors	Sums of Factors
-32, 1	-31
-16, 2	-14
-8, 4	-4

The numbers we need are -8 and 4. Thus, $y^2 - 4y - 32 = (y - 8)(y + 4)$. We must not forget to include the common factor:

$32y + 4y^2 - y^3 = -y(y - 8)(y + 4)$, or

$y(-y + 8)(y + 4)$, or $y(8 - y)(4 + y)$

25. $x^4 + 11x^2 - 80$

First make a substitution. We let $u = x^2$, so $u^2 = x^4$. Then we consider $u^2 + 11u - 80$. We look for pairs of factors of -80, one positive and one negative, such that the positive factor has the larger absolute value and the sum of the factors is 11.

Pairs of Factors	Sums of Factors
80, -1	79
40, -2	38
20, -4	16
16, -5	11
10, -8	2

The numbers we need are 16 and -5. Then $u^2 + 11u - 80 = (u + 16)(u - 5)$. Replacing u by x^2 we obtain the factorization of the original trinomial: $(x^2 + 16)(x^2 - 5)$.

27. $x^2 - 3x + 7$

There are no factors of 7 whose sum is -3. This trinomial is not factorable into binomials with integer coefficients.

29. $x^2 + 12xy + 27y^2$

We look for numbers p and q such that $x^2 + 12xy + 27y^2 = (x + py)(x + qy)$. Our thinking is much the same as if we were factoring $x^2 + 12x + 27$. We look for factors of 27 whose sum is 12. Those factors are 9 and 3. Then

$$x^2 + 12xy + 27y^2 = (x + 9y)(x + 3y).$$

31. $x^4 + 50x^2 + 49$

Substitute u for x^2 (and hence u^2 for x^4). Consider $u^2 + 50u + 49$. We look for a pair of positive factors of 49 whose sum is 50.

Pairs of Factors	Sums of Factors
7, 7	14
1, 49	50

The numbers we need are 1 and 49. Then $u^2 + 50u + 49 = (u + 1)(u + 49)$. Replacing u by x^2 we have

$$x^4 + 50x^2 + 49 = (x^2 + 1)(x^2 + 49).$$

33. $x^6 + 11x^3 + 18$

Substitute u for x^3 (and hence u^2 for x^6). Consider $u^2 + 11u + 18$. We look for two numbers whose product is 18 and whose sum is 11. Since both 18 and 11 are positive, we need consider only positive factors.

Pairs of Factors	Sums of Factors
1, 18	19
2, 9	11
3, 6	9

The numbers we need are 2 and 9. Then $u^2 + 11u + 18 = (u + 2)(u + 9)$. Replacing u by x^3 we obtain the factorization of the original trinomial: $(x^3 + 2)(x^3 + 9)$.

35. $x^8 - 11x^4 + 24$

Substitute u for x^4 (and hence u^2 for x^8). Consider $u^2 - 11u + 24$. Since the constant term, 24, is positive and the coefficient of the middle term, -11, is negative, we look for a factorization of 24 in which both factors are negative. Their sum must be -11.

Pairs of Factors	Sums of Factors
$-1, -24$	-25
$-2, -12$	-14
$-3, -8$	-11
$-4, -6$	-10

The numbers we need are -3 and -8. Then $u^2 - 11u + 24 = (u-3)(u-8)$. Replacing u by x^4 we obtain the factorization of the original trinomial: $(x^4 - 3)(x^4 - 8)$.

37. $3x^2 - 14x - 5$

We will use the FOIL method.

1) There is no common factor (other than 1 or -1).

2) We factor the first term, $3x^2$. The factors are $3x$ and x. We have this possibility:
$$(3x+\quad)(x+\quad)$$

3) Next we factor the last term, -5. The possibilities are $-1, 5$ and $1, -5$.

4) We look for combinations of factors from steps (2) and (3) such that the sum of their products is the middle term, $-14x$. We try the possibilities:
$$(3x - 1)(x + 5) = 3x^2 + 14x - 5$$
$$(3x + 1)(x - 5) = 3x^2 - 14x - 5$$

The factorization is $(3x + 1)(x - 5)$.

39. $10y^3 + y^2 - 21y$

We will use the grouping method.

1) Look for a common factor. We factor out y:
$$y(10y^2 + y - 21)$$

2) Factor the trinomial $10y^2 + y - 21$. Multiply the leading coefficient, 10, and the constant, -21.
$$10(-21) = -210$$

3) Look for a factorization of -210 in which the sum of the factors is the coefficient of the middle term, 1.

Pairs of Factors	Sums of Factors
$-1,\quad 210$	209
$1,\ -210$	-209
$-2,\quad 105$	103
$2,\ -105$	-103
$-3,\quad 70$	67
$3,\ -70$	-67
$-5,\quad 42$	37
$5,\ -42$	-37
$-6,\quad 35$	29
$6,\ -35$	-29
$-7,\quad 30$	23
$7,\ -30$	-23
$-10,\quad 21$	11
$10,\ -21$	-11
$-14,\quad 15$	1 $\leftarrow -14+15=1$
$14,\ -15$	-1

4) Next, split the middle term, y, as follows:
$$y = -14y + 15y$$

5) Factor by grouping:
$$10y^2 + y - 21 = 10y^2 - 14y + 15y - 21$$
$$= 2y(5y - 7) + 3(5y - 7)$$
$$= (2y + 3)(5y - 7)$$

We must include the common factor to get a factorization of the original trinomial:
$$10y^3 + y^2 - 21y = y(2y + 3)(5y - 7)$$

41. $3c^2 - 20c + 32$

We will use the FOIL method.

1) There is no common factor(other than 1 or -1).

2) Factor the first term, $3c^2$. The factors are $3c$ and c. We have this possibility:
$$(3c+\quad)(c+\quad)$$

3) Next we factor the last term, 32. The possibilities are $1, 32$ and $-1, -32$ and $2, 16$ and $-2, -16$ and $4, 8$ and $-4, -8$.

4) We look for a combination of factors from steps (2) and (3) such that the sum of their products is the middle term, $-20c$. Trial and error leads us to the correct factorization, $(3c - 8)(c - 4)$.

43. $35y^2 + 34y + 8$

We will use the grouping method.

1) There is no common factor (other than 1 or -1).

2) Multiply the leading coefficient, 35, and the constant, 8: $35(8) = 280$

3) Try to factor 280 so the sum of the factors is 34. We need only consider pairs of positive factors since 280 and 34 are both positive.

Pairs of Factors	Sums of Factors
280, 1	281
140, 2	142
70, 4	74
56, 5	61
40, 7	47
28, 10	38
20, 14	34

4) Split $34y$ as follows:
$$34y = 20y + 14y$$

5) Factor by grouping:
$$35y^2 + 34y + 8 = 35y^2 + 20y + 14y + 8$$
$$= 5y(7y + 4) + 2(7y + 4)$$
$$= (7y + 4)(5y + 2)$$

45. $4t + 10t^2 - 6 = 10t^2 + 4t - 6$

We will use the FOIL method.

1) Factor out the common factor, 2:
$$2(5t^2 + 2t - 3)$$

2) Now we factor out the trinomial $5t^2 + 2t - 3$.

Factor the first term, $5t^2$. The factors are $5t$ and t. We have this possibility:

$(5t+\quad)(t+\quad)$

3) Factor the last term, -3. The factors are $1, -3$ and $-1, 3$.

4) Look for factors in steps (2) and (3) such that the sum of the products is the middle term, $2t$. Trial and error leads us to the correct factorization:
$5t^2 + 2t - 3 = (5t - 3)(t + 1)$

We must include the common factor to get a factorization of the original trinomial:

$4t + 10t^2 - 6 = 2(5t - 3)(t + 1)$

47. $8x^2 - 16 - 28x = 8x^2 - 28x - 16$

We will use the grouping method.

1) Factor out the common factor, 4:

$4(2x^2 - 7x - 4)$

2) Now we factor the trinomial $2x^2 - 7x - 4$. Multiply the leading coefficient, 2, and the constant, -4:
$2(-4) = -8$

3) Factor -8 so the sum of the factors is -7. We need only consider parts of factors in which the negative factor has the larger absolute value, since their sum is negative.

Pairs of Factors	Sums of Factors
$-4, \quad 2$	-2
$-8, \quad 1$	-7

4) Split $-7x$ as follows:

$-7x = -8x + x$

5) Factor by grouping:
$$2x^2 - 7x - 4 = 2x^2 - 8x + x - 4$$
$$= 2x(x - 4) + (x - 4)$$
$$= (x - 4)(2x + 1)$$

We must include the common factor to get a factorization of the original trinomial:

$8x^2 - 16 - 28x = 4(x - 4)(2x + 1)$

49. $12x^3 - 31x^2 + 20x$

We will use the FOIL method.

1) Factor out the common factor, x:

$x(12x^2 - 31x + 20)$

2) We now factor the trinomial $12x^2 - 31x + 20$. Factor the first term, $12x^2$. The factors are $12x, x$ and $6x, 2x$ and $4x, 3x$. We have these possibilities:
$(12x+\quad)(x+\quad)$, $(6x+\quad)(2x+\quad)$, $(4x+\quad)(3x+\quad)$

3) Factor the last term, 20. The factors are 20, 1 and $-20, -1$ and 10, 2 and $-10, -2$ and 5, 4 and $-5, -4$.

4) Look for factors in steps (2) and (3) such that the sum of the products is the middle term, $-31x$. Trial and error leads us to the correct factorization:
$12x^2 - 31x + 20 = (4x - 5)(3x - 4)$

We must include the common factor to get a factorization of the original trinomial:

$12x^3 - 31x^2 + 20x = x(4x - 5)(3x - 4)$

51. $14x^4 - 19x^3 - 3x^2$

We will use the grouping method.

1) Factor out the common factor, x^2:

$x^2(14x^2 - 19x - 3)$

2) Now we factor the trinomial $14x^2 - 19x - 3$. Multiply the leading coefficient, 14, and the constant, -3:
$14(-3) = -42$

3) Factor -42 so the sum of the factors is -19. We need only consider pairs of factors in which the negative factor has the larger absolute value, since the sum is negative.

Pairs of Factors	Sums of Factors
$-42, \quad 1$	-41
$-21, \quad 2$	-19
$-14, \quad 3$	-11
$-7, \quad 6$	-1

4) Split $-19x$ as follows:

$-19x = -21x + 2x$

5) Factor by grouping:
$$14x^2 - 19x - 3 = 14x^2 - 21x + 2x - 3$$
$$= 7x(2x - 3) + 2x - 3$$
$$= (2x - 3)(7x + 1)$$

We must include the common factor to get a factorization of the original trinomial:

$14x^4 - 19x^3 - 3x^2 = x^2(2x - 3)(7x + 1)$

53. $3a^2 - a - 4$

We will use the FOIL method.

1) There is no common factor (other than 1 or -1).

2) Factor the first term, $3a^2$. The factors are $3a$ and a. We have this possibility: $(3a+\quad)(a+\quad)$

3) Factor the last term, -4. The factors are $4, -1$ and $-4, 1$ and $2, -2$.

4) Look for factors in steps (2) and (3) such that the sum of the products is the middle term, $-a$. Trial and error leads us to the correct factorization:
$(3a - 4)(a + 1)$

55. $9x^2 + 15x + 4$

We will use the grouping method.

1) There is no common factor (other than 1 or -1).

2) Multiply the leading coefficient and the constant:
$9(4) = 36$

⌐ Factor 36 so the sum of the factors is 15. We need only consider pairs of positive factors since 36 and 15 are both positive.

Pairs of Factors		Sums of Factors
36,	1	37
18,	2	20
12,	3	15
9,	4	13
6,	6	12

4) Split $15x$ as follows:

$$15x = 12x + 3x$$

5) Factor by grouping:

$$9x^2 + 15x + 4 = 9x^2 + 12x + 3x + 4$$
$$= 3x(3x + 4) + 3x + 4$$
$$= (3x + 4)(3x + 1)$$

57. $3 + 35z - 12z^2 = -12z^2 + 35z + 3$

We will use the FOIL method.

1) Factor out -1 so the leading coefficient is positive:
$-(12z^2 - 35z - 3)$

2) Now we factor the trinomial $12z^2 - 35z - 3$. Factor the first term, $12z^2$. The factors are $12z$, z and $6z$, $2z$ and $4z$, $3z$. We have these possibilities: $(12z+\quad)(z+\quad)$, $(6z+\quad)(2z+\quad)$, $(4z+\quad)(3z+\quad)$

3) Factor the last term, -3. The factors are 3, -1 and -3, 1.

4) Look for factors in steps (2) and (3) such that the sum of the products is the middle term, $-35z$. Trial and error leads us to the correct factorization: $(12z + 1)(z - 3)$

We must include the common factor to get a factorization of the original trinomial:

$$3 + 35z - 12z^2 = -(12z + 1)(z - 3), \text{ or}$$
$$(12z + 1)(-z + 3), \text{ or } (1 + 12z)(3 - z)$$

59. $-4t^2 - 4t + 15$

We will use the grouping method.

1) Factor out -1 so the leading coefficient is positive:
$-(4t^2 + 4t - 15)$

2) Now we factor the trinomial $4t^2 + 4t - 15$. Multiply the leading coefficient and the constant: $4(-15) = -60$

3) Factor -60 so the sum of the factors is 4. The desired factorization is $10(-6)$.

4) Split $4t$ as follows:

$$4t = 10t - 6t$$

5) Factor by grouping:

$$4t^2 + 4t - 15 = 4t^2 + 10t - 6t - 15$$
$$= 2t(2t + 5) - 3(2t + 5)$$
$$= (2t + 5)(2t - 3)$$

We must include the common factor to get a factorization of the original trinomial:

$$-4t^2 - 4t + 15 = -(2t + 5)(2t - 3), \text{ or}$$
$$(2t + 5)(-2t + 3)$$

61. $3x^3 - 5x^2 - 2x$

We will use the FOIL method.

1) Factor out the common factor, x:
$x(3x^2 - 5x - 2)$

2) Now we factor the trinomial $3x^2 - 5x - 2$. Factor the first term, $3x^2$. The factors are $3x$ and x. We have this possibility: $(3x+\quad)(x+\quad)$

3) Factor the last term, -2. The factors are 2, -1 and -2, 1.

4) Look for factors in steps (2) and (3) such that the sum of the products is the middle term, $-5x$. Trial and error leads us to the correct factorization: $(3x + 1)(x - 2)$

We must include the common factor to get a factorization of the original trinomial:

$$3x^3 - 5x^2 - 2x = x(3x + 1)(x - 2)$$

63. $24x^2 - 2 - 47x = 24x^2 - 47x - 2$

We will use the grouping method.

1) There is no common factor (other than 1 or -1).

2) Multiply the leading coefficient and the constant:
$24(-2) = -48$

3) Factor -48 so the sum of the factors is -47. The desired factorization is $-48 \cdot 1$.

4) Split $-47x$ as follows:

$$-47x = -48x + x$$

5) Factor by grouping:

$$24x^2 - 47x - 2 = 24x^2 - 48x + x - 2$$
$$= 24x(x - 2) + (x - 2)$$
$$= (x - 2)(24x + 1)$$

65. $21x^2 + 37x + 12$

We will use the FOIL method.

1) There is no common factor (other than 1 or -1).

2) Factor the first term $21x^2$. The factors are $21x$, x and $7x$, $3x$. We have these possibilities: $(21x+\quad)(x+\quad)$ and $(7x+\quad)(3x+\quad)$.

3) Factor the last term, 12. The factors are 12, 1 and -12, -1 and 6, 2 and -6, -2 and 4, 3 and -4, -3.

4) Look for factors in steps (2) and (3) such that the sum of the products is the middle term, $37x$. Trial and error leads us to the correct factorization: $(7x + 3)(3x + 4)$

67. $40x^4 + 16x^2 - 12$

We will use the grouping method.

1) Factor out the common factor, 4.

$$4(10x^4 + 4x^2 - 3)$$

Now we will factor the trinomial $10x^4 + 4x^2 - 3$. Substitute u for x^2 (and u^2 for x^4), and factor $10u^2 + 4u - 3$.

2) Multiply the leading coefficient and the constant: $10(-3) = -30$

3) Factor -30 so the sum of the factors is 4. This cannot be done. The trinomial $10u^2 + 4u - 3$ cannot be factored into binomials with integer coefficients. We have

$$40x^4 + 16x^2 - 12 = 4(10x^4 + 4x^2 - 3)$$

69. $12a^2 - 17ab + 6b^2$

We will use the FOIL method. (Our thinking is much the same as if we were factoring $12a^2 - 17a + 6$.)

1) There is no common factor (other than 1 or -1).

2) Factor the first term, $12a^2$. The factors are $12a$, a and $6a$, $2a$ and $4a$, $3a$. We have these possibilities: $(12a+\ \ )(a+\ \ )$ and $(6a+\ \ )(2a+\ \ )$ and $(4a+\ \ )(3a+\ \ )$.

3) Factor the last term, $6b^2$. The factors are $6b$, b and $-6b$, $-b$ and $3b$, $2b$ and $-3b$, $-2b$.

4) Look for factors in steps (2) and (3) such that the sum of the products is the middle term, $-17ab$. Trial and error leads us to the correct factorization: $(4a - 3b)(3a - 2b)$

71. $2x^2 + xy - 6y^2$

We will use the grouping method.

1) There is no common factor (other than 1 or -1).

2) Multiply the coefficients of the first and last terms: $2(-6) = -12$

3) Factor -12 so the sum of the factors is 1. The desired factorization is $4(-3)$.

4) Split xy as follows:

$$xy = 4xy - 3xy$$

5) Factor by grouping:

$$2x^2 + xy - 6y^2 = 2x^2 + 4xy - 3xy - 6y^2$$
$$= 2x(x + 2y) - 3y(x + 2y)$$
$$= (x + 2y)(2x - 3y)$$

73. $6x^2 - 29xy + 28y^2$

We will use the FOIL method.

1) There is no common factor (other than 1 or -1).

2) Factor the first term, $6x^2$. The factors are $6x$, x and $3x$, $2x$. We have these possibilities: $(6x+\ \ )(x+\ \ )$ and $(3x+\ \ )(2x+\ \ )$.

3) Factor the last term, $28y^2$. The factors are $28y$, y and $-28y$, $-y$ and $14y$, $2y$ and $-14y$, $-2y$ and $7y$, $4y$ and $-7y$, $-4y$.

4) Look for factors in steps (2) and (3) such that the sum of the products is the middle term, $-29xy$. Trial and error leads us to the correct factorization: $(3x - 4y)(2x - 7y)$

75. $9x^2 - 30xy + 25y^2$

We will use the grouping method.

1) There is no common factor (other than 1 or -1).

2) Multiply the coefficients of the first and last terms: $9(25) = 225$

3) Factor 225 so the sum of the factors is -30. The desired factorization is $-15(-15)$.

4) Split $-30xy$ as follows:

$$-30xy = -15xy - 15xy$$

5) Factor by grouping:

$$9x^2 - 30xy + 25y^2 = 9x^2 - 15xy - 15xy + 25y^2$$
$$= 3x(3x - 5y) - 5y(3x - 5y)$$
$$= (3x - 5y)(3x - 5y)$$

77. $3x^6 + 4x^3 - 4$

We will use the FOIL method.

1) There is no common factor (other than 1 or -1). Substitute u for x^3 (and hence u^2 for x^6). We factor $3u^2 + 4u - 4$.

2) Factor the first term, $3u^2$. The factors are $3u$ and u. We have this possibility: $(3u+\ \ )(u+\ \ )$

3) Factor the last term, -4. The factors are -1, 4 and 1, -4 and -2, 2.

4) Look for factors in steps (2) and (3) such that the sum of the products is the middle term, $4u$. Trial and error leads us to the correct factorization of $3u^2 + 4u - 4$: $(3u - 2)(u + 2)$. Replacing u with x^3 we have the factorization of the original trinomial: $(3x^3 - 2)(x^3 + 2)$.

79. a) $h = -16(0)^2 + 80(0) + 224 = 224$ ft

$h = -16(1)^2 + 80(1) + 224 = 288$ ft

$h = -16(3)^2 + 80(3) + 224 = 320$ ft

$h = -16(4)^2 + 80(4) + 224 = 288$ ft

$h = -16(6)^2 + 80(6) + 224 = 128$ ft

$h = -16t^2 + 80t + 224$

We will use the grouping method.

1) Factor out -16 so the leading coefficient is positive: $-16(t^2 - 5t - 14)$

2) Factor the trinomial $t^2 - 5t - 14$. Multiply the leading coefficient and the constant: $1(-14) = -14$

3) Factor -14 so the sum of the factors is -5. The desired factorization is $-7 \cdot 2$.

4) Split $-5t$ as follows:
$$-5t = -7t + 2t$$

5) Factor by grouping:
$$t^2 - 5t - 14 = t^2 - 7t + 2t - 14$$
$$= t(t - 7) + 2(t - 7)$$
$$= (t - 7)(t + 2)$$

We must include the common factor to get a factorization of the original trinomial.
$$h = -16(t - 7)(t + 2)$$

81. $x + 2y - z = 0,$

$4x + 2y + 5z = 6,$

$2x - y + z = 5$

We will solve the system of equations using matrices. First we write a matrix using only the constants.

$$\begin{bmatrix} 1 & 2 & -1 & | & 0 \\ 4 & 2 & 5 & | & 6 \\ 2 & -1 & 1 & | & 5 \end{bmatrix} \quad \begin{matrix} \text{(P1)} \\ \text{(P2)} \\ \text{(P3)} \end{matrix}$$

Multiply row 1 by -4 and add it to row 2.

Multiply row 1 by -2 and add it to row 3.

$$\begin{bmatrix} 1 & 2 & -1 & | & 0 \\ 0 & -6 & 9 & | & 6 \\ 0 & -5 & 3 & | & 5 \end{bmatrix}$$

Multiply row 3 by 6.

$$\begin{bmatrix} 1 & 2 & -1 & | & 0 \\ 0 & -6 & 9 & | & 6 \\ 0 & -30 & 18 & | & 30 \end{bmatrix}$$

Multiply row 2 by -5 and add it to row 3.

$$\begin{bmatrix} 1 & 2 & -1 & | & 0 \\ 0 & -6 & 9 & | & 6 \\ 0 & 0 & -27 & | & 0 \end{bmatrix}$$

Reinserting the variables, we have

$x + 2y - z = 0,$ (P1)

$- 6y + 9z = 6,$ (P2)

$- 27z = 0$ (P3)

Solve (P3) for z.

$$-27z = 0$$
$$z = 0$$

Back-substitute 0 for z in (P2) and solve for y.

$$-6y + 9 \cdot 0 = 6$$
$$-6y + 0 = 6$$
$$-6y = 6$$
$$y = -1$$

Back-substitute -1 for y and 0 for z in (P1) and solve for x.

$$x + 2(-1) - 0 = 0$$
$$x - 2 = 0$$
$$x = 2$$

The solution is $(2, -1, 0)$.

83. $|10x - 3| \geq 27$

$10x - 3 \geq 27 \quad or \quad 10x - 3 \leq -27$

$10x \geq 30 \quad or \quad 10x \leq -24$

$x \geq 3 \quad or \quad x \leq -\dfrac{24}{10}$

$x \geq 3 \quad or \quad x \leq -\dfrac{12}{5}$

The solution set is $\left\{ x \middle| x \leq -\dfrac{12}{5} \text{ or } x \geq 3 \right\}$.

85. $|7 - 4x| < 32$

$-32 < 7 - 4x < 32$

$-39 < -4x < 25$

$\dfrac{39}{4} > x > -\dfrac{25}{4}$ Dividing by -4 and reversing the inequality symbol

The solution set is $\left\{ x \middle| \dfrac{39}{4} > x > -\dfrac{25}{4} \right\}$, or

$\left\{ x \middle| -\dfrac{25}{4} < x < \dfrac{39}{4} \right\}$.

87. $p^2q^2 + 7pq + 12$

The factorization will be of the form $(pq + \quad)(pq + \quad)$. We look for factors of 12 whose sum is 7. The factors we need are 4 and 3. The factorization is $(pq + 4)(pq + 3)$.

89. $x^2 - \dfrac{4}{25} + \dfrac{3}{5}x = x^2 + \dfrac{3}{5}x - \dfrac{4}{25}$

We look for factors of $-\dfrac{4}{25}$ whose sum is $\dfrac{3}{5}$. The factors are $\dfrac{4}{5}$ and $-\dfrac{1}{5}$. The factorization is $\left(x + \dfrac{4}{5} \right)\left(x - \dfrac{1}{5} \right)$.

91. $y^2 + 0.4y - 0.05$

We look for factors of -0.05 whose sum is 0.4. The factors are -0.1 and 0.5. The factorization is $(y - 0.1)(y + 0.5)$.

93. $7a^2b^2 + 6 + 13ab = 7a^2b^2 + 13ab + 6$

We will use the grouping method. There is no common factor (other than 1 or -1). Multiply the leading coefficient and the constant: $7(6) = 42$. Factor 42 so the sum of the factors is 13. The desired factorization is $6 \cdot 7$. Split the middle term and factor by grouping.

$$7a^2b^2 + 13ab + 6 = 7a^2b^2 + 6ab + 7ab + 6$$
$$= ab(7ab + 6) + 7ab + 6$$
$$= (7ab + 6)(ab + 1)$$

95. $3x^2 + 12x - 495$

Factor out the common factor, 3.

$$3(x^2 + 4x - 165)$$

Now factor $x^2 + 4x - 165$. Find factors of -165 whose sum is 4. The factors are -11 and 15. Then $x^2 + 4x - 165 = (x - 11)(x + 15)$, and $3x^2 + 12x - 495 = 3(x - 11)(x + 15)$.

97. $216x + 78x^2 + 6x^3 = 6x^3 + 78x^2 + 216x$

Factor out the common factor, $6x$.

$$6x(x^2 + 13x + 36)$$

Now factor $x^2 + 13x + 36$. Look for factors of 36 whose sum is 13. The factors are 9 and 4. Then $x^2 + 13x + 36 = (x + 9)(x + 4)$, and $6x^3 + 78x^2 + 216x = 6x(x + 9)(x + 4)$.

99. $x^{2a} + 5x^a - 24$

$x^{2a} = (x^a)^2$, so the factorization is of the form $(x^a + \ )(x^a + \ )$.

Look for factors of -24 whose sum is 5. The factors are 8 and -3. Then the factorization is $(x^a + 8)(x^a - 3)$.

Exercise Set 5.5

1. $x^2 - 4x + 4 = (x - 2)^2$ Find the square terms and write their square roots with a minus sign between them.

3. $y^2 + 18y + 81 = (y + 9)^2$ Find the square terms and write their square roots with a minus sign between them.

5. $x^2 + 1 + 2x = x^2 + 2x + 1$ Writing in descending order
$$= (x + 1)^2$$ Factoring the trinomial square

7. $9y^2 + 12y + 4 = (3y + 2)^2$ Find the square terms and write their square roots with a minus sign between them.

9. $-18y^2 + y^3 + 81y = y^3 - 18y^2 + 81y$ Writing in descending order
$$= y(y^2 - 18y + 81)$$ Removing the common factor
$$= y(y - 9)^2$$ Factoring the trinomial square

11. $12a^2 + 36a + 27 = 3(4a^2 + 12a + 9)$ Removing the common factor
$$= 3(2a + 3)^2$$ Factoring the trinomial square

13. $2x^2 - 40x + 200 = 2(x^2 - 20x + 100)$
$$= 2(x - 10)^2$$

15. $1 - 8d + 16d^2 = (1 - 4d)^2$,   Find the square terms or $(4d - 1)^2$ and write their square roots with a minus sign between them.

17. $y^4 - 8y^2 + 16 = (y^2 - 4)^2$ Find the square terms and write their square roots with a minus sign between them.

$$= [(y + 2)(y - 2)]^2$$ Factoring the difference of squares
$$= (y + 2)^2(y - 2)^2$$

19. $0.25x^2 + 0.30x + 0.09 = (0.5x + 0.3)^2$ Find the square terms and write their square roots with a plus sign between them.

21. $p^2 - 2pq + q^2 = (p - q)^2$

23. $a^2 + 4ab + 4b^2 = (a + 2b)^2$

25. $25a^2 - 30ab + 9b^2 = (5a - 3b)^2$

27. $y^6 + 26y^3 + 169 = (y^3 + 13)^2$ Find the square terms and write their square roots with a plus sign between them

29. $16x^{10} - 8x^5 + 1 = (4x^5 - 1)^2$ $[16x^{10} = (4x^5)^2]$

31. $x^4 + 2y^2y^2 + y^4 = (x^2 + y^2)^2$

33. $x^2 - 16 = x^2 - 4^2 = (x + 4)(x - 4)$

35. $p^2 - 49 = p^2 - 7^2 = (p + 7)(p - 7)$

37. $p^2q^2 - 25 = (pq)^2 - 5^2 = (pq + 5)(pq - 5)$

39. $6x^2 - 6y^2$
$$= 6(x^2 - y^2)$$ Removing the common factor
$$= 6(x + y)(x - y)$$ Factoring the difference of squares

41. $4xy^4 - 4xz^4$
$$= 4x(y^4 - z^4)$$ Removing the common factor
$$= 4x[(y^2)^2 - (z^2)^2]$$
$$= 4x(y^2 + z^2)(y^2 - z^2)$$ Factoring the difference of squares
$$= 4x(y^2 + z^2)(y + z)(y - z)$$ Factoring $y^2 - z^2$

43. $4a^3 - 49a = a(4a^2 - 49)$
$$= a[(2a)^2 - 7^2]$$
$$= a(2a + 7)(2a - 7)$$

$$x^8 - 3y^8 = 3(x^8 - y^8)$$
$$= 3[(x^4)^2 - (y^4)^2]$$
$$= 3(x^4 + y^4)(x^4 - y^4)$$
$$= 3(x^4 + y^4)[(x^2)^2 - (y^2)^2]$$
$$= 3(x^4 + y^4)(x^2 + y^2)(x^2 - y^2)$$
$$= 3(x^4 + y^4)(x^2 + y^2)(x + y)(x - y)$$

47. $9a^4 - 25a^2b^4 = a^2(9a^2 - 25b^4)$
$$= a^2[(3a)^2 - (5b^2)^2]$$
$$= a^2(3a + 5b^2)(3a - 5b^2)$$

49. $\dfrac{1}{36} - z^2 = \left(\dfrac{1}{6}\right)^2 - z^2 = \left(\dfrac{1}{6} + z\right)\left(\dfrac{1}{6} - z\right)$

51. $0.04x^2 - 0.09y^2 = (0.2x)^2 - (0.3y)^2$
$$= (0.2x + 0.3y)(0.2x - 0.3y)$$

53. $\quad m^3 - 7m^2 - 4m + 28$
$= m^2(m - 7) - 4(m - 7)$ Factoring by grouping
$= (m^2 - 4)(m - 7)$
$= (m + 2)(m - 2)(m - 7)$ Factoring the difference of squares

55. $\quad a^3 - ab^2 - 2a^2 + 2b^2$
$= a(a^2 - b^2) - 2(a^2 - b^2)$ Factoring by grouping
$= (a - 2)(a^2 - b^2)$
$= (a - 2)(a + b)(a - b)$ Factoring the difference of squares

57. $(a + b)^2 - 100 = (a + b)^2 - 10^2$
$$= (a + b + 10)(a + b - 10)$$

59. $\quad a^2 + 2ab + b^2 - 9$
$= (a^2 + 2ab + b^2) - 9$ Grouping as a difference of squares
$= (a + b)^2 - 3^2$
$= (a + b + 3)(a + b - 3)$

61. $\quad r^2 - 2r + 1 - 4s^2$
$= (r^2 - 2r + 1) - 4s^2$ Grouping as a difference of squares
$= (r - 1)^2 - (2s)^2$
$= (r - 1 + 2s)(r - 1 - 2s)$

63. $\quad 2m^2 + 4mn + 2n^2 - 50b^2$
$= 2(m^2 + 2mn + n^2 - 25b^2)$ Removing the common factor
$= 2[(m^2 + 2mn + n^2) - 25b^2]$ Grouping as a difference of squares
$= 2[(m + n)^2 - (5b)^2]$
$= 2(m + n + 5b)(m + n - 5b)$

65. $9 - (a^2 + 2ab + b^2) = 9 - (a + b)^2$
$$= [3 + (a + b)][3 - (a + b)]$$
$$= (3 + a + b)(3 - a - b)$$

67. Graph: $y + 1 = 2x$

To find the x-intercept we can cover up the y-term and look at the rest of the equation. We have $1 = 2x$ or $\dfrac{1}{2} = x$. The x-intercept is $\left(\dfrac{1}{2}, 0\right)$.

To find the y-intercept we can cover up the x-term and proceed as above. We have $y + 1 = 0$ or $y = -1$. The y-intercept is $(0, -1)$. We plot these points and draw the line.

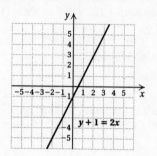

We can use a third point as a check. We choose $x = 2$ and solve for y.
$$y + 1 = 2 \cdot 2$$
$$y + 1 = 4$$
$$y = 3$$

We plot $(2, 3)$ and note that it is on the line.

69. Graph: $2x - y = 4$

To find the x-intercept we can cover up the y-term and look at the rest of the equation. We have $2x = 4$ or $x = 2$. The x-intercept is $(2, 0)$.

To find the y-intercept we can cover up the x-term and proceed as above. We have $-y = 4$ or $y = -4$. The y-intercept is $(0, -4)$. We plot these points and draw the line.

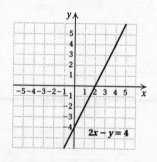

We can use a third point as a check. We choose $x = 4$ and solve for y.
$$2 \cdot 4 - y = 4$$
$$8 - y = 4$$
$$-y = -4$$
$$y = 4$$

We plot $(4, 4)$ and note that it is on the line.

71. $\frac{1}{49}p^2 - \frac{8}{63}p + \frac{16}{81} = \left(\frac{1}{7}p - \frac{4}{9}\right)^2$ Find the square terms and write their square roots with a plus sign between them

73. $9x^{2n} - 6x^n + 1 = (3x^n)^2 - 6x^n + 1$
$$= (3x^n - 1)^2$$

75. $y^8 - 256 = (y^4)^2 - 16^2$
$$= (y^4 + 16)(y^4 - 16)$$
$$= (y^4 + 16)[(y^2)^2 - 4^2]$$
$$= (y^4 + 16)(y^2 + 4)(y^2 - 4)$$
$$= (y^4 + 16)(y^2 + 4)(y + 2)(y - 2)$$

Exercise Set 5.6

1. $z^3 + 27 = z^3 + 3^3$
$$= (z + 3)(z^2 - 3z + 9)$$
$$A^3 + B^3 = (A + B)(A^2 - AB + B^2)$$

3. $x^3 - 1 = x^3 - 1^3$
$$= (x - 1)(x^2 + x + 1)$$
$$A^3 - B^3 = (A - B)(A^2 + AB + B^2)$$

5. $y^3 + 125 = y^3 + 5^3$
$$= (y + 5)(y^2 - 5y + 25)$$
$$A^3 + B^3 = (A + B)(A^2 - AB + B^2)$$

7. $8a^3 + 1 = (2a)^3 + 1^3$
$$= (2a + 1)(4a^2 - 2a + 1)$$
$$A^3 + B^3 = (A + B)(A^2 - AB + B^2)$$

9. $y^3 - 8 = y^3 - 2^3$
$$= (y - 2)(y^2 + 2y + 4)$$
$$A^3 - B^3 = (A - B)(A^2 + AB + B^2)$$

11. $8 - 27b^3 = 2^3 - (3b)^3$
$$= (2 - 3b)(4 + 6b + 9b^2)$$

13. $64y^3 + 1 = (4y)^3 + 1^3$
$$= (4y + 1)(16y^2 - 4y + 1)$$

15. $8x^3 + 27 = (2x)^3 + 3^3$
$$= (2x + 3)(4x^2 - 6x + 9)$$

17. $a^3 - b^3 = (a - b)(a^2 + ab + b^2)$

19. $a^3 + \frac{1}{8} = a^3 + \left(\frac{1}{2}\right)^3$
$$= \left(a + \frac{1}{2}\right)\left(a^2 - \frac{1}{2}a + \frac{1}{4}\right)$$

21. $2y^3 - 128 = 2(y^3 - 64)$
$$= 2(y^3 - 4^3)$$
$$= 2(y - 4)(y^2 + 4y + 16)$$

23. $24a^3 + 3 = 3(8a^3 + 1)$
$$= 3[(2a)^3 + 1^3]$$
$$= 3(2a + 1)(4a^2 - 2a + 1)$$

25. $rs^3 + 64r = r(s^3 + 64)$
$$= r(s^3 + 4^3)$$
$$= r(s + 4)(s^2 - 4s + 16)$$

27. $5x^3 - 40z^3 = 5(x^3 - 8z^3)$
$$= 5[x^3 - (2z)^3]$$
$$= 5(x - 2z)(x^2 + 2xz + 4z^2)$$

29. $x^3 + 0.001 = x^3 + (0.1)^3$
$$= (x + 0.1)(x^2 - 0.1x + 0.01)$$

31. $64x^6 - 8t^6 = 8(8x^6 - t^6)$
$$= 8[(2x^2)^3 - (t^2)^3]$$
$$= 8(2x^2 - t^2)(4x^4 + 2x^2t^2 + t^4)$$

33. $2y^4 - 128y = 2y(y^3 - 64)$
$$= 2y(y^3 - 4^3)$$
$$= 2y(y - 4)(y^2 + 4y + 16)$$

35. $z^6 - 1$
$$= (z^3)^2 - 1^2 \qquad \text{Writing as a difference of squares}$$
$$= (z^3 + 1)(z^3 - 1) \qquad \text{Factoring a difference of squares}$$
$$= (z + 1)(z^2 - z + 1)(z - 1)(z^2 + z + 1)$$
$$\text{Factoring a sum and a difference of cubes}$$

37. $t^6 + 64y^6 = (t^2)^3 + (4y^2)^3$
$$= (t^2 + 4y^2)(t^4 - 4t^2y^2 + 16y^4)$$

39. Graph: $\qquad 5x = 10 - 2y$
$$2y + 5x = 10 \qquad \text{Rewriting}$$

To find the x-intercept, we cover up the y-term and consider the rest of the equation. We have $5x = 10$, or $x = 2$. The x-intercept is $(2, 0)$. To find the y-intercept, we cover up the x-term and consider the rest of the equation. We have $2y = 10$, or $y = 5$. The y-intercept is $(0, 5)$. We plot these points and draw the line.

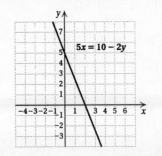

We find a third point as a check. We let $x = 4$ and solve for y:

$$2y + 5 \cdot 4 = 10$$
$$2y + 20 = 10$$
$$2y = -10$$
$$y = -5$$

We plot the point $(4, -5)$ and note that it is on the line.

41. Graph: $y + x = -3$

To find the x-intercept, we cover up the y-term and consider the rest of the equation. We have $x = -3$. The x-intercept is $(-3, 0)$. To find the y-intercept, we cover up the x-term and consider the rest of the equation. We have $y = -3$. The y-intercept is $(0, -3)$. We plot these points and draw the line.

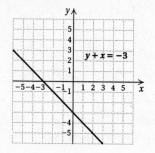

We find a third point as a check. We let $x = 1$ and solve for y:

$$y + 1 = -3$$
$$y = -4$$

We plot the point $(1, -4)$ and note that it is on the line.

43. $|5x - 6| \leq 39$
$$-39 \leq 5x - 6 \leq 39$$
$$-33 \leq 5x \leq 45$$
$$-\frac{33}{5} \leq x \leq 9$$

The solution set is $\left\{ x \,\middle|\, -\dfrac{33}{5} \leq x \leq 9 \right\}$.

45. $(a+b)^3 = (-2+3)^3 = 1^3 = 1$

$a^3 + b^3 = (-2)^3 + (3)^3 = -8 + 27 = 19$

$(a+b)(a^2 - ab + b^2)$
$= (-2+3)[(-2)^2 - (-2)(3) + (3)^2]$
$= 1(4 + 6 + 9)$
$= 19$

$(a+b)(a^2 + ab + b^2)$
$= (-2+3)[(-2)^2 + (-2)(3) + (3)^2]$
$= 1(4 - 6 + 9)$
$= 7$

$(a+b)(a+b)(a+b)$
$= (-2+3)(-2+3)(-2+3)$
$= 1 \cdot 1 \cdot 1$
$= 1$

47. $x^{6a} + y^{3b} = (x^{2a})^3 + (y^b)^3$
$\qquad = (x^{2a} + y^b)(x^{4a} - x^{2a}y^b + y^{2b})$

49. $3x^{3a} + 24y^{3b} = 3(x^{3a} + 8y^{3b})$
$\qquad = 3[(x^a)^3 + (2y^b)^3]$
$\qquad = 3(x^a + 2y^b)(x^{2a} - 2x^a y^b + 4y^{2b})$

51. $\dfrac{1}{24}x^3y^3 + \dfrac{1}{3}z^3 = \dfrac{1}{3}\left(\dfrac{1}{8}x^3y^3 + z^3\right)$
$\qquad = \dfrac{1}{3}\left[\left(\dfrac{1}{2}xy\right)^3 + z^3\right]$
$\qquad = \dfrac{1}{3}\left(\dfrac{1}{2}xy + z\right)\left(\dfrac{1}{4}x^2y^2 - \dfrac{1}{2}xyz + z^2\right)$

53. $(x+y)^3 - x^3$
$= [(x+y) - x][(x+y)^2 + x(x+y) + x^2]$
$= (x+y-x)(x^2 + 2xy + y^2 + x^2 + xy + x^2)$
$= y(3x^2 + 3xy + y^2)$

55. $(a+2)^3 - (a-2)^3$
$= [(a+2) - (a-2)][(a+2)^2 + (a+2)(a-2) + (a-2)^2]$
$= (a+2-a+2)(a^2 + 4a + 4 + a^2 - 4 + a^2 - 4a + 4)$
$= 4(3a^2 + 4)$

Exercise Set 5.7

1. $\quad y^2 - 225$
$= y^2 - 15^2 \quad$ Difference of squares
$= (y + 15)(y - 15)$

3. $\quad 2x^2 + 11x + 12$
$= (2x + 3)(x + 4) \qquad$ FOIL or grouping method

5. $\quad 5x^4 - 20$
$= 5(x^4 - 4) \qquad$ Removing the common factor
$= 5[(x^2)^2 - 2^2] \qquad$ Difference of squares
$= 5(x^2 + 2)(x^2 - 2)$

7. $\quad p^2 + 36 + 12p$
$= p^2 + 12p + 36 \quad$ Trinomial square
$= (p + 6)^2$

9. $\quad 2x^2 - 10x - 132$
$= 2(x^2 - 5x - 66)$
$= 2(x - 11)(x + 6) \quad$ Trial and error

11. $\quad 9x^2 - 25y^2$
$= (3x)^2 - (5y)^2 \quad$ Difference of squares
$= (3x + 5y)(3x - 5y)$

13. $m^6 - 1$

$= (m^3)^2 - 1^2$ Difference of squares

$= (m^3 + 1)(m^3 - 1)$ Sum and difference of cubes

$= (m + 1)(m^2 - m + 1)(m - 1)(m^2 + m + 1)$

15. $x^2 + 6x - y^2 + 9$

$= x^2 + 6x + 9 - y^2$

$= (x + 3)^2 - y^2$ Difference of squares

$= [(x + 3) + y][(x + 3) - y]$

$= (x + 3 + y)(x + 3 - y)$

17. $250x^3 - 128y^3$

$= 2(125x^3 - 64y^3)$

$= 2[(5x)^3 - (4y)^3]$ Difference of cubes

$= 2(5x - 4y)(25x^2 + 20xy + 16y^2)$

19. $8m^3 + m^6 - 20$

$= m^6 + 8m^3 - 20$

$= (m^3)^2 + 8m^3 - 20$

$= (m^3 - 2)(m^3 + 10)$ Trial and error

21. $ac + cd - ab - bd$

$= c(a + d) - b(a + d)$ Factoring by grouping

$= (c - b)(a + d)$

23. $50b^2 - 5ab - a^2$

$= (5b - a)(10b + a)$ FOIL or grouping method

25. $-7x^2 + 2x^3 + 4x - 14$

$= 2x^3 - 7x^2 + 4x - 14$

$= x^2(2x - 7) + 2(2x - 7)$ Factoring by grouping

$= (x^2 + 2)(2x - 7)$

27. $2x^3 + 6x^2 - 8x - 24$

$= 2(x^3 + 3x^2 - 4x - 12)$

$= 2[x^2(x + 3) - 4(x + 3)]$ Factoring by grouping

$= 2(x^2 - 4)(x + 3)$ Difference of squares

$= 2(x + 2)(x - 2)(x + 3)$

29. $16x^3 + 54y^3$

$= 2(8x^3 + 27y^3)$

$= 2[(2x)^3 + (3y)^3]$ Sum of cubes

$= 2(2x + 3y)(4x^2 - 6xy + 9y^2)$

31. $36y^2 - 35 + 12y$

$= 36y^2 + 12y - 35$

$= (6y - 5)(6y + 7)$ FOIL or grouping method

33. $a^8 - b^8$ Difference of squares

$= (a^4 + b^4)(a^4 - b^4)$ Difference of squares

$= (a^4 + b^4)(a^2 + b^2)(a^2 - b^2)$ Difference of squares

$= (a^4 + b^4)(a^2 + b^2)(a + b)(a - b)$

35. $a^3b - 16ab^3$

$= ab(a^2 - 16b^2)$ Difference of squares

$= ab(a + 4b)(a - 4b)$

37. $\dfrac{1}{16}x^2 - \dfrac{1}{6}xy^2 + \dfrac{1}{9}y^4$

$= \dfrac{1}{16}x^2 - \dfrac{1}{6}xy^2 + \dfrac{1}{9}(y^2)^2$ Trinomial square

$= \left(\dfrac{1}{4}x - \dfrac{1}{3}y^2\right)^2$

39. $5x^3 - 5x^2y - 5xy^2 + 5y^3$

$= 5(x^3 - x^2y - xy^2 + y^3)$

$= 5[x^2(x - y) - y^2(x - y)]$ Factoring by grouping

$= 5(x^2 - y^2)(x - y)$

$= 5(x + y)(x - y)(x - y)$ Factoring the difference of squares

$= 5(x + y)(x - y)^2$

41. $42ab + 27a^2b^2 + 8$

$= 27a^2b^2 + 42ab + 8$

$= (9ab + 2)(3ab + 4)$ FOIL or grouping method

43. $8y^4 - 125y$

$= y(8y^3 - 125)$

$= y[(2y)^3 - 5^3]$ Difference of cubes

$= y(2y - 5)(4y^2 + 10y + 25)$

45. $2x - 3y + 4z = 10,$ (1)

$\quad 4x + 6y - 4z = -5,$ (2)

$\quad -8x - 9y + 8z = -2$ (3)

We will eliminate z.

$\quad 2x - 3y + 4z = 10$ (1)

$\quad \underline{4x + 6y - 4z = -5}$ (2)

$\quad 6x + 3y \qquad = 5$ (4)   Adding

$\quad 8x + 12y - 8z = -10$ Multiplying (2) by 2

$\quad \underline{-8x - 9y + 8z = -2}$ (3)

$\quad 3y \qquad = -12$ (5)   Adding

We can solve Equation (5) for y.

$\quad 3y = -12$

$\quad y = -4$

Substitute -4 for y in Equation (4) and solve for x.

$\quad 6x + 3(-4) = 5$

$\quad 6x - 12 = 5$

$\quad 6x = 17$

$\quad x = \dfrac{17}{6}$

Finally, substitute $\dfrac{17}{6}$ for x and -4 for y in one of the original equations and solve for z. We choose Equation (1).

$$2\left(\frac{17}{6}\right) - 3(-4) + 4z = 10$$

$$\frac{17}{3} + 12 + 4z = 10$$

$$4z = -\frac{23}{3}$$

$$z = -\frac{23}{12}$$

The triple $\left(\frac{17}{6}, -4, -\frac{23}{12}\right)$ checks. It is the solution.

47. $|3x - 5| = 16$

$3x - 5 = 16 \ \ or \ \ 3x - 5 = -16$

$3x = 21 \ \ or \ \ \ \ \ 3x = -11$

$x = 7 \ \ or \ \ \ \ \ \ \ x = -\frac{11}{3}$

The solution set is $\left\{7, -\frac{11}{3}\right\}$.

49. $|3x - 5| < 16$

$-16 < 3x - 5 < 16$

$-11 < 3x < 21$

$-\frac{11}{3} < x < 7$

The solution set is $\left\{x \ \middle| -\frac{11}{3} < x < 7\right\}$.

51. $30y^4 - 97xy^2 + 60x^2$

$= (5y^2 - 12x)(6y^2 - 5x)$ FOIL or grouping method

53. $5x^3 - \frac{5}{27}$

$= 5\left(x^3 - \frac{1}{27}\right)$

$= 5\left[x^3 - \left(\frac{1}{3}\right)^3\right]$ Difference of cubes

$= 5\left(x - \frac{1}{3}\right)\left(x^2 + \frac{1}{3}x + \frac{1}{9}\right)$

55. $(x - p)^2 - p^2$ Difference of squares

$= (x - p + p)(x - p - p)$

$= x(x - 2p)$

57. $(y - 1)^4 - (y - 1)^2$

$= (y - 1)^2[(y - 1)^2 - 1]$

    Removing the common factor

$= (y - 1)^2[(y - 1) + 1][(y - 1) - 1]$

    Factoring the difference of squares

$= (y - 1)^2(y)(y - 2)$, or $y(y - 1)^2(y - 2)$

59. $4x^2 + 4xy + y^2 - r^2 + 6rs - 9s^2$

$= (4x^2 + 4xy + y^2) - 1(r^2 - 6rs + 9s^2)$ Grouping

$= (2x + y)^2 - (r - 3s)^2$ Difference of squares

$= [(2x + y) + (r - 3s)][(2x + y) - (r - 3s)]$

$= (2x + y + r - 3s)(2x + y - r + 3s)$

61. $c^{2w+1} + 2c^{w+1} + c$

$= c^{2w} \cdot c + 2c^w \cdot c + c$

$= c(c^{2w} + 2c^w + 1)$

$= c[(c^w)^2 + 2(c^w) + 1]$ Trinomial square

$= c(c^w + 1)^2$

63. $3(x + 1)^2 + 9(x + 1) - 12$

$= 3[(x + 1)^2 + 3(x + 1) - 4]$

$= 3[(x + 1) + 4][(x + 1) - 1]$ Factor $u^2 + 3u - 4$

                           where $u = x + 1$

$= 3(x + 5)(x)$, or

$3x(x + 5)$

Exercise Set 5.8

1. $x^2 + 3x = 28$

$x^2 + 3x - 28 = 0$ Getting 0 on one side

$(x + 7)(x - 4) = 0$ Factoring

$x + 7 = 0 \ \ \ or \ \ \ x - 4 = 0$ Principle of zero

                                       products

$x = -7 \ \ or \ \ \ \ \ \ x = 4$

The solutions are -7 and 4.

3. $y^2 + 9 = 6y$

$y^2 - 6y + 9 = 0$ Getting 0 on one side

$(y - 3)(y - 3) = 0$ Factoring

$y - 3 = 0 \ \ or \ \ y - 3 = 0$ Principle of zero

                                     products

$y = 3 \ \ or \ \ \ \ \ \ y = 3$

There is only one solution, 3.

5. $x^2 + 20x + 100 = 0$

$(x + 10)(x + 10) = 0$ Factoring

$x + 10 = 0 \ \ \ or \ \ \ x + 10 = 0$ Principle of zero

                                         products

$x = -10 \ \ or \ \ \ \ \ \ x = -10$

There is only one solution, -10.

7. $9x + x^2 + 20 = 0$

$x^2 + 9x + 20 = 0$ Changing order

$(x + 5)(x + 4) = 0$ Factoring

$x + 5 = 0 \ \ \ or \ \ x + 4 = 0$ Principle of zero

                                     products

$x = -5 \ \ or \ \ \ \ \ \ x = -4$

The solutions are -5 and -4.

9. $x^2 + 8x = 0$

$x(x + 8) = 0$ Factoring

$x = 0 \ \ or \ \ x + 8 = 0$ Principle of zero

                               products

$x = 0 \ \ or \ \ \ \ \ \ x = -8$

The solutions are 0 and -8.

11. $x^2 - 25 = 0$

$(x + 5)(x - 5) = 0$ Factoring

$x + 5 = 0$ or $x - 5 = 0$ Principle of zero products

$x = -5$ or $x = 5$

The solutions are -5 and 5.

13. $z^2 = 144$

$z^2 - 144 = 0$ Getting 0 on one side

$(z + 12)(z - 12) = 0$ Factoring

$z + 12 = 0$ or $z - 12 = 0$ Principle of zero products

$z = -12$ or $z = 12$

The solutions are -12 and 12.

15. $y^2 + 2y = 63$

$y^2 + 2y - 63 = 0$

$(y + 9)(y - 7) = 0$

$y + 9 = 0$ or $y - 7 = 0$

$y = -9$ or $y = 7$

The solutions are -9 and 7.

17. $32 + 4x - x^2 = 0$

$0 = x^2 - 4x - 32$

$0 = (x - 8)(x + 4)$

$x - 8 = 0$ or $x + 4 = 0$

$x = 8$ or $x = -4$

The solutions are 8 and -4.

19. $3b^2 + 8b + 4 = 0$

$(3b + 2)(b + 2) = 0$

$3b + 2 = 0$ or $b + 2 = 0$

$3b = -2$ or $b = -2$

$b = -\dfrac{2}{3}$ or $b = -2$

The solutions are $-\dfrac{2}{3}$ and -2.

21. $8y^2 - 10y + 3 = 0$

$(4y - 3)(2y - 1) = 0$

$4y - 3 = 0$ or $2y - 1 = 0$

$4y = 3$ or $2y = 1$

$y = \dfrac{3}{4}$ or $y = \dfrac{1}{2}$

The solutions are $\dfrac{3}{4}$ and $\dfrac{1}{2}$.

23. $6z - z^2 = 0$

$0 = z^2 - 6z$

$0 = z(z - 6)$

$z = 0$ or $z - 6 = 0$

$z = 0$ or $z = 6$

The solutions are 0 and 6.

25. $12z^2 + z = 6$

$12z^2 + z - 6 = 0$

$(4z + 3)(3z - 2) = 0$

$4z + 3 = 0$ or $3z - 2 = 0$

$4z = -3$ or $3z = 2$

$z = -\dfrac{3}{4}$ or $z = \dfrac{2}{3}$

The solutions are $-\dfrac{3}{4}$ and $\dfrac{2}{3}$.

27. $7x^2 - 7 = 0$

$7(x^2 - 1) = 0$

$7(x + 1)(x - 1) = 0$

$x + 1 = 0$ or $x - 1 = 0$

$x = -1$ or $x = 1$

The solutions are -1 and 1.

29. $21r^2 + r - 10 = 0$

$(3r - 2)(7r + 5) = 0$

$3r - 2 = 0$ or $7r + 5 = 0$

$3r = 2$ or $7r = -5$

$r = \dfrac{2}{3}$ or $r = -\dfrac{5}{7}$

The solutions are $\dfrac{2}{3}$ and $-\dfrac{5}{7}$.

31. $15y^2 = 3y$

$15y^2 - 3y = 0$

$3y(5y - 1) = 0$

$3y = 0$ or $5y - 1 = 0$

$y = 0$ or $5y = 1$

$y = 0$ or $y = \dfrac{1}{5}$

The solutions are 0 and $\dfrac{1}{5}$.

33. $14 = x(x - 5)$

$14 = x^2 - 5x$

$0 = x^2 - 5x - 14$ Getting 0 on one side

$0 = (x - 7)(x + 2)$

$x - 7 = 0$ or $x + 2 = 0$

$x = 7$ or $x = -2$

The solutions are 7 and -2.

35. $2x^3 - 2x^2 = 12x$

$2x^3 - 2x^2 - 12x = 0$

$2x(x^2 - x - 6) = 0$

$2x(x - 3)(x + 2) = 0$

$2x = 0$ or $x - 3 = 0$ or $x + 2 = 0$

$x = 0$ or $x = 3$ or $x = -2$

The solutions are 0, 3, and -2.

37.
$$2x^3 = 128x$$
$$2x^3 - 128x = 0$$
$$2x(x^2 - 64) = 0$$
$$2x(x + 8)(x - 8) = 0$$

$x = 0$ or $x + 8 = 0$ or $x - 8 = 0$
$x = 0$ or $x = -8$ or $x = 8$

The solutions are 0, −8, and 8.

39. *Familiarize*. Let $x =$ the number.

Translate.

The square of a number | plus | the number | is | 132.
x^2 | $+$ | x | $=$ | 132

Solve. We solve the equation.
$$x^2 + x = 132$$
$$x^2 + x - 132 = 0$$
$$(x + 12)(x - 11) = 0$$

$x + 12 = 0$ or $x - 11 = 0$
$x = -12$ or $x = 11$

Check. For −12: The square of −12 plus −12 is $144 - 12$, or 132.

For 11: The square of 11 plus 11 is $121 + 11$, or 132. Both numbers check.

State. The number is −12 or 11.

41. *Familiarize*. We will use $N = n^2 - n$, where $n =$ the number of teams in the league and $N =$ the total number of games played.

Translate. We substitute 56 for N.
$$56 = n^2 - n$$

Solve. We solve the equation.
$$n^2 - n = 56$$
$$n^2 - n - 56 = 0$$
$$(n - 8)(n + 7) = 0$$

$n - 8 = 0$ or $n + 7 = 0$
$n = 8$ or $n = -7$

Check. Since the number of teams cannot be negative, −7 cannot be a solution. But $8^2 - 8 = 56$, so 8 checks.

State. There are 8 teams in the league.

43. *Familiarize*. We make a drawing. We let $l =$ the length of the flower bed and $w =$ the width.

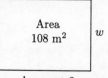

l, or $w + 3$

Recall the formula for the area of a rectangle: $A = lw$.

Translate. We translate to a system of equations.
$$108 = lw, \quad \text{Substituting 108 for } A$$
$$l = w + 3$$

Solve. We will use the system of equations to find an equation in one variable. We substitute $w + 3$ for l in the first equation.
$$108 = (w + 3)w$$
$$108 = w^2 + 3w$$
$$0 = w^2 + 3w - 108$$
$$0 = (w + 12)(w - 9)$$

$w + 12 = 0$ or $w - 9 = 0$
$w = -12$ or $w = 9$

Check. Width cannot be negative, so −12 cannot be a solution. If the width is 9 and the length is 3 m longer, or 12, then the area will be $12 \cdot 9$, or 108 m². We have a solution.

State. The length is 12 m, and the width is 9 m.

45. *Familiarize*. We make a drawing. We let $l =$ the length and $w =$ the width. When both the length and width are increased by 15 yd, they become $l + 15$ and $w + 15$, respectively.

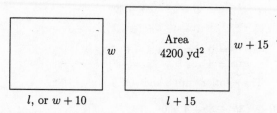

l, or $w + 10$ $l + 15$

Recall the formula for the area of a rectangle: $A = lw$

Translate. We translate to a system of equations.
$$l = w + 10,$$
$$4200 = (l + 15)(w + 15) \quad \text{Substituting in the area formula}$$

Solve. We will use the system of equations to find an equation in one variable. We substitute $w + 10$ for l in the second equation.
$$4200 = (w + 10 + 15)(w + 15)$$
$$4200 = (w + 25)(w + 15)$$
$$4200 = w^2 + 40w + 375$$
$$0 = w^2 + 40w - 3825$$
$$0 = (w + 85)(w - 45)$$

$w + 85 = 0$ or $w - 45 = 0$
$w = -85$ or $w = 45$

Check. Width cannot be negative, so −85 cannot be the solution. If the width is 45 and the length is 10 yd greater, or 55 yd, then the length and width of the larger rectangle are $55 + 15$, or 70, and $45 + 15$, or 60, respectively. The area of the larger rectangle is $70 \cdot 60$, or 4200 yd². We have a solution.

State. The length is 55 yd, and the width is 45 yd.

47. Familiarize. Let $x = $ the first positive odd integer. Then $x + 2 = $ the next positive odd integer.

Translate.

Square of first integer	plus	Square of second integer		is	202.
↓	↓	↓		↓	↓
x^2	$+$	$(x+2)^2$		$=$	202

Solve. We solve the equation:

$$x^2 + (x+2)^2 = 202$$
$$x^2 + x^2 + 4x + 4 = 202$$
$$2x^2 + 4x - 198 = 0$$
$$2(x^2 + 2x - 99) = 0$$
$$2(x + 11)(x - 9) = 0$$
$$x + 11 = 0 \quad \text{or} \quad x - 9 = 0$$
$$x = -11 \quad \text{or} \quad x = 9$$

Check. We only check 9 since the problem asks for consecutive positive odd integers. If $x = 9$, then $x + 2 = 11$, and 9 and 11 are consecutive positive odd integers. The sum of the squares of 9 and 11 is $81 + 121$, or 202. The numbers check.

State. The integers are 9 and 11.

49. Familiarize. Let $h = $ the height of the triangle. Then $h + 9 = $ the base. Recall that the formula for the area of a triangle is $A = \frac{1}{2} \times \text{base} \times \text{height}$.

Translate.

Area	is	56 ft².
↓	↓	↓
$\frac{1}{2}(h+9)h$	$=$	56

Solve. We solve the equation:

$$\frac{1}{2}(h+9)h = 56$$
$$(h+9)h = 112 \qquad \text{Multiplying by 2}$$
$$h^2 + 9h = 112$$
$$h^2 + 9h - 112 = 0$$
$$(h+16)(h-7) = 0$$
$$h + 16 = 0 \quad \text{or} \quad h - 7 = 0$$
$$h = -16 \quad \text{or} \quad h = 7$$

Check. We only check 7, since height cannot be negative. If the height is 7 ft, the base is $7 + 9$, or 16 ft and the area is $\frac{1}{2} \cdot 16 \cdot 7$, or 56 ft². We have a solution.

State. The height is 7 ft, and the base is 16 ft.

51. Familiarize. Let $x = $ the length of a side of the square. Then the perimeter is $x + x + x + x$, or $4x$, and the area is $x \cdot x$, or x^2.

Translate.

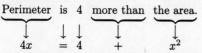

Perimeter	is	4	more than	the area.
↓	↓ ↓		↓	↓
$4x$	$=$ 4		$+$	x^2

Solve. We solve the equation:

$$4x = 4 + x^2$$
$$0 = x^2 - 4x + 4$$
$$0 = (x-2)(x-2)$$
$$x - 2 = 0 \quad \text{or} \quad x - 2 = 0$$
$$x = 2 \quad \text{or} \quad x = 2$$

Check. If the length of a side is 2, the perimeter or 8, the area is $2 \cdot 2$ or 4, and 8 is four more than value checks.

State. The length of a side is 2.

53. Familiarize. Let x represent the first integer, $x + 1$ second, and $x + 2$ the third.

Translate.

First	·	Third	−	Second	=	1	+	10 · Third
x	·	$(x+2)$	−	$(x+1)$	$=$	1	+	$10(x+2)$

Solve. We solve the equation:

$$x(x+2) - (x+1) = 1 + 10(x+2)$$
$$x^2 + 2x - x - 1 = 1 + 10x + 20$$
$$x^2 - 9x - 22 = 0$$
$$(x-11)(x+2) = 0$$
$$x - 11 = 0 \quad \text{or} \quad x + 2 = 0$$
$$x = 11 \quad \text{or} \quad x = -2$$

Check. If $x = 11$, the consecutive integers are 11, 12, and 13.

First · Third - Second	$1 + 10 \cdot$ Third
$11 \cdot 13 - 12$	$1 + 10 \cdot 13$
$143 - 12$	$1 + 130$
131	131

If $x = -2$, the consecutive integers are -2, -1, and 0.

First · Third - Second	$1 + 10 \cdot$ Third
$-2 \cdot 0 - (-1)$	$1 + 10 \cdot 0$
$0 + 1$	$1 + 0$
1	1

Both sets of integers check.

State. The three consecutive integers can be 11, 12, and 13 or -2, -1, and 0.

55. Familiarize. Let $x = $ the other leg of the triangle. Then $x + 1 = $ the hypotenuse.

Translate. We use the Pythagorean theorem.

$$a^2 + b^2 = c^2$$
$$9^2 + x^2 = (x+1)^2$$

Solve. We solve the equation:

$$9^2 + x^2 = (x+1)^2$$
$$81 + x^2 = x^2 + 2x + 1$$
$$80 = 2x$$
$$40 = x$$

heck. When $x = 40$, then $x + 1 = 41$, and $9^2 + 40^2 = 81 = 41^2$. The numbers check.

State. The other sides have lengths of 40 m and 41 m.

Familiarize. We will use the equation
$h = -16t^2 + 80t + 224$.

Translate.

$$\underbrace{\text{Height}}_{\downarrow} \quad \text{is} \quad \underbrace{0 \text{ ft.}}_{\downarrow} $$
$$-16t^2 + 80t + 224 \quad = \quad 0$$

Solve. We solve the equation:
$$-16(t^2 - 5t - 14) = 0$$
$$-16(t - 7)(t + 2) = 0$$
$$t - 7 = 0 \quad \text{or} \quad t + 2 = 0$$
$$t = 7 \quad \text{or} \quad t = -2$$

Check. The number -2 is not a solution, since time cannot be negative in this application. When $t = 7$, $h = -16 \cdot 7^2 + 80 \cdot 7 + 224 = 0$. We have a solution.

State. The object reaches the ground after 7 sec.

59. $\quad 2x - 14y + 10z = 100, \quad (1)$
$$\quad\quad\quad 5y - 8z = 80, \quad (2)$$
$$\quad\quad\quad\quad\quad 4z = 64 \quad (3)$$

Solve equation (3) for z:
$$4z = 64$$
$$z = 16$$

Substitute 16 for z in equation (2) and solve for y:
$$5y - 8(16) = 80$$
$$5y - 128 = 80$$
$$5y = 208$$
$$y = \frac{208}{5}$$

Substitute $\frac{208}{5}$ for y and 16 for z in equation (1) and solve for x:
$$2x - 14\left(\frac{208}{5}\right) + 10(16) = 100$$
$$2x - \frac{2912}{5} + 160 = 100$$
$$2x - \frac{2112}{5} = 100$$
$$2x = \frac{2612}{5}$$
$$x = \frac{1306}{5}$$

The solution is $\left(\frac{1306}{5}, \frac{208}{5}, 16\right)$.

61. Graph: $y = -x$

We find several ordered pairs, plot them, and draw the line.

x	y $y = -x$	(x, y)
-3	3	$(-3, 3)$
0	0	$(0, 0)$
2	-2	$(2, -2)$

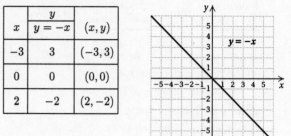

63. Graph: $3x - 2y = -6$

To find the x-intercept, we cover up the y-term and consider the rest of the equation. We have $3x = -6$ or $x = -2$. The x-intercept is $(-2, 0)$. To find the y-intercept, we cover up the x-term and consider the rest of the equation. We have $-2y = -6$ or $y = 3$. The y-intercept is $(0, 3)$. We plot these points and draw the line.

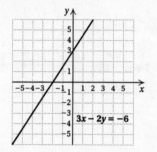

We find a third point as a check. We let $x = 2$ and solve for y:
$$3 \cdot 2 - 2y = -6$$
$$6 - 2y = -6$$
$$-2y = -12$$
$$y = 6$$

We plot the point $(2, 6)$ and note that it is on the line.

65. $\quad\quad x(x + 8) = 16(x - 1)$
$$x^2 + 8x = 16x - 16$$
$$x^2 - 8x + 16 = 0$$
$$(x - 4)(x - 4) = 0$$
$$x - 4 = 0 \quad \text{or} \quad x - 4 = 0$$
$$x = 4 \quad \text{or} \quad x = 4$$

The solution is 4.

67. $\quad\quad (a - 5)^2 = 36$
$$a^2 - 10a + 25 = 36$$
$$a^2 - 10a - 11 = 0$$
$$(a - 11)(a + 1) = 0$$
$$a - 11 = 0 \quad \text{or} \quad a + 1 = 0$$
$$a = 11 \quad \text{or} \quad a = -1$$

The solutions are 11 and -1.

69. $(3x^2 - 7x - 20)(x - 5) = 0$

$(3x + 5)(x - 4)(x - 5) = 0$

$3x + 5 = 0 \quad \text{or} \quad x - 4 = 0 \quad \text{or} \quad x - 5 = 0$
$3x = -5 \quad \text{or} \qquad x = 4 \quad \text{or} \qquad x = 5$
$x = -\dfrac{5}{3} \quad \text{or} \qquad x = 4 \quad \text{or} \qquad x = 5$

The solutions are $-\dfrac{5}{3}$, 4, and 5.

71. $\qquad 2x^3 + 6x^2 = 8x + 24$

$2x^3 + 6x^2 - 8x - 24 = 0$

$2(x^3 + 3x^2 - 4x - 12) = 0$

$2[x^2(x + 3) - 4(x + 3)] = 0$

$2(x + 3)(x^2 - 4) = 0$

$2(x + 3)(x + 2)(x - 2) = 0$

$x + 3 = 0 \quad \text{or} \quad x + 2 = 0 \quad \text{or} \quad x - 2 = 0$
$x = -3 \quad \text{or} \qquad x = -2 \quad \text{or} \qquad x = 2$

The solutions are -3, -2, and 2.

73. Familiarize. Let $x =$ one of the numbers. Then $17 - x =$ the other number.

Translate.

The sum of the squares of the numbers is 205.

$$x^2 + (17 - x)^2 \qquad = \quad 205$$

Solve. We solve the equation.

$x^2 + (17 - x)^2 = 205$

$x^2 + 289 - 34x + x^2 = 205$

$2x^2 - 34x + 289 = 205$

$2x^2 - 34x + 84 = 0$

$2(x^2 - 17x + 42) = 0$

$2(x - 3)(x - 14) = 0$

$x - 3 = 0 \quad \text{or} \quad x - 14 = 0$
$x = 3 \quad \text{or} \qquad x = 14$

Check. If one number is 3, then the other is $17 - 3$, or 14. If one number is 14, then the other is $17 - 14$, or 3. In either case, we have the numbers 3 and 14. Their sum is 17. Also, $3^2 + 14^2 = 9 + 196 = 205$. The numbers check.

State. The numbers are 3 and 14.

75. Familiarize. Let x represent the length of the hypotenuse. Then $x - 3$ and $x - 6$ represent the lengths of the legs of the triangle. We will first find the lengths of the legs (that is, the base and the height of the triangle). Then we will find the area of the triangle using the formula Area $= \dfrac{1}{2} \times$ base $\times$ height.

Translate. We use the Pythagorean equation.

$$a^2 + b^2 = c^2$$
$$(x - 3)^2 + (x - 6)^2 = x^2$$

Solve. We solve the equation:

$$(x - 3)^2 + (x - 6)^2 = x^2$$
$$x^2 - 6x + 9 + x^2 - 12x + 36 = x^2$$
$$2x^2 - 18x + 45 = x^2$$
$$x^2 - 18x + 45 = 0$$
$$(x - 15)(x - 3) = 0$$

$x - 15 = 0 \quad \text{or} \quad x - 3 = 0$
$x = 15 \quad \text{or} \qquad x = 3$

Check. If $x = 3$, then $x - 3 = 0$ and $x - 6 = -3$. Thus, 3 cannot be a solution since the lengths of the legs must be positive. If $x = 15$, $x - 3 = 12$ and $x - 6 = 9$, and $12^2 + 9^2 = 225 = 15^2$. These numbers check. We find the area of the triangle:

$$A = \frac{1}{2}bh = \frac{1}{2}(12)(9) = 54$$

State. The area of the triangle is 54 cm^2.

Exercise Set 5.9

1. $\dfrac{45y^7 - 20y^4 + 15y^2}{5y^2}$

$= \dfrac{45y^7}{5y^2} - \dfrac{20y^4}{5y^2} + \dfrac{15y^2}{5y^2}$

$= 9y^5 - 4y^2 + 3$

3. $(32a^4b^3 + 14a^3b^2 - 22a^2b) \div 2a^2b$

$= \dfrac{32a^4b^3 + 14a^3b^2 - 22a^2b}{2a^2b}$

$= \dfrac{32a^4b^3}{2a^2b} + \dfrac{14a^3b^2}{2a^2b} - \dfrac{22a^2b}{2a^2b}$

$= 16a^2b^2 + 7ab - 11$

5.
$$\begin{array}{r} x + 7 \\ x + 3 \overline{\smash{\big)}\ x^2 + 10x + 21} \\ \underline{x^2 + 3x} \\ 7x + 21 \\ \underline{7x + 21} \\ 0 \end{array}$$
$(x^2 + 10x) - (x^2 + 3x) = 7x$

The answer is $x + 7$.

7.
$$\begin{array}{r} a - 12 \\ a + 4 \overline{\smash{\big)}\ a^2 - 8a - 16} \\ \underline{a^2 + 4a} \\ -12a - 16 \\ \underline{-12a - 48} \\ 32 \end{array}$$
$(a^2 - 8a) - (a^2 + 4a) = -12a$
$(-12a - 16) - (-12a - 48) = 32$

The answer is $a - 12$, R 32, or $a - 12 + \dfrac{32}{a + 4}$.

9.
$$\begin{array}{r} x + 2 \\ x + 5 \overline{\smash{\big)}\ x^2 + 7x + 14} \\ \underline{x^2 + 5x} \\ 2x + 14 \\ \underline{2x + 10} \\ 4 \end{array}$$
$(x^2 + 7x) - (x^2 + 5x) = 2x$
$(2x + 14) - (2x + 10) = 4$

The answer is $x + 2$, R 4, or $x + 2 + \dfrac{4}{x + 5}$.

11.

$$
\begin{array}{r}
2y^2 - y + 2 \\
2y + 4\overline{\smash{\big)}\,4y^3+6y^2 + 0y + 14} \\
\underline{4y^3+8y^2} \\
-2y^2 + 0y \\
\underline{-2y^2 - 4y} \\
4y + 14 \\
\underline{4y + 8} \\
6
\end{array}
$$

The answer is $2y^2 - y + 2$, R 6, or $2y^2 - y + 2 + \dfrac{6}{2y + 4}$.

13.

$$
\begin{array}{r}
2y^2+ 2y - 1 \\
5y - 2\overline{\smash{\big)}\,10y^3+ 6y^2 - 9y + 10} \\
\underline{10y^3 - 4y^2} \\
10y^2 - 9y \\
\underline{10y^2 - 4y} \\
- 5y + 10 \\
\underline{- 5y + 2} \\
8
\end{array}
$$

The answer is $2y^2 + 2y - 1$, R 8, or $2y^2 + 2y - 1 + \dfrac{8}{5y - 2}$.

15.

$$
\begin{array}{r}
2x^2 - x - 9 \\
x^2 + 2\overline{\smash{\big)}\,2x^4 - x^3 -5x^2 + x - 6} \\
\underline{2x^4 +4x^2} \\
- x^3 -9x^2 + x \\
\underline{- x^3 - 2x} \\
-9x^2 + 3x - 6 \\
\underline{-9x^2 - 18} \\
3x + 12
\end{array}
$$

The answer is $2x^2 - x - 9$, R $(3x + 12)$, or
$2x^2 - x - 9 + \dfrac{3x + 12}{x^2 + 2}$.

17.

$$
\begin{array}{r}
2x^3 + 5x^2+ 17x + 51 \\
x^2 - 3x\overline{\smash{\big)}\,2x^5 - x^4+ 2x^3 + 0x^2 - x} \\
\underline{2x^5 - 6x^4} \\
5x^4+ 2x^3 \\
\underline{5x^4 -15x^3} \\
17x^3 + 0x^2 \\
\underline{17x^3 - 51x^2} \\
51x^2 - x \\
\underline{51x^2 - 153x} \\
152x
\end{array}
$$

The answer is $2x^3 + 5x^2 + 17x + 51$, R $152x$, or
$2x^3 + 5x^2 + 17x + 51 + \dfrac{152x}{x^2 - 3x}$.

19. $(x^3 - 2x^2 + 2x - 5) \div (x - 1)$

$$
\begin{array}{r|rrrr}
1 & 1 & -2 & 2 & -5 \\
  &   & 1 & -1 & 1 \\
\hline
  & 1 & -1 & 1 & \!\!-4
\end{array}
$$

The answer is $x^2 - x + 1$, R -4, or $x^2 - x + 1 + \dfrac{-4}{x - 1}$.

21. $(a^2 + 11a - 19) \div (a + 4) =$

$(a^2 + 11a - 19) \div [a - (-4)]$

$$
\begin{array}{r|rrr}
-4 & 1 & 11 & -19 \\
   &   & -4 & -28 \\
\hline
   & 1 & 7 & \!\!-47
\end{array}
$$

The answer is $a + 7$, R -47, or $a + 7 + \dfrac{-47}{a + 4}$.

23. $(x^3 - 7x^2 - 13x + 3) \div (x - 2)$

$$
\begin{array}{r|rrrr}
2 & 1 & -7 & -13 & 3 \\
  &   & 2 & -10 & -46 \\
\hline
  & 1 & -5 & -23 & \!\!-43
\end{array}
$$

The answer is $x^2-5x-23$, R -43, or $x^2 - 5x - 23 + \dfrac{-43}{x - 2}$.

25. $(3x^3 + 7x^2 - 4x + 3) \div (x + 3) =$

$(3x^3 + 7x^2 - 4x + 3) \div [x - (-3)]$

$$
\begin{array}{r|rrrr}
-3 & 3 & 7 & -4 & 3 \\
   &   & -9 & 6 & -6 \\
\hline
   & 3 & -2 & 2 & \!\!-3
\end{array}
$$

The answer is $3x^2-2x+2$, R -3, or $3x^2 - 2x + 2 + \dfrac{-3}{x + 3}$.

27. $(y^3 - 3y + 10) \div (y - 2) =$

$(y^3 + 0y^2 - 3y + 10) \div (y - 2)$

$$
\begin{array}{r|rrrr}
2 & 1 & 0 & -3 & 10 \\
  &   & 2 & 4 & 2 \\
\hline
  & 1 & 2 & 1 & \!\!12
\end{array}
$$

The answer is $y^2 + 2y + 1$, R 12, or $y^2 + 2y + 1 + \dfrac{12}{y - 2}$.

29. $(3x^4 - 25x^2 - 18) \div (x - 3) =$

$(3x^4 + 0x^3 - 25x^2 + 0x - 18) \div (x - 3)$

$$
\begin{array}{r|rrrrr}
3 & 3 & 0 & -25 & 0 & -18 \\
  &   & 9 & 27 & 6 & 18 \\
\hline
  & 3 & 9 & 2 & 6 & \!\!0
\end{array}
$$

The answer is $3x^3 + 9x^2 + 2x + 6$.

31. $(x^3 - 8) \div (x - 2) = (x^3 + 0x^2 + 0x - 8) \div (x - 2)$

$$
\begin{array}{r|rrrr}
2 & 1 & 0 & 0 & -8 \\
  &   & 2 & 4 & 8 \\
\hline
  & 1 & 2 & 4 & \!\!0
\end{array}
$$

The answer is $x^2 + 2x + 4$.

33. $(y^4 - 16) \div (y - 2) =$

$(y^4 + 0y^3 + 0y^2 + 0y - 16) \div (y - 2)$

$$
\begin{array}{r|rrrrr}
2 & 1 & 0 & 0 & 0 & -16 \\
  &   & 2 & 4 & 8 & 16 \\
\hline
  & 1 & 2 & 4 & 8 & \!\!0
\end{array}
$$

The answer is $y^3 + 2y^2 + 4y + 8$.

35. Graph: $2x - 3y = 6$

To find the x-intercept we cover up the y-term and consider the rest of the equation. We have $2x = 6$, or $x = 3$. The x-intercept is $(3, 0)$. To find the y-intercept we cover up the x-term and consider the rest of the equation. We have $-3y = 6$, or $y = -2$. The y-intercept is $(0, -2)$. We plot these points and draw the graph.

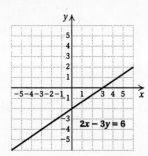

We find a third point as a check. We let $x = 6$ and solve for y.

$$2 \cdot 6 - 3y = 6$$
$$12 - 3y = 6$$
$$-3y = -6$$
$$y = 2$$

We plot the point $(6, 2)$ and see that it is on the line.

37. Graph: $y = -2$

Since x is missing, any number for x will do. Thus all ordered pairs $(x, -2)$ are solutions. The graph is parallel to the x-axis.

x	y
-2	-2
0	-2
3	-2

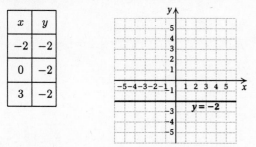

39. $|x - 5| < 10$

$$-10 < x - 5 < 10$$
$$-5 < x < 15$$

The solution set is $\{x \mid -5 < x < 15\}$.

41.
$$
\begin{array}{r}
2x^4 - 2x^3 + 5x^2 - 4x - 1 \\
x^2 + x + 1 \overline{\smash{\big)}\ 2x^6 + 0x^5 + 5x^4 - x^3 + 0x^2 + 0x + 1} \\
\underline{2x^6 + 2x^5 + 2x^4} \\
-2x^5 + 3x^4 - x^3 \\
\underline{-2x^5 - 2x^4 - 2x^3} \\
5x^4 + x^3 + 0x^2 \\
\underline{5x^4 + 5x^3 + 5x^2} \\
-4x^3 - 5x^2 + 0x \\
\underline{-4x^3 - 4x^2 - 4x} \\
-x^2 + 4x + 1 \\
\underline{-x^2 - x - 1} \\
5x + 2
\end{array}
$$

The answer is $2x^4 - 2x^3 + 5x^2 - 4x - 1$, R $(5x + 2)$, or
$$2x^4 - 2x^3 + 5x^2 - 4x - 1 + \frac{5x + 2}{x^2 + x + 1}.$$

43.
$$
\begin{array}{r}
x^2 + 2y \\
x^2 - xy + y^2 \overline{\smash{\big)}\ x^4 - x^3y + x^2y^2 + 2x^2y - 2xy^2 + 2y^3} \\
\underline{x^4 - x^3y + x^2y^2} \\
0 + 2x^2y - 2xy^2 + 2y^3 \\
\underline{2x^2y - 2xy^2 + 2y^3} \\
0
\end{array}
$$

The answer is $x^2 + 2y$.

45.
$$
\begin{array}{r}
a^6 - a^5b + a^4b^2 - a^3b^3 + a^2b^4 - ab^5 + b^6 \\
a + b \overline{\smash{\big)}\ a^7 \qquad\qquad\qquad\qquad\qquad + b^7} \\
\underline{a^7 + a^6b} \\
-a^6b \\
\underline{-a^6b - a^5b^2} \\
a^5b^2 \\
\underline{a^5b^2 + a^4b^3} \\
-a^4b^3 \\
\underline{-a^4b^3 - a^3b^4} \\
a^3b^4 \\
\underline{a^3b^4 + a^2b^5} \\
-a^2b^5 \\
\underline{-a^2b^5 - ab^6} \\
ab^6 + b^7 \\
\underline{ab^6 + b^7} \\
0
\end{array}
$$

The answer is $a^6 - a^5b + a^4b^2 - a^3b^3 + a^2b^4 - ab^5 + b^6$.

Chapter 6

Rational Expressions and Equations

Exercise Set 6.1

---

1. $\dfrac{5t^2 - 64}{3t + 17}$

We set the denominator equal to 0 and solve.

$$3t + 17 = 0$$
$$3t = -17$$
$$t = -\frac{17}{3}$$

The expression is undefined for the replacement number $-\dfrac{17}{3}$.

3. $\dfrac{y^3 - y^2 + y + 2}{y^2 + 12y + 35}$

We set the denominator equal to 0 and solve.

$$y^2 + 12y + 35 = 0$$
$$(y + 5)(y + 7) = 0$$
$$y + 5 = 0 \quad \text{or} \quad y + 7 = 0$$
$$y = -5 \quad \text{or} \quad y = -7$$

The expression is undefined for the replacement numbers -5 and -7.

5. $\dfrac{7x}{7x} \cdot \dfrac{x + 2}{x + 8} = \dfrac{7x(x + 2)}{7x(x + 8)}$ Multiplying numerators and multiplying denominators

7. $\dfrac{q - 5}{q + 3} \cdot \dfrac{q + 5}{q + 5} = \dfrac{(q - 5)(q + 5)}{(q + 3)(q + 5)}$ Multiplying numerators and multiplying denominators

9. $\dfrac{15y^5}{5y^4} = \dfrac{3 \cdot 5 \cdot y^4 \cdot y}{5 \cdot y^4 \cdot 1}$ Factoring the numerator and the denominator

$= \dfrac{5y^4}{5y^4} \cdot \dfrac{3y}{1}$ Factoring the rational expression

$= 1 \cdot 3y \qquad \dfrac{5y^4}{5y^4} = 1$

$= 3y$ Removing a factor of 1

11. $\dfrac{16p^3}{24p^7} = \dfrac{8p^3 \cdot 2}{8p^3 \cdot 3p^4}$ Factoring the numerator and the denominator

$= \dfrac{8p^3}{8p^3} \cdot \dfrac{2}{3p^4}$ Factoring the rational expression

$= 1 \cdot \dfrac{2}{3p^4} \qquad \dfrac{8p^3}{8p^3} = 1$

$= \dfrac{2}{3p^4}$ Removing a factor of 1

13. $\dfrac{9a - 27}{9} = \dfrac{9(a - 3)}{9 \cdot 1}$ Factoring the numerator and the denominator

$= \dfrac{9}{9} \cdot \dfrac{a - 3}{1}$

$= \dfrac{a - 3}{1}$ Removing a factor of 1

$= a - 3$

15. $\dfrac{12x - 15}{21} = \dfrac{3(4x - 5)}{3 \cdot 7}$ Factoring the numerator and the denominator

$= \dfrac{3}{3} \cdot \dfrac{4x - 5}{7}$

$= \dfrac{4x - 5}{7}$ Removing a factor of 1

17. $\dfrac{4y - 12}{4y + 12} = \dfrac{4(y - 3)}{4(y + 3)} = \dfrac{4}{4} \cdot \dfrac{y - 3}{y + 3} = \dfrac{y - 3}{y + 3}$

19. $\dfrac{t^2 - 16}{t^2 - 8t + 16} = \dfrac{(t + 4)(t - 4)}{(t - 4)(t - 4)} = \dfrac{t + 4}{t - 4} \cdot \dfrac{t - 4}{t - 4} = \dfrac{t + 4}{t - 4}$

21. $\dfrac{x^2 - 9x + 8}{x^2 + 3x - 4} = \dfrac{(x - 8)(x - 1)}{(x + 4)(x - 1)} = \dfrac{x - 8}{x + 4} \cdot \dfrac{x - 1}{x - 1} = \dfrac{x - 8}{x + 4}$

23. $\dfrac{w^3 - z^3}{w^2 - z^2} = \dfrac{(w - z)(w^2 + wz + z^2)}{(w + z)(w - z)} =$

$\dfrac{w - z}{w - z} \cdot \dfrac{w^2 + wz + z^2}{w + z} = \dfrac{w^2 + wz + z^2}{w + z}$

25. $\dfrac{x^4}{3x + 6} \cdot \dfrac{5x + 10}{5x^7}$

$= \dfrac{x^4(5x + 10)}{(3x + 6)(5x^7)}$ Multiplying the numerators and the denominators

$= \dfrac{x^4(5)(x + 2)}{3(x + 2)(5)(x^4)(x^3)}$ Factoring the numerator and the denominator

$= \dfrac{x^4(5)(x + 2)(1)}{3(x + 2)(5)(x^4)(x^3)}$ Removing a factor of 1: $\dfrac{(x^4)(5)(x + 2)}{(x + 2)(5)(x^4)} = 1$

$= \dfrac{1}{3x^3}$ Simplifying

27. $\dfrac{x^2-16}{x^2}\cdot\dfrac{x^2-4x}{x^2-x-12}$

$=\dfrac{(x^2-16)(x^2-4x)}{x^2(x^2-x-12)}$ Multiplying the numerators and the denominators

$=\dfrac{(x+4)(x-4)(x)(x-4)}{x\cdot x(x-4)(x+3)}$ Factoring the numerator and the denominator

$=\dfrac{(x+4)(x-4)(\not x)(x-4)}{\not x\cdot x(x-4)(x+3)}$ Removing a factor of 1

$=\dfrac{(x+4)(x-4)}{x(x+3)}$

29. $\dfrac{y^2-16}{2y+6}\cdot\dfrac{y+3}{y-4}=\dfrac{(y^2-16)(y+3)}{(2y+6)(y-4)}$

$=\dfrac{(y+4)(y-4)(y+3)}{2(y+3)(y-4)}$

$=\dfrac{(y+4)(y-4)(y+3)}{2(y+3)(y-4)}$

$=\dfrac{y+4}{2}$

31. $\dfrac{x^2-2x-35}{2x^3-3x^2}\cdot\dfrac{4x^3-9x}{7x-49}$

$=\dfrac{(x^2-2x-35)(4x^3-9x)}{(2x^3-3x^2)(7x-49)}$

$=\dfrac{(x-7)(x+5)(x)(2x+3)(2x-3)}{x\cdot x(2x-3)(7)(x-7)}$

$=\dfrac{(x-7)(x+5)(\not x)(2x+3)(2x-3)}{\not x\cdot x(2x-3)(7)(x-7)}$

$=\dfrac{(x+5)(2x+3)}{7x}$

33. $\dfrac{c^3+8}{c^2-4}\cdot\dfrac{c^2-4c+4}{c^2-2c+4}$

$=\dfrac{(c^3+8)(c^2-4c+4)}{(c^2-4)(c^2-2c+4)}$

$=\dfrac{(c+2)(c^2-2c+4)(c-2)(c-2)}{(c+2)(c-2)(c^2-2c+4)\cdot 1}$

$=\dfrac{(c+2)(c^2-2c+4)(c-2)}{(c+2)(c^2-2c+4)(c-2)}\cdot\dfrac{c-2}{1}$

$=\dfrac{c-2}{1}$

$=c-2$

35. $\dfrac{x^2-y^2}{x^3-y^3}\cdot\dfrac{x^2+xy+y^2}{x^2+2xy+y^2}$

$=\dfrac{(x^2-y^2)(x^2+xy+y^2)}{(x^3-y^3)(x^2+2xy+y^2)}$

$=\dfrac{(x+y)(x-y)(x^2+xy+y^2)\cdot 1}{(x-y)(x^2+xy+y^2)(x+y)(x+y)}$

$=\dfrac{(x+y)(x-y)(x^2+xy+y^2)}{(x+y)(x-y)(x^2+xy+y^2)}\cdot\dfrac{1}{x+y}$

$=\dfrac{1}{x+y}$

37. $\dfrac{12x^8}{3y^4}\div\dfrac{16x^3}{6y}$

$=\dfrac{12x^8}{3y^4}\cdot\dfrac{6y}{16x^3}$ Multiplying by the reciprocal of the divisor

$=\dfrac{12x^8(6y)}{3y^4(16x^3)}$ Multiplying the numerators and the denominators

$=\dfrac{3\cdot 4\cdot x^3\cdot x^5\cdot 2\cdot 3\cdot y}{3\cdot y\cdot y^3\cdot 4\cdot 2\cdot 2\cdot x^3}$ Factoring the numerator and the denominator

$=\dfrac{\not 3\cdot\not 4\cdot\not{x^3}\cdot x^5\cdot\not 2\cdot 3\cdot\not y}{\not 3\cdot\not y\cdot y^3\cdot\not 4\cdot\not 2\cdot 2\cdot\not{x^3}}$ Removing a factor of 1

$=\dfrac{3x^5}{2y^3}$

39. $\dfrac{3y+15}{y}\div\dfrac{y+5}{y}=\dfrac{3y+15}{y}\cdot\dfrac{y}{y+5}$

$=\dfrac{(3y+15)(y)}{y(y+5)}$

$=\dfrac{3(y+5)(y)}{y(y+5)\cdot 1}$

$=\dfrac{3(y+5)(\not y)}{\not y(y+5)\cdot 1}$

$=\dfrac{3}{1}$

$=3$

41. $\dfrac{y^2-9}{y}\div\dfrac{y+3}{y+2}=\dfrac{y^2-9}{y}\cdot\dfrac{y+2}{y+3}$

$=\dfrac{(y^2-9)(y+2)}{y(y+3)}$

$=\dfrac{(y+3)(y-3)(y+2)}{y(y+3)}$

$=\dfrac{(y+3)(y-3)(y+2)}{y(y+3)}$

$=\dfrac{(y-3)(y+2)}{y}$

43. $\dfrac{4a^2-1}{a^2-4}\div\dfrac{2a-1}{a-2}=\dfrac{4a^2-1}{a^2-4}\cdot\dfrac{a-2}{2a-1}$

$=\dfrac{(4a^2-1)(a-2)}{(a^2-4)(2a-1)}$

$=\dfrac{(2a+1)(2a-1)(a-2)}{(a+2)(a-2)(2a-1)}$

$=\dfrac{(2a+1)(2a-1)(a-2)}{(a+2)(a-2)(2a-1)}$

$=\dfrac{2a+1}{a+2}$

45. $\dfrac{x^2 - 16}{x^2 - 10x + 25} \div \dfrac{3x - 12}{x^2 - 3x - 10}$

$= \dfrac{x^2 - 16}{x^2 - 10x + 25} \cdot \dfrac{x^2 - 3x - 10}{3x - 12}$

$= \dfrac{(x^2 - 16)(x^2 - 3x - 10)}{(x^2 - 10x + 25)(3x - 12)}$

$= \dfrac{(x + 4)(x - 4)(x - 5)(x + 2)}{(x - 5)(x - 5)(3)(x - 4)}$

$= \dfrac{(x + 4)\cancel{(x - 4)}\cancel{(x - 5)}(x + 2)}{\cancel{(x - 5)}(x - 5)(3)\cancel{(x - 4)}}$

$= \dfrac{(x + 4)(x + 2)}{3(x - 5)}$

47. $\dfrac{y^3 + 3y}{y^2 - 9} \div \dfrac{y^2 + 5y - 14}{y^2 + 4y - 21}$

$= \dfrac{y^3 + 3y}{y^2 - 9} \cdot \dfrac{y^2 + 4y - 21}{y^2 + 5y - 14}$

$= \dfrac{(y^3 + 3y)(y^2 + 4y - 21)}{(y^2 - 9)(y^2 + 5y - 14)}$

$= \dfrac{y(y^2 + 3)(y + 7)(y - 3)}{(y + 3)(y - 3)(y + 7)(y - 2)}$

$= \dfrac{y(y^2 + 3)\cancel{(y + 7)}\cancel{(y - 3)}}{(y + 3)\cancel{(y - 3)}\cancel{(y + 7)}(y - 2)}$

$= \dfrac{y(y^2 + 3)}{(y + 3)(y - 2)}$

49. $\dfrac{x^3 - 64}{x^3 + 64} \div \dfrac{x^2 - 16}{x^2 - 4x + 16}$

$= \dfrac{x^3 - 64}{x^3 + 64} \cdot \dfrac{x^2 - 4x + 16}{x^2 - 16}$

$= \dfrac{(x^3 - 64)(x^2 - 4x + 16)}{(x^3 + 64)(x^2 - 16)}$

$= \dfrac{(x - 4)(x^2 + 4x + 16)(x^2 - 4x + 16)}{(x + 4)(x^2 - 4x + 16)(x + 4)(x - 4)}$

$= \dfrac{(x - 4)(x^2 + 4x + 16)}{(x - 4)(x^2 - 4x + 16)} \cdot \dfrac{x^2 + 4x + 16}{(x + 4)(x + 4)}$

$= \dfrac{x^2 + 4x + 16}{(x + 4)(x + 4)},$ or $\dfrac{x^2 + 4x + 16}{(x + 4)^2}$

51. $\left[\dfrac{r^2 - 4s^2}{r + 2s} \div (r + 2s) \right] \cdot \dfrac{2s}{r - 2s}$

$= \left[\dfrac{r^2 - 4s^2}{r + 2s} \cdot \dfrac{1}{r + 2s} \right] \cdot \dfrac{2s}{r - 2s}$

$= \dfrac{(r^2 - 4s^2)(1)(2s)}{(r + 2s)(r + 2s)(r - 2s)}$

$= \dfrac{(r + 2s)(r - 2s)(2s)}{(r + 2s)(r + 2s)(r - 2s)}$

$= \dfrac{\cancel{(r + 2s)}\cancel{(r - 2s)}(2s)}{\cancel{(r + 2s)}(r + 2s)\cancel{(r - 2s)}}$

$= \dfrac{2s}{r + 2s}$

53. *Familiarize*. Let $x =$ the number of field goals and $y =$ the number of three-point baskets the player scores. Then

she scores $2x$ points for field goals and $3y$ points for three-point baskets.

Translate.

The total number of baskets made is 15.

$$x + y = 15$$

The total numbe of points is 33.

$$2x + 3y = 33$$

We have a system of equations:

$$x + y = 15,$$
$$2x + 3y = 33$$

Solve. Solving the system of equations, we get $(12, 3)$.

Check. If the player makes 12 field goals and 3 three-point shots she makes a total of 15 baskets. She scores $2 \cdot 12$, or 24 points, from field goals and $3 \cdot 3$, or 9 points from three-point baskets, so the total points are 33. These values check.

State. The player makes 12 field goals and 3 three-point shots.

55. *Familiarize* We make a drawing. Let $l =$ the length and $w =$ the width of the rectangle.

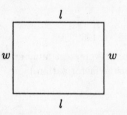

Recall the formula for the perimeter of a rectangle: $P = 2l + 2w$.

Translate.

The length is 5 yd greater than the width.

$$l = 5 + w$$

The perimeter is 4910 yd.

$$2l + 2w = 4910$$

We have a system of equations:

$$l = 5 + w,$$
$$2l + 2w = 4910$$

Solve. Solving the system of equations, we get $(1230, 1225)$.

Check. The length, 1230 yd, is 5 yd greater than 1225 yd, the width. The perimeter is $2 \cdot 1230 + 2 \cdot 1225$, or 4910 yd. These values check.

State. The length of the field is 1230 yd, and the width is 1225 yd.

57. $8 - x^3 = 2^3 - x^3 = (2 - x)(4 + 2x + x^2)$

59.
$$\frac{x(x+1)-2(x+3)}{(x+1)(x+2)(x+3)} = \frac{x^2+x-2x-6}{(x+1)(x+2)(x+3)}$$
$$= \frac{x^2-x-6}{(x+1)(x+2)(x+3)}$$
$$= \frac{(x-3)(x+2)}{(x+1)(x+2)(x+3)}$$
$$= \frac{(x-3)\cancel{(x+2)}}{(x+1)\cancel{(x+2)}(x+3)}$$
$$= \frac{x-3}{(x+1)(x+3)}$$

61.
$$\frac{m^2-t^2}{m^2+t^2+m+t+2mt} = \frac{m^2-t^2}{(m^2+2mt+t^2)+(m+t)}$$
$$= \frac{(m+t)(m-t)}{(m+t)^2+(m+t)}$$
$$= \frac{(m+t)(m-t)}{(m+t)[(m+t)+1]}$$
$$= \frac{\cancel{(m+t)}(m-t)}{\cancel{(m+t)}(m+t+1)}$$
$$= \frac{m-t}{m+t+1}$$

Exercise Set 6.2

1. $15 = 3 \cdot 5$

$40 = 2 \cdot 2 \cdot 2 \cdot 5$

$LCM = 2 \cdot 2 \cdot 2 \cdot 3 \cdot 5$, or 120

(We used each factor the greatest number of times that it occurs in any one prime factorization.)

3. $18 = 2 \cdot 3 \cdot 3$

$48 = 2 \cdot 2 \cdot 2 \cdot 2 \cdot 3$

$LCM = 2 \cdot 2 \cdot 2 \cdot 2 \cdot 3 \cdot 3$, or 144

5. $30 = 2 \cdot 3 \cdot 5$

$105 = 3 \cdot 5 \cdot 7$

$LCM = 2 \cdot 3 \cdot 5 \cdot 7$, or 210

7. $9 = 3 \cdot 3$

$15 = 3 \cdot 5$

$5 = 5$

$LCM = 3 \cdot 3 \cdot 5$, or 45

9.
$$\frac{5}{6} + \frac{4}{15} = \frac{5}{2 \cdot 3} + \frac{4}{3 \cdot 5}, \quad LCD = 2 \cdot 3 \cdot 5, \text{ or } 30$$
$$= \frac{5}{2 \cdot 3} \cdot \frac{5}{5} + \frac{4}{3 \cdot 5} \cdot \frac{2}{2}$$
$$= \frac{25}{2 \cdot 3 \cdot 5} + \frac{8}{2 \cdot 3 \cdot 5}$$
$$= \frac{33}{2 \cdot 3 \cdot 5} = \frac{\cancel{3} \cdot 11}{2 \cdot \cancel{3} \cdot 5}$$
$$= \frac{11}{10}$$

11.
$$\frac{7}{36} + \frac{1}{24}$$
$$= \frac{7}{2 \cdot 2 \cdot 3 \cdot 3} + \frac{1}{2 \cdot 2 \cdot 2 \cdot 3}, \quad LCD = 2 \cdot 2 \cdot 2 \cdot 3 \cdot 3, \text{ or } 72$$
$$= \frac{7}{2 \cdot 2 \cdot 3 \cdot 3} \cdot \frac{2}{2} + \frac{1}{2 \cdot 2 \cdot 2 \cdot 3} \cdot \frac{3}{3}$$
$$= \frac{14}{2 \cdot 2 \cdot 2 \cdot 3 \cdot 3} + \frac{3}{2 \cdot 2 \cdot 2 \cdot 3 \cdot 3}$$
$$= \frac{17}{2 \cdot 2 \cdot 2 \cdot 3 \cdot 3}$$
$$= \frac{17}{72}$$

13.
$$\frac{3}{4} + \frac{7}{30} + \frac{1}{16}$$
$$= \frac{3}{2 \cdot 2} + \frac{7}{2 \cdot 3 \cdot 5} + \frac{1}{2 \cdot 2 \cdot 2 \cdot 2}, \quad LCD = 2 \cdot 2 \cdot 2 \cdot 2 \cdot 3 \cdot 5$$
$$= \frac{3}{2 \cdot 2} \cdot \frac{2 \cdot 2 \cdot 3 \cdot 5}{2 \cdot 2 \cdot 3 \cdot 5} + \frac{7}{2 \cdot 3 \cdot 5} \cdot \frac{2 \cdot 2 \cdot 2}{2 \cdot 2 \cdot 2} +$$
$$\frac{1}{2 \cdot 2 \cdot 2 \cdot 2} \cdot \frac{3 \cdot 5}{3 \cdot 5}$$
$$= \frac{180}{2 \cdot 2 \cdot 2 \cdot 2 \cdot 3 \cdot 5} + \frac{56}{2 \cdot 2 \cdot 2 \cdot 2 \cdot 3 \cdot 5} + \frac{15}{2 \cdot 2 \cdot 2 \cdot 2 \cdot 3 \cdot 5}$$
$$= \frac{251}{2 \cdot 2 \cdot 2 \cdot 2 \cdot 3 \cdot 5}$$
$$= \frac{251}{240}$$

15. $21x^2y = 3 \cdot 7 \cdot x \cdot x \cdot y$

$7xy = 7 \cdot x \cdot y$

$LCM = 3 \cdot 7 \cdot x \cdot x \cdot y$, or $21x^2y$

17. $y^2 - 100 = (y+10)(y-10)$

$10y + 100 = 10(y+10)$

$LCM = 10(y+10)(y-10)$

19. $15ab^2 = 3 \cdot 5 \cdot a \cdot b \cdot b$

$3ab = 3 \cdot a \cdot b$

$10a^3b = 2 \cdot 5 \cdot a \cdot a \cdot a \cdot b$

$LCM = 2 \cdot 3 \cdot 5 \cdot a \cdot a \cdot a \cdot b \cdot b$, or $30a^3b^2$

21. $5y - 15 = 5(y-3)$

$y^2 - 6y + 9 = (y-3)(y-3)$

$LCM = 5(y-3)(y-3)$, or $5(y-3)^2$

23. $y^2 - 25 = (y+5)(y-5)$

$5 - y$

We can use $y - 5$ from the prime factorization of $y^2 - 25$ or $5 - y$ from the second expression, but not both.

$LCM = (y+5)(y-5)$, or $(y+5)(5-y)$

25. $2r^2 - 5r - 12 = (2r+3)(r-4)$

$3r^2 - 13r + 4 = (3r-1)(r-4)$

$r^2 - 16 = (r+4)(r-4)$

$LCM = (2r+3)(3r-1)(r+4)(r-4)$

27. $x^5 + 4x^3 = x^3(x^2 + 4) = x \cdot x \cdot x(x^2 + 4)$

$x^3 - 4x^2 + 4x = x(x^2 - 4x + 4) = x(x-2)(x-2)$

LCM $= x \cdot x \cdot x(x-2)(x-2)(x^2+4)$, or $x^3(x-2)(x-2)(x^2+4)$

29. $x^5 - 2x^4 + x^3 = x^3(x^2 - 2x + 1) = x \cdot x \cdot x(x-1)(x-1)$

$2x^3 + 2x = 2x(x^2 + 1)$

$5x + 5 = 5(x + 1)$

LCM $= 2 \cdot 5 \cdot x \cdot x \cdot x(x-1)(x-1)(x+1)(x^2+1)$, or $10x^3(x-1)(x-1)(x+1)(x^2+1)$

31. $\dfrac{x - 2y}{x + y} + \dfrac{x + 9y}{x + y}$

$= \dfrac{x - 2y + x + 9y}{x + y}$ Adding the numerators

$= \dfrac{2x + 7y}{x + y}$

33. $\dfrac{4y + 2}{y - 2} - \dfrac{y - 3}{y - 2}$

$= \dfrac{4y + 2 - (y - 3)}{y - 2}$ Subtracting numerators

$= \dfrac{4y + 2 - y + 3}{y - 2}$

$= \dfrac{3y + 5}{y - 2}$

35. $\dfrac{a^2}{a - b} + \dfrac{b^2}{b - a}$

$= \dfrac{a^2}{a - b} + \dfrac{b^2}{b - a} \cdot \dfrac{-1}{-1}$ Multiplying by 1, using $\dfrac{-1}{-1}$

$= \dfrac{a^2}{a - b} + \dfrac{-b^2}{a - b}$

$= \dfrac{a^2 - b^2}{a - b}$ Adding numerators

$= \dfrac{(a + b)(a - b)}{a - b}$ Factoring the numerator

$= \dfrac{(a + b)(a - b)}{1(a - b)}$ Removing a factor of 1

$= \dfrac{a + b}{1}$

$= a + b$

37. $\dfrac{6}{y} - \dfrac{7}{-y}$

$= \dfrac{6}{y} - \dfrac{7}{-y} \cdot \dfrac{-1}{-1}$ Multiplying by 1, using $\dfrac{-1}{-1}$

$= \dfrac{6}{y} - \dfrac{-7}{y}$

$= \dfrac{6 - (-7)}{y}$ Subtracting numerators

$= \dfrac{13}{y}$

39. $\dfrac{4a - 2}{a^2 - 49} + \dfrac{5 + 3a}{49 - a^2}$

$= \dfrac{4a - 2}{a^2 - 49} + \dfrac{5 + 3a}{49 - a^2} \cdot \dfrac{-1}{-1}$ Multiplying by 1, using $\dfrac{-1}{-1}$

$= \dfrac{4a - 2}{a^2 - 49} + \dfrac{-5 - 3a}{a^2 - 49}$

$= \dfrac{4a - 2 - 5 - 3a}{a^2 - 49}$ Adding numerators

$= \dfrac{a - 7}{a^2 - 49}$

$= \dfrac{a - 7}{(a + 7)(a - 7)}$ Factoring

$= \dfrac{(a - 7) \cdot 1}{(a + 7)(a - 7)}$ Removing a factor of 1

$= \dfrac{1}{a + 7}$

41. $\dfrac{y - 2}{y + 4} + \dfrac{y + 3}{y - 5}$ LCD $= (y + 4)(y - 5)$

$= \dfrac{y - 2}{y + 4} \cdot \dfrac{y - 5}{y - 5} + \dfrac{y + 3}{y - 5} \cdot \dfrac{y + 4}{y + 4}$

$= \dfrac{(y^2 - 7y + 10) + (y^2 + 7y + 12)}{(y + 4)(y - 5)}$

$= \dfrac{2y^2 + 22}{(y + 4)(y - 5)}$, or $\dfrac{2y^2 + 22}{y^2 - y - 20}$

43. $\dfrac{4xy}{x^2 - y^2} + \dfrac{x - y}{x + y}$

$= \dfrac{4xy}{(x + y)(x - y)} + \dfrac{x - y}{x + y}$ LCD $= (x + y)(x - y)$

$= \dfrac{4xy}{(x + y)(x - y)} + \dfrac{x - y}{x + y} \cdot \dfrac{x - y}{x - y}$

$= \dfrac{4xy + x^2 - 2xy + y^2}{(x + y)(x - y)}$

$= \dfrac{x^2 + 2xy + y^2}{(x + y)(x - y)} = \dfrac{(x + y)(x + y)}{(x + y)(x - y)}$

$= \dfrac{(x + y)(x + y)}{(x + y)(x - y)} = \dfrac{x + y}{x - y}$

45. $\dfrac{9x + 2}{3x^2 - 2x - 8} + \dfrac{7}{3x^2 + x - 4}$

$= \dfrac{9x + 2}{(3x + 4)(x - 2)} + \dfrac{7}{(3x + 4)(x - 1)}$

LCD $= (3x + 4)(x - 2)(x - 1)$

$= \dfrac{9x + 2}{(3x + 4)(x - 2)} \cdot \dfrac{x - 1}{x - 1} + \dfrac{7}{(3x + 4)(x - 1)} \cdot \dfrac{x - 2}{x - 2}$

$= \dfrac{9x^2 - 7x - 2 + 7x - 14}{(3x + 4)(x - 2)(x - 1)}$

$= \dfrac{9x^2 - 16}{(3x + 4)(x - 2)(x - 1)} = \dfrac{(3x + 4)(3x - 4)}{(3x + 4)(x - 2)(x - 1)}$

$= \dfrac{(3x + 4)(3x - 4)}{(3x + 4)(x - 2)(x - 1)}$

$= \dfrac{3x - 4}{(x - 2)(x - 1)}$, or $\dfrac{3x - 4}{x^2 - 3x + 2}$

47. $\dfrac{4}{x+1} + \dfrac{x+2}{x^2-1} + \dfrac{3}{x-1}$

$= \dfrac{4}{x+1} + \dfrac{x+2}{(x+1)(x-1)} + \dfrac{3}{x-1}$

$\qquad\qquad\qquad$ LCD $= (x+1)(x-1)$

$= \dfrac{4}{x+1} \cdot \dfrac{x-1}{x-1} + \dfrac{x+2}{(x+1)(x-1)} + \dfrac{3}{x-1} \cdot \dfrac{x+1}{x+1}$

$= \dfrac{4x-4+x+2+3x+3}{(x+1)(x-1)}$

$= \dfrac{8x+1}{(x+1)(x-1)},$ or $\dfrac{8x+1}{x^2-1}$

49. $\dfrac{x-1}{3x+15} - \dfrac{x+3}{5x+25}$

$= \dfrac{x-1}{3(x+5)} - \dfrac{x+3}{5(x+5)}$

$\qquad\qquad\quad$ LCD $= 3 \cdot 5(x+5),$ or $15(x+5)$

$= \dfrac{x-1}{3(x+5)} \cdot \dfrac{5}{5} - \dfrac{x+3}{5(x+5)} \cdot \dfrac{3}{3}$

$= \dfrac{5x-5-(3x+9)}{15(x+5)}$

$= \dfrac{5x-5-3x-9}{15(x+5)}$

$= \dfrac{2x-14}{15(x+5)},$ or $\dfrac{2x-14}{15x+75}$

51. $\dfrac{5ab}{a^2-b^2} - \dfrac{a-b}{a+b}$

$= \dfrac{5ab}{(a+b)(a-b)} - \dfrac{a-b}{a+b} \qquad$ LCD $= (a+b)(a-b)$

$= \dfrac{5ab}{(a+b)(a-b)} - \dfrac{a-b}{a+b} \cdot \dfrac{a-b}{a-b}$

$= \dfrac{5ab-(a^2-2ab+b^2)}{(a+b)(a-b)}$

$= \dfrac{5ab-a^2+2ab-b^2}{(a+b)(a-b)}$

$= \dfrac{-a^2+7ab-b^2}{(a+b)(a-b)},$ or $\dfrac{-a^2+7ab-b^2}{a^2-b^2}$

53. $\dfrac{3y}{y^2-7y+10} - \dfrac{2y}{y^2-8y+15}$

$= \dfrac{3y}{(y-5)(y-2)} - \dfrac{2y}{(y-5)(y-3)}$

$\qquad\qquad\quad$ LCD $= (y-5)(y-2)(y-3)$

$= \dfrac{3y}{(y-5)(y-2)} \cdot \dfrac{y-3}{y-3} - \dfrac{2y}{(y-5)(y-3)} \cdot \dfrac{y-2}{y-2}$

$= \dfrac{3y^2-9y-(2y^2-4y)}{(y-5)(y-2)(y-3)}$

$= \dfrac{3y^2-9y-2y^2+4y}{(y-5)(y-2)(y-3)}$

$= \dfrac{y^2-5y}{(y-5)(y-2)(y-3)} = \dfrac{y(y-5)}{(y-5)(y-2)(y-3)}$

$= \dfrac{y(y-5)}{(y-5)(y-2)(y-3)}$

$= \dfrac{y}{(y-2)(y-3)},$ or $\dfrac{y}{y^2-5y+6}$

55. $\dfrac{y}{y^2-y-20} + \dfrac{2}{y+4}$

$= \dfrac{y}{(y-5)(y+4)} + \dfrac{2}{y+4} \qquad$ LCD $= (y-5)(y+4)$

$= \dfrac{y}{(y-5)(y+4)} + \dfrac{2}{y+4} \cdot \dfrac{y-5}{y-5}$

$= \dfrac{y+2y-10}{(y-5)(y+4)}$

$= \dfrac{3y-10}{(y-5)(y+4)},$ or $\dfrac{3y-10}{y^2-y-20}$

57. $\dfrac{3y+2}{y^2+5y-24} + \dfrac{7}{y^2+4y-32}$

$= \dfrac{3y+2}{(y+8)(y-3)} + \dfrac{7}{(y+8)(y-4)}$

$\qquad\qquad\quad$ LCD $= (y+8)(y-3)(y-4)$

$= \dfrac{3y+2}{(y+8)(y-3)} \cdot \dfrac{y-4}{y-4} + \dfrac{7}{(y+8)(y-4)} \cdot \dfrac{y-3}{y-3}$

$= \dfrac{3y^2-10y-8+7y-21}{(y+8)(y-3)(y-4)}$

$= \dfrac{3y^2-3y-29}{(y+8)(y-3)(y-4)}$

59. $\dfrac{3x-1}{x^2+2x-3} - \dfrac{x+4}{x^2-9}$

$= \dfrac{3x-1}{(x+3)(x-1)} - \dfrac{x+4}{(x+3)(x-3)}$

$\qquad\qquad\quad$ LCD $= (x+3)(x-1)(x-3)$

$= \dfrac{3x-1}{(x+3)(x-1)} \cdot \dfrac{x-3}{x-3} - \dfrac{x+4}{(x+3)(x-3)} \cdot \dfrac{x-1}{x-1}$

$= \dfrac{3x^2-10x+3-(x^2+3x-4)}{(x+3)(x-1)(x-3)}$

$= \dfrac{3x^2-10x+3-x^2-3x+4}{(x+3)(x-1)(x-3)}$

$= \dfrac{2x^2-13x+7}{(x+3)(x-1)(x-3)}$

61. $\dfrac{1}{x+1} - \dfrac{x}{x-2} + \dfrac{x^2+2}{x^2-x-2}$

$= \dfrac{1}{x+1} - \dfrac{x}{x-2} + \dfrac{x^2+2}{(x-2)(x+1)}$

$\qquad \text{LCD} = (x+1)(x-2)$

$= \dfrac{1}{x+1} \cdot \dfrac{x-2}{x-2} - \dfrac{x}{x-2} \cdot \dfrac{x+1}{x+1} + \dfrac{x^2+2}{(x-2)(x+1)}$

$= \dfrac{x-2-(x^2+x)+x^2+2}{(x+1)(x-2)}$

$= \dfrac{x-2-x^2-x+x^2+2}{(x+1)(x-2)}$

$= \dfrac{0}{(x+1)(x-2)}$

$= 0$

63. $\dfrac{x-1}{x-2} - \dfrac{x+1}{x+2} + \dfrac{x-6}{x^2-4}$

$= \dfrac{x-1}{x-2} - \dfrac{x+1}{x+2} + \dfrac{x-6}{(x+2)(x-2)}$

$\qquad \text{LCD} = (x-2)(x+2)$

$= \dfrac{x-1}{x-2} \cdot \dfrac{x+2}{x+2} - \dfrac{x+1}{x+2} \cdot \dfrac{x-2}{x-2} + \dfrac{x-6}{(x+2)(x-2)}$

$= \dfrac{(x^2+x-2)-(x^2-x-2)+(x-6)}{(x-2)(x+2)}$

$= \dfrac{x^2+x-2-x^2+x+2+x-6}{(x-2)(x+2)}$

$= \dfrac{3x-6}{(x-2)(x+2)}$

$= \dfrac{3(x-2)}{(x-2)(x+2)}$

$= \dfrac{3\cancel{(x-2)}}{\cancel{(x-2)}(x+2)}$

$= \dfrac{3}{x+2}$

65. $\dfrac{4x}{x^2-1} + \dfrac{3x}{1-x} - \dfrac{4}{x-1}$

$= \dfrac{4x}{x^2-1} + \dfrac{3x}{1-x} \cdot \dfrac{-1}{-1} - \dfrac{4}{x-1}$

$= \dfrac{4x}{(x+1)(x-1)} + \dfrac{-3x}{x-1} - \dfrac{4}{x-1}$

$\qquad \text{LCD} = (x+1)(x-1)$

$= \dfrac{4x}{(x+1)(x-1)} + \dfrac{-3x}{x-1} \cdot \dfrac{x+1}{x+1} - \dfrac{4}{x-1} \cdot \dfrac{x+1}{x+1}$

$= \dfrac{4x-3x^2-3x-4x-4}{(x+1)(x-1)}$

$= \dfrac{-3x^2-3x-4}{x^2-1}$

67. $\dfrac{5}{3-2x} - \dfrac{3}{2x-3} + \dfrac{x-3}{2x^2-x-3}$

$= \dfrac{5}{3-2x} \cdot \dfrac{-1}{-1} - \dfrac{3}{2x-3} + \dfrac{x-3}{2x^2-x-3}$

$= \dfrac{-5}{2x-3} - \dfrac{3}{2x-3} + \dfrac{x-3}{(2x-3)(x+1)}$

$\qquad \text{LCD} = (2x-3)(x+1)$

$= \dfrac{-5}{2x-3} \cdot \dfrac{x+1}{x+1} - \dfrac{3}{2x-3} \cdot \dfrac{x+1}{x+1} + \dfrac{x-3}{(2x-3)(x+1)}$

$= \dfrac{-5x-5-3x-3+x-3}{(2x-3)(x+1)}$

$= \dfrac{-7x-11}{2x^2-x-3}$

69. $\dfrac{3}{2c-1} - \dfrac{1}{c+2} + \dfrac{5}{2c^2+3c-2}$

$= \dfrac{3}{2c-1} - \dfrac{1}{c+2} + \dfrac{5}{(2c-1)(c+2)}$

$\qquad \text{LCD} = (2c-1)(c+2)$

$= \dfrac{3}{2c-1} \cdot \dfrac{c+2}{c+2} - \dfrac{1}{c+2} \cdot \dfrac{2c-1}{2c-1} + \dfrac{5}{(2c-1)(c+2)}$

$= \dfrac{3c+6-2c+1+5}{(2c-1)(c+2)}$

$= \dfrac{c+12}{2c^2+3c-2}$

71. $\dfrac{1}{x+y} + \dfrac{1}{y-x} - \dfrac{2x}{x^2-y^2}$

$= \dfrac{1}{x+y} + \dfrac{1}{y-x} \cdot \dfrac{-1}{-1} - \dfrac{2x}{x^2-y^2}$

$= \dfrac{1}{x+y} + \dfrac{-1}{x-y} - \dfrac{2x}{(x+y)(x-y)}$

$\qquad \text{LCD} = (x+y)(x-y)$

$= \dfrac{1}{x+y} \cdot \dfrac{x-y}{x-y} + \dfrac{-1}{x-y} \cdot \dfrac{x+y}{x+y} - \dfrac{2x}{(x+y)(x-y)}$

$= \dfrac{x-y-x-y-2x}{(x+y)(x-y)}$

$= \dfrac{-2x-2y}{(x+y)(x-y)} = \dfrac{-2(x+y)}{(x+y)(x-y)}$

$= \dfrac{-2\cancel{(x+y)}}{\cancel{(x+y)}(x-y)}$

$= \dfrac{-2}{x-y}, \text{ or } \dfrac{2}{y-x}$

73. Graph: $2x-3y > 6$

We first graph the line $2x-3y = 6$. The intercepts are $(0,-2)$ and $(3,0)$. We draw the line dashed since the inequality symbol is $>$. To determine which half-plane to shade, we consider a test point not on the line. We try $(0,0)$:

$$\begin{array}{c|c} \multicolumn{2}{c}{2x-3y > 6} \\ \hline 2 \cdot 0 - 3 \cdot 0 & 6 \\ 0 & \text{FALSE} \end{array}$$

Since $0 > 6$ is false, we shade the half-plane that does not contain $(0,0)$.

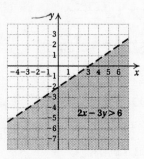

75. Graph: $5x + 3y \leq 15$

We first graph the line $5x + 3y = 15$. The intercepts are $(3,0)$ and $(0,5)$. We draw the line solid since the inequality symbol is $\leq$. To determine which half-plane to shade, we consider a test point not on the line. We try $(0,0)$:

$$\frac{5x + 3y \leq 15}{5 \cdot 0 + 3 \cdot 0 \mid 15}$$
$$0 \mid$$
$$\text{TRUE}$$

Since $0 \leq 15$ is true, we shade the half-plane that contains $(0,0)$.

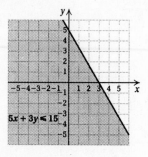

77. $x^2 - 25 = x^2 - 5^2$ Difference of squares
$$= (x+5)(x-5)$$

79. $18 = 2 \cdot 3 \cdot 3$
$42 = 2 \cdot 3 \cdot 7$
$82 = 2 \cdot 41$
$120 = 2 \cdot 2 \cdot 2 \cdot 3 \cdot 5$
$300 = 2 \cdot 2 \cdot 3 \cdot 5 \cdot 5$
$700 = 2 \cdot 2 \cdot 5 \cdot 5 \cdot 7$
LCM $= 2 \cdot 2 \cdot 2 \cdot 3 \cdot 3 \cdot 5 \cdot 5 \cdot 7 \cdot 41$, or $516{,}600$

81. $2a^3b^7 = 2 \cdot a \cdot a \cdot a \cdot b \cdot b \cdot b \cdot b \cdot b \cdot b \cdot b$

Other expression $= ?$

LCM $= 8a^4b^7$, or $2 \cdot 2 \cdot 2 \cdot a \cdot a \cdot a \cdot a \cdot b \cdot b \cdot b \cdot b \cdot b \cdot b \cdot b$

The other expression must contain three factors of 2 and four factors of a, or $8a^4$. It may also contain from 0 to 7 factors of b. The possibilities are:

$8a^4$, $8a^4b$, $8a^4b^2$, $8a^4b^3$, $8a^4b^4$, $8a^4b^5$, $8a^4b^6$, or $8a^4b^7$.

83.
$$\frac{x+y+1}{y-(x+1)} + \frac{x+y-1}{x-(y-1)} - \frac{x-y-1}{1-(y-x)}$$
$$= \frac{x+y+1}{-x+y-1} + \frac{x+y-1}{x-y+1} - \frac{x-y-1}{x-y+1}$$
$$= \frac{x+y+1}{-x+y-1} \cdot \frac{-1}{-1} + \frac{x+y-1}{x-y+1} - \frac{x-y-1}{x-y+1}$$
$$= \frac{-x-y-1}{x-y+1} + \frac{x+y-1}{x-y+1} - \frac{x-y-1}{x-y+1}$$
$$= \frac{-x-y-1+x+y-1-(x-y-1)}{x-y+1}$$
$$= \frac{-x-y-1+x+y-1-x+y+1}{x-y+1}$$
$$= \frac{-x+y-1}{x-y+1}$$
$$= \frac{-(x-y+1)}{x-y+1}$$
$$= \frac{-1 \cdot (x-y+1)}{1 \cdot (x-y+1)}$$
$$= \frac{-1}{1} \cdot \frac{x-y+1}{x-y+1}$$
$$= -1$$

85.
$$\frac{x}{x^4 - y^4} - \frac{1}{x^2 + 2xy + y^2}$$
$$= \frac{x}{(x^2+y^2)(x+y)(x-y)} - \frac{1}{(x+y)(x+y)}$$
$$= \frac{x}{(x^2+y^2)(x+y)(x-y)} \cdot \frac{x+y}{x+y} -$$
$$\frac{1}{(x+y)(x+y)} \cdot \frac{(x^2+y^2)(x-y)}{(x^2+y^2)(x-y)}$$
$$= \frac{x^2+xy}{(x^2+y^2)(x+y)(x-y)(x+y)} -$$
$$\frac{x^3 - x^2y + xy^2 - y^3}{(x^2+y^2)(x+y)(x-y)(x+y)}$$
$$= \frac{x^2+xy-(x^3-x^2y+xy^2-y^3)}{(x^2+y^2)(x+y)(x-y)(x+y)}$$
$$= \frac{x^2+xy-x^3+x^2y-xy^2+y^3}{(x^2+y^2)(x+y)(x-y)(x+y)}$$

Exercise Set 6.3

1.
$$\frac{y}{10} = \frac{2}{5} + \frac{3}{8}, \quad \text{LCM is } 40$$
$$40 \cdot \frac{y}{10} = 40 \cdot \left(\frac{2}{5} + \frac{3}{8}\right) \quad \text{Multiplying by the LCM}$$
$$4y = 40 \cdot \frac{2}{5} + 40 \cdot \frac{3}{8} \quad \text{Removing parentheses}$$
$$4y = 16 + 15$$
$$4y = 31$$
$$y = \frac{31}{4}$$

Check: $\dfrac{y}{10} = \dfrac{2}{5} + \dfrac{3}{8}$

$$
\begin{array}{c|c}
\dfrac{\frac{31}{4}}{10} & \dfrac{2}{5} + \dfrac{3}{8} \\[2mm]
\dfrac{31}{4} \cdot \dfrac{1}{10} & \dfrac{16}{40} + \dfrac{15}{40} \\[2mm]
\dfrac{31}{40} & \dfrac{31}{40} \qquad \text{TRUE}
\end{array}
$$

The solution is $\dfrac{31}{4}$.

3. $\qquad \dfrac{1}{4} - \dfrac{5}{6} = \dfrac{1}{a}$, LCM is $12a$

$12a \cdot \left(\dfrac{1}{4} - \dfrac{5}{6} \right) = 12a \cdot \dfrac{1}{a}$ Multiplying by the LCM

$12a \cdot \dfrac{1}{4} - 12a \cdot \dfrac{5}{6} = 12$

$3a - 10a = 12$

$-7a = 12$

$a = -\dfrac{12}{7}$

Check: $\dfrac{1}{4} - \dfrac{5}{6} = \dfrac{1}{a}$

$$
\begin{array}{c|c}
\dfrac{1}{4} - \dfrac{5}{6} & \dfrac{1}{-\frac{12}{7}} \\[3mm]
\dfrac{3}{12} - \dfrac{10}{12} & 1 \cdot \left(-\dfrac{7}{12} \right) \\[3mm]
-\dfrac{7}{12} & -\dfrac{7}{12} \qquad \text{TRUE}
\end{array}
$$

The solution is $-\dfrac{12}{7}$.

5. $\qquad \dfrac{x}{3} - \dfrac{x}{4} = 12$, LCM is 12

$12 \cdot \left(\dfrac{x}{3} - \dfrac{x}{4} \right) = 12 \cdot 12$

$12 \cdot \dfrac{x}{3} - 12 \cdot \dfrac{x}{4} = 144$

$4x - 3x = 144$

$x = 144$

Check: $\dfrac{x}{3} - \dfrac{x}{4} = 12$

$$
\begin{array}{c|c}
\dfrac{144}{3} - \dfrac{144}{4} & 12 \\[3mm]
48 - 36 & \\[2mm]
12 & \text{TRUE}
\end{array}
$$

The solution is 144.

7. $\qquad x + \dfrac{8}{x} = -9$, LCM is x

$x \left(x + \dfrac{8}{x} \right) = x(-9)$

$x \cdot x + x \cdot \dfrac{8}{x} = -9x$

$x^2 + 8 = -9x$

$x^2 + 9x + 8 = 0$

$(x + 1)(x + 8) = 0$

$x + 1 = 0 \quad \text{or} \quad x + 8 = 0$ Principle of zero products

$x = -1 \quad \text{or} \qquad x = -8$

Check:

For -1:

$x + \dfrac{8}{x} = -9$

$$
\begin{array}{c|c}
-1 + \dfrac{8}{-1} & -9 \\[3mm]
-1 - 8 & \\[2mm]
-9 & \text{TRUE}
\end{array}
$$

For -8:

$x + \dfrac{8}{x} = -9$

$$
\begin{array}{c|c}
-8 + \dfrac{8}{-8} & -9 \\[3mm]
-8 - 1 & \\[2mm]
-9 & \text{TRUE}
\end{array}
$$

The solutions are -1 and -8.

9. $\qquad \dfrac{3}{y} + \dfrac{7}{y} = 5$, LCM is y

$y \left(\dfrac{3}{y} + \dfrac{7}{y} \right) = y \cdot 5$

$y \cdot \dfrac{3}{y} + y \cdot \dfrac{7}{y} = 5y$

$3 + 7 = 5y$

$10 = 5y$

$2 = y$

Check: $\dfrac{3}{y} + \dfrac{7}{y} = 5$

$$
\begin{array}{c|c}
\dfrac{3}{2} + \dfrac{7}{2} & 5 \\[3mm]
\dfrac{10}{2} & \\[2mm]
5 & \text{TRUE}
\end{array}
$$

The solution is 2.

11. $\qquad \dfrac{1}{2} = \dfrac{z - 5}{z + 1}$, LCM is $2(z + 1)$

$2(z + 1) \cdot \dfrac{1}{2} = 2(z + 1) \cdot \dfrac{z - 5}{z + 1}$

$z + 1 = 2(z - 5)$

$z + 1 = 2z - 10$

$11 = z$

Check: $\dfrac{1}{2} = \dfrac{z-5}{z+1}$

$\dfrac{1}{2}$	$\dfrac{11-5}{11+1}$
	$\dfrac{6}{12}$
	$\dfrac{1}{2}$ TRUE

The solution is 11.

13. $\qquad \dfrac{3}{y+1} = \dfrac{2}{y-3},$ LCM is $(y+1)(y-3)$

$$(y+1)(y-3) \cdot \dfrac{3}{y+1} = (y+1)(y-3) \cdot \dfrac{2}{y-3}$$

$$3(y-3) = 2(y+1)$$

$$3y - 9 = 2y + 2$$

$$y = 11$$

Check: $\qquad \dfrac{3}{y+1} = \dfrac{2}{y-3}$

$\dfrac{3}{11+1}$	$\dfrac{2}{11-3}$
$\dfrac{3}{12}$	$\dfrac{2}{8}$
$\dfrac{1}{4}$	$\dfrac{1}{4}$ TRUE

The solution is 11.

15. $\qquad \dfrac{y-1}{y-3} = \dfrac{2}{y-3},$ LCM is $y-3$

$$(y-3) \cdot \dfrac{y-1}{y-3} = (y-3) \cdot \dfrac{2}{y-3}$$

$$y - 1 = 2$$

$$y = 3$$

Check: $\dfrac{y-1}{y-3} = \dfrac{2}{y-3}$

$\dfrac{3-1}{3-3}$	$\dfrac{2}{3-3}$
$\dfrac{2}{0}$	$\dfrac{2}{0}$ UNDEFINED

We know that 3 is not a solution of the original equation, because it results in division by 0. The equation has no solution.

17. $\qquad \dfrac{x+1}{x} = \dfrac{3}{2},$ LCM is $2x$

$$2x \cdot \dfrac{x+1}{x} = 2x \cdot \dfrac{3}{2}$$

$$2(x+1) = x \cdot 3$$

$$2x + 2 = 3x$$

$$2 = x$$

Check: $\dfrac{x+1}{x} = \dfrac{3}{2}$

$\dfrac{2+1}{2}$	$\dfrac{3}{2}$
$\dfrac{3}{2}$	TRUE

The solution is 2.

19. $\qquad 10 = \dfrac{5}{x} - \dfrac{6}{x} + \dfrac{8}{x},$ LCM is x

$$x \cdot 10 = x\left(\dfrac{5}{x} - \dfrac{6}{x} + \dfrac{8}{x}\right)$$

$$10x = x \cdot \dfrac{5}{x} - x \cdot \dfrac{6}{x} + x \cdot \dfrac{8}{x}$$

$$10x = 5 - 6 + 8$$

$$10x = 7$$

$$x = \dfrac{7}{10}$$

Check: $\qquad 10 = \dfrac{5}{x} - \dfrac{6}{x} + \dfrac{8}{x}$

10	$\dfrac{5}{\frac{7}{10}} - \dfrac{6}{\frac{7}{10}} + \dfrac{8}{\frac{7}{10}}$
	$5 \cdot \dfrac{10}{7} - 6 \cdot \dfrac{10}{7} + 8 \cdot \dfrac{10}{7}$
	$\dfrac{50}{7} - \dfrac{60}{7} + \dfrac{80}{7}$
	$\dfrac{70}{7}$
	10 TRUE

The solution is $\dfrac{7}{10}$.

21. $\qquad \dfrac{1}{2} - \dfrac{4}{9x} = \dfrac{4}{9} - \dfrac{1}{6x},$ LCM is $18x$

$$18x\left(\dfrac{1}{2} - \dfrac{4}{9x}\right) = 18x\left(\dfrac{4}{9} - \dfrac{1}{6x}\right)$$

$$18x \cdot \dfrac{1}{2} - 18x \cdot \dfrac{4}{9x} = 18x \cdot \dfrac{4}{9} - 18x \cdot \dfrac{1}{6x}$$

$$9x - 8 = 8x - 3$$

$$x = 5$$

Since 5 checks, it is the solution.

23. $\qquad \dfrac{60}{x} - \dfrac{60}{x-5} = \dfrac{2}{x},$ LCM is $x(x-5)$

$$x(x-5)\left(\dfrac{60}{x} - \dfrac{60}{x-5}\right) = x(x-5) \cdot \dfrac{2}{x}$$

$$60(x-5) - 60x = 2(x-5)$$

$$60x - 300 - 60x = 2x - 10$$

$$-300 = 2x - 10$$

$$-290 = 2x$$

$$-145 = x$$

Since -145 checks, it is the solution.

25. $\dfrac{7}{5x-2} = \dfrac{5}{4x}$, LCM is $4x(5x-2)$

$$4x(5x-2) \cdot \dfrac{7}{5x-2} = 4x(5x-2) \cdot \dfrac{5}{4x}$$

$$4x \cdot 7 = 5(5x-2)$$

$$28x = 25x - 10$$

$$3x = -10$$

$$x = -\dfrac{10}{3}$$

Since $-\dfrac{10}{3}$ checks, it is the solution.

27. $$\dfrac{x}{x-2} + \dfrac{x}{x^2-4} = \dfrac{x+3}{x+2}$$

$$\dfrac{x}{x-2} + \dfrac{x}{(x+2)(x-2)} = \dfrac{x+3}{x+2}$$

LCM is $(x+2)(x-2)$

$$(x+2)(x-2)\Big(\dfrac{x}{x-2} + \dfrac{x}{(x+2)(x-2)}\Big) =$$

$$(x+2)(x-2) \cdot \dfrac{x+3}{x+2}$$

$$x(x+2) + x =$$

$$(x-2)(x+3)$$

$$x^2 + 2x + x = x^2 + x - 6$$

$$3x = x - 6$$

$$2x = -6$$

$$x = -3$$

Since -3 checks, it is the solution.

29. $$\dfrac{6}{x^2-4x+3} - \dfrac{1}{x-3} = \dfrac{1}{4x-4}$$

$$\dfrac{6}{(x-3)(x-1)} - \dfrac{1}{x-3} = \dfrac{1}{4(x-1)}$$

LCM is $4(x-3)(x-1)$

$$4(x-3)(x-1)\Big(\dfrac{6}{(x-3)(x-1)} - \dfrac{1}{x-3}\Big) =$$

$$4(x-3)(x-1) \cdot \dfrac{1}{4(x-1)}$$

$$4 \cdot 6 - 4(x-1) = x-3$$

$$24 - 4x + 4 \quad x - 3$$

$$-5x = -31$$

$$x = \dfrac{31}{5}$$

Since $\dfrac{31}{5}$ checks, it is the solution.

31. $$\dfrac{5}{y+3} = \dfrac{1}{4y^2-36} + \dfrac{2}{y-3}$$

$$\dfrac{5}{y+3} = \dfrac{1}{4(y+3)(y-3)} + \dfrac{2}{y-3}$$

LCM is $4(y+3)(y-3)$

$$4(y+3)(y-3) \cdot \dfrac{5}{y+3} =$$

$$4(y+3)(y-3)\Big(\dfrac{1}{4(y+3)(y-3)} + \dfrac{2}{y-3}\Big)$$

$$4 \cdot 5(y-3) = 1 + 4 \cdot 2(y+3)$$

$$20y - 60 = 1 + 8y + 24$$

$$12y = 85$$

$$y = \dfrac{85}{12}$$

Since $\dfrac{85}{12}$ checks, it is the solution.

33. $$\dfrac{a}{2a-6} - \dfrac{3}{a^2-6a+9} = \dfrac{a-2}{3a-9}$$

$$\dfrac{a}{2(a-3)} - \dfrac{3}{(a-3)(a-3)} = \dfrac{a-2}{3(a-3)}$$

LCM is $2 \cdot 3(a-3)(a-3)$

$$6(a-3)(a-3)\Big(\dfrac{a}{2(a-3)} - \dfrac{3}{(a-3)(a-3)}\Big) =$$

$$6(a-3)(a-3) \cdot \dfrac{a-2}{3(a-3)}$$

$$3a(a-3) - 6 \cdot 3 = 2(a-3)(a-2)$$

$$3a^2 - 9a - 18 = 2(a^2 - 5a + 6)$$

$$3a^2 - 9a - 18 = 2a^2 - 10a + 12$$

$$a^2 + a - 30 = 0$$

$$(a+6)(a-5) = 0$$

$$a + 6 = 0 \quad \text{or} \quad a - 5 = 0$$

$$a = -6 \quad \text{or} \quad a = 5$$

Both -6 and 5 check. The solutions are -6 and 5.

35. $$\dfrac{2x+3}{x-1} = \dfrac{10}{x^2-1} + \dfrac{2x-3}{x+1}$$

$$\dfrac{2x+3}{x-1} = \dfrac{10}{(x+1)(x-1)} + \dfrac{2x-3}{x+1}$$

LCM is $(x+1)(x-1)$

$$(x+1)(x-1) \cdot \dfrac{2x+3}{x-1} =$$

$$(x+1)(x-1)\Big(\dfrac{10}{(x+1)(x-1)} + \dfrac{2x-3}{x+1}\Big)$$

$$(x+1)(2x+3) = 10 + (x-1)(2x-3)$$

$$2x^2 + 5x + 3 = 10 + 2x^2 - 5x + 3$$

$$5x + 3 = 13 - 5x$$

$$10x = 10$$

$$x = 1$$

We know that 1 is not a solution of the original equation, because it results in division by 0. The equation has no solution.

37. $\dfrac{4}{x+3} + \dfrac{7}{x^2-3x+9} = \dfrac{108}{x^3+27}$

Note: $x^3 + 27 = (x+3)(x^2-3x+9)$

Thus the LCM is $(x+3)(x^2-3x+9)$.

$$(x+3)(x^2-3x+9)\left(\dfrac{4}{x+3} + \dfrac{7}{x^2-3x+9}\right) =$$

$$(x+3)(x^2-3x+9)\left(\dfrac{108}{(x+3)(x^2-3x+9)}\right)$$

$$4(x^2-3x+9) + 7(x+3) = 108$$

$$4x^2 - 12x + 36 + 7x + 21 = 108$$

$$4x^2 - 5x - 51 = 0$$

$$(4x-17)(x+3) = 0$$

$$4x - 17 = 0 \quad \text{or} \quad x + 3 = 0$$

$$4x = 17 \quad \text{or} \qquad x = -3$$

$$x = \dfrac{17}{4} \quad \text{or} \qquad x = -3$$

We know that -3 is not a solution of the original equation, because it results in division by 0. Since $\dfrac{17}{4}$ checks, it is the solution.

39. $\dfrac{5x}{x-7} - \dfrac{35}{x+7} = \dfrac{490}{x^2-49}$

$$\dfrac{5x}{x-7} - \dfrac{35}{x+7} = \dfrac{490}{(x+7)(x-7)}$$

$$\text{LCM is } (x+7)(x-7)$$

$$(x+7)(x-7)\left(\dfrac{5x}{x-7} - \dfrac{35}{x+7}\right) =$$

$$(x+7)(x-7)\cdot\dfrac{490}{(x+7)(x-7)}$$

$$5x(x+7) - 35(x-7) = 490$$

$$5x^2 + 35x - 35x + 245 = 490$$

$$5x^2 + 245 = 490$$

$$5x^2 - 245 = 0$$

$$5(x^2-49) = 0$$

$$5(x+7)(x-7) = 0$$

$$x + 7 = 0 \quad \text{or} \quad x - 7 = 0$$

$$x = -7 \quad \text{or} \qquad x = 7$$

The numbers -7 and 7 are possible solutions. Each makes a denominator 0, so the equation has no solution.

41. $\dfrac{x-1}{3} + \dfrac{6x+1}{15} + \dfrac{2(x-2)}{13-7x} = \dfrac{2(x+2)}{5}$,

$$\text{LCM is } 15(13-7x)$$

$$15(13-7x)\left(\dfrac{x-1}{3} + \dfrac{6x+1}{15} + \dfrac{2(x-2)}{13-7x}\right) =$$

$$15(13-7x)\cdot\dfrac{2(x+2)}{5}$$

$$5(13-7x)(x-1) + (13-7x)(6x+1) + 15\cdot 2(x-2) =$$
$$3(13-7x)(2)(x+2)$$

$$5(-7x^2+20x-13) + (-42x^2+71x+13) + 30x - 60 =$$
$$6(-7x^2-x+26)$$

$$-35x^2+100x-65 - 42x^2+71x+13 + 30x - 60 =$$
$$-42x^2 - 6x + 156$$

$$-77x^2 + 201x - 112 = -42x^2 - 6x + 156$$

$$0 = 35x^2 - 207x + 268$$

$$0 = (35x-67)(x-4)$$

$$35x - 67 = 0 \quad \text{or} \quad x - 4 = 0$$

$$35x = 67 \quad \text{or} \qquad x = 4$$

$$x = \dfrac{67}{35} \quad \text{or} \qquad x = 4$$

Both numbers check. The solutions are $\dfrac{67}{35}$ and 4.

43. $4x^2 - 5x - 51$

We will use the FOIL method.

1) There are no common factors (other than 1 or -1).

2) Factor the first term, $4x^2$. The factors are x, $4x$ and $2x$, $2x$. We have these possibilities:

$$(x+\ \ )(4x+\ \ ), \ (2x+\ \ )(2x+\ \ )$$

3) Factor the last term, -51. The factors are -1, 51 and 1, -51 and -3, 17 and 3, -17.

4) Look for factors in steps (2) and (3) such that the sum of the products is the middle term, $-5x$. Trial and error leads us to the correct factorization: $(x+3)(4x-17)$

45. $\quad 1 - t^6$

$$= 1^2 - (t^3)^2$$

$$= (1+t^3)(1-t^3)$$

$$= (1+t)(1-t+t^2)(1-t)(1+t+t^2)$$

47. $\quad a^3 - 8b^3$

$$= a^3 - (2b)^3$$

$$= (a-2b)(a^2+2ab+4b^2)$$

49. $\left(\dfrac{1}{1+x} + \dfrac{x}{1-x}\right) \div \left(\dfrac{x}{1+x} - \dfrac{1}{1-x}\right) = -1$

First we add inside each set of parentheses.

$$\dfrac{1}{1+x} + \dfrac{x}{1-x} = \dfrac{1}{1+x} \cdot \dfrac{1-x}{1-x} + \dfrac{x}{1-x} \cdot \dfrac{1+x}{1+x}$$

$$= \dfrac{1-x}{(1+x)(1-x)} + \dfrac{x(1+x)}{(1+x)(1-x)}$$

$$= \dfrac{1-x}{(1+x)(1-x)} + \dfrac{x+x^2}{(1+x)(1-x)}$$

$$= \dfrac{1-x+x+x^2}{(1+x)(1-x)}$$

$$= \dfrac{1+x^2}{(1+x)(1-x)}$$

$$\dfrac{x}{1+x} - \dfrac{1}{1-x} = \dfrac{x}{1+x} \cdot \dfrac{1-x}{1-x} - \dfrac{1}{1-x} \cdot \dfrac{1+x}{1+x}$$

$$= \dfrac{x(1-x)}{(1+x)(1-x)} - \dfrac{1+x}{(1+x)(1-x)}$$

$$= \dfrac{x-x^2}{(1+x)(1-x)} - \dfrac{1+x}{(1+x)(1-x)}$$

$$= \dfrac{x-x^2-(1+x)}{(1+x)(1-x)}$$

$$= \dfrac{x-x^2-1-x}{(1+x)(1-x)}$$

$$= \dfrac{-x^2-1}{(1+x)(1-x)}, \text{ or }$$

$$= -\dfrac{x^2+1}{(1+x)(1-x)}$$

Then the equation becomes

$$\dfrac{1+x^2}{(1+x)(1-x)} \div \left(-\dfrac{x^2+1}{(1+x)(x-1)}\right) = -1$$

$$\dfrac{1+x^2}{(1+x)(1-x)} \cdot \left(-\dfrac{(1+x)(1-x)}{x^2+1}\right) = -1$$

$$-\dfrac{(1+x^2)(1+x)(1-x)}{(1+x)(1-x)(x^2+1)} = -1$$

$$-1 = -1$$

We know that 1 and -1 cannot be solutions of the original equation, because they result in division by 0. Since $-1 = -1$ is true, all real numbers except 1 and -1 are solutions.

51.

$$\dfrac{7}{x-9} - \dfrac{7}{x} = \dfrac{63}{x^2-9x}$$

$$\dfrac{7}{x-9} - \dfrac{7}{x} = \dfrac{63}{x(x-9)}, \text{ LCM is } x(x-9)$$

$$x(x-9)\left(\dfrac{7}{x-9} - \dfrac{7}{x}\right) = x(x-9) \cdot \dfrac{63}{x(x-9)}$$

$$7x - 7(x-9) = 63$$

$$7x - 7x + 63 = 63$$

$$63 = 63$$

Since $63 = 63$ is true, all real numbers except 0 and 9 (which result in division by 0) are solutions of the equation.

Exercise Set 6.4

1. Familiarize. Let $x =$ the number. Then $\dfrac{1}{x} =$ its reciprocal.

Translate. We reword the problem.

A number plus 5 times its reciprocal is -6.

$$x + 5 \cdot \dfrac{1}{x} = -6$$

Solve. We solve the equation.

$$x + \dfrac{5}{x} = -6, \text{ LCM is } x$$

$$x\left(x + \dfrac{5}{x}\right) = x(-6)$$

$$x \cdot x + x \cdot \dfrac{5}{x} = -6x$$

$$x^2 + 5 = -6x$$

$$x^2 + 6x + 5 = 0$$

$$(x+5)(x+1) = 0$$

$$x + 5 = 0 \quad \text{or} \quad x + 1 = 0$$

$$x = -5 \quad \text{or} \quad x = -1$$

Check. If the number is -5, its reciprocal is $-\dfrac{1}{5}$ and $-5 + 5\left(-\dfrac{1}{5}\right) = -5 - 1 = -6$. If the number is -1, its reciprocal is also -1 and $-1 + 5(-1) = -1 - 5 = -6$. Both values check.

State. The number is -5 or -1.

3. Familiarize. Let $x =$ the number. Then $\dfrac{1}{x} =$ its reciprocal.

Translate. We reword the problem.

The reciprocal of 8 plus the reciprocal of 9 is the reciprocal of what number?

$$\dfrac{1}{8} + \dfrac{1}{9} = \dfrac{1}{x}$$

Solve. We solve the equation.

$$\dfrac{1}{8} + \dfrac{1}{9} = \dfrac{1}{x}, \text{ LCM is } 72x$$

$$72x\left(\dfrac{1}{8} + \dfrac{1}{9}\right) = 72x \cdot \dfrac{1}{x}$$

$$72x \cdot \dfrac{1}{8} + 72x \cdot \dfrac{1}{9} = 72$$

$$9x + 8x = 72$$

$$17x = 72$$

$$x = \dfrac{72}{17}$$

Check. $\dfrac{1}{8} + \dfrac{1}{9} = \dfrac{9}{72} + \dfrac{8}{72} = \dfrac{17}{72}$, which is the reciprocal of $\dfrac{72}{17}$.

State. The number is $\frac{72}{17}$.

5. *Familiarize*. Let $t =$ the time it takes them, working together, to fill the order.

Translate. Using the work principle, we get the following equation:

$$\frac{t}{5} + \frac{t}{9} = 1$$

Solve. We solve the equation.

$$\frac{t}{5} + \frac{t}{9} = 1, \text{ LCM is } 45$$

$$45\left(\frac{t}{5} + \frac{t}{9}\right) = 45 \cdot 1$$

$$45 \cdot \frac{t}{5} + 45 \cdot \frac{t}{9} = 45$$

$$9t + 5t = 45$$

$$14t = 45$$

$$t = \frac{45}{14}, \text{ or } 3\frac{3}{14}$$

Check. We verify the work principle.

$$\frac{\frac{45}{14}}{5} + \frac{\frac{45}{14}}{9} = \frac{45}{14} \cdot \frac{1}{5} + \frac{45}{14} \cdot \frac{1}{9} = \frac{9}{14} + \frac{5}{14} = \frac{14}{14} = 1$$

State. It will take them $3\frac{3}{14}$ hr, working together.

7. *Familiarize*. Let $t =$ the time it will take to fill the pool using both the pipe and the hose.

Translate. Using the work principle, we get the following equation:

$$\frac{t}{12} + \frac{t}{30} = 1$$

Solve. We solve the equation.

$$\frac{t}{12} + \frac{t}{30} = 1, \text{ LCM is } 60$$

$$60\left(\frac{t}{12} + \frac{t}{30}\right) = 60 \cdot 1$$

$$\frac{60t}{12} + \frac{60t}{30} = 60$$

$$5t + 2t = 60$$

$$7t = 60$$

$$t = \frac{60}{7}, \text{ or } 8\frac{4}{7}$$

Check. We verify the work principle.

$$\frac{\frac{60}{7}}{12} + \frac{\frac{60}{7}}{30} = \frac{60}{7} \cdot \frac{1}{12} + \frac{60}{7} \cdot \frac{1}{30} = \frac{5}{7} + \frac{2}{7} = 1$$

State. It will take them $8\frac{4}{7}$ hr to fill the pool using both the pipe and the hose.

9. *Familiarize*. Let $t =$ the time it will take to print the order if both presses are used.

Translate. Using the work principle, we get the following equation:

$$\frac{t}{4.5} + \frac{t}{5.5} = 1$$

Solve. We solve the equation.

$$\frac{t}{4.5} + \frac{t}{5.5} = 1, \text{ LCM is } (4.5)(5.5)$$

$$(4.5)(5.5)\left(\frac{t}{4.5} + \frac{t}{5.5}\right) = (4.5)(5.5)(1)$$

$$(4.5)(5.5) \cdot \frac{t}{4.5} + (4.5)(5.5) \cdot \frac{t}{5.5} = 24.75$$

$$5.5t + 4.5t = 24.75$$

$$10t = 24.75$$

$$t = 2.475, \text{ or } 2\frac{19}{40}$$

Check. We verify the work principle.

$$\frac{2.475}{4.5} + \frac{2.475}{5.5} = 0.55 + 0.45 = 1$$

State. It will take 2.475 hr, or $2\frac{19}{40}$ hr, to print the order if both presses are used.

11. *Familiarize*. Let $a =$ the number of days it takes Juan to paint the house. Then $4a =$ the number of days it takes Ariel to paint the house.

Translate. Using the work principle, we get the following equation:

$$\frac{8}{a} + \frac{8}{4a} = 1, \text{ or } \frac{8}{a} + \frac{2}{a} = 1$$

Solve. We solve the equation.

$$\frac{8}{a} + \frac{2}{a} = 1$$

$$\frac{10}{a} = 1 \qquad \text{Adding}$$

$$10 = a \qquad \text{Multiplying by } a$$

Check. If it takes Juan 10 days to paint the house, then it takes Ariel $4 \cdot 10$, or 40 days. We verify the work principle.

$$\frac{8}{10} + \frac{8}{40} = \frac{4}{5} + \frac{1}{5} = \frac{5}{5} = 1$$

State. It would take Juan 10 days to paint the house alone, and it would take Ariel 40 days.

13. Let $d =$ the distance a black racer snake can travel in 3 hr. We translate to a proportion and solve for d

$$\begin{array}{l}\text{Distance} \rightarrow \\ \text{Time} \rightarrow\end{array} \frac{46}{20} = \frac{d}{3} \begin{array}{l}\leftarrow \text{Distance} \\ \leftarrow \text{Time}\end{array}$$

$$3 \cdot \frac{46}{20} = 3 \cdot \frac{d}{3} \qquad \text{Multiplying by 3}$$

$$\frac{138}{20} = d$$

$$\frac{69}{10} = d, \text{ or}$$

$$6\frac{9}{10} = d$$

The black racer will travel $6\frac{9}{10}$ km in 3 hr.

15. Let h = the number of hits the player will have. We translate to a proportion and solve for h.

$$\text{Hits} \rightarrow \frac{120}{300} = \frac{h}{500} \leftarrow \text{Hits}$$
$$\text{At bats} \rightarrow \quad \quad \leftarrow \text{At bats}$$

$$500 \cdot \frac{120}{300} = 500 \cdot \frac{h}{500} \quad \text{Multiplying by 500}$$

$$\frac{120 \cdot 500}{300} = h$$

$$200 = h$$

The player will get 200 hits.

17. Let g = the number of grams of hemoglobin contained in 32 cm^3 of the blood. We translate to a proportion and solve for g.

$$\text{Specimen} \rightarrow \frac{10}{1.2} = \frac{32}{g} \leftarrow \text{Specimen}$$
$$\text{Hemoglobin} \rightarrow \quad \quad \leftarrow \text{Hemoglobin}$$

$$1.2g \cdot \frac{10}{1.2} = 1.2g \cdot \frac{32}{g} \quad \text{Multiplying by } 1.2g$$

$$g \cdot 10 = 1.2 \cdot 32$$

$$g = \frac{1.2 \cdot 32}{10}$$

$$g = 3.84$$

The blood contains 3.84 g of hemoglobin.

19. Let T = the number of trout in the lake. We translate to a proportion and solve for T.

$$\begin{array}{l} \text{Trout tagged} \\ \text{originally} \rightarrow \\ \text{Trout} \rightarrow \\ \text{in lake} \end{array} \frac{112}{T} = \frac{32}{82} \begin{array}{l} \leftarrow \text{Tagged trout} \\ \leftarrow \text{caught later} \\ \leftarrow \text{Trout caught} \\ \text{later} \end{array}$$

$$82T \cdot \frac{112}{T} = 82T \cdot \frac{32}{82}$$

$$82 \cdot 112 = 32 \cdot T$$

$$\frac{82 \cdot 112}{32} = T$$

$$287 = T$$

There are 287 trout in the lake.

21. a) Let w = the number of tons the rocket will weigh on Mars. We translate to a proportion and solve for w.

$$\text{Weight on Mars} \rightarrow \frac{0.4}{1} = \frac{w}{12} \leftarrow \text{Weight on Mars}$$
$$\text{Weight on earth} \rightarrow \quad \quad \leftarrow \text{Weight on earth}$$

$$0.4 \cdot 12 = 1 \cdot w \quad \text{Using cross-}$$
$$\text{products}$$
$$4.8 = w$$

A 12-T rocket will weigh 4.8 T on Mars.

b) Let w = the number of pounds the astronaut will weigh on Mars. We translate to a proportion and solve for w.

$$\text{Weight on Mars} \rightarrow \frac{0.4}{1} = \frac{w}{120} \leftarrow \text{Weight on Mars}$$
$$\text{Weight on earth} \rightarrow \quad \quad \leftarrow \text{Weight on earth}$$

$$0.4 \cdot 120 = 1 \cdot w \quad \text{Using cross-}$$
$$\text{products}$$
$$48 = w$$

A 120-lb astronaut will weigh 48 lb on Mars.

23. Let x = the number that is added to each of the given numbers. We translate to a proportion and solve for x.

$$\frac{1+x}{2+x} = \frac{3+x}{5+x}$$

$$(1+x)(5+x) = (2+x)(3+x) \quad \text{Using cross-products}$$

$$5 + 6x + x^2 = 6 + 5x + x^2$$

$$x = 1$$

The number is 1.

25. *Familiarize* We first make a drawing. We let r = the speed of the boat in still water. Then $r - 3$ = the speed upstream and $r + 3$ = the speed downstream.

Upstream 4 miles $r - 3$ mph

10 miles $r + 3$ mph Downstream

We organize the information in a table. The time is the same both upstream and downstream so we use t for each time.

	Distance	Speed	Time
Upstream	4	$r - 3$	t
Downstream	10	$r + 3$	t

Translate. Using $t = \dfrac{d}{r}$ we get two equations from the rows of the table.

$$t = \frac{4}{r - 3} \quad \text{and} \quad t = \frac{10}{r + 3}$$

Solve. Since both rational expressions represent the same time, t, we can set them equal to each other and solve.

$$\frac{4}{r - 3} = \frac{10}{r + 3}, \quad \text{LCM is } (r-3)(r+3)$$

$$(r-3)(r+3) \cdot \frac{4}{r-3} = (r-3)(r+3) \cdot \frac{10}{r+3}$$

$$4(r + 3) = 10(r - 3)$$

$$4r + 12 = 10r - 30$$

$$42 = 6r$$

$$7 = r$$

Check. If $r = 7$ mph, the $r - 3$ is 4 mph and $r + 3$ is 10 mph. The time upstream is $\dfrac{4}{4}$, or 1 hour. The time downstream is $\dfrac{10}{10}$, or 1 hour. The times are the same. The values check.

State. The speed of the boat in still water is 7 mph.

27. *Familiarize* We first make a drawing. We let r = the speed of the freight train. Then $r + 14$ = the speed of the passenger train.

Passenger train $r + 14$ 400 mi

Freight train r 330 mi

We organize the information in a table.

Train	Distance	Speed	Time
Passenger	400	$r + 14$	t
Freight	330	r	t

Translate. Using $t = \dfrac{d}{r}$ we get two equations from the rows of the table.

$$t = \frac{400}{r + 14} \text{ and } t = \frac{330}{r}$$

Solve. Set the rational expressions equal to each other and solve.

$$\frac{400}{r + 14} = \frac{330}{r}, \text{ LCM is } r(r + 14)$$

$$r(r + 14) \cdot \frac{400}{r + 14} = r(r + 14) \cdot \frac{330}{r}$$

$$400r = 330(r + 14)$$

$$400r = 330r + 4620$$

$$70r = 4620$$

$$r = 66$$

Check. If the speed of the freight train is 66 mph, then the speed of the passenger train is $66 + 14$, or 80 mph. The time for the freight train is $\dfrac{330}{66}$, or 5 hr. The time for the passenger train is $\dfrac{400}{80}$, or 5 hr. Since the times are the same, the answer checks.

State. The speed of the passenger train is 80 mph; the speed of the freight train is 66 mph.

29. **Familiarize** We first make a drawing. We let $r =$ the speed of the river and $t =$ the time.

Downstream $\qquad$ $2 + r$ $\qquad$ 4 km

$\qquad$ 1 km $\quad$ $2 - r$ $\quad$ Upstream

We organize the information in a table.

	Distance	Speed	Time
Downstream	4	$2 + r$	t
Upstream	1	$2 - r$	t

Translate. Using $t = \dfrac{d}{r}$ we get two equations from the rows of the table.

$$t = \frac{4}{2 + r} \text{ and } t = \frac{1}{2 - r}$$

Solve. Set the rational expressions equal to each other and solve.

$$\frac{4}{2 + r} = \frac{1}{2 - r}, \text{ LCM is } (2 + r)(2 - r)$$

$$(2 + r)(2 - r) \cdot \frac{4}{2 + r} = (2 + r)(2 - r) \cdot \frac{1}{2 - r}$$

$$4(2 - r)r = 2 + r$$

$$8 - 4r = 2 + r$$

$$6 = 5r$$

$$\frac{6}{5} = r, \text{ or}$$

$$1\frac{1}{5} = r$$

Check. If the speed of the river is $\dfrac{6}{5}$ km/h, then the paddleboat's speed downstream is $2 + \dfrac{6}{5}$, or $\dfrac{16}{5}$ km/h and the speed upstream is $2 - \dfrac{6}{5}$, or $\dfrac{4}{5}$ km/h. The time downstream is $\dfrac{4}{16/5}$, or $\dfrac{5}{4}$ hr. The time upstream is $\dfrac{1}{4/5}$, or $\dfrac{5}{4}$ hr. Since the times are the same, the answer checks.

State. The speed of the river is $\dfrac{6}{5}$ km/h, or $1\dfrac{1}{5}$ km/h.

31. Graph: $x - 4y \geq 8$

First graph the line $x - 4y = 8$. The intercepts are $(8, 0)$ and $(0, -2)$. We draw the line solid since the inequality symbol is $\geq$. To determine which half-plane to shade we consider a test point not on the line. We try $(0, 0)$:

$$\frac{x - 4y \geq 8}{\begin{array}{c|c} 0 - 4 \cdot 0 & 8 \\ 0 & \text{FALSE} \end{array}}$$

Since $0 \geq 8$ is false, we shade the half-plane that does not contain $(0, 0)$.

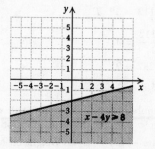

33. Graph: $x \geq 3$, or $x + 0y \geq 3$

First graph the line $x = 3$ in the plane. We draw the line solid since the inequality symbol is $\geq$. To determine which half-plane to shade we consider a test point not on the line. We try $(0, 0)$:

$$\frac{x + 0y \geq 3}{\begin{array}{c|c} 0 + 0 \cdot 0 & 3 \\ 0 & \text{FALSE} \end{array}}$$

Since $0 \geq 3$ is false, we shade the half-plane that does not contain $(0, 0)$.

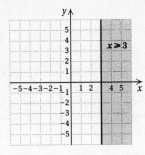

35. Graph: $2x - 5y < 10$

First graph the line $2x - 5y = 10$. The intercepts are $(5, 0)$ and $(0, -2)$. We draw the line dashed since the inequality symbol is $<$. To determine which half-plane to shade we consider a test point not on the line. We try $(0, 0)$:

$$\frac{2x - 5y < 10}{2 \cdot 0 - 5 \cdot 0 \;\bigg|\; 10}$$
$$0 \;\bigg|\; \text{TRUE}$$

Since $0 < 10$ is true, we shade the half-plane that contains $(0, 0)$.

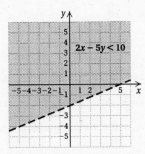

37. *Familiarize* Let d = the distance from the woman's house to work. The trip at 40 mph takes 2 min, or $\frac{1}{30}$ hr, longer than the trip at 45 mph. Let t = the time the trip takes at 45 mph. We organize the information in a table.

	Distance	Speed	Time
Faster trip	d	45	t
Slower trip	d	40	$t + \frac{1}{30}$

Translate. Using $t = \frac{d}{r}$ we get two equations from the rows of the table.

$$t = \frac{d}{45} \text{ and } t + \frac{1}{30} = \frac{d}{40}, \text{ or}$$
$$t = \frac{d}{45} \text{ and } t = \frac{d}{40} - \frac{1}{30}$$

Solve. We set the rational expressions equal to each other and solve.

$$\frac{d}{45} = \frac{d}{40} - \frac{1}{30}, \text{ LCM is } 360$$
$$360 \cdot \frac{d}{45} = 360\left(\frac{d}{40} - \frac{1}{30}\right)$$
$$8d = 9d - 12$$
$$12 = d$$

Check. At 40 mph, the woman drives 12 mi in $\frac{12}{40}$, or $\frac{3}{10}$ hr, or 18 min. At 45 mph, the woman drives 12 mi in $\frac{12}{45}$, or $\frac{4}{15}$ hr, or 16 min. The time at 40 mph is 2 min longer than at 45 mph, so the answer checks.

State. The woman lives 12 mi from work.

39. *Familiarize* Let t_1 = the time the airplane flies away from the airport, t_2 = the time the plane flies back to the airport, and d = the distance flown in one direction. The plane's speed against the wind is $240 - 40$, or 200 mph, and the speed with the wind is $240 + 40$, or 280 mph. We organize the information in a table.

	Distance	Speed	Time
Away from airport	d	200	t_1
Return to airport	d	280	t_2

Translate. Using $t = \frac{d}{r}$ we get two equations from the rows of the table.

$$t_1 = \frac{d}{200} \text{ and } t_2 = \frac{d}{280}$$

We know that the total time is 6 hr, so we have $t_1 + t_2 = 6$, or $\frac{d}{200} + \frac{d}{280} = 6$.

Solve. We solve the equation.

$$\frac{d}{200} + \frac{d}{280} = 6, \text{ LCM is } 1400$$
$$1400\left(\frac{d}{200} + \frac{d}{280}\right) = 1400 \cdot 6$$
$$7d + 5d = 8400$$
$$12d = 8400$$
$$d = 700$$

Check. Flying 700 mi away from the airport (against the wind), the plane travels for $\frac{700}{200}$, or $3\frac{1}{2}$ hr. Returning 700 mi to the airport (with the wind), the plane travels for $\frac{700}{280}$, or $2\frac{1}{2}$ hr. The total time is $3\frac{1}{2} + 2\frac{1}{2}$, or 6 hr. The answer checks.

State. The airplane can fly 700 mi away from the airport and return to the airport without refueling. Thus, the airplane can make a 1400 mi round-trip without refueling.

41. *Familiarize* Express the position of the hands in terms of minute units on the face of the clock. At 10:30 the hour hand is at $\frac{10.5}{12}$ hr$\times\frac{60 \text{ min}}{1 \text{ hr}}$, or 52.5 minutes, and the minute hand is at 30 minutes. The rate of the minute hand is 12 times the rate of the hour hand. (When the minute hand moves 60 minutes, the hour hand moves 5 minutes.) Let $t=$ the number of minutes after 10:30 that the hands will first be perpendicular. After t minutes the minute hand has moved t units, and the hour hand has moved $\frac{t}{12}$ units. The position of the hour hand will be 15 units "ahead" of the position of the minute hand when they are first perpendicular.

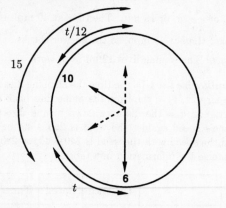

Translate.

$$
\begin{array}{ccccc}
\underbrace{\text{Position of}}_{\text{hour hand}} & \text{is} & \underbrace{\text{position of}}_{\text{minute hand}} & \text{plus} & \text{15 min.} \\
\text{after } t \text{ min} & & \text{after } t \text{ min} & & \\
\downarrow & \downarrow & \downarrow & \downarrow & \downarrow \\
52.5 + \dfrac{t}{12} & = & 30 + t & + & 15
\end{array}
$$

Solve. We solve the equation.

$$52.5 + \frac{t}{12} = 30 + t + 15$$

$$52.5 + \frac{t}{12} = 45 + t, \text{ LCM is 12}$$

$$12\left(52.5 + \frac{t}{12}\right) = 12(45 + t)$$

$$630 + t = 540 + 12t$$

$$90 = 11t$$

$$\frac{90}{11} = t, \text{ or}$$

$$8\frac{2}{11} = t$$

Check. At $\frac{90}{11}$ min after 10:30, the position of the hour hand is at $52.5 + \frac{90/11}{12}$, or $53\frac{2}{11}$ min. The minute hand is at $30 + \frac{90}{11}$, or $38\frac{2}{11}$ min. The hour hand is 15 minutes ahead of the minute hand so the hands are perpendicular. The answer checks.

State. After 10:30 the hands of a clock will first be perpendicular in $8\frac{2}{11}$ min. The time is $10:38\frac{2}{11}$, or $21\frac{9}{11}$ min before 11:00.

43. *Familiarize* We let $x =$ the speed of the current and $3x =$ the speed of the boat. Then the speed up the river is $3x - x$, or $2x$, and the speed down the river is $3x + x$, or $4x$. The total distance is 100 km; thus the distance each way is 50 km. Using $t = \frac{d}{r}$, we can use $\frac{50}{2x}$ for the time up the river and $\frac{50}{4x}$ for the time down the river.

Translate. Since the total of the times is 10 hours, we have the following equation.

$$\frac{50}{2x} + \frac{50}{4x} = 10$$

Solve. We solve the equation. The LCM is $4x$.

$$4x\left(\frac{50}{2x} + \frac{50}{4x}\right) = 4x \cdot 10$$

$$100 + 50 = 40x$$

$$150 = 40x$$

$$\frac{15}{4} = x, \text{ or}$$

$$3\frac{3}{4} = x$$

Check. If the speed of the current is $\frac{15}{4}$ km/h, then the speed of the boat is $3 \cdot \frac{15}{4}$, or $\frac{45}{4}$. The speed up the river is $\frac{45}{4} - \frac{15}{4}$, or $\frac{15}{2}$ km/h, and the time traveling up the river is $50 \div \frac{15}{2}$, or $6\frac{2}{3}$ hr. The speed down the river is $\frac{45}{4} + \frac{15}{4}$, or 15 km/h, and the time traveling down the river is $50 \div 15$, or $3\frac{1}{3}$ hr. The total time for the trip is $6\frac{2}{3} + 3\frac{1}{3}$, or 10 hr. The value checks.

State. The speed of the current is $3\frac{3}{4}$ km/h.

45. *Familiarize* Trucks A, B, and C, working together, move a load of sand in t hours. Then Truck A, alone, can move a load of sand in $t + 1$ hours; Truck B, alone, can move a load of sand in $t + 6$ hours; and Truck C, alone can move a load of sand in $t + t$, or $2t$ hours.

Translate. Using the work principle, we get the following equation:

$$\frac{t}{t+1} + \frac{t}{t+6} + \frac{t}{2t} = 1, \text{ or}$$

$$\frac{t}{t+1} + \frac{t}{t+6} + \frac{1}{2} = 1$$

Solve. We solve the equation.

$$\frac{t}{t+1} + \frac{t}{t+6} = \frac{1}{2} \qquad \text{Subtracting } \frac{1}{2}$$

$$2(t+1)(t+6)\left(\frac{t}{t+1} + \frac{t}{t+6}\right) = 2(t+1)(t+6)\cdot\frac{1}{2}$$

$$2t(t+6) + 2t(t+1) = (t+1)(t+6)$$

$$2t^2 + 12t + 2t^2 + 2t = t^2 + 7t + 6$$

$$4t^2 + 14t = t^2 + 7t + 6$$

$$3t^2 + 7t - 6 = 0$$

$$(3t-2)(t+3) = 0$$

$$3t - 2 = 0 \quad \text{or} \quad t + 3 = 0$$

$$3t = 2 \quad \text{or} \quad t = -3$$

$$t = \frac{2}{3} \quad \text{or} \quad t = -3$$

Check. Since time cannot be negative, we need only check $\frac{2}{3}$. If $t = \frac{2}{3}$, then working alone it takes A $\frac{2}{3} + 1$, or $\frac{5}{3}$ hr, B $\frac{2}{3} + 6$, or $\frac{20}{3}$ hr, and C $2 \cdot \frac{2}{3}$, or $\frac{4}{3}$ hr. We verify the work principle.

$$\frac{\frac{2}{3}}{\frac{5}{3}} + \frac{\frac{2}{3}}{\frac{20}{3}} + \frac{\frac{2}{3}}{\frac{4}{3}}$$

$$= \frac{2}{3}\cdot\frac{3}{5} + \frac{2}{3}\cdot\frac{3}{20} + \frac{2}{3}\cdot\frac{3}{4}$$

$$= \frac{2}{5} + \frac{1}{10} + \frac{1}{2} = \frac{4}{10} + \frac{1}{10} + \frac{5}{10}$$

$$= \frac{10}{10} = 1$$

State. Working together, it takes $\frac{2}{3}$ hour.

Exercise Set 6.5

1.
$$\frac{W_1}{W_2} = \frac{d_1}{d_2}$$

$$W_2 d_2 \cdot \frac{W_1}{W_2} = W_2 d_2 \cdot \frac{d_1}{d_2} \qquad \text{Multiplying by the LCM}$$

$$d_2 W_1 = W_2 d_1$$

$$\frac{d_2 W_1}{d_1} = W_2 \qquad \text{Dividing by } d_1$$

3.
$$\frac{1}{R} = \frac{1}{r_1} + \frac{1}{r_2}$$

$$Rr_1 r_2 \cdot \frac{1}{R} = Rr_1 r_2 \left(\frac{1}{r_1} + \frac{1}{r_2}\right) \qquad \begin{array}{l}\text{Multiplying by}\\ \text{the LCM}\end{array}$$

$$r_1 r_2 = Rr_2 + Rr_1$$

$$r_1 r_2 - Rr_2 = Rr_1 \qquad \text{Subtracting } Rr_2$$

$$r_2(r_1 - R) = Rr_1 \qquad \text{Factoring}$$

$$r_2 = \frac{Rr_1}{r_1 - R} \qquad \text{Dividing by } r_1 - R$$

5.
$$s = \frac{(v_1 + v_2)t}{2}$$

$$2 \cdot s = 2 \cdot \frac{(v_1 + v_2)t}{2} \qquad \text{Multiplying by 2}$$

$$2s = (v_1 + v_2)t$$

$$\frac{2s}{v_1 + v_2} = t \qquad \text{Dividing by } v_1 + v_2$$

7.
$$R = \frac{gs}{g+s}$$

$$(g+s) \cdot R = (g+s) \cdot \frac{gs}{g+s} \qquad \text{Multiplying by } g+s$$

$$R(g+s) = gs$$

$$Rg + Rs = gs \qquad \text{Removing parentheses}$$

$$Rg = gs - Rs \qquad \text{Subtracting } Rs$$

$$Rg = s(g - R) \qquad \text{Factoring}$$

$$\frac{Rg}{g-R} = s \qquad \text{Dividing by } g-R$$

9.
$$\frac{1}{p} + \frac{1}{q} = \frac{1}{f}$$

$$pqf\left(\frac{1}{p} + \frac{1}{q}\right) = pqf \cdot \frac{1}{f} \qquad \text{Multiplying by } pqf$$

$$pqf \cdot \frac{1}{p} + pqf \cdot \frac{1}{q} = pq$$

$$qf + pf = pq$$

$$qf = pq - pf \qquad \text{Subtracting } pf$$

$$qf = p(q - f) \qquad \text{Factoring}$$

$$\frac{qf}{q-f} = p \qquad \text{Dividing by } q-f$$

11.
$$\frac{t}{a} + \frac{t}{b} = 1$$

$$ab\left(\frac{t}{a} + \frac{t}{b}\right) = ab \cdot 1 \qquad \text{Multiplying by } ab$$

$$ab \cdot \frac{t}{a} + ab \cdot \frac{t}{b} = ab \qquad \text{Removing parentheses}$$

$$bt + at = ab$$

$$bt = ab - at \qquad \text{Subtracting } at$$

$$bt = a(b - t) \qquad \text{Factoring}$$

$$\frac{bt}{b-t} = a \qquad \text{Dividing by } b-t$$

13.
$$I = \frac{nE}{E+nr}$$

$$(E+nr)I = (E+nr) \cdot \frac{nE}{E+nr} \qquad \begin{array}{l}\text{Multiplying by}\\ E+nr\end{array}$$

$$EI + nrI = nE$$

$$nrI = nE - EI \qquad \text{Subtracting } EI$$

$$nrI = E(n - I) \qquad \text{Factoring}$$

$$\frac{nrI}{n-I} = E \qquad \text{Dividing by } n-I$$

15.
$$S = \frac{H}{m(t_1 - t_2)}$$

$$S \cdot m(t_1 - t_2) = \frac{H}{m(t_1 - t_2)} \cdot m(t_1 - t_2) \quad \text{Multiplying by } m(t_1 - t_2)$$

$$Sm(t_1 - t_2) = H$$

17.
$$\frac{E}{e} = \frac{R+r}{r}$$

$$er \cdot \frac{E}{e} = er \cdot \frac{R+r}{r} \quad \text{Multiplying by } er$$

$$Er = eR + er$$

$$Er - er = eR \quad\quad \text{Subtracting } er$$

$$r(E - e) = eR \quad\quad \text{Factoring}$$

$$r = \frac{eR}{E - e} \quad\quad \text{Dividing by } E - e$$

19.
$$V = \frac{1}{3}\pi h^2 (3R - h)$$

$$\frac{3}{\pi h^2} \cdot V = \frac{3}{\pi h^2} \cdot \frac{1}{3}\pi h^2 (3R - h) \quad \text{Multiplying by } \frac{3}{\pi h^2}$$

$$\frac{3V}{\pi h^2} = 3R - h$$

$$\frac{3V}{\pi h^2} + h = 3R \quad\quad \text{Adding } h$$

$$\frac{1}{3}\left(\frac{3V}{\pi h^2} + h\right) = \frac{1}{3} \cdot 3R \quad \text{Multiplying by } \frac{1}{3}$$

$$\frac{V}{\pi h^2} + \frac{h}{3} = R, \text{ or}$$

$$\frac{3V + \pi h^3}{3\pi h^2} = R$$

21.
$$A = P(1 + rt)$$

$$A = P + Prt \quad\quad \text{Removing parentheses}$$

$$A - P = Prt \quad\quad \text{Subtracting } P$$

$$\frac{A - P}{Pr} = t \quad\quad \text{Dividing by } Pr$$

23. Graph: $6x - y < 6$

First graph the line $6x - y = 6$. The intercepts are $(1, 0)$ and $(0, -6)$. We draw the line dashed since the inequality symbol is $<$. Since the ordered pair $(0, 0)$ is a solution of the inequality ($6 \cdot 0 - 0 < 6$ is true), we shade the half-plane containing $(0, 0)$.

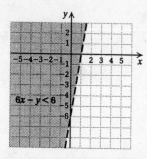

25. $p^3 - 27q^3 = p^3 - (3q)^3 = (p - 3q)(p^2 + 3pq + 9q^2)$

27. $t^3 + 8b^3 = t^3 + (2b)^3 = (t + 2b)(t^2 - 2bt + 4b^2)$

29.
$$\frac{1}{R} = \frac{1}{r_1} + \frac{1}{r_2} + \frac{1}{r_3}$$

$$Rr_1r_2r_3 \cdot \frac{1}{R} = Rr_1r_2r_3\left(\frac{1}{r_1} + \frac{1}{r_2} + \frac{1}{r_3}\right)$$

$$r_1r_2r_3 = Rr_2r_3 + Rr_1r_3 + Rr_1r_2$$

$$r_1r_2r_3 = R(r_2r_3 + r_1r_3 + r_1r_2)$$

$$\frac{r_1r_2r_3}{r_2r_3 + r_1r_3 + r_1r_2} = R$$

Exercise Set 6.6

1.
$$\frac{\frac{1}{a} + 2}{\frac{1}{a} - 1}$$

The LCM of all the denominators is a. We multiply by 1 using a/a.

$$\frac{\frac{1}{a} + 2}{\frac{1}{a} - 1} = \frac{\frac{1}{a} + 2}{\frac{1}{a} - 1} \cdot \frac{a}{a}$$

$$= \frac{\left(\frac{1}{a} + 2\right) \cdot a}{\left(\frac{1}{a} - 1\right) \cdot a}$$

$$= \frac{\frac{1}{a} \cdot a + 2a}{\frac{1}{a} \cdot a - a}$$

$$= \frac{1 + 2a}{1 - a}$$

3.
$$\frac{x - \frac{1}{x}}{x + \frac{1}{x}}$$

The LCM of all the denominators is x. We multiply by 1 using x/x.

$$\frac{x - \frac{1}{x}}{x + \frac{1}{x}} = \frac{x - \frac{1}{x}}{x + \frac{1}{x}} \cdot \frac{x}{x}$$

$$= \frac{\left(x - \frac{1}{x}\right) \cdot x}{\left(x + \frac{1}{x}\right) \cdot x}$$

$$= \frac{x \cdot x - \frac{1}{x} \cdot x}{x \cdot x + \frac{1}{x} \cdot x}$$

$$= \frac{x^2 - 1}{x^2 + 1}$$

5. $\dfrac{\dfrac{3}{x} + \dfrac{4}{y}}{\dfrac{4}{x} - \dfrac{3}{y}}$

The LCM of all the denominators is xy. We multiply by 1 using $\dfrac{xy}{xy}$.

$$\dfrac{\dfrac{3}{x} + \dfrac{4}{y}}{\dfrac{4}{x} - \dfrac{3}{y}} = \dfrac{\dfrac{3}{x} + \dfrac{4}{y}}{\dfrac{4}{x} - \dfrac{3}{y}} \cdot \dfrac{xy}{xy}$$

$$= \dfrac{\left(\dfrac{3}{x} + \dfrac{4}{y}\right) \cdot xy}{\left(\dfrac{4}{x} - \dfrac{3}{y}\right) \cdot xy}$$

$$= \dfrac{\dfrac{3}{x} \cdot xy + \dfrac{4}{y} \cdot xy}{\dfrac{4}{x} \cdot xy - \dfrac{3}{y} \cdot xy}$$

$$= \dfrac{3y + 4x}{4y - 3x}$$

7. $= \dfrac{\dfrac{9x^2 - y^2}{xy}}{\dfrac{3x - y}{y}}$

$= \dfrac{9x^2 - y^2}{xy} \cdot \dfrac{y}{3x - y}$ Multiplying by the reciprocal of the divisor

$= \dfrac{(9x^2 - y^2)(y)}{xy(3x - y)}$

$= \dfrac{(3x + y)(3x - y)(y)}{xy\,(3x - y)}$

$= \dfrac{3x + y}{x}$

9. $\dfrac{a - \dfrac{3a}{b}}{b - \dfrac{b}{a}}$

$= \dfrac{a - \dfrac{3a}{b}}{b - \dfrac{b}{a}} \cdot \dfrac{ab}{ab}$ Using the LCM of the denominators

$= \dfrac{\left(a - \dfrac{3a}{b}\right) \cdot ab}{\left(b - \dfrac{b}{a}\right) \cdot ab}$

$= \dfrac{a(ab) - \dfrac{3a}{b} \cdot ab}{b(ab) - \dfrac{b}{a} \cdot ab}$

$= \dfrac{a^2 b - 3a^2}{ab^2 - b^2}$

$= \dfrac{a^2(b - 3)}{b^2(a - 1)}$

11. $\dfrac{\dfrac{1}{a} + \dfrac{1}{b}}{\dfrac{a^2 - b^2}{ab}}$

$= \dfrac{\dfrac{1}{a} + \dfrac{1}{b}}{\dfrac{a^2 - b^2}{ab}} \cdot \dfrac{ab}{ab}$ Using the LCM of the denominators

$= \dfrac{\left(\dfrac{1}{a} + \dfrac{1}{b}\right) \cdot ab}{\left(\dfrac{a^2 - b^2}{ab}\right) \cdot ab}$

$= \dfrac{\dfrac{1}{a} \cdot ab + \dfrac{1}{b} \cdot ab}{\dfrac{a^2 - b^2}{ab} \cdot ab}$

$= \dfrac{b + a}{a^2 - b^2} = \dfrac{b + a}{(a + b)(a - b)}$

$= \dfrac{a + b}{a + b} \cdot \dfrac{1}{a - b}$ $(b + a = a + b)$

$= \dfrac{1}{a - b}$

13. $\dfrac{\dfrac{1}{x + h} - \dfrac{1}{x}}{h}$

$= \dfrac{\dfrac{1}{x + h} \cdot \dfrac{x}{x} - \dfrac{1}{x} \cdot \dfrac{x + h}{x + h}}{h}$ Adding in the numerator

$= \dfrac{\dfrac{x - x - h}{x(x + h)}}{h} = \dfrac{\dfrac{-h}{x(x + h)}}{h}$

$= \dfrac{-h}{x(x + h)} \cdot \dfrac{1}{h}$ Multiplying by the reciprocal of the divisor

$= \dfrac{h}{h} \cdot \dfrac{-1}{x(x + h)}$ Removing a factor of 1

$= \dfrac{-1}{x(x + h)}$, or $-\dfrac{1}{x(x + h)}$

15. $\dfrac{\dfrac{x^2 - x - 12}{x^2 - 2x - 15}}{\dfrac{x^2 + 8x + 12}{x^2 - 5x - 14}}$

$= \dfrac{x^2 - x - 12}{x^2 - 2x - 15} \cdot \dfrac{x^2 - 5x - 14}{x^2 + 8x + 12}$ Multiplying by the reciprocal of the divisor

$= \dfrac{(x^2 - x - 12)(x^2 - 5x - 14)}{(x^2 - 2x - 15)(x^2 + 8x + 12)}$

$= \dfrac{(x + 3)(x - 4)(x - 7)(x + 2)}{(x - 5)(x + 3)(x + 6)(x + 2)}$ Removing a factor of 1

$= \dfrac{(x - 4)(x - 7)}{(x - 5)(x + 6)}$

17. $\dfrac{\dfrac{1}{x+2}+\dfrac{4}{x-3}}{\dfrac{2}{x-3}-\dfrac{7}{x+2}}$

$=\dfrac{\dfrac{1}{x+2}+\dfrac{4}{x-3}}{\dfrac{2}{x-3}-\dfrac{7}{x+2}}\cdot\dfrac{(x+2)(x-3)}{(x+2)(x-3)}$ Using the LCM

   of the denominators

$=\dfrac{\dfrac{1}{x+2}\cdot(x+2)(x-3)+\dfrac{4}{x-3}\cdot(x+2)(x-3)}{\dfrac{2}{x-3}\cdot(x+2)(x-3)-\dfrac{7}{x+2}\cdot(x+2)(x-3)}$

$=\dfrac{x-3+4(x+2)}{2(x+2)-7(x-3)}$

$=\dfrac{x-3+4x+8}{2x+4-7x+21}$

$=\dfrac{5x+5}{-5x+25}$

$=\dfrac{\cancel{5}\,(x+1)}{\cancel{5}\,(-x+5)}$ Removing a factor of 1

$=\dfrac{x+1}{-x+5}$, or $\dfrac{x+1}{5-x}$

19. $\dfrac{\dfrac{6}{x^2-4}-\dfrac{5}{x+2}}{\dfrac{7}{x^2-4}-\dfrac{4}{x-2}}$

$=\dfrac{\dfrac{6}{(x+2)(x-2)}-\dfrac{5}{x+2}\cdot\dfrac{x-2}{x-2}}{\dfrac{7}{(x+2)(x-2)}-\dfrac{4}{x-2}\cdot\dfrac{(x+2)}{(x+2)}}$

   Finding the LCM of the denominators and multiplying by 1

$=\dfrac{\dfrac{6}{(x+2)(x-2)}-\dfrac{5x-10}{(x+2)(x-2)}}{\dfrac{7}{(x+2)(x-2)}-\dfrac{4x+8}{(x+2)(x-2)}}$

$=\dfrac{\dfrac{6-5x+10}{(x+2)(x-2)}}{\dfrac{7-4x-8}{(x+2)(x-2)}}$ Subtracting in the numerator and in the denominator

$=\dfrac{\dfrac{16-5x}{(x+2)(x-2)}}{\dfrac{-1-4x}{(x+2)(x-2)}}$

$=\dfrac{16-5x}{(x+2)(x-2)}\cdot\dfrac{(x+2)(x-2)}{-1-4x}$

   Multiplying by the reciprocal of the denominator

$=\dfrac{(16-5x)(\cancel{x+2})(\cancel{x-2})}{(\cancel{x+2})(\cancel{x-2})(-1-4x)}$ Removing a factor of 1

$=\dfrac{16-5x}{-1-4x}$

$=\dfrac{\cancel{-1}\cdot(5x-16)}{\cancel{-1}(4x+1)}$ Removing a factor of 1

$=\dfrac{5x-16}{4x+1}$

21. $\dfrac{\dfrac{1}{z^2}-\dfrac{1}{w^2}}{\dfrac{1}{z^3}+\dfrac{1}{w^3}}$

$=\dfrac{\dfrac{1}{z^2}-\dfrac{1}{w^2}}{\dfrac{1}{z^3}+\dfrac{1}{w^3}}\cdot\dfrac{z^3w^3}{z^3w^3}$ Multiplying by the LCM of the denominators

$=\dfrac{\left(\dfrac{1}{z^2}-\dfrac{1}{w^2}\right)\cdot z^3w^3}{\left(\dfrac{1}{z^3}+\dfrac{1}{w^3}\right)\cdot z^3w^3}$

$=\dfrac{zw^3-z^3w}{w^3+z^3}$

$=\dfrac{zw(w^2-z^2)}{(w+z)(w^2-wz+z^2)}$

$=\dfrac{zw(\cancel{w+z})(w-z)}{(\cancel{w+z})(w^2-wz+z^2)}$

$=\dfrac{zw(w-z)}{w^2-wz+z^2}$

23. $4x^3+20x^2+6x$

$=2x(2x^2+10x+3)$ Factoring out the largest common factor

The trinomial $2x^2+10x+3$ cannot be factored as a product of binomials so we have the complete factorization.

25. $y^3+8=y^3+2^3=(y+2)(y^2-2y+4)$

27. $1000x^3+1=(10x)^3+1^3=(10x+1)(100x^2-10x+1)$

29. $\dfrac{5x^{-1}-5y^{-1}+10x^{-1}y^{-1}}{6x^{-1}-6y^{-1}+12x^{-1}y^{-1}}$

$=\dfrac{\dfrac{5}{x}-\dfrac{5}{y}+\dfrac{10}{xy}}{\dfrac{6}{x}-\dfrac{6}{y}+\dfrac{12}{xy}}$

$=\dfrac{\dfrac{5}{x}-\dfrac{5}{y}+\dfrac{10}{xy}}{\dfrac{6}{x}-\dfrac{6}{y}+\dfrac{12}{xy}}\cdot\dfrac{xy}{xy}$

$=\dfrac{5y-5x+10}{6y-6x+12}$

$=\dfrac{5(y-x+2)}{6(y-x+2)}$

$=\dfrac{5}{6}\cdot\dfrac{y-x+2}{y-x+2}$

$=\dfrac{5}{6}$

31.
$$\frac{(a^2b^{-1} + b^2a^{-1})(a^{-2} - b^{-2})}{(a^2 - ab + b^2)(a^{-2} + 2a^{-1}b^{-1} + b^{-2})}$$

$$= \frac{\left(\dfrac{a^2}{b} + \dfrac{b^2}{a}\right)\left(\dfrac{1}{a^2} - \dfrac{1}{b^2}\right)}{(a^2 - ab + b^2)\left(\dfrac{1}{a^2} + \dfrac{2}{ab} + \dfrac{1}{b^2}\right)}$$

$$= \frac{\left(\dfrac{a^3 + b^3}{ab}\right)\left(\dfrac{b^2 - a^2}{a^2b^2}\right)}{(a^2 - ab + b^2)\left(\dfrac{b^2 + 2ab + a^2}{a^2b^2}\right)}$$
Adding inside parentheses

$$= \frac{\dfrac{(a^3 + b^3)(b^2 - a^2)}{ab \cdot a^2b^2}}{\dfrac{(a^2 - ab + b^2)(b^2 + 2ab + a^2)}{a^2b^2}}$$

$$= \frac{(a^3 + b^3)(b^2 - a^2)}{ab \cdot a^2b^2} \cdot \frac{a^2b^2}{(a^2 - ab + b^2)(b^2 + 2ab + a^2)}$$

$$= \frac{(a+b)(a^2 - ab + b^2)(b+a)(b-a)}{a^3b^3} \cdot \frac{a^2b^2}{(a^2 - ab + b^2)(b+a)^2}$$

$$= \frac{(a+b)^2(a^2 - ab + b^2)(a^2b^2)}{(a+b)^2(a^2 - ab + b^2)(a^2b^2)} \cdot \frac{b-a}{ab}$$

$$= \frac{b - a}{ab}$$

33. The reciprocal of $\dfrac{1 - \dfrac{1}{a}}{a - 1}$ is $\dfrac{a - 1}{1 - \dfrac{1}{a}}$.

$$\frac{a - 1}{1 - \dfrac{1}{a}}$$

$$= \frac{a - 1}{\dfrac{a - 1}{a}}$$ Adding in the denominator

$$= \frac{a - 1}{1} \cdot \frac{a}{a - 1}$$ Multiplying by the reciprocal of the divisor

$$= \frac{(a-1)(a)}{(1)(a-1)}$$

$$= a$$

Exercise Set 6.7

1. $y = kx$

$40 = k \cdot 8$ Substituting

$5 = k$ Solving for k

The variation constant is 5.

The equation of variation is $y = 5x$.

3. $y = kx$

$4 = k \cdot 30$ Substituting

$\dfrac{4}{30} = k$, or Solving for k

$\dfrac{2}{15} = k$ Simplifying

The variation constant is $\dfrac{2}{15}$.

The equation of variation is $y = \dfrac{2}{15}x$.

5. $y = kx$

$0.9 = k \cdot 0.4$ Substituting

$\dfrac{0.9}{0.4} = k$, or

$\dfrac{9}{4} = k$

The variation constant is $\dfrac{9}{4}$.

The equation of variation is $y = \dfrac{9}{4}x$.

7. Let p = the number of people using the cans.

$N = kp$ N varies directly as p.

$60,000 = k \cdot 250$ Substituting

$\dfrac{60,000}{250} = k$ Solving for k

$240 = k$ Variation constant

$N = 240p$ Equation of variation

$N = 240(1,007,000)$ Substituting

$N = 241,680,000$

In Dallas 241,680,000 cans are used each year.

9. $d = kw$ d varies directly as w.

$40 = k \cdot 3$ Substituting

$\dfrac{40}{3} = k$ Variation constant

$d = \dfrac{40}{3}w$ Equation of variation

$d = \dfrac{40}{3} \cdot 5$ Substituting

$d = \dfrac{200}{3}$, or $66\dfrac{2}{3}$

The spring is stretched $66\dfrac{2}{3}$ cm by a 5-kg barbell.

11. Let F = the number of grams of fat and w = the weight.

$F = kw$ F varies directly as w.

$60 = k \cdot 120$ Substituting

$\dfrac{60}{120} = k$, or Solving for k

$\dfrac{1}{2} = k$ Variation constant

$$F = \frac{1}{2}w \qquad \text{Equation of variation}$$

$$F = \frac{1}{2} \cdot 180 \qquad \text{Substituting}$$

$$F = 90$$

The maximum daily fat intake for a person weighing 180 lb is 90 g.

13.
$$y = \frac{k}{x}$$

$$14 = \frac{k}{7} \qquad \text{Substituting}$$

$$7 \cdot 14 = k \qquad \text{Solving for } k$$

$$98 = k$$

The variation constant is 98.

The equation of variation is $y = \dfrac{98}{x}$.

15.
$$y = \frac{k}{x}$$

$$3 = \frac{k}{12} \qquad \text{Substituting}$$

$$12 \cdot 3 = k \qquad \text{Solving for } k$$

$$36 = k$$

The variation constant is 36.

The equation of variation is $y = \dfrac{36}{x}$.

17.
$$y = \frac{k}{x}$$

$$0.1 = \frac{k}{0.5} \qquad \text{Substituting}$$

$$0.5(0.1) = k \qquad \text{Solving for } k$$

$$0.05 = k$$

The variation constant is 0.05.

The equation of variation is $y = \dfrac{0.05}{x}$.

19.
$$P = \frac{k}{W} \qquad P \text{ varies inversely as } W.$$

$$330 = \frac{k}{3.2} \qquad \text{Substituting}$$

$$1056 = k \qquad \text{Solving for } k, \text{ the variation constant}$$

$$P = \frac{1056}{W} \qquad \text{Equation of variation}$$

$$550 = \frac{1056}{W} \qquad \text{Substituting}$$

$$550W = 1056 \qquad \text{Multiplying by } W$$

$$W = \frac{1056}{550} \qquad \text{Dividing by 550}$$

$$W = 1.92 \qquad \text{Simplifying}$$

A tone with a pitch of 550 vibrations per second has a wavelength of 1.92 ft.

21.
$$W = \frac{k}{F} \qquad W \text{ varies inversely as } F.$$

$$300 = \frac{k}{1200} \qquad \text{Substituting}$$

$$360,000 = k \qquad \text{Variation constant}$$

$$W = \frac{360,000}{F} \qquad \text{Equation of variation}$$

$$W = \frac{360,000}{800} \qquad \text{Substituting}$$

$$W = 450$$

A wave with a frequency of 800 kHz has a length of 450 m.

23.
$$y = kx^2$$

$$0.15 = k(0.1)^2 \qquad \text{Substituting}$$

$$0.15 = 0.01k$$

$$\frac{0.15}{0.01} = k$$

$$15 = k$$

The equation of variation is $y = 15x^2$.

25.
$$y = \frac{k}{x^2}$$

$$0.15 = \frac{k}{(0.1)^2} \qquad \text{Substituting}$$

$$0.15 = \frac{k}{0.01}$$

$$0.15(0.01) = k$$

$$0.0015 = k$$

The equation of variation is $y = \dfrac{0.0015}{x^2}$.

27.
$$y = kxz$$

$$56 = k \cdot 7 \cdot 8 \qquad \text{Substituting}$$

$$56 = 56k$$

$$1 = k$$

The equation of variation is $y = xz$.

29.
$$y = kxz^2$$

$$105 = k \cdot 14 \cdot 5^2 \qquad \text{Substituting}$$

$$105 = 350k$$

$$\frac{105}{350} = k$$

$$\frac{3}{10} = k$$

The equation of variation is $y = \dfrac{3}{10}xz^2$.

31.
$$y = k\frac{xz}{wp}$$

$$\frac{3}{28} = k\frac{3 \cdot 10}{7 \cdot 8} \quad \text{Substituting}$$

$$\frac{3}{28} = k \cdot \frac{30}{56}$$

$$\frac{3}{28} \cdot \frac{56}{30} = k$$

$$\frac{1}{5} = k$$

The equation of variation is $y = \frac{xz}{5wp}$.

33.
$$d = kr^2$$

$$200 = k \cdot 60^2 \quad \text{Substituting}$$

$$200 = 3600k$$

$$\frac{200}{3600} = k$$

$$\frac{1}{18} = k$$

The equation of variation is $d = \frac{1}{18}r^2$.

Substitute 80 for r and find d.

$$d = \frac{1}{18} \cdot 80^2 = \frac{6400}{18} = 355\frac{5}{9}$$

It will take $355\frac{5}{9}$ ft to stop when traveling 80 mph.

35. $W = \dfrac{k}{d^2}$

We first find k.

$$220 = \frac{k}{(3978)^2} \quad \text{Substituting}$$

$$220 = \frac{k}{15,824,484}$$

$$3,481,386,480 = k$$

The equation of variation is $W = \dfrac{3,481,386,480}{d^2}$.

Substitute 3978 + 200, or 4178 for d and find W.

$$W = \frac{3,481,386,480}{(4178)^2} = \frac{3,481,386,480}{17,455,684} \approx 199.4$$

When the astronaut is 200 km above the surface of the earth, his weight is about 199.4 lb.

37. $E = \dfrac{kR}{I}$

We first find k.

$$2.89 = \frac{k \cdot 78}{243} \quad \text{Substituting}$$

$$2.89\left(\frac{243}{78}\right) = k \quad \text{Multiplying by } \frac{243}{78}$$

$$9 \approx k$$

The equation of variation is $E = \dfrac{9R}{I}$.

Substitute 2.89 for E and 300 for I and solve R.

$$2.89 = \frac{9R}{300}$$

$$2.89\left(\frac{300}{9}\right) = R \quad \text{Multiplying by } \frac{300}{9}$$

$$96 \approx R \quad \text{Rounding}$$

Jim Abbott would have given up 96 earned runs if he had pitched 300 innings.

39. $Q = kd^2$

We first find k.

$$225 = k \cdot 5^2$$

$$225 = 25k$$

$$9 = k$$

The equation of variation is $Q = 9d^2$.

Substitute 9 for d and compute Q.

$$Q = 9 \cdot 9^2$$

$$Q = 9 \cdot 81$$

$$Q = 729$$

729 gallons of water are emptied by a pipe that is 9 in. in diameter.

41. *Familiarize*. Let x = the number of correct answers and y = the number of incorrect answers. Then $2x$ points are awarded for the correct answers, and $\frac{1}{2} \cdot y$, or $\frac{y}{2}$, points are deducted for the incorrect answers.

Translate.

$$\underbrace{\text{The total number of answers}}_{x + y} \ \underbrace{\text{is}}_{=} \ \underbrace{75.}_{75}$$

$$\underbrace{\text{The total score}}_{2x - \frac{y}{2}} \ \underbrace{\text{is}}_{=} \ \underbrace{100.}_{100}$$

We have a system of equations:

$$x + y = 75,$$
$$2x - \frac{y}{2} = 100$$

Solve. Solving the system, we get $(55, 20)$.

Check. If there are 55 correct answers and 20 incorrect answers, the total number of answers is 75. Also, $2 \cdot 55$, or 110, points are awarded for the correct answers, and $\frac{1}{2} \cdot 20$, or 10, points are deducted for the incorrect answers. The score is $110 - 10$, or 100. The solution checks.

State. There were 55 correct answers and 20 incorrect answers.

43.
$$I = kP \quad I \text{ varies directly as } P.$$

$$1665 = k \cdot 9000 \quad \text{Substituting}$$

$$\frac{1665}{9000} = k$$

$$0.185 = k \quad \text{Variation constant}$$

The equation of variation is $I = 0.185P$.

45. We are told $A = kd^2$, and we know $A = \pi r^2$ so we have:

$$kd^2 = \pi r^2$$

$$kd^2 = \pi \left(\frac{d}{2}\right)^2 \qquad r = \frac{d}{2}$$

$$kd^2 = \frac{\pi d^2}{4}$$

$$k = \frac{\pi}{4} \qquad \text{Variation constant}$$

47. $Q = \dfrac{kp^2}{q^3}$

Q varies directly as the square of p and inversely as the cube of q.

Chapter 7

Radical Expressions and Equations

Exercise Set 7.1

1. The square roots of 16 are 4 and -4, because $4^2 = 16$ and $(-4)^2 = 16$.

3. The square roots of 144 are 12 and -12, because $12^2 = 144$ and $(-12)^2 = 144$.

5. The square roots of 400 are 20 and -20, because $20^2 = 400$ and $(-20)^2 = 400$.

7. $-\sqrt{\dfrac{49}{36}} = -\dfrac{7}{6}$ Since $\sqrt{\dfrac{49}{36}} = \dfrac{7}{6}$, $-\sqrt{\dfrac{49}{36}} = -\dfrac{7}{6}$.

9. $\sqrt{196} = 14$ Remember, $\sqrt{}$ indicates the principle square root.

11. $-\sqrt{\dfrac{16}{81}} = -\dfrac{4}{9}$ Since $\sqrt{\dfrac{16}{81}} = \dfrac{4}{9}$, $-\sqrt{\dfrac{16}{81}} = -\dfrac{4}{9}$.

13. $\sqrt{0.04} = 0.2$

15. $-\sqrt{0.0009} = -0.03$

17. $9\sqrt{y^2 + 16}$

The radicand is the expression written under the radical sign, $y^2 + 16$.

19. $x^4 y^5 \sqrt{\dfrac{x}{y-1}}$

The radicand is the expression written under the radical sign, $\dfrac{x}{y-1}$.

21. $\sqrt{16x^2} = \sqrt{(4x)^2} = |4x| = 4|x|$

(The absolute value is used to ensure that the principal square root is nonnegative.)

23. $\sqrt{(-12c)^2} = |-12c| = |-12| \cdot |c| = 12|c|$

(The absolute value is used to ensure that the principal square root is nonnegative.)

25. $\sqrt{(p+3)^2} = |p+3|$

(The absolute value is used to ensure that the principal square root is nonnegative.)

27. $\sqrt{x^2 - 4x + 4} = \sqrt{(x-2)^2} = |x-2|$

(The absolute value is used to ensure that the principal square root is nonnegative.)

29. $\sqrt{4x^2 + 28x + 49} = \sqrt{(2x+7)^2} = |2x+7|$

(The absolute value is used to ensure that the principal square root is nonnegative.)

31. $\sqrt[3]{27} = 3$ $[3^3 = 27]$

33. $\sqrt[3]{-64x^3} = -4x$ $[(-4x)^3 = -64x^3]$

35. $\sqrt[3]{-216} = -6$ $[(-6)^3 = -216]$

37. $\sqrt[3]{0.343(x+1)^3} = 0.7(x+1)$

$[(0.7(x+1))^3 = 0.343(x+1)^3]$

39. $\sqrt[4]{625} = 5$ Since $5^4 = 625$

41. $\sqrt[5]{-1} = -1$ Since $(-1)^5 = -1$

43. $\sqrt[5]{-\dfrac{32}{243}} = -\dfrac{2}{3}$ Since $\left(-\dfrac{2}{3}\right)^5 = -\dfrac{32}{243}$

45. $\sqrt[6]{x^6} = |x|$

The index is even so we use absolute-value notation.

47. $\sqrt[4]{(5a)^4} = |5a| = 5|a|$

The index is even so we use absolute-value notation.

49. $\sqrt[10]{(-6)^{10}} = |-6| = 6$

51. $\sqrt[414]{(a+b)^{414}} = |a+b|$

The index is even so we use absolute-value notation.

53. $\sqrt[7]{y^7} = y$

We do not use absolute-value notation when the index is odd.

55. $\sqrt[5]{(x-2)^5} = x-2$

We do not use absolute-value notation when the index is odd.

57. $x^2 + x - 2 = 0$

$(x+2)(x-1) = 0$ Factoring

$x + 2 = 0$ or $x - 1 = 0$ Principle of zero products

$x = -2$ or $x = 1$

The solutions are -2 and 1.

59. $4x^2 - 49 = 0$

$(2x+7)(2x-7) = 0$ Factoring

$2x + 7 = 0$ or $2x - 7 = 0$ Principle of zero products

$2x = -7$ or $2x = 7$

$x = -\dfrac{7}{2}$ or $x = \dfrac{7}{2}$

The solutions are $-\dfrac{7}{2}$ and $\dfrac{7}{2}$.

61.
$$3x^2 + x = 10$$
$$3x^2 + x - 10 = 0$$
$$(3x - 5)(x + 2) = 0$$
$$3x - 5 = 0 \quad \text{or} \quad x + 2 = 0$$
$$3x = 5 \quad \text{or} \quad x = -2$$
$$x = \frac{5}{3} \quad \text{or} \quad x = -2$$

The solutions are $\frac{5}{3}$ and -2.

63. $N = 2.5\sqrt{A}$

a) $N = 2.5\sqrt{25} = 2.5(5) = 12.5 \approx 13$

b) $N = 2.5\sqrt{36} = 2.5(6) = 15$

c) $N = 2.5\sqrt{49} = 2.5(7) = 17.5 \approx 18$

d) $N = 2.5\sqrt{64} = 2.5(8) = 20$

Exercise Set 7.2

1. $\sqrt{24} = \sqrt{4 \cdot 6} = \sqrt{4}\,\sqrt{6} = 2\sqrt{6}$

3. $\sqrt{90} = \sqrt{9 \cdot 10} = \sqrt{9}\,\sqrt{10} = 3\sqrt{10}$

5. $\sqrt[3]{250} = \sqrt[3]{125 \cdot 2} = \sqrt[3]{125}\,\sqrt[3]{2} = 5\sqrt[3]{2}$

7. $\sqrt{180x^4} = \sqrt{36 \cdot 5 \cdot x^4} = \sqrt{36x^4}\,\sqrt{5} = 6x^2\sqrt{5}$

9. $\sqrt[3]{54x^8} = \sqrt[3]{27 \cdot 2 \cdot x^6 \cdot x^2} = \sqrt[3]{27x^6}\,\sqrt[3]{2x^2} = 3x^2\sqrt[3]{2x^2}$

11. $\sqrt[3]{80t^8} = \sqrt[3]{8 \cdot 10 \cdot t^6 \cdot t^2} = \sqrt[3]{8t^6}\,\sqrt[3]{10t^2} = 2t^2\sqrt[3]{10t^2}$

13. $\sqrt[4]{80} = \sqrt[4]{16 \cdot 5} = \sqrt[4]{16}\,\sqrt[4]{5} = 2\sqrt[4]{5}$

15. $\sqrt[4]{243x^8y^{10}} = \sqrt[4]{81 \cdot 3 \cdot x^8 \cdot y^8 \cdot y^2} = \sqrt[4]{81x^8y^8}\,\sqrt[4]{3y^2} = 3x^2y^2\sqrt[4]{3y^2}$

17. $\sqrt[3]{(p+q)^4} = \sqrt[3]{(p+q)^3(p+q)} = \sqrt[3]{(p+q)^3}\,\sqrt[3]{p+q} = (p+q)\sqrt[3]{p+q}$

19. $\sqrt[3]{-24x^4y^5} = \sqrt[3]{-8 \cdot 3 \cdot x^3 \cdot x \cdot y^3 \cdot y^2} = \sqrt[3]{-8x^3y^3}\,\sqrt[3]{3xy^2} = -2xy\sqrt[3]{3xy^2}$

21. $\sqrt[5]{96x^7y^{15}} = \sqrt[5]{32 \cdot 3 \cdot x^5 \cdot x^2 \cdot y^{15}} = \sqrt[5]{32x^5y^{15}}\,\sqrt[5]{3x^2} = 2xy^3\sqrt[5]{3x^2}$

23. $\sqrt{15}\,\sqrt{6} = \sqrt{15 \cdot 6} = \sqrt{90}$
$$= \sqrt{9 \cdot 10} = \sqrt{9}\,\sqrt{10} = 3\sqrt{10}$$

25. $\sqrt[3]{3}\,\sqrt[3]{18} = \sqrt[3]{3 \cdot 18} = \sqrt[3]{54}$
$$= \sqrt[3]{27 \cdot 2} = \sqrt[3]{27}\,\sqrt[3]{2} = 3\sqrt[3]{2}$$

27. $\sqrt{45}\,\sqrt{60} = \sqrt{45 \cdot 60} = \sqrt{2700}$
$$= \sqrt{900 \cdot 3} = \sqrt{900}\,\sqrt{3} = 30\sqrt{3}$$

29. $\sqrt{5b^3}\,\sqrt{10c^4} = \sqrt{5b^3 \cdot 10c^4}$
$$= \sqrt{50b^3c^4}$$
$$= \sqrt{25 \cdot 2 \cdot b^2 \cdot b \cdot c^4}$$
$$= \sqrt{25b^2c^4}\,\sqrt{2b}$$
$$= 5bc^2\sqrt{2b}$$

31. $\sqrt[3]{y^4}\,\sqrt[3]{16y^5} = \sqrt[3]{y^4 \cdot 16y^5}$
$$= \sqrt[3]{16y^9}$$
$$= \sqrt[3]{8 \cdot 2 \cdot y^9}$$
$$= \sqrt[3]{8y^9}\,\sqrt[3]{2}$$
$$= 2y^3\sqrt[3]{2}$$

33. $\sqrt[3]{(c+6)^4}\,\sqrt[3]{(c+6)^2} = \sqrt[3]{(c+6)^4(c+6)^2}$
$$= \sqrt[3]{(c+6)^6}$$
$$= (c+6)^2$$

35. $\sqrt{12a^3b}\,\sqrt{8a^4b^2} = \sqrt{12a^3b \cdot 8a^4b^2}$
$$= \sqrt{96a^7b^3}$$
$$= \sqrt{16 \cdot 6 \cdot a^6 \cdot a \cdot b^2 \cdot b}$$
$$= \sqrt{16a^6b^2}\,\sqrt{6ab}$$
$$= 4a^3b\sqrt{6ab}$$

37. $\sqrt[4]{16} \cdot \sqrt[4]{64} = \sqrt[4]{16 \cdot 64} = \sqrt[4]{1024} = \sqrt[4]{256 \cdot 4} = \sqrt[4]{256}\,\sqrt[4]{4} = 4\sqrt[4]{4}$

39. $\sqrt[4]{10a^3} \cdot \sqrt[4]{8a^2} = \sqrt[4]{10a^3 \cdot 8a^2} = \sqrt[4]{80a^5} = \sqrt[4]{16 \cdot 5 \cdot a^4 \cdot a} = \sqrt[4]{16a^4}\,\sqrt[4]{5a} = 2a\sqrt[4]{5a}$

41. $\sqrt{30x^3y^4}\,\sqrt{18x^2y^5} = \sqrt{30x^3y^4 \cdot 18x^2y^5} = \sqrt{540x^5y^9} = \sqrt{36 \cdot 15 \cdot x^4 \cdot x \cdot y^8 \cdot y} = \sqrt{36x^4y^8}\,\sqrt{15xy} = 6x^2y^4\sqrt{15xy}$

43. $\sqrt[5]{a^3(b-c)^7}\,\sqrt[5]{a^8(b-c)^{11}} =$
$\sqrt[5]{a^3(b-c)^7 \cdot a^8(b-c)^{11}} = \sqrt[5]{a^{11}(b-c)^{18}} =$
$\sqrt[5]{a^{10} \cdot a \cdot (b-c)^{15} \cdot (b-c)^3} =$
$\sqrt[5]{a^{10}(b-c)^{15}}\,\sqrt[5]{a(b-c)^3} = a^2(b-c)^3\sqrt[5]{a(b-c)^3}$

45. $\sqrt{140} \approx 11.83215957 \approx 11.832$

47. $\dfrac{8 + \sqrt{480}}{4} \approx \dfrac{8 + 21.9089023}{4} \approx \dfrac{29.9089023}{4}$
$$\approx 7.477225575 \approx 7.477$$

49. $\dfrac{16 - \sqrt{48}}{20} \approx \dfrac{16 - 6.92820323}{20} \approx \dfrac{9.07179677}{20}$
$$\approx 0.453589838 \approx 0.454$$

51. $\dfrac{24 + \sqrt{128}}{8} \approx \dfrac{24 + 11.3137085}{8} \approx \dfrac{35.3137085}{8}$
$$\approx 4.414213562 \approx 4.414$$

53. $r = 2\sqrt{5L}$

a) $r = 2\sqrt{5 \cdot 20}$

$= 2\sqrt{100}$

$= 2 \cdot 10 = 20$ mph

b) $r = 2\sqrt{5 \cdot 70}$

$= 2\sqrt{350}$

$\approx 2 \times 18.708$ Using a calculator

≈ 37.4 mph Multiplying and rounding

c) $r = 2\sqrt{5 \cdot 90}$

$= 2\sqrt{450}$

$\approx 2 \times 21.213$ Using a calculator

≈ 42.4 mph Multiplying and rounding

55. $x^2 - \dfrac{2}{3}x = 0$

$x\left(x - \dfrac{2}{3}\right) = 0$ Factoring

$x = 0$ or $x - \dfrac{2}{3} = 0$ Principle of zero products

$x = 0$ or $x = \dfrac{2}{3}$

The solutions are 0 and $\dfrac{2}{3}$.

57. $\dfrac{x^2}{x-2} = \dfrac{4}{x-2}$, LCM is $x - 2$

$(x-2)\dfrac{x^2}{x-2} = (x-2) \cdot \dfrac{4}{x-2}$

$x^2 = 4$

$x^2 - 4 = 0$

$(x+2)(x-2) = 0$

$x + 2 = 0$ or $x - 2 = 0$

$x = -2$ or $x = 2$

2 cannot be a solution, because it makes the denominators 0; -2 checks and is the solution.

59. $\dfrac{x^3 - y^3}{x+y} \cdot \dfrac{x^2 - y^2}{x^2 + xy + y^2}$

$= \dfrac{(x^3 - y^3)(x^2 - y^2)}{(x+y)(x^2 + xy + y^2)}$

$= \dfrac{(x-y)(x^2 + xy + y^2)(x+y)(x-y)}{(x+y)(x^2 + xy + y^2)(1)}$

$= \dfrac{(x+y)(x^2 + xy + y^2)}{(x+y)(x^2 + xy + y^2)} \cdot \dfrac{(x-y)(x-y)}{1}$ Removing a factor of 1

$= (x-y)^2$

61. $\sqrt{968}\,\sqrt{1014} = \sqrt{968 \cdot 1014}$

$= \sqrt{4 \cdot 121 \cdot 2 \cdot 169 \cdot 2 \cdot 3}$

$= \sqrt{4 \cdot 121 \cdot 2 \cdot 2 \cdot 169 \cdot 3}$

$= \sqrt{4}\,\sqrt{121}\,\sqrt{2 \cdot 2}\,\sqrt{169}\,\sqrt{3}$

$= 2 \cdot 11 \cdot 2 \cdot 13\sqrt{3}$

$= 572\sqrt{3}$

63. $\sqrt[3]{2x^{t+3}} \cdot \sqrt[3]{32x^t} = 4x^9$

$\sqrt[3]{2x^{t+3} \cdot 32x^t} = 4x^9$

$\sqrt[3]{64x^{2t+3}} = 4x^9$ Adding exponents

$\sqrt[3]{64 \cdot x^{2t} \cdot x^3} = 4x^9$

$\sqrt[3]{64}\,\sqrt[3]{x^3}\,\sqrt[3]{x^{2t}} = 4x^9$

$4x\sqrt[3]{x^{2t}} = 4x^9$

$\sqrt[3]{x^{2t}} = x^8$ Dividing by $4x$, assuming $x \neq 0$

Now, if $\sqrt[3]{x^{2t}} = x^8$, then it must be true that $x^{2t} = x^{24}$. Since the bases are the same, the exponents must be the same, so we have:

$2t = 24$

$t = 12$

The solution is 12.

Exercise Set 7.3

1. $\dfrac{\sqrt{35q}}{\sqrt{7q}} = \sqrt{\dfrac{35q}{7q}} = \sqrt{5}$

3. $\dfrac{\sqrt[3]{54}}{\sqrt[3]{2}} = \sqrt[3]{\dfrac{54}{2}} = \sqrt[3]{27} = 3$

5. $\dfrac{\sqrt{56xy^3}}{\sqrt{8x}} = \sqrt{\dfrac{56xy^3}{8x}} = \sqrt{7y^3} = \sqrt{y^2 \cdot 7y} =$

$\sqrt{y^2}\,\sqrt{7y} = y\sqrt{7y}$

7. $\dfrac{\sqrt[3]{96a^4b^2}}{\sqrt[3]{12a^2b}} = \sqrt[3]{\dfrac{96a^4b^2}{12a^2b}} = \sqrt[3]{8a^2b} = \sqrt[3]{8}\,\sqrt[3]{a^2b} = 2\sqrt[3]{a^2b}$

9. $\dfrac{\sqrt{128xy}}{2\sqrt{2}} = \dfrac{1}{2}\dfrac{\sqrt{128xy}}{\sqrt{2}} = \dfrac{1}{2}\sqrt{\dfrac{128xy}{2}} = \dfrac{1}{2}\sqrt{64xy} =$

$\dfrac{1}{2}\sqrt{64}\,\sqrt{xy} = \dfrac{1}{2} \cdot 8\sqrt{xy} = 4\sqrt{xy}$

11. $\dfrac{\sqrt[4]{48x^9y^{13}}}{\sqrt[4]{3xy^5}} = \sqrt[4]{\dfrac{48x^9y^{13}}{3xy^5}} = \sqrt[4]{16x^8y^8} = 2x^2y^2$

13. $\dfrac{\sqrt{x^3 - y^3}}{\sqrt{x-y}} = \sqrt{\dfrac{x^3 - y^3}{x-y}} = \sqrt{\dfrac{(x-y)(x^2 + xy + y^2)}{x-y}} =$

$\sqrt{x^2 + xy + y^2}$

15. $\sqrt{\dfrac{16}{49}} = \dfrac{\sqrt{16}}{\sqrt{49}} = \dfrac{4}{7}$

17. $\sqrt[3]{\dfrac{125}{27}} = \dfrac{\sqrt[3]{125}}{\sqrt[3]{27}} = \dfrac{5}{3}$

19. $\sqrt{\dfrac{25}{y^2}} = \dfrac{\sqrt{25}}{\sqrt{y^2}} = \dfrac{5}{y}$

21. $\sqrt{\dfrac{25y^3}{x^4}} = \dfrac{\sqrt{25y^3}}{\sqrt{x^4}} = \dfrac{\sqrt{25y^2 \cdot y}}{\sqrt{x^4}} = \dfrac{\sqrt{25y^2}\,\sqrt{y}}{\sqrt{x^4}} = \dfrac{5y\sqrt{y}}{x^2}$

23. $\sqrt[3]{\dfrac{8x^5}{27y^3}} = \dfrac{\sqrt[3]{8x^5}}{\sqrt[3]{27y^3}} = \dfrac{\sqrt[3]{8x^3 \cdot x^2}}{\sqrt[3]{27y^3}} = \dfrac{\sqrt[3]{8x^3}\,\sqrt[3]{x^2}}{\sqrt[3]{27y^3}} = \dfrac{2x\sqrt[3]{x^2}}{3y}$

25. $\sqrt[4]{\dfrac{81x^4}{16}} = \dfrac{\sqrt[4]{81x^4}}{\sqrt[4]{16}} = \dfrac{3x}{2}$

27. $\sqrt[4]{\dfrac{p^5q^8}{r^{12}}} = \dfrac{\sqrt[4]{p^5q^8}}{\sqrt[4]{r^{12}}} = \dfrac{\sqrt[4]{p^4 \cdot p \cdot q^8}}{\sqrt[4]{r^{12}}} = \dfrac{\sqrt[4]{p^4q^8}\,\sqrt[4]{p}}{\sqrt[4]{r^{12}}} =$

$\dfrac{pq^2\sqrt[4]{p}}{r^3}$

29. $\sqrt[5]{\dfrac{32x^8}{y^{10}}} = \dfrac{\sqrt[5]{32x^8}}{\sqrt[5]{y^{10}}} = \dfrac{\sqrt[5]{32 \cdot x^5 \cdot x^3}}{\sqrt[5]{y^{10}}} = \dfrac{\sqrt[5]{32x^5}\,\sqrt[5]{x^3}}{\sqrt[5]{y^{10}}} =$

$\dfrac{2x\sqrt[5]{x^3}}{y^2}$

31. $\sqrt[6]{\dfrac{x^{13}}{y^6z^{12}}} = \dfrac{\sqrt[6]{x^{13}}}{\sqrt[6]{y^6z^{12}}} = \dfrac{\sqrt[6]{x^{12} \cdot x}}{\sqrt[6]{y^6z^{12}}} = \dfrac{\sqrt[6]{x^{12}}\,\sqrt[6]{x}}{\sqrt[6]{y^6z^{12}}} = \dfrac{x^2\sqrt[6]{x}}{yz^2}$

33. a) $\sqrt{(6a)^3} = \sqrt{6^3a^2} = \sqrt{6^2a^2}\,\sqrt{6a} = 6a\sqrt{6a}$

   b) $(\sqrt{6a})^3 = \sqrt{6a}\,\sqrt{6a}\,\sqrt{6a} = 6a\sqrt{6a}$

35. a) $(\sqrt[3]{16b^2})^2 = \sqrt[3]{16b^2}\,\sqrt[3]{16b^2} = \sqrt[3]{256b^4} =$

   $\sqrt[3]{64 \cdot 4 \cdot b^3 \cdot b} = 4b\sqrt[3]{4b}$

   b) $\sqrt[3]{(16b^2)^2} = \sqrt[3]{256b^4} = 4b\sqrt[3]{4b}$, as in part (a)

37. a) $\sqrt{(18a^2b)^3} = \sqrt{5832a^6b^3} = \sqrt{2916a^6b^2 \cdot 2b} =$

   $54a^3b\sqrt{2b}$

   b) $(\sqrt{18a^2b})^3 = \sqrt{18a^2b}\,\sqrt{18a^2b}\,\sqrt{18a^2b} =$

   $18a^2b\sqrt{18a^2b} = 18a^2b\sqrt{9a^2 \cdot 2b} = 18a^2b \cdot 3a\sqrt{2b} =$

   $54a^3b\sqrt{2b}$

39. a) $(\sqrt[3]{12c^2d})^2 = \sqrt[3]{12c^2d} \cdot \sqrt[3]{12c^2d} = \sqrt[3]{144c^4d^2} =$

   $\sqrt[3]{8 \cdot 18 \cdot c^3 \cdot c \cdot d^2} = 2c\sqrt[3]{18cd^2}$

   b) $\sqrt[3]{(12c^2d)^2} = \sqrt[3]{144c^4d^2} = 2c\sqrt[3]{18cd^2}$, as in part (a)

41. a) $\sqrt[3]{(7x^2y)^2} = \sqrt[3]{49x^4y^2} = \sqrt[3]{49 \cdot x^3 \cdot x \cdot y^2} =$

   $x\sqrt[3]{49xy^2}$

   b) $(\sqrt[3]{7x^2y})^2 = \sqrt[3]{7x^2y} \cdot \sqrt[3]{7x^2y} = \sqrt[3]{49x^4y} =$

   $x\sqrt[3]{49xy^2}$, as in part (a)

43. a) $(\sqrt[4]{81xy^5})^2 = \sqrt[4]{81xy^5} \cdot \sqrt[4]{81xy^5} = \sqrt[4]{6561x^2y^{10}} =$

   $\sqrt[4]{6561 \cdot x^2 \cdot y^8 \cdot y^2} = 9y^2\sqrt[4]{x^2y^2}$

   b) $\sqrt[4]{(81xy^5)^2} = \sqrt[4]{6561x^2y^{10}} = 9y^2\sqrt[4]{x^2y^2}$, as in

   part (a)

45.

$\dfrac{12x}{x-4} - \dfrac{3x^2}{x+4} = \dfrac{384}{x^2-16}$

$\dfrac{12x}{x-4} - \dfrac{3x^2}{x+4} = \dfrac{384}{(x+4)(x-4)}$,

LCM is $(x+4)(x-4)$.

$(x+4)(x-4)\left[\dfrac{12x}{x-4} - \dfrac{3x^2}{x+4}\right] = (x+4)(x-4) \cdot \dfrac{384}{(x+4)(x-4)}$

$12x(x+4) - 3x^2(x-4) = 384$

$12x^2 + 48x - 3x^3 + 12x^2 = 384$

$-3x^3 + 24x^2 + 48x - 384 = 0$

$-3(x^3 - 8x^2 - 16x + 128) = 0$

$-3[x^2(x-8) - 16(x-8)] = 0$

$-3(x-8)(x^2-16) = 0$

$-3(x-8)(x+4)(x-4) = 0$

$x - 8 = 0$ or $x + 4 = 0$ or $x - 4 = 0$

$x = 8$ or $x = -4$ or $x = 4$

Check: For 8:

$$\dfrac{12x}{x-4} - \dfrac{3x^2}{x+4} = \dfrac{384}{x^2-16}$$

$$\begin{array}{c|c} \dfrac{12 \cdot 8}{8-4} - \dfrac{3 \cdot 8^2}{8+4} & \dfrac{384}{8^2-16} \\[2mm] \dfrac{96}{4} - \dfrac{192}{12} & \dfrac{384}{48} \\[2mm] 24 - 16 & 8 \\[1mm] 8 & \text{TRUE} \end{array}$$

8 is a solution.

For -4:

$$\dfrac{12x}{x-4} - \dfrac{3x^2}{x+4} = \dfrac{384}{x^2-16}$$

$$\begin{array}{c|c} \dfrac{12(-4)}{-4-4} - \dfrac{3(-4)^2}{-4+4} & \dfrac{384}{(-4)^2-16} \\[2mm] \dfrac{-48}{-8} - \dfrac{48}{0} & \dfrac{384}{16-16} \quad \text{UNDEFINED} \end{array}$$

-4 is not a solution.

For 4:

$$\dfrac{12x}{x-4} - \dfrac{3x^2}{x+4} = \dfrac{384}{x^2-16}$$

$$\begin{array}{c|c} \dfrac{12 \cdot 4}{4-4} - \dfrac{3 \cdot 4^2}{4+4} & \dfrac{384}{4^2-16} \\[2mm] \dfrac{48}{0} - \dfrac{48}{8} & \dfrac{384}{16-16} \quad \text{UNDEFINED} \end{array}$$

4 is not a solution.

The solution is 8.

47. $\dfrac{x+1}{2} - \dfrac{x-3}{3} = 3$,   LCM is 6

$$6\left(\dfrac{x+1}{2} - \dfrac{x-3}{3}\right) = 6 \cdot 3$$

$$6 \cdot \dfrac{x+1}{2} - 6 \cdot \dfrac{x-3}{3} = 18$$

$$3(x+1) - 2(x-3) = 18$$

$$3x + 3 - 2x + 6 = 18$$

$$x + 9 = 18$$

$$x = 9$$

Check: $\dfrac{x+1}{2} - \dfrac{x-3}{3} = 3$

$$\begin{array}{c|c} \dfrac{9+1}{2} - \dfrac{9-3}{3} & 3 \\[2mm] \dfrac{10}{2} - \dfrac{6}{3} & \\[2mm] 5 - 2 & \\[2mm] 3 & \text{TRUE} \end{array}$$

The solution is 9.

49.
$$\dfrac{4x}{x+5} + \dfrac{20}{x} = \dfrac{100}{x^2 + 5x}$$

$$\dfrac{4x}{x+5} + \dfrac{20}{x} = \dfrac{100}{x(x+5)},$$

$$\text{LCM is } x(x+5)$$

$$x(x+5)\left(\dfrac{4x}{x+5} + \dfrac{20}{x}\right) = x(x+5) \cdot \dfrac{100}{x(x+5)}$$

$$x(x+5) \cdot \dfrac{4x}{x+5} + x(x+5) \cdot \dfrac{20}{x} = 100$$

$$x \cdot 4x + 20(x+5) = 100$$

$$4x^2 + 20x + 100 = 100$$

$$4x^2 + 20x = 0$$

$$4x(x+5) = 0$$

$$4x = 0 \quad \text{or} \quad x + 5 = 0$$

$$x = 0 \quad \text{or} \qquad x = -5$$

Neither value checks. (Each make a denominator 0.) The equation has no solution.

51. a) Substitute 65 for L and 3.14 for π.

$$T = 2\pi\sqrt{\dfrac{L}{980}}$$

$$\approx 2(3.14)\sqrt{\dfrac{65}{980}}$$

$$\approx 2(3.14)(0.257539376) \quad \text{Using a calculator}$$

$$\approx 1.62 \qquad\qquad\qquad \text{Rounding}$$

The period of a 65 cm pendulum is about 1.62 sec.

b) Substitute 98 for L and 3.14 for π.

$$T = 2\pi\sqrt{\dfrac{L}{980}}$$

$$\approx 2(3.14)\sqrt{\dfrac{98}{980}}$$

$$\approx 2(3.14)(0.316227766) \quad \text{Using a calculator}$$

$$\approx 1.99 \qquad\qquad\qquad \text{Rounding}$$

The period of a 98 cm pendulum is about 1.99 sec.

c) Substitute 120 for L and 3.14 for π.

$$T = 2\pi\sqrt{\dfrac{L}{980}}$$

$$\approx 2(3.14)\sqrt{\dfrac{120}{980}}$$

$$\approx 2(3.14)(0.349927106) \quad \text{Using a calculator}$$

$$\approx 2.20 \qquad\qquad\qquad \text{Rounding}$$

The period of a 120 cm pendulum is about 2.20 sec.

53.
$$\dfrac{(\sqrt[3]{81mn^2})^2}{(\sqrt[3]{mn})^2} = \dfrac{\sqrt[3]{(81mn^2)^2}}{\sqrt[3]{(mn)^2}}$$

$$= \dfrac{\sqrt[3]{6561m^2n^4}}{\sqrt[3]{m^2n^2}}$$

$$= \sqrt[3]{\dfrac{6561m^2n^4}{m^2n^2}}$$

$$= \sqrt[3]{6561n^2}$$

$$= \sqrt[3]{729 \cdot 9n^2}$$

$$= \sqrt[3]{729}\,\sqrt[3]{9n^2}$$

$$= 9\sqrt[3]{9n^2}$$

Exercise Set 7.4

1. $7\sqrt{5} + 4\sqrt{5} = (7+4)\sqrt{5}$ Factoring out $\sqrt{5}$
$$= 11\sqrt{5}$$

3. $6\sqrt[3]{7} - 5\sqrt[3]{7} = (6-5)\sqrt[3]{7}$ Factoring out $\sqrt[3]{7}$
$$= \sqrt[3]{7}$$

5. $4\sqrt[3]{y} + 9\sqrt[3]{y} = (4+9)\sqrt[3]{y} = 13\sqrt[3]{y}$

7. $5\sqrt{6} - 9\sqrt{6} - 4\sqrt{6} = (5-9-4)\sqrt{6} = -8\sqrt{6}$

9. $4\sqrt[3]{3} - \sqrt{5} + 2\sqrt[3]{3} + \sqrt{5} =$
$$(4+2)\sqrt[3]{3} + (-1+1)\sqrt{5} = 6\sqrt[3]{3}$$

11. $8\sqrt{27} - 3\sqrt{3} = 8\sqrt{9 \cdot 3} - 3\sqrt{3}$ $\Big\}$ Factoring the
$$= 8\sqrt{9} \cdot \sqrt{3} - 3\sqrt{3} \quad \text{first radical}$$
$$= 8 \cdot 3\sqrt{3} - 3\sqrt{3} \quad \text{Taking the square root}$$
$$= 24\sqrt{3} - 3\sqrt{3}$$
$$= (24-3)\sqrt{3} \quad \text{Factoring out } \sqrt{3}$$
$$= 21\sqrt{3}$$

13. $8\sqrt{45} + 7\sqrt{20} = 8\sqrt{9 \cdot 5} + 7\sqrt{4 \cdot 5}$ $\left.\begin{array}{l}\text{Factoring}\\\text{the}\\\text{radicals}\end{array}\right\}$

$= 8\sqrt{9} \cdot \sqrt{5} + 7\sqrt{4} \cdot \sqrt{5}$

$= 8 \cdot 3\sqrt{5} + 7 \cdot 2\sqrt{5}$ Taking the square roots

$= 24\sqrt{5} + 14\sqrt{5}$

$= (24 + 14)\sqrt{5}$ Factoring out $\sqrt{5}$

$= 38\sqrt{5}$

15. $18\sqrt{72} + 2\sqrt{98} = 18\sqrt{36 \cdot 2} + 2\sqrt{49 \cdot 2} =$
$18\sqrt{36} \cdot \sqrt{2} + 2\sqrt{49} \cdot \sqrt{2} = 18 \cdot 6\sqrt{2} + 2 \cdot 7\sqrt{2} =$
$108\sqrt{2} + 14\sqrt{2} = (108 + 14)\sqrt{2} = 122\sqrt{2}$

17. $3\sqrt[3]{16} + \sqrt[3]{54} = 3\sqrt[3]{8 \cdot 2} + \sqrt[3]{27 \cdot 2} =$
$3\sqrt[3]{8} \cdot \sqrt[3]{2} + \sqrt[3]{27} \cdot \sqrt[3]{2} = 3 \cdot 2\sqrt[3]{2} + 3\sqrt[3]{2} =$
$6\sqrt[3]{2} + 3\sqrt[3]{2} = (6 + 3)\sqrt[3]{2} = 9\sqrt[3]{2}$

19. $2\sqrt{128} - \sqrt{18} + 4\sqrt{32} =$
$2\sqrt{64 \cdot 2} - \sqrt{9 \cdot 2} + 4\sqrt{16 \cdot 2} =$
$2\sqrt{64} \cdot \sqrt{2} - \sqrt{9} \cdot \sqrt{2} + 4\sqrt{16} \cdot \sqrt{2} =$
$2 \cdot 8\sqrt{2} - 3\sqrt{2} + 4 \cdot 4\sqrt{2} = 16\sqrt{2} - 3\sqrt{2} + 16\sqrt{2} =$
$(16 - 3 + 16)\sqrt{2} = 29\sqrt{2}$

21. $\sqrt{5a} + 2\sqrt{45a^3} = \sqrt{5a} + 2\sqrt{9a^2 \cdot 5a} =$
$\sqrt{5a} + 2\sqrt{9a^2} \cdot \sqrt{5a} = \sqrt{5a} + 2 \cdot 3a\sqrt{5a} =$
$\sqrt{5a} + 6a\sqrt{5a} = (1 + 6a)\sqrt{5a}$

23. $\sqrt[3]{24x} - \sqrt[3]{3x^4} = \sqrt[3]{8 \cdot 3x} - \sqrt[3]{x^3 \cdot 3x} =$
$\sqrt[3]{8} \cdot \sqrt[3]{3x} - \sqrt[3]{x^3} \cdot \sqrt[3]{3x} = 2\sqrt[3]{3x} - x\sqrt[3]{3x} =$
$(2 - x)\sqrt[3]{3x}$

25. $\sqrt{24y - 24} + \sqrt{6y - 6} = \sqrt{4(6y - 6)} + \sqrt{6y - 6} =$
$\sqrt{4} \cdot \sqrt{6y - 6} + \sqrt{6y - 6} = 2\sqrt{6y - 6} + \sqrt{6y - 6} =$
$(2 + 1)\sqrt{6y - 6} = 3\sqrt{6y - 6}$

27. $\sqrt{x^3 - x^2} + \sqrt{9x - 9} = \sqrt{x^2(x - 1)} + \sqrt{9(x - 1)} =$
$\sqrt{x^2} \cdot \sqrt{x - 1} + \sqrt{9} \cdot \sqrt{x - 1} = x\sqrt{x - 1} + 3\sqrt{x - 1} =$
$(x + 3)\sqrt{x - 1}$

29. $5\sqrt[3]{32} - \sqrt[3]{108} + 2\sqrt[3]{256} =$
$5\sqrt[3]{8 \cdot 4} - \sqrt[3]{27 \cdot 4} + 2\sqrt[3]{64 \cdot 4} =$
$5\sqrt[3]{8} \cdot \sqrt[3]{4} - \sqrt[3]{27} \cdot \sqrt[3]{4} + 2\sqrt[3]{64} \cdot \sqrt[3]{4} =$
$5 \cdot 2\sqrt[3]{4} - 3\sqrt[3]{4} + 2 \cdot 4\sqrt[3]{4} = 10\sqrt[3]{4} - 3\sqrt[3]{4} + 8\sqrt[3]{4} =$
$(10 - 3 + 8)\sqrt[3]{4} = 15\sqrt[3]{4}$

31. $\sqrt{5}(4 - 2\sqrt{5}) = \sqrt{5} \cdot 4 - 2(\sqrt{5})^2$ Distributive law
$= 4\sqrt{5} - 2 \cdot 5$
$= 4\sqrt{5} - 10$

33. $\sqrt{3}(\sqrt{2} - \sqrt{7}) = \sqrt{3}\,\sqrt{2} - \sqrt{3}\,\sqrt{7}$ Distributive law
$= \sqrt{6} - \sqrt{21}$

35. $\sqrt{3}(2\sqrt{5} - 3\sqrt{4}) = \sqrt{3}(2\sqrt{5} - 3 \cdot 2) =$
$\sqrt{3} \cdot 2\sqrt{5} - \sqrt{3} \cdot 6 = 2\sqrt{15} - 6\sqrt{3}$

37. $\sqrt[3]{2}(\sqrt[3]{4} - 2\sqrt[3]{32}) = \sqrt[3]{2} \cdot \sqrt[3]{4} - \sqrt[3]{2} \cdot 2\sqrt[3]{32} =$
$\sqrt[3]{8} - 2\sqrt[3]{64} = 2 - 2 \cdot 4 = 2 - 8 = -6$

39. $\sqrt[3]{a}(\sqrt[3]{2a^2} + \sqrt[3]{16a^2}) = \sqrt[3]{a} \cdot \sqrt[3]{2a^2} + \sqrt[3]{a} \cdot \sqrt[3]{16a^2} =$
$\sqrt[3]{2a^3} + \sqrt[3]{16a^3} = \sqrt[3]{a^3 \cdot 2} + \sqrt[3]{8a^3 \cdot 2} = a\sqrt[3]{2} + 2a\sqrt[3]{2} =$
$3a\sqrt[3]{2}$

41. $(\sqrt{3} - \sqrt{2})(\sqrt{3} + \sqrt{2}) = (\sqrt{3})^2 - (\sqrt{2})^2 = 3 - 2 = 1$

43. $(\sqrt{8} + 2\sqrt{5})(\sqrt{8} - 2\sqrt{5}) = (\sqrt{8})^2 - (2\sqrt{5})^2 =$
$8 - 4 \cdot 5 = 8 - 20 = -12$

45. $(7 + \sqrt{5})(7 - \sqrt{5}) = 7^2 - (\sqrt{5})^2 = 49 - 5 = 44$

47. $(2 - \sqrt{3})(2 + \sqrt{3}) = 2^2 - (\sqrt{3})^2 = 4 - 3 = 1$

49. $(\sqrt{8} + \sqrt{5})(\sqrt{8} - \sqrt{5}) = (\sqrt{8})^2 - (\sqrt{5})^2 = 8 - 5 = 3$

51. $(3 + 2\sqrt{7})(3 - 2\sqrt{7}) = 3^2 - (2\sqrt{7})^2 =$
$9 - 4 \cdot 7 = 9 - 28 = -19$

53. $(\sqrt{a} + \sqrt{b})(\sqrt{a} - \sqrt{b}) = (\sqrt{a})^2 - (\sqrt{b})^2 = a - b$

55. $(3 - \sqrt{5})(2 + \sqrt{5})$
$= 3 \cdot 2 + 3\sqrt{5} - 2\sqrt{5} - (\sqrt{5})^2$ Using FOIL
$= 6 + 3\sqrt{5} - 2\sqrt{5} - 5$
$= 1 + \sqrt{5}$ Simplifying

57. $(\sqrt{3} + 1)(2\sqrt{3} + 1)$
$= \sqrt{3} \cdot 2\sqrt{3} + \sqrt{3} \cdot 1 + 1 \cdot 2\sqrt{3} + 1^2$ Using FOIL
$= 2 \cdot 3 + \sqrt{3} + 2\sqrt{3} + 1$
$= 7 + 3\sqrt{3}$ Simplifying

59. $(2\sqrt{7} - 4\sqrt{2})(3\sqrt{7} + 6\sqrt{2}) =$
$2\sqrt{7} \cdot 3\sqrt{7} + 2\sqrt{7} \cdot 6\sqrt{2} - 4\sqrt{2} \cdot 3\sqrt{7} - 4\sqrt{2} \cdot 6\sqrt{2} =$
$6 \cdot 7 + 12\sqrt{14} - 12\sqrt{14} - 24 \cdot 2 =$
$42 + 12\sqrt{14} - 12\sqrt{14} - 48 = -6$

61. $(\sqrt{a} + \sqrt{2})(\sqrt{a} + \sqrt{3}) =$
$(\sqrt{a})^2 + \sqrt{a} \cdot \sqrt{3} + \sqrt{2} \cdot \sqrt{a} + \sqrt{2} \cdot \sqrt{3} =$
$a + \sqrt{3a} + \sqrt{2a} + \sqrt{6}$

63. $(2\sqrt[3]{3} + \sqrt[3]{2})(\sqrt[3]{3} - 2\sqrt[3]{2}) =$
$2\sqrt[3]{3} \cdot \sqrt[3]{3} - 2\sqrt[3]{3} \cdot 2\sqrt[3]{2} + \sqrt[3]{2} \cdot \sqrt[3]{3} - \sqrt[3]{2} \cdot 2\sqrt[3]{2} =$
$2\sqrt[3]{9} - 4\sqrt[3]{6} + \sqrt[3]{6} - 2\sqrt[3]{4} = 2\sqrt[3]{9} - 3\sqrt[3]{6} - 2\sqrt[3]{4}$

65. $(2 + \sqrt{3})^2 = 2^2 + 4\sqrt{3} + (\sqrt{3})^2$ Squaring a binomial
$= 4 + 4\sqrt{3} + 3$
$= 7 + 4\sqrt{3}$

67. $(\sqrt[5]{9} - \sqrt[5]{3})(\sqrt[5]{8} + \sqrt[5]{27})$

$= \sqrt[5]{9} \cdot \sqrt[5]{8} + \sqrt[5]{9} \cdot \sqrt[5]{27} - \sqrt[5]{3} \cdot \sqrt[5]{8} - \sqrt[5]{3} \cdot \sqrt[5]{27}$

$\qquad\qquad\qquad\qquad\qquad$ Using FOIL

$= \sqrt[5]{72} + \sqrt[5]{243} - \sqrt[5]{24} - \sqrt[5]{81}$

$= \sqrt[5]{72} + 3 - \sqrt[5]{24} - \sqrt[5]{81}$

69. $\dfrac{x^3 + 4x}{x^2 - 16} \div \dfrac{x^2 + 8x + 15}{x^2 + x - 20}$

$= \dfrac{x^3 + 4x}{x^2 - 16} \cdot \dfrac{x^2 + x - 20}{x^2 + 8x + 15}$

$= \dfrac{(x^3 + 4x)(x^2 + x - 20)}{(x^2 - 16)(x^2 + 8x + 15)}$

$= \dfrac{x(x^2 + 4)(x + 5)(x - 4)}{(x + 4)(x - 4)(x + 3)(x + 5)}$

$= \dfrac{x(x^2 + 4)\cancel{(x + 5)}\cancel{(x - 4)}}{(x + 4)\cancel{(x - 4)}(x + 3)\cancel{(x + 5)}}$

$= \dfrac{x(x^2 + 4)}{(x + 4)(x + 3)}$

71. $\dfrac{a^3 + 8}{a^2 - 4} \cdot \dfrac{a^2 - 4a + 4}{a^2 - 2a + 4}$

$= \dfrac{(a^3 + 8)(a^2 - 4a + 4)}{(a^2 - 4)(a^2 - 2a + 4)}$

$= \dfrac{(a + 2)(a^2 - 2a + 4)(a - 2)(a - 2)}{(a + 2)(a - 2)(a^2 - 2a + 4)(1)}$

$= \dfrac{(a + 2)(a^2 - 2a + 4)(a - 2)}{(a + 2)(a^2 - 2a + 4)(a - 2)} \cdot \dfrac{a - 2}{1}$

$= a - 2$

73. $\sqrt{9 + 3\sqrt{5}}\sqrt{9 - 3\sqrt{5}} = \sqrt{(9 + 3\sqrt{5})(9 - 3\sqrt{5})} =$

$\sqrt{81 - 9 \cdot 5} = \sqrt{81 - 45} = \sqrt{36} = 6$

75. $(\sqrt{3} + \sqrt{5} - \sqrt{6})^2 = [(\sqrt{3} + \sqrt{5}) - \sqrt{6}]^2 =$

$(\sqrt{3} + \sqrt{5})^2 - 2(\sqrt{3} + \sqrt{5})(\sqrt{6}) + (\sqrt{6})^2 =$

$3 + 2\sqrt{15} + 5 - 2\sqrt{18} - 2\sqrt{30} + 6 =$

$14 + 2\sqrt{15} - 2\sqrt{9 \cdot 2} - 2\sqrt{30} =$

$14 + 2\sqrt{15} - 6\sqrt{2} - 2\sqrt{30}$

77. $(\sqrt[3]{9} - 2)(\sqrt[3]{9} + 4)$

$= \sqrt[3]{81} + 2\sqrt[3]{9} - 8$

$= \sqrt[3]{27 \cdot 3} + 2\sqrt[3]{9} - 8$

$= 3\sqrt[3]{3} + 2\sqrt[3]{9} - 8$

Exercise Set 7.5

1. $\sqrt{\dfrac{5}{3}} = \sqrt{\dfrac{5}{3} \cdot \dfrac{3}{3}} = \sqrt{\dfrac{15}{9}} = \dfrac{\sqrt{15}}{\sqrt{9}} = \dfrac{\sqrt{15}}{3}$

3. $\sqrt{\dfrac{11}{2}} = \sqrt{\dfrac{11}{2} \cdot \dfrac{2}{2}} = \sqrt{\dfrac{22}{4}} = \dfrac{\sqrt{22}}{\sqrt{4}} = \dfrac{\sqrt{22}}{2}$

5. $\dfrac{2\sqrt{3}}{7\sqrt{5}} = \dfrac{2\sqrt{3}}{7\sqrt{5}} \cdot \dfrac{\sqrt{5}}{\sqrt{5}} = \dfrac{2\sqrt{15}}{7\sqrt{5^2}} = \dfrac{2\sqrt{15}}{7 \cdot 5} = \dfrac{2\sqrt{15}}{35}$

7. $\sqrt[3]{\dfrac{16}{9}} = \sqrt[3]{\dfrac{16}{9} \cdot \dfrac{3}{3}} = \sqrt[3]{\dfrac{48}{27}} = \dfrac{\sqrt[3]{8 \cdot 6}}{\sqrt[3]{27}} = \dfrac{2\sqrt[3]{6}}{3}$

9. $\dfrac{\sqrt[3]{3a}}{\sqrt[3]{5c}} = \dfrac{\sqrt[3]{3a}}{\sqrt[3]{5c}} \cdot \dfrac{\sqrt[3]{5^2 c^2}}{\sqrt[3]{5^2 c^2}} = \dfrac{\sqrt[3]{75ac^2}}{\sqrt[3]{5^3 c^3}} = \dfrac{\sqrt[3]{75ac^2}}{5c}$

11. $\dfrac{\sqrt[3]{2y^4}}{\sqrt[3]{6x^4}} = \dfrac{\sqrt[3]{2y^4}}{\sqrt[3]{6x^4}} \cdot \dfrac{\sqrt[3]{6^2 x^2}}{\sqrt[3]{6^2 x^2}} = \dfrac{\sqrt[3]{72x^2 y^4}}{\sqrt[3]{6^3 x^6}} = \dfrac{\sqrt[3]{8y^3 \cdot 9x^2 y}}{6x^2} =$

$\dfrac{2y\sqrt[3]{9x^2 y}}{6x^2} = \dfrac{y\sqrt[3]{9x^2 y}}{3x^2}$

13. $\dfrac{1}{\sqrt[4]{st}} = \dfrac{1}{\sqrt[4]{st}} \cdot \dfrac{\sqrt[4]{s^3 t^3}}{\sqrt[4]{s^3 t^3}} = \dfrac{\sqrt[4]{s^3 t^3}}{\sqrt[4]{s^4 t^4}} = \dfrac{\sqrt[4]{s^3 t^3}}{st}$

15. $\sqrt{\dfrac{3x}{20}} = \sqrt{\dfrac{3x}{20} \cdot \dfrac{5}{5}} = \sqrt{\dfrac{15x}{100}} = \dfrac{\sqrt{15x}}{\sqrt{100}} = \dfrac{\sqrt{15x}}{10}$

17. $\sqrt[3]{\dfrac{4}{5x^5 y^2}} = \sqrt[3]{\dfrac{4}{5x^5 y^2} \cdot \dfrac{25xy}{25xy}} = \sqrt[3]{\dfrac{100xy}{125x^6 y^3}} =$

$\dfrac{\sqrt[3]{100xy}}{\sqrt[3]{125x^6 y^3}} = \dfrac{\sqrt[3]{100xy}}{5x^2 y}$

19. $\sqrt[4]{\dfrac{1}{8x^7 y^3}} = \sqrt[4]{\dfrac{1}{8x^7 y^3} \cdot \dfrac{2xy}{2xy}} = \sqrt[4]{\dfrac{2xy}{16x^8 y^4}} = \dfrac{\sqrt[4]{2xy}}{\sqrt[4]{16x^8 y^4}} =$

$\dfrac{\sqrt[4]{2xy}}{2x^2 y}$

21. $\dfrac{2x}{\sqrt[5]{18x^8 y^6}} = \dfrac{2x}{\sqrt[5]{2 \cdot 3^2 \cdot x^8 y^6}} =$

$\dfrac{2x}{\sqrt[5]{2 \cdot 3^2 \cdot x^8 y^6}} \cdot \dfrac{\sqrt[5]{2^4 \cdot 3^3 \cdot x^8 y^6}}{\sqrt[5]{2^4 \cdot 3^3 \cdot x^8 y^6}} = \dfrac{2x\sqrt[5]{2^4 \cdot 3^3 \cdot x^8 y^6}}{\sqrt[5]{2^5 \cdot 3^5 \cdot x^{10} y^{10}}} =$

$\dfrac{2x\sqrt[5]{432x^2 y^4}}{2 \cdot 3 \cdot x^2 y^2} = \dfrac{2x\sqrt[5]{432x^2 y^4}}{2 \cdot 3 \cdot x \cdot x \cdot y \cdot y} = \dfrac{2\cancel{x}\sqrt[5]{432x^2 y^4}}{\cancel{2} \cdot 3 \cdot \cancel{x} \cdot x \cdot y \cdot y} =$

$\dfrac{\sqrt[5]{432x^2 y^4}}{3xy^2}$

23. $\dfrac{\sqrt{3}}{\sqrt{5x}} = \dfrac{\sqrt{3}}{\sqrt{5x}} \cdot \dfrac{\sqrt{3}}{\sqrt{3}} = \dfrac{\sqrt{9}}{\sqrt{15x}} = \dfrac{3}{\sqrt{15x}}$

25. $\sqrt{\dfrac{14}{21}} = \sqrt{\dfrac{2}{3}} = \sqrt{\dfrac{2}{3} \cdot \dfrac{2}{2}} = \sqrt{\dfrac{4}{6}} = \dfrac{\sqrt{4}}{\sqrt{6}} = \dfrac{2}{\sqrt{6}}$

27. $\dfrac{4\sqrt{13}}{3\sqrt{7}} = \dfrac{4\sqrt{13}}{3\sqrt{7}} \cdot \dfrac{\sqrt{13}}{\sqrt{13}} = \dfrac{4\sqrt{169}}{3\sqrt{91}} = \dfrac{4 \cdot 13}{3\sqrt{91}} = \dfrac{52}{3\sqrt{91}}$

29. $\dfrac{\sqrt[3]{7}}{\sqrt[3]{2}} = \dfrac{\sqrt[3]{7}}{\sqrt[3]{2}} \cdot \dfrac{\sqrt[3]{7^2}}{\sqrt[3]{7^2}} = \dfrac{\sqrt[3]{7^3}}{\sqrt[3]{98}} = \dfrac{7}{\sqrt[3]{98}}$

31. $\sqrt{\dfrac{7x}{3y}} = \sqrt{\dfrac{7x}{3y} \cdot \dfrac{7x}{7x}} = \dfrac{\sqrt{(7x)^2}}{\sqrt{21xy}} = \dfrac{7x}{\sqrt{21xy}}$

33. $\dfrac{\sqrt[3]{5y^4}}{\sqrt[3]{6x^5}} = \dfrac{\sqrt[3]{5y^4}}{\sqrt[3]{6x^5}} \cdot \dfrac{\sqrt[3]{5^2 y^2}}{\sqrt[3]{5^2 y^2}} = \dfrac{\sqrt[3]{5^3 y^6}}{\sqrt[3]{150x^5 y^2}} = \dfrac{5y^2}{x\sqrt[3]{150x^2 y^2}}$

35. $\dfrac{\sqrt{ab}}{3} = \dfrac{\sqrt{ab}}{3} \cdot \dfrac{\sqrt{ab}}{\sqrt{ab}} = \dfrac{ab}{3\sqrt{ab}}$

37. $\dfrac{9}{6 - \sqrt{10}} = \dfrac{9}{6 - \sqrt{10}} \cdot \dfrac{6 + \sqrt{10}}{6 + \sqrt{10}} = \dfrac{9(6 + \sqrt{10})}{6^2 - (\sqrt{10})^2} =$

$\dfrac{9(6 + \sqrt{10})}{36 - 10} = \dfrac{54 + 9\sqrt{10}}{26}$

39. $\dfrac{-4\sqrt{7}}{\sqrt{5} - \sqrt{3}} = \dfrac{-4\sqrt{7}}{\sqrt{5} - \sqrt{3}} \cdot \dfrac{\sqrt{5} + \sqrt{3}}{\sqrt{5} + \sqrt{3}} =$

$\dfrac{-4\sqrt{7}(\sqrt{5} + \sqrt{3})}{(\sqrt{5})^2 - (\sqrt{3})^2} = \dfrac{-4\sqrt{7}(\sqrt{5} + \sqrt{3})}{5 - 3} =$

$\dfrac{-4\sqrt{7}(\sqrt{5} + \sqrt{3})}{2} = -2\sqrt{7}(\sqrt{5} + \sqrt{3}) = -2\sqrt{35} - 2\sqrt{21}$

41. $\dfrac{\sqrt{5} - 2\sqrt{6}}{\sqrt{3} - 4\sqrt{5}} = \dfrac{\sqrt{5} - 2\sqrt{6}}{\sqrt{3} - 4\sqrt{5}} \cdot \dfrac{\sqrt{3} + 4\sqrt{5}}{\sqrt{3} + 4\sqrt{5}} =$

$\dfrac{\sqrt{15} + 4 \cdot 5 - 2\sqrt{18} - 8\sqrt{30}}{(\sqrt{3})^2 - (4\sqrt{5})^2} =$

$\dfrac{\sqrt{15} + 20 - 2\sqrt{9 \cdot 2} - 8\sqrt{30}}{3 - 16 \cdot 5} =$

$\dfrac{\sqrt{15} + 20 - 6\sqrt{2} - 8\sqrt{30}}{-77}$, or $-\dfrac{\sqrt{15} + 20 - 6\sqrt{2} - 8\sqrt{30}}{77}$

43. $\dfrac{\sqrt{x} - \sqrt{y}}{\sqrt{x} + \sqrt{y}} = \dfrac{\sqrt{x} - \sqrt{y}}{\sqrt{x} + \sqrt{y}} \cdot \dfrac{\sqrt{x} - \sqrt{y}}{\sqrt{x} - \sqrt{y}} =$

$\dfrac{x - \sqrt{xy} - \sqrt{xy} + y}{x - y} = \dfrac{x - 2\sqrt{xy} + y}{x - y}$

45. $\dfrac{5\sqrt{3} - 3\sqrt{2}}{3\sqrt{2} - 2\sqrt{3}} = \dfrac{5\sqrt{3} - 3\sqrt{2}}{3\sqrt{2} - 2\sqrt{3}} \cdot \dfrac{3\sqrt{2} + 2\sqrt{3}}{3\sqrt{2} + 2\sqrt{3}} =$

$\dfrac{15\sqrt{6} + 10 \cdot 3 - 9 \cdot 2 - 6\sqrt{6}}{9 \cdot 2 - 4 \cdot 3} = \dfrac{12 + 9\sqrt{6}}{6} =$

$\dfrac{3(4 + 3\sqrt{6})}{3 \cdot 2} = \dfrac{4 + 3\sqrt{6}}{2}$

47. $\dfrac{\sqrt{x} - 2\sqrt{y}}{2\sqrt{x} + \sqrt{y}} = \dfrac{\sqrt{x} - 2\sqrt{y}}{2\sqrt{x} + \sqrt{y}} \cdot \dfrac{2\sqrt{x} - \sqrt{y}}{2\sqrt{x} - \sqrt{y}} =$

$\dfrac{2(\sqrt{x})^2 - \sqrt{xy} - 4\sqrt{xy} + 2(\sqrt{y})^2}{(2\sqrt{x})^2 - (\sqrt{y})^2} = \dfrac{2x - 5\sqrt{xy} + 2y}{4x - y}$

49. $\dfrac{\sqrt{2} - 5}{7} = \dfrac{\sqrt{2} - 5}{7} \cdot \dfrac{\sqrt{2} + 5}{\sqrt{2} + 5} = \dfrac{(\sqrt{2})^2 - 5^2}{7(\sqrt{2} + 5)} =$

$\dfrac{2 - 25}{7(\sqrt{2} + 5)} = \dfrac{-23}{7\sqrt{2} + 35}$, or $-\dfrac{23}{7\sqrt{2} + 35}$

51. $\dfrac{\sqrt{3} - 5}{\sqrt{2} + 5} = \dfrac{\sqrt{3} - 5}{\sqrt{2} + 5} \cdot \dfrac{\sqrt{3} + 5}{\sqrt{3} + 5} =$

$\dfrac{3 - 25}{\sqrt{6} + 5\sqrt{2} + 5\sqrt{3} + 25} = \dfrac{-22}{\sqrt{6} + 5\sqrt{2} + 5\sqrt{3} + 25}$, or

$-\dfrac{22}{\sqrt{6} + 5\sqrt{2} + 5\sqrt{3} + 25}$

53. $\dfrac{\sqrt{x} - \sqrt{y}}{\sqrt{x} + \sqrt{y}} = \dfrac{\sqrt{x} - \sqrt{y}}{\sqrt{x} + \sqrt{y}} \cdot \dfrac{\sqrt{x} + \sqrt{y}}{\sqrt{x} + \sqrt{y}} =$

$\dfrac{x - y}{x + \sqrt{xy} + \sqrt{xy} + y} = \dfrac{x - y}{x + 2\sqrt{xy} + y}$

55. $\dfrac{4\sqrt{6} - 5\sqrt{3}}{2\sqrt{3} + 7\sqrt{6}} = \dfrac{4\sqrt{6} - 5\sqrt{3}}{2\sqrt{3} + 7\sqrt{6}} \cdot \dfrac{4\sqrt{6} + 5\sqrt{3}}{4\sqrt{6} + 5\sqrt{3}} =$

$\dfrac{16 \cdot 6 - 25 \cdot 3}{8\sqrt{18} + 10 \cdot 3 + 28 \cdot 6 + 35\sqrt{18}} =$

$\dfrac{96 - 75}{43\sqrt{18} + 30 + 168} = \dfrac{21}{43\sqrt{9 \cdot 2} + 198} =$

$\dfrac{21}{43 \cdot 3\sqrt{2} + 198} = \dfrac{3 \cdot 7}{3(43\sqrt{2} + 66)} = \dfrac{7}{43\sqrt{2} + 66}$

57. $\dfrac{\sqrt{2} + 3\sqrt{x}}{\sqrt{2} - \sqrt{x}} = \dfrac{\sqrt{2} + 3\sqrt{x}}{\sqrt{2} - \sqrt{x}} \cdot \dfrac{\sqrt{2} - 3\sqrt{x}}{\sqrt{2} - 3\sqrt{x}} =$

$\dfrac{(\sqrt{2})^2 - (3\sqrt{x})^2}{(\sqrt{2})^2 - 3\sqrt{2x} - \sqrt{2x} + 3(\sqrt{x})^2} = \dfrac{2 - 9x}{2 - 4\sqrt{2x} + 3x}$

59. $\dfrac{a\sqrt{b} + c}{\sqrt{b} + c} = \dfrac{a\sqrt{b} + c}{\sqrt{b} + c} \cdot \dfrac{a\sqrt{b} - c}{a\sqrt{b} - c} =$

$\dfrac{(a\sqrt{b})^2 - c^2}{a(\sqrt{b})^2 - c\sqrt{b} + ac\sqrt{b} - c^2} =$

$\dfrac{a^2 b - c^2}{ab - c\sqrt{b} + ac\sqrt{b} - c^2}$

61. $\dfrac{1}{2} - \dfrac{1}{3} = \dfrac{1}{t}$, LCM is $6t$

$6t\left(\dfrac{1}{2} - \dfrac{1}{3}\right) = 6t\left(\dfrac{1}{t}\right)$

$3t - 2t = 6$

$t = 6$

Check:

$$\dfrac{1}{2} - \dfrac{1}{3} = \dfrac{1}{t}$$

$\dfrac{1}{2} - \dfrac{1}{3}$	$\dfrac{1}{6}$
$\dfrac{3}{6} - \dfrac{2}{6}$	
$\dfrac{1}{6}$	TRUE

The solution is 6.

63. $\dfrac{1}{x^3 - y^2} \div \dfrac{1}{(x - y)(x^2 + xy + y^2)}$

$= \dfrac{1}{(x - y)(x^2 + xy + y^2)} \cdot \dfrac{(x - y)(x^2 + xy + y^2)}{1}$

$= \dfrac{(x - y)(x^2 + xy + y^2)}{(x - y)(x^2 + xy + y^2)}$

$= 1$

65.
$$\frac{\sqrt{5}+\sqrt{10}-\sqrt{6}}{\sqrt{50}} = \frac{\sqrt{5}+\sqrt{10}-\sqrt{6}}{\sqrt{50}} \cdot \frac{\sqrt{2}}{\sqrt{2}}$$
$$= \frac{\sqrt{10}+\sqrt{20}-\sqrt{12}}{\sqrt{100}}$$
$$= \frac{\sqrt{10}+2\sqrt{5}-2\sqrt{3}}{10}$$

67.
$$\frac{b+\sqrt{b}}{1+b+\sqrt{b}} = \frac{b+\sqrt{b}}{(1+b)+\sqrt{b}} \cdot \frac{(1+b)-\sqrt{b}}{(1+b)-\sqrt{b}}$$
$$= \frac{(b+\sqrt{b})(1+b-\sqrt{b})}{(1+b)^2-(\sqrt{b})^2}$$
$$= \frac{b+b^2-b\sqrt{b}+\sqrt{b}+b\sqrt{b}-b}{1+2b+b^2-b}$$
$$= \frac{b^2+\sqrt{b}}{b^2+b+1}$$

69.
$$\frac{36a^2b}{\sqrt[3]{6a^2b}} = \frac{36a^2b}{\sqrt[3]{6a^2b}} \cdot \frac{\sqrt[3]{6^2ab^2}}{\sqrt[3]{6^2ab^2}}$$
$$= \frac{36a^2b\sqrt[3]{6^2ab^2}}{\sqrt[3]{6^3a^3b^3}}$$
$$= \frac{36a^2b\sqrt[3]{6^2ab^2}}{6ab}$$
$$= 6a\sqrt[3]{36ab^2}$$

71.
$$\frac{\sqrt{x+6}-5}{\sqrt{x+6}+5} = \frac{\sqrt{x+6}-5}{\sqrt{x+6}+5} \cdot \frac{\sqrt{x+6}+5}{\sqrt{x+6}+5}$$
$$= \frac{(x+6)-25}{(x+6)+10\sqrt{x+6}+25}$$
$$= \frac{x-19}{x+10\sqrt{x+6}+31}$$

73.
$$\sqrt{a^2-3}-\frac{a^2}{\sqrt{a^2-3}}$$
$$= \sqrt{a^2-3}-\frac{a^2}{\sqrt{a^2-3}} \cdot \frac{\sqrt{a^2-3}}{\sqrt{a^2-3}}$$
$$= \sqrt{a^2-3}-\frac{a^2\sqrt{a^2-3}}{a^2-3}$$
$$= \sqrt{a^2-3} \cdot \frac{a^2-3}{a^2-3}-\frac{a^2\sqrt{a^2-3}}{a^2-3}$$
$$= \frac{a^2\sqrt{a^2-3}-3\sqrt{a^2-3}-a^2\sqrt{a^2-3}}{a^2-3}$$
$$= \frac{-3\sqrt{a^2-3}}{a^2-3}, \text{ or } -\frac{3\sqrt{a^2-3}}{a^2-3}$$

75.
$$\frac{\frac{1}{\sqrt{w}}-\sqrt{w}}{\frac{\sqrt{w}+1}{\sqrt{w}}} = \frac{\frac{1}{\sqrt{w}}-\sqrt{w}}{\frac{\sqrt{w}+1}{\sqrt{w}}} \cdot \frac{\sqrt{w}}{\sqrt{w}} = \frac{1-w}{\sqrt{w}+1} =$$
$$\frac{1-w}{\sqrt{w}+1} \cdot \frac{\sqrt{w}-1}{\sqrt{w}-1} = \frac{\sqrt{w}-1-w\sqrt{w}+w}{w-1} =$$
$$\frac{(w-1)-\sqrt{w}(w-1)}{w-1} = \frac{(w-1)(1-\sqrt{w})}{w-1} = 1-\sqrt{w}$$

Exercise Set 7.6

1. $y^{1/7} = \sqrt[7]{y}$

3. $(8)^{1/3} = \sqrt[3]{8} = 2$

5. $(a^3b^3)^{1/5} = \sqrt[5]{a^3b^3}$

7. $16^{3/4} = \sqrt[4]{16^3} = (\sqrt[4]{16})^3 = 2^3 = 8$

9. $\sqrt[3]{18} = 18^{1/3}$

11. $\sqrt[5]{xy^2z} = (xy^2z)^{1/5}$

13. $(\sqrt{3mn})^3 = (3mn)^{3/2}$

15. $(\sqrt[7]{8x^2y})^5 = (8x^2y)^{5/7}$

17. $x^{-1/4} = \dfrac{1}{x^{1/4}}$

19. $\dfrac{1}{x^{-2/3}} = x^{2/3}$

21. $5^{3/4} \cdot 5^{1/8} = 5^{3/4+1/8} = 5^{6/8+1/8} = 5^{7/8}$

23. $\dfrac{7^{5/8}}{7^{3/8}} = 7^{5/8-3/8} = 7^{2/8} = 7^{1/4}$

25. $\dfrac{4.9^{3/5}}{4.9^{1/4}} = 4.9^{3/5-1/4} = 4.9^{12/20-5/20} = 4.9^{7/20}$

27. $(6^{3/8})^{2/7} = 6^{3/8 \cdot 2/7} = 6^{6/56} = 6^{3/28}$

29. $a^{2/3} \cdot a^{5/6} = a^{2/3+5/6} = a^{4/6+5/6} = a^{9/6} = a^{3/2}$

31. $(a^{2/3} \cdot b^{5/8})^4 = (a^{2/3})^4(b^{5/8})^4 = a^{8/3}b^{20/8} = a^{8/3}b^{5/2}$

33. $\sqrt[6]{a^4} = a^{4/6} = a^{2/3} = \sqrt[3]{a^2}$

35. $\sqrt[3]{8y^6} = (2^3y^6)^{1/3} = 2^{3/3}y^{6/3} = 2y^2$

37. $\sqrt[4]{32} = \sqrt[4]{2^5} = 2^{5/4} = 2^{4/4} \cdot 2^{1/4} = 2\sqrt[4]{2}$

39. $\sqrt[6]{4x^2} = (2^2x^2)^{1/6} = 2^{2/6}x^{2/6}$
$$= 2^{1/3}x^{1/3} = (2x)^{1/3} = \sqrt[3]{2x}$$

41. $\sqrt[5]{32c^{10}d^{15}} = (2^5c^{10}d^{15})^{1/5} = 2^{5/5}c^{10/5}d^{15/5}$
$$= 2c^2d^3$$

43. $\sqrt[6]{\dfrac{m^{12}n^{24}}{64}} = \left(\dfrac{m^{12}n^{24}}{2^6}\right)^{1/6} = \dfrac{m^{12/6}n^{24/6}}{2^{6/6}} = \dfrac{m^2n^4}{2}$

45. $\sqrt[8]{r^4s^2} = (r^4s^2)^{1/8} = r^{4/8}s^{2/8} = r^{2/4}s^{1/4}$
$$= (r^2s)^{1/4} = \sqrt[4]{r^2s}$$

47. $\sqrt[3]{27a^3b^9} = (3^3a^3b^9)^{1/3} = 3^{3/3}a^{3/3}b^{9/3} = 3ab^3$

49. $\sqrt[5]{32x^{15}y^{40}} = (2^5x^{15}y^{40})^{1/5} = 2^{5/5}x^{15/5}y^{40/5} = 2x^3y^8$

51. $\sqrt[4]{64p^{12}q^{32}} = (2^6p^{12}q^{32})^{1/4} = 2^{6/4}p^{12/4}q^{32/4} =$
$$2^{3/2}p^3q^8 = 2 \cdot 2^{1/2}p^3q^8 = 2\sqrt{2}p^3q^8$$

53. $\sqrt[3]{3}\,\sqrt{3} = 3^{1/3}3^{1/2} = 3^{2/6}3^{3/6} = (3^2 \cdot 3^3)^{1/6} =$
$(9 \cdot 27)^{1/6} = (243)^{1/6} = \sqrt[6]{243}$

55. $\sqrt[4]{5} \cdot \sqrt[5]{7} = 5^{1/4} \cdot 7^{1/5} = 5^{5/20} \cdot 7^{4/20} = (5^5 \cdot 7^4)^{1/20} =$
$\sqrt[20]{5^5 \cdot 7^4}$

57. $\sqrt[3]{y}\,\sqrt[5]{3y} = y^{1/3}(3y)^{1/5} = y^{5/15}(3y)^{3/15} =$
$[y^5(3y)^3]^{1/15} = \sqrt[15]{y^5(3y)^3} = \sqrt[15]{y^5 \cdot 3^3 \cdot y^3} =$
$\sqrt[15]{3^3y^8} = \sqrt[15]{27y^8}$

59. $\sqrt[4]{3x}\,\sqrt{y+4} = (3x)^{1/4}(y+4)^{1/2} = (3x)^{1/4}(y+4)^{2/4} =$
$[3x(y+4)^2]^{1/4} = \sqrt[4]{3x(y+4)^2} = \sqrt[4]{3x(y^2+8y+16)} =$
$\sqrt[4]{3xy^2 + 24xy + 48x}$

61. $\dfrac{\sqrt[3]{(a+b)^2}}{\sqrt{(a+b)}} = \dfrac{(a+b)^{2/3}}{(a+b)^{1/2}} = \dfrac{(a+b)^{4/6}}{(a+b)^{3/6}} = (a+b)^{4/6-3/6} =$
$(a+b)^{1/6} = \sqrt[6]{a+b}$

63. $a^{2/3} \cdot b^{3/4} = a^{8/12} \cdot b^{9/12} = (a^8b^9)^{1/12} = \sqrt[12]{a^8b^9}$

65. $\dfrac{x^{8/15} \cdot y^{7/5}}{x^{1/3} \cdot y^{-1/5}} = x^{8/15-1/3}y^{7/5-(-1/5)} =$
$x^{8/15-5/15}y^{7/5+1/5} = x^{3/15}y^{8/5} = x^{1/5}y^{8/5} = (xy^8)^{1/5} =$
$\sqrt[5]{xy^8}$, or $y\sqrt[5]{xy^3}$

67. $\sqrt[4]{x^3y^5} \cdot \sqrt{xy} = (x^3y^5)^{1/4}(xy)^{1/2} = x^{3/4}y^{5/4}x^{1/2}y^{1/2} =$
$x^{3/4+1/2}y^{5/4+1/2} = x^{5/4}y^{7/4} = (x^5y^7)^{1/4} = \sqrt[4]{x^5y^7}$, or
$xy\sqrt[4]{xy^3}$

69. $\sqrt{a^4b^3c^4}\,\sqrt[3]{ab^2c} = (a^4b^3c^4)^{1/2}(ab^2c)^{1/3}$
$= a^{4/2}b^{3/2}c^{4/2}a^{1/3}b^{2/3}c^{1/3}$
$= a^{4/2+1/3}b^{3/2+2/3}c^{4/2+1/3}$
$= a^{14/6}b^{13/6}c^{14/6}$
$= (a^{14}b^{13}c^{14})^{1/6}$
$= \sqrt[6]{a^{14}b^{13}c^{14}}$, or $a^2b^2c^2\sqrt[6]{a^2bc^2}$

71. $\left(\dfrac{c^{-4/5}d^{5/9}}{c^{3/10}d^{1/6}}\right)^3 = (c^{-4/5-3/10}d^{5/9-1/6})^3 =$
$(c^{-8/10-3/10}d^{10/18-3/18})^3 = (c^{-11/10}d^{7/18})^3 =$
$c^{-33/10}d^{7/6} = c^{-99/30}d^{35/30} = (c^{-99}d^{35})^{1/30} =$
$\left(\dfrac{d^{35}}{c^{99}}\right)^{1/30} = \sqrt[30]{\dfrac{d^{35}}{c^{99}}} = \sqrt[30]{\dfrac{d^{30}}{c^{90}} \cdot \dfrac{d^5}{c^9}} = \dfrac{d}{c^3}\sqrt[30]{\dfrac{d^5}{c^9}}$

73. $\sqrt{x^5\sqrt[3]{x^4}} = \sqrt{x^5x^{4/3}} = (x^{19/3})^{1/2} = x^{19/6} =$
$\sqrt[6]{x^{19}} = x^3\sqrt[6]{x}$

75. $\sqrt[4]{\sqrt[3]{8x^3y^6}} = \sqrt[4]{(2^3x^3y^6)^{1/3}} = \sqrt[4]{2^{3/3}x^{3/3}y^{6/3}} = \sqrt[4]{2xy^2}$

77. $\sqrt[12]{p^2+2pq+q^2} = \sqrt[12]{(p+q)^2} = [(p+q)^2]^{1/12} =$
$(p+q)^{2/12} = (p+q)^{1/6} = \sqrt[6]{p+q}$

79. $\left[\sqrt[10]{\sqrt[5]{x^{15}}}\right]^5 \left[\sqrt[5]{\sqrt[10]{x^{15}}}\right]^5$
$= \left[\left((x^{15})^{1/5}\right)^{1/10}\right]^5 \left[\left((x^{15})^{1/10}\right)^{1/5}\right]^5$
$= x^{3/2}x^{3/2}$ 			Multiplying exponents
$= x^3$

Exercise Set 7.7

1. $\sqrt{2x-3} = 4$
$(\sqrt{2x-3})^2 = 4^2$ 	Principle of powers
$2x - 3 = 16$
$2x = 19$
$x = \dfrac{19}{2}$

Check: 	$\begin{array}{c|c} \sqrt{2x-3} = 4 \\ \hline \sqrt{2 \cdot \dfrac{19}{2} - 3} & 4 \\ \sqrt{19-3} & \\ \sqrt{16} & \\ 4 & \text{TRUE} \end{array}$

The solution is $\dfrac{19}{2}$.

3. $\sqrt{6x} + 1 = 8$
$\sqrt{6x} = 7$ 		Subtracting to isolate the radical
$(\sqrt{6x})^2 = 7^2$ 	Principle of powers
$6x = 49$
$x = \dfrac{49}{6}$

Check: 	$\begin{array}{c|c} \sqrt{6x} + 1 = 8 \\ \hline \sqrt{6 \cdot \dfrac{49}{6} + 1} & 8 \\ \sqrt{49} + 1 & \\ 7 + 1 & \\ 8 & \text{TRUE} \end{array}$

The solution is $\dfrac{49}{6}$.

5. $\sqrt{y+7} - 4 = 4$
$\sqrt{y+7} = 8$ 		Adding to isolate the radical
$(\sqrt{y+7})^2 = 8^2$ 	Principle of powers
$y + 7 = 64$
$y = 57$

Check: $\dfrac{\sqrt{y+7}-4=4}{}$

$\quad\sqrt{57+7}-4 \quad\big|\quad 4$

$\qquad \sqrt{64}-4$

$\qquad\quad 8-4$

$\qquad\qquad\quad 4 \quad\big|\quad$ TRUE

The solution is 57.

7. $\quad \sqrt{5y+8}=10$

$\quad (\sqrt{5y+8})^2 = 10^2 \quad$ Principle of powers

$\qquad 5y+8 = 100$

$\qquad\quad 5y = 92$

$\qquad\quad\ y = \dfrac{92}{5}$

Check: $\dfrac{\sqrt{5y+8}=10}{}$

$\quad \sqrt{5\cdot\dfrac{92}{5}+8} \quad\big|\quad 10$

$\qquad\quad \sqrt{92+8}$

$\qquad\qquad \sqrt{100}$

$\qquad\qquad\quad 10 \quad\big|\quad$ TRUE

The solution is $\dfrac{92}{5}$.

9. $\quad \sqrt[3]{x} = -1$

$\quad (\sqrt[3]{x})^3 = (-1)^3 \quad$ Principle of powers

$\qquad x = -1$

Check: $\dfrac{\sqrt[3]{x}=-1}{}$

$\quad \sqrt[3]{-1} \quad\big|\quad -1$

$\qquad -1 \quad\big|\quad$ TRUE

The solution is -1.

11. $\quad \sqrt{x+2} = -4$

$\quad (\sqrt{x+2})^2 = (-4)^2$

$\qquad x+2 = 16$

$\qquad\quad x = 14$

Check: $\dfrac{\sqrt{x+2}=-4}{}$

$\quad \sqrt{14+2} \quad\big|\quad -4$

$\qquad \sqrt{16}$

$\qquad\quad 4 \quad\big|\quad$ FALSE

The number 14 does not check. The equation has no solution. We might have observed at the outset that this equation has no solution because the principle square root of a number is never negative.

13. $\quad \sqrt[3]{x+5} = 2$

$\quad (\sqrt[3]{x+5})^3 = 2^3$

$\qquad x+5 = 8$

$\qquad\quad x = 3$

Check: $\dfrac{\sqrt[3]{x+5}=2}{}$

$\quad \sqrt[3]{3+5} \quad\big|\quad 2$

$\qquad \sqrt[3]{8}$

$\qquad\quad 2 \quad\big|\quad$ TRUE

The solution is 3.

15. $\quad \sqrt[4]{y-3} = 2$

$\quad (\sqrt[4]{y-3})^4 = 2^4$

$\qquad y-3 = 16$

$\qquad\quad y = 19$

Check: $\dfrac{\sqrt[4]{y-3}=2}{}$

$\quad \sqrt[4]{19-3} \quad\big|\quad 2$

$\qquad \sqrt[4]{16}$

$\qquad\quad 2 \quad\big|\quad$ TRUE

The solution is 19.

17. $\quad \sqrt[3]{6x+9}+8 = 5$

$\qquad \sqrt[3]{6x+9} = -3$

$\quad (\sqrt[3]{6x+9})^3 = (-3)^3$

$\qquad 6x+9 = -27$

$\qquad\quad 6x = -36$

$\qquad\quad\ x = -6$

Check: $\dfrac{\sqrt[3]{6x+9}+8=5}{}$

$\quad \sqrt[3]{6(-6)+9}+8 \quad\big|\quad 5$

$\qquad \sqrt[3]{-27}+8$

$\qquad\quad -3+8$

$\qquad\qquad 5 \quad\big|\quad$ TRUE

The solution is -6.

19. $\qquad 8 = \dfrac{1}{\sqrt{x}}$

$\quad 8\cdot\sqrt{x} = \dfrac{1}{\sqrt{x}}\cdot\sqrt{x}$

$\qquad 8\sqrt{x} = 1$

$\quad (8\sqrt{x})^2 = 1^2$

$\qquad 64x = 1$

$\qquad\quad x = \dfrac{1}{64}$

Check: $8 = \dfrac{1}{\sqrt{x}}$

$$
\begin{array}{c|c}
8 & \dfrac{1}{\sqrt{\dfrac{1}{64}}} \\[6ex]
 & \dfrac{1}{\dfrac{1}{8}} \\[4ex]
 & 8 \qquad \text{TRUE}
\end{array}
$$

The solution is $\dfrac{1}{64}$.

21. $\sqrt{3y+1} = \sqrt{2y+6}$

$(\sqrt{3y+1})^2 = (\sqrt{2y+6})^2$

$\quad 3y+1 = 2y+6$

$\qquad\quad y = 5$

Check: $\begin{array}{c|c} \sqrt{3y+1} = & \sqrt{2y+6} \\ \hline \sqrt{3\cdot5+1} & \sqrt{2\cdot5+6} \\ \sqrt{16} & \sqrt{16} \ \text{TRUE} \end{array}$

The solution is 5.

23. $\sqrt{y-5} + \sqrt{y} = 5$

$\quad \sqrt{y-5} = 5 - \sqrt{y}$ Isolating one radical

$(\sqrt{y-5})^2 = (5 - \sqrt{y})^2$

$\quad y-5 = 25 - 10\sqrt{y} + y$

$\quad 10\sqrt{y} = 30$ Isolating the remaining radical

$\quad \sqrt{y} = 3$ Dividing by 10

$(\sqrt{y})^2 = 3^2$

$\quad y = 9$

The number 9 checks, so it is the solution.

25. $3 + \sqrt{z-6} = \sqrt{z+9}$

$(3 + \sqrt{z-6})^2 = (\sqrt{z+9})^2$

$9 + 6\sqrt{z-6} + z - 6 = z + 9$

$6\sqrt{z-6} = 6$

$\sqrt{z-6} = 1$ Dividing by 6

$(\sqrt{z-6})^2 = 1^2$

$z - 6 = 1$

$z = 7$

The number 7 checks, so it is the solution.

27. $\sqrt{20-x} + 8 = \sqrt{9-x} + 11$

$\sqrt{20-x} = \sqrt{9-x} + 3$ Isolating one radical

$(\sqrt{20-x})^2 = (\sqrt{9-x} + 3)^2$

$20 - x = 9 - x + 6\sqrt{9-x} + 9$

$2 = 6\sqrt{9-x}$ Isolating the remaining radical

$1 = 3\sqrt{9-x}$ Dividing by 2

$1^2 = (3\sqrt{9-x})^2$

$1 = 9(9-x)$

$1 = 81 - 9x$

$9x = 80$

$x = \dfrac{80}{9}$

The number $\dfrac{80}{9}$ checks, so it is the solution.

29. $\sqrt{4y+1} - \sqrt{y-2} = 3$

$\sqrt{4y+1} = 3 + \sqrt{y-2}$ Isolating one radical

$(\sqrt{4y+1})^2 = (3 + \sqrt{y-2})^2$

$4y + 1 = 9 + 6\sqrt{y-2} + y - 2$

$3y - 6 = 6\sqrt{y-2}$ Isolating the remaining radical

$y - 2 = 2\sqrt{y-2}$ Multiplying by $\dfrac{1}{3}$

$(y-2)^2 = (2\sqrt{y-2})^2$

$y^2 - 4y + 4 = 4(y-2)$

$y^2 - 4y + 4 = 4y - 8$

$y^2 - 8y + 12 = 0$

$(y-6)(y-2) = 0$

$y - 6 = 0 \quad \text{or} \quad y - 2 = 0$

$y = 6 \quad \text{or} \quad\quad y = 2$

The numbers 6 and 2 check, so they are the solutions.

31. $\sqrt{x+2} + \sqrt{3x+4} = 2$

$\sqrt{x+2} = 2 - \sqrt{3x+4}$ Isolating one radical

$(\sqrt{x+2})^2 = (2 - \sqrt{3x-4})^2$

$x + 2 = 4 - 4\sqrt{3x+4} + 3x + 4$

$-2x - 6 = -4\sqrt{3x+4}$ Isolating the remaining radical

$x + 3 = 2\sqrt{3x+4}$ Dividing by 2

$(x+3)^2 = (2\sqrt{3x+4})^2$

$x^2 + 6x + 9 = 4(3x+4)$

$x^2 + 6x + 9 = 12x + 16$

$x^2 - 6x - 7 = 0$

$(x-7)(x+1) = 0$

$x - 7 = 0 \quad \text{or} \quad x + 1 = 0$

$x = 7 \quad \text{or} \quad\quad x = -1$

<u>Check</u>: For 7:

$$\frac{\sqrt{x+2} + \sqrt{3x+4} = 2}{}$$

$$\begin{array}{c|c} \sqrt{7+2} + \sqrt{3 \cdot 7 + 4} & 2 \\ \sqrt{9} + \sqrt{25} & \\ 8 & \text{FALSE} \end{array}$$

<u>Check</u>: For −1:

$$\frac{\sqrt{x+2} + \sqrt{3x+4} = 2}{}$$

$$\begin{array}{c|c} \sqrt{-1+2} + \sqrt{3(-1)+4} & 2 \\ \sqrt{1} + \sqrt{1} & \\ 2 & \text{TRUE} \end{array}$$

Since −1 checks but 7 does not, the solution is −1.

33. $\sqrt{3x-5} + \sqrt{2x+3} + 1 = 0$

$$\sqrt{3x-5} + 1 = -\sqrt{2x+3}$$
$$(\sqrt{3x-5}+1)^2 = (-\sqrt{2x+3})^2$$
$$3x - 5 + 2\sqrt{3x-5} + 1 = 2x + 3$$
$$2\sqrt{3x-5} = -x + 7$$
$$(2\sqrt{3x-5})^2 = (-x+7)^2$$
$$4(3x-5) = x^2 - 14x + 49$$
$$12x - 20 = x^2 - 14x + 49$$
$$0 = x^2 - 26x + 69$$
$$0 = (x-23)(x-3)$$

$$x - 23 = 0 \quad \text{or} \quad x - 3 = 0$$
$$x = 23 \quad \text{or} \quad x = 3$$

Neither number checks. There is no solution. (At the outset we might have observed that there is no solution since the sum on the left side of the equation must be at least 1.)

35. $2\sqrt{t-1} - \sqrt{3t-1} = 0$

$$2\sqrt{t-1} = \sqrt{3t-1}$$
$$(2\sqrt{t-1})^2 = (\sqrt{3t-1})^2$$
$$4(t-1) = 3t - 1$$
$$4t - 4 = 3t - 1$$
$$t = 3$$

Since 3 checks, it is the solution.

37. $x^2 + 2.8x = 0$

$x(x + 2.8) = 0$

$x = 0 \quad \text{or} \quad x + 2.8 = 0$

$x = 0 \quad \text{or} \quad x = -2.8$

The solutions are 0 and −2.8.

39. $x^2 - 64 = 0$

$(x+8)(x-8) = 0$

$x + 8 = 0 \quad \text{or} \quad x - 8 = 0$

$x = -8 \quad \text{or} \quad x = 8$

The solutions are −8 and 8.

41. $6x^2 = x + 1$

$$6x^2 - x - 1 = 0$$
$$(3x+1)(2x-1) = 0$$
$$3x + 1 = 0 \quad \text{or} \quad 2x - 1 = 0$$
$$3x = -1 \quad \text{or} \quad 2x = 1$$
$$x = -\frac{1}{3} \quad \text{or} \quad x = \frac{1}{2}$$

The solutions are $-\frac{1}{3}$ and $\frac{1}{2}$.

43. $\sqrt[3]{\frac{z}{4}} - 10 = 2$

$$\sqrt[3]{\frac{z}{4}} = 12$$
$$\left(\sqrt[3]{\frac{z}{4}}\right)^3 = 12^3$$
$$\frac{z}{4} = 1728$$
$$z = 6912$$

The number 6912 checks, so it is the solution.

45. $\sqrt{\sqrt{y+49} - \sqrt{y}} = \sqrt{7}$

$$\left(\sqrt{\sqrt{y+49} - \sqrt{y}}\right)^2 = (\sqrt{7})^2$$
$$\sqrt{y+49} - \sqrt{y} = 7$$
$$\sqrt{y+49} = 7 + \sqrt{y}$$
$$(\sqrt{y+49})^2 = (7+\sqrt{y})^2$$
$$y + 49 = 49 + 14\sqrt{y} + y$$
$$0 = 14\sqrt{y}$$
$$0 = \sqrt{y}$$
$$0^2 = (\sqrt{y})^2$$
$$0 = y$$

The number 0 checks and is the solution.

47. $\sqrt{\sqrt{x^2+9x+34}} = 2$

$$\left(\sqrt{\sqrt{x^2+9x+34}}\right)^2 = 2^2$$
$$\sqrt{x^2+9x+34} = 4$$
$$(\sqrt{x^2+9x+34})^2 = 4^2$$
$$x^2 + 9x + 34 = 16$$
$$x^2 + 9x + 18 = 0$$
$$(x+6)(x+3) = 0$$

$$x + 6 = 0 \quad \text{or} \quad x + 3 = 0$$
$$x = -6 \quad \text{or} \quad x = -3$$

Both values check. The solutions are −6 and −3.

49.
$$\sqrt{x-2} - \sqrt{x+2} + 2 = 0$$
$$\sqrt{x-2} + 2 = \sqrt{x+2}$$
$$(\sqrt{x-2} + 2)^2 = (\sqrt{x+2})^2$$
$$(x-2) + 4\sqrt{x-2} + 4 = x+2$$
$$4\sqrt{x-2} = 0$$
$$\sqrt{x-2} = 0$$
$$(\sqrt{x-2})^2 = 0^2$$
$$x - 2 = 0$$
$$x = 2$$

The number 2 checks, so it is the solution.

51.
$$\sqrt{a^2 + 30a} = a + \sqrt{5a}$$
$$(\sqrt{a^2 + 30a})^2 = (a + \sqrt{5a})^2$$
$$a^2 + 30a = a^2 + 2a\sqrt{5a} + 5a$$
$$25a = 2a\sqrt{5a}$$
$$(25a)^2 = (2a\sqrt{5a})^2$$
$$625a^2 = 4a^2 \cdot 5a$$
$$625a^2 = 20a^3$$
$$0 = 20a^3 - 625a^2$$
$$0 = 5a^2(4a - 125)$$
$$5a^2 = 0 \quad \text{or} \quad 4a - 125 = 0$$
$$a^2 = 0 \quad \text{or} \qquad 4a = 125$$
$$a = 0 \quad \text{or} \qquad a = \frac{125}{4}$$

Both values check. The solutions are 0 and $\frac{125}{4}$.

53.
$$\frac{x-1}{\sqrt{x^2 + 3x + 6}} = \frac{1}{4},$$
$$\text{LCM} = 4\sqrt{x^2 + 3x + 6}$$
$$4\sqrt{x^2 + 3x + 6} \cdot \frac{x-1}{\sqrt{x^2 + 3x + 6}} = 4\sqrt{x^2 + 3x + 6} \cdot \frac{1}{4}$$
$$4x - 4 = \sqrt{x^2 + 3x + 6}$$
$$16x^2 - 32x + 16 = x^2 + 3x + 6$$

$$\text{Squaring both sides}$$
$$15x^2 - 35x + 10 = 0$$
$$3x^2 - 7x + 2 = 0 \qquad \text{Dividing by 5}$$
$$(3x - 1)(x - 2) = 0$$
$$3x - 1 = 0 \quad \text{or} \quad x - 2 = 0$$
$$3x = 1 \quad \text{or} \qquad x = 2$$
$$x = \frac{1}{3} \quad \text{or} \qquad x = 2$$

The number 2 checks but $\frac{1}{3}$ does not. The solution is 2.

55.
$$\sqrt{y^2 + 6} + y - 3 = 0$$
$$\sqrt{y^2 + 6} = 3 - y$$
$$(\sqrt{y^2 + 6})^2 = (3 - y)^2$$
$$y^2 + 6 = 9 - 6y + y^2$$
$$-3 = -6y$$
$$\frac{1}{2} = y$$

The number $\frac{1}{2}$ checks and is the solution.

57.
$$\sqrt{y+1} - \sqrt{2y-5} = \sqrt{y-2}$$
$$(\sqrt{y+1} - \sqrt{2y-5})^2 = (\sqrt{y-2})^2$$
$$y + 1 - 2\sqrt{(y+1)(2y-5)} + 2y - 5 = y - 2$$
$$-2\sqrt{2y^2 - 3y - 5} = -2y + 2$$
$$\sqrt{2y^2 - 3y - 5} = y - 1$$
$$\text{Dividing by } -2$$
$$(\sqrt{2y^2 - 3y - 5})^2 = (y - 1)^2$$
$$2y^2 - 3y - 5 = y^2 - 2y + 1$$
$$y^2 - y - 6 = 0$$
$$(y - 3)(y + 2) = 0$$
$$y - 3 = 0 \quad \text{or} \quad y + 2 = 0$$
$$y = 3 \quad \text{or} \qquad y = -2$$

The number 3 checks but -2 does not. The solution is 3.

Exercise Set 7.8

1. $a = 3, \quad b = 5$
Find c.
$$c^2 = a^2 + b^2 \qquad \text{Pythagorean equation}$$
$$c^2 = 3^2 + 5^2 \qquad \text{Substituting}$$
$$c^2 = 9 + 25$$
$$c^2 = 34$$
$$c = \sqrt{34} \qquad \text{Exact answer}$$
$$c \approx 5.831 \qquad \text{Approximation}$$

3. $a = 15, \quad b = 15$
Find c.
$$c^2 = a^2 + b^2 \qquad \text{Pythagorean equation}$$
$$c^2 = 15^2 + 15^2 \qquad \text{Substituting}$$
$$c^2 = 225 + 225$$
$$c^2 = 450$$
$$c = \sqrt{450} \qquad \text{Exact answer}$$
$$c \approx 21.213 \qquad \text{Approximation}$$

5. $b = 12, \quad c = 13$
Find a.
$$a^2 + b^2 = c^2 \qquad \text{Pythagorean equation}$$
$$a^2 + 12^2 = 13^2 \qquad \text{Substituting}$$
$$a^2 + 144 = 169$$
$$a^2 = 25$$
$$a = 5$$

7. $c = 7, \quad a = \sqrt{6}$

Find b.

$c^2 = a^2 + b^2$ Pythagorean equation

$7^2 = (\sqrt{6})^2 + b^2$ Substituting

$49 = 6 + b^2$

$43 = b^2$

$\sqrt{43} = b$ Exact answer

$6.557 \approx b$ Approximation

9. $b = 1, \quad c = \sqrt{13}$

Find a.

$a^2 + b^2 = c^2$ Pythagorean equation

$a^2 + 1^2 = (\sqrt{13})^2$ Substituting

$a^2 + 1 = 13$

$a^2 = 12$

$a = \sqrt{12}$ Exact answer

$a \approx 3.464$ Approximation

11. $a = 1, \quad c = \sqrt{n}$

Find b.

$a^2 + b^2 = c^2$

$1^2 + b^2 = (\sqrt{n})^2$

$1 + b^2 = n$

$b^2 = n - 1$

$b = \sqrt{n - 1}$

13. We make a drawing and let $d =$ the length of the guy wire.

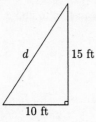

We use the Pythagorean equation to find d.

$d^2 = 10^2 + 15^2$

$d^2 = 100 + 225$

$d^2 = 325$

$d = \sqrt{325}$

$d \approx 18.028$

The wire is $\sqrt{325}$, or about 18.028 ft long.

15. We add labels to the drawing in the text. We let $h =$ the height of the bulge.

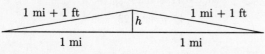

Note that 1 mi = 5280 ft, so 1 mi + 1 ft = 5280 + 1, or 5281 ft.

We use the Pythagorean equation to find h.

$5281^2 = 5280^2 + h^2$

$27,888,961 = 27,878,400 + h^2$

$10,561 = h^2$

$\sqrt{10,561} = h$

$102.767 \approx h$

The bulge is $\sqrt{10,561}$, or about 102.767 ft high.

17. We add some labels to the drawing in the text.

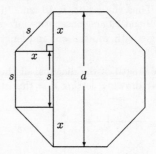

Note that $d = s + 2x$. We use the Pythagorean equation to find x.

$x^2 + x^2 = s^2$

$2x^2 = s^2$

$x^2 = \dfrac{s^2}{2}$

$x = \sqrt{\dfrac{s^2}{2}}$

$x = \dfrac{s}{\sqrt{2}}$

$x = \dfrac{s\sqrt{2}}{2}$ Rationalizing the denominator

Then $d = s + 2x = s + 2\left(\dfrac{s\sqrt{2}}{2}\right) = s + s\sqrt{2}.$

19. $L = \dfrac{0.000169 d^{2.27}}{h}$

$L = \dfrac{0.000169(200)^{2.27}}{4}$

≈ 7.1

The length of the letters should be about 7.1 ft.

21. We make a drawing. Let $x =$ the width of the rectangle. Then $x + 1 =$ the length.

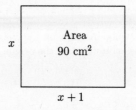

We first find the length and width of the rectangle. Recall the formula for the area of a rectangle, $A = lw$. We substitute 90 for A, $x+1$ for l, and x for w in this formula and solve for x.

$$90 = (x+1)x$$
$$90 = x^2 + x$$
$$0 = x^2 + x - 90$$
$$0 = (x+10)(x-9)$$
$$x + 10 = 0 \quad \text{or} \quad x - 9 = 0$$
$$x = -10 \quad \text{or} \quad x = 9$$

Since the width cannot be negative, we know that the width is 9 cm. Thus the length is 10 cm. (These numbers check since 9 and 10 are consecutive integers and the area of a rectangle with width 9 cm and length 10 cm is $10 \cdot 9$, or 90 cm^2.)

Now we find the length of the diagonal of the rectangle. We make another drawing, letting d = the length of the diagonal.

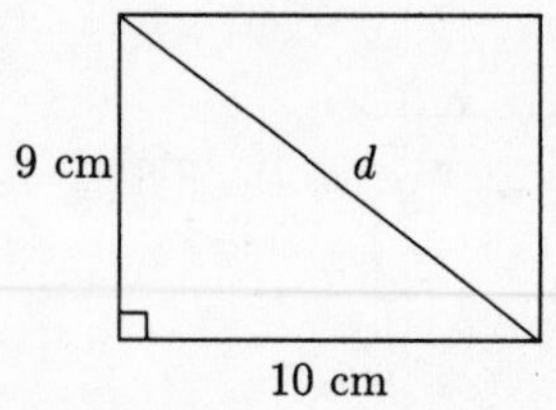

We use the Pythagorean equation to find d.

$$d^2 = 9^2 + 10^2$$
$$d^2 = 81 + 100$$
$$d^2 = 181$$
$$d = \sqrt{181}$$
$$d \approx 13.454$$

The length of the diagonal is $\sqrt{181}$, or about 13.454 cm.

23. We use the drawing in the text, replacing w with 16 in.

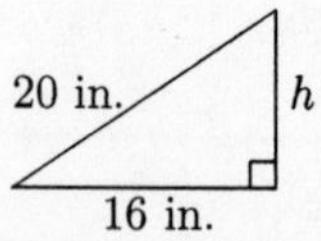

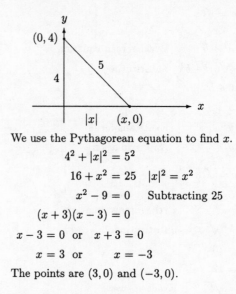

We use the Pythagorean equation to find h.

$$h^2 + 16^2 = 20^2$$
$$h^2 + 256 = 400$$
$$h^2 = 144$$
$$h = 12$$

The height is 12 in.

25. We first make a drawing. A point on the x-axis has coordinates $(x, 0)$ and is $|x|$ units from the origin.

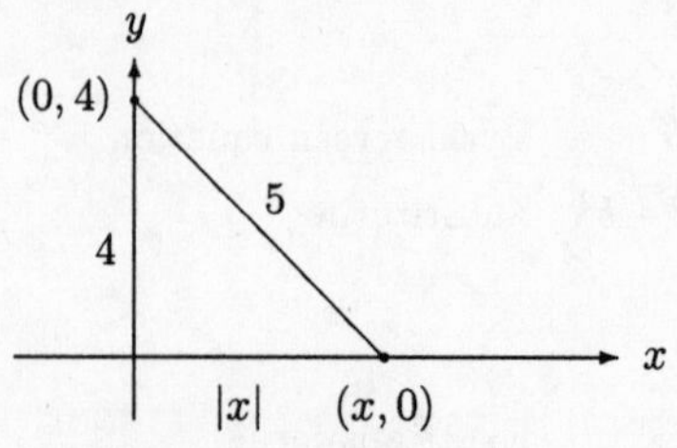

We use the Pythagorean equation to find x.

$$4^2 + |x|^2 = 5^2$$
$$16 + x^2 = 25 \qquad |x|^2 = x^2$$
$$x^2 - 9 = 0 \qquad \text{Subtracting 25}$$
$$(x+3)(x-3) = 0$$
$$x - 3 = 0 \quad \text{or} \quad x + 3 = 0$$
$$x = 3 \quad \text{or} \quad x = -3$$

The points are $(3, 0)$ and $(-3, 0)$.

27. We make a drawing. We let d = the length the diagonal should be

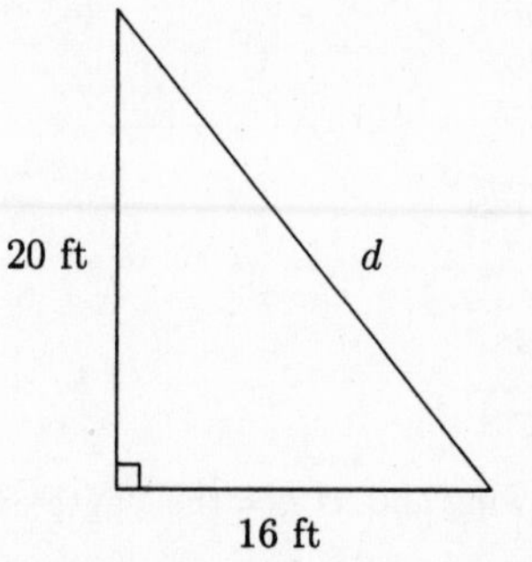

We use the Pythagorean equation to find d.

$$d^2 = 20^2 + 16^2$$
$$d^2 = 400 + 256$$
$$d^2 = 656$$
$$d = \sqrt{656}$$
$$d \approx 25.612$$

The diagonal should be $\sqrt{656}$, or about 25.612 ft long.

29. $x^2 - 11x + 24 = 0$

$$(x-8)(x-3) = 0$$
$$x - 8 = 0 \quad \text{or} \quad x - 3 = 0$$
$$x = 8 \quad \text{or} \quad x = 3$$

The solutions are 8 and 3.

31. $\dfrac{x-5}{x-7} = \dfrac{4}{3}$, LCM is $3(x-7)$

$$3(x-7) \cdot \frac{x-5}{x-7} = 3(x-7) \cdot \frac{4}{3}$$
$$3(x-5) = 4(x-7)$$
$$3x - 15 = 4x - 28$$
$$13 = x$$

The number 13 checks and is the solution.

33. $\dfrac{x-1}{x-3} = \dfrac{6}{x-3}$, LCM is $x-3$

$(x-3) \cdot \dfrac{x-1}{x-3} = (x-3) \cdot \dfrac{6}{x-3}$

$x - 1 = 6$

$x = 7$

The number 7 checks and is the solution.

35. If the area of square $PQRS$ is 100 ft^2, then each side measures 10 ft. If A, B, C, and D are midpoints, then each of the segments PB, BQ, QC, CS, SD, DR, RA, and AP measures 5 ft. We can label the figure with additional information.

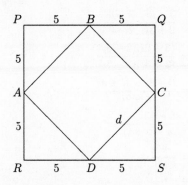

We label a side of square $ABCD$ with d. Then we use the Pythagorean property.

$5^2 + 5^2 = d^2$

$25 + 25 = d^2$

$50 = d^2$

$\sqrt{50} = d$

If a side of square $ABCD$ is $\sqrt{50}$ ft, then its area is $\sqrt{50} \cdot \sqrt{50}$, or 50 ft^2.

Exercise Set 7.9

1. $\sqrt{-35} = \sqrt{-1 \cdot 35} = \sqrt{-1} \cdot \sqrt{35} = i\sqrt{35}$, or $\sqrt{35}i$

3. $\sqrt{-16} = \sqrt{-1 \cdot 16} = \sqrt{-1} \cdot \sqrt{16} = i \cdot 4 = 4i$

5. $-\sqrt{-12} = -\sqrt{-1 \cdot 12} = -\sqrt{-1} \cdot \sqrt{12} = -i \cdot 2\sqrt{3} = -2\sqrt{3}i$, or $-2i\sqrt{3}$

7. $\sqrt{-3} = \sqrt{-1 \cdot 3} = \sqrt{-1} \cdot \sqrt{3} = i\sqrt{3}$, or $\sqrt{3}i$

9. $\sqrt{-81} = \sqrt{-1 \cdot 81} = \sqrt{-1} \cdot \sqrt{81} = i \cdot 9 = 9i$

11. $\sqrt{-98} = \sqrt{-1 \cdot 98} = \sqrt{-1} \cdot \sqrt{98} = i \cdot 7\sqrt{2} = 7\sqrt{2}i$, or $7i\sqrt{2}$

13. $-\sqrt{-49} = -\sqrt{-1 \cdot 49} = -\sqrt{-1} \cdot \sqrt{49} = -i \cdot 7 = -7i$

15. $4 - \sqrt{-60} = 4 - \sqrt{-1 \cdot 60} = 4 - \sqrt{-1} \cdot \sqrt{60} = 4 - i \cdot 2\sqrt{15} = 4 - 2\sqrt{15}i$, or $4 - 2i\sqrt{15}$

17. $\sqrt{-4} + \sqrt{-12} = \sqrt{-1 \cdot 4} + \sqrt{-1 \cdot 12} = \sqrt{-1} \cdot \sqrt{4} + \sqrt{-1} \cdot \sqrt{12} = i \cdot 2 + i \cdot 2\sqrt{3} = (2 + 2\sqrt{3})i$

19. $(7 + 2i) + (5 - 6i)$

$= (7 + 5) + (2 - 6)i$ Collecting like terms

$= 12 - 4i$

21. $(4 - 3i) + (5 - 2i)$

$= (4 + 5) + (-3 - 2)i$ Collecting like terms

$= 9 - 5i$

23. $(9 - i) + (-2 + 5i) = (9 - 2) + (-1 + 5)i$

$= 7 + 4i$

25. $(6 - i) - (10 + 3i) = (6 - 10) + (-1 - 3)i$

$= -4 - 4i$

27. $(4 - 2i) - (5 - 3i) = (4 - 5) + [-2 - (-3)]i$

$= -1 + i$

29. $(9 + 5i) - (-2 - i) = [9 - (-2)] + [5 - (-1)]i$

$= 11 + 6i$

31. $\sqrt{-36} \cdot \sqrt{-9} = \sqrt{-1} \cdot \sqrt{36} \cdot \sqrt{-1} \cdot \sqrt{9}$

$= i \cdot 6 \cdot i \cdot 3$

$= i^2 \cdot 18$

$= -1 \cdot 18 \qquad i^2 = -1$

$= -18$

33. $\sqrt{-7} \cdot \sqrt{-2} = \sqrt{-1} \cdot \sqrt{7} \cdot \sqrt{-1} \cdot \sqrt{2}$

$= i \cdot \sqrt{7} \cdot i \cdot \sqrt{2}$

$= i^2(\sqrt{14})$

$= -1(\sqrt{14}) \qquad i^2 = -1$

$= -\sqrt{14}$

35. $-3i \cdot 7i = -21 \cdot i^2$

$= -21(-1) \qquad i^2 = -1$

$= 21$

37. $-3i(-8 - 2i) = -3i(-8) - 3i(-2i)$

$= 24i + 6i^2$

$= 24i + 6(-1) \qquad i^2 = -1$

$= 24i - 6$

$= -6 + 24i$

39. $(3 + 2i)(1 + i)$

$= 3 + 3i + 2i + 2i^2$ Using FOIL

$= 3 + 3i + 2i - 2 \qquad i^2 = -1$

$= 1 + 5i$

41. $(2 + 3i)(6 - 2i)$

$= 12 - 4i + 18i - 6i^2$ Using FOIL

$= 12 - 4i + 18i + 6 \qquad i^2 = -1$

$= 18 + 14i$

43. $(6-5i)(3+4i) = 18 + 24i - 15i - 20i^2$
$$= 18 + 24i - 15i + 20$$
$$= 38 + 9i$$

45. $(7-2i)(2-6i) = 14 - 42i - 4i + 12i^2$
$$= 14 - 42i - 4i - 12$$
$$= 2 - 46i$$

47. $(3-2i)^2 = 3^2 - 2 \cdot 3 \cdot 2i + (2i)^2$ Squaring a binomial
$$= 9 - 12i + 4i^2$$
$$= 9 - 12i - 4 \qquad i^2 = -1$$
$$= 5 - 12i$$

49. $(1+5i)^2$
$$= 1^2 + 2 \cdot 1 \cdot 5i + (5i)^2 \quad \text{Squaring a binomial}$$
$$= 1 + 10i + 25i^2$$
$$= 1 + 10i - 25 \qquad i^2 = -1$$
$$= -24 + 10i$$

51. $(-2+3i)^2 = 4 - 12i + 9i^2 = 4 - 12i - 9 =$
$-5 - 12i$

53. $i^7 = i^6 \cdot i = (i^2)^3 \cdot i = (-1)^3 \cdot i = -1 \cdot i = -i$

55. $i^{24} = (i^2)^{12} = (-1)^{12} = 1$

57. $i^{42} = (i^2)^{21} = (-1)^{21} = -1$

59. $i^9 = (i^2)^4 \cdot i = (-1)^4 \cdot i = 1 \cdot i = i$

61. $i^6 = (i^2)^3 = (-1)^3 = -1$

63. $(5i)^3 = 5^3 \cdot i^3 = 125 \cdot i^2 \cdot i = 125(-1)(i) = -125i$

65. $7 + i^4 = 7 + (i^2)^2 = 7 + (-1)^2 = 7 + 1 = 8$

67. $i^{28} - 23i = (i^2)^{14} - 23i = (-1)^{14} - 23i = 1 - 23i$

69. $i^2 + i^4 = -1 + (i^2)^2 = -1 + (-1)^2 = -1 + 1 = 0$

71. $i^5 + i^7 = i^4 \cdot i + i^6 \cdot i = (i^2)^2 \cdot i + (i^2)^3 \cdot i =$
$(-1)^2 \cdot i + (-1)^3 \cdot i = 1 \cdot i + (-1)i = i - i = 0$

73. $1 + i + i^2 + i^3 + i^4 = 1 + i + i^2 + i^2 \cdot i + (i^2)^2$
$$= 1 + i + (-1) + (-1) \cdot i + (-1)^2$$
$$= 1 + i - 1 - i + 1$$
$$= 1$$

75. $5 - \sqrt{-64} = 5 - \sqrt{-1} \cdot \sqrt{64} = 5 - i \cdot 8 = 5 - 8i$

77. $\dfrac{8 - \sqrt{-24}}{4} = \dfrac{8 - \sqrt{-1} \cdot \sqrt{24}}{4} = \dfrac{8 - i \cdot 2\sqrt{6}}{4} =$
$\dfrac{2(4 - i\sqrt{6})}{2 \cdot 2} = \dfrac{\cancel{2}(4 - i\sqrt{6})}{\cancel{2} \cdot 2} = \dfrac{4 - i\sqrt{6}}{2} = 2 - \dfrac{\sqrt{6}}{2}i$

79. $\dfrac{4 + 3i}{3 - i} = \dfrac{4 + 3i}{3 - i} \cdot \dfrac{3 + i}{3 + i}$
$$= \dfrac{(4 + 3i)(3 + i)}{(3 - i)(3 + i)}$$
$$= \dfrac{12 + 4i + 9i + 3i^2}{9 - i^2}$$
$$= \dfrac{12 + 13i - 3}{9 - (-1)}$$
$$= \dfrac{9 + 13i}{10}$$
$$= \dfrac{9}{10} + \dfrac{13}{10}i$$

81. $\dfrac{3 - 2i}{2 + 3i} = \dfrac{3 - 2i}{2 + 3i} \cdot \dfrac{2 - 3i}{2 - 3i}$
$$= \dfrac{(3 - 2i)(2 - 3i)}{(2 + 3i)(2 - 3i)}$$
$$= \dfrac{6 - 9i - 4i + 6i^2}{4 - 9i^2}$$
$$= \dfrac{6 - 13i - 6}{4 - 9(-1)}$$
$$= \dfrac{-13i}{13}$$
$$= -i$$

83. $\dfrac{8 - 3i}{7i} = \dfrac{8 - 3i}{7i} \cdot \dfrac{-7i}{-7i}$
$$= \dfrac{-56i + 21i^2}{-49i^2}$$
$$= \dfrac{-21 - 56i}{49}$$
$$= -\dfrac{21}{49} - \dfrac{56}{49}i$$
$$= -\dfrac{3}{7} - \dfrac{8}{7}i$$

85. $\dfrac{4}{3 + i} = \dfrac{4}{3 + i} \cdot \dfrac{3 - i}{3 - i}$
$$= \dfrac{12 - 4i}{9 - i^2}$$
$$= \dfrac{12 - 4i}{9 - (-1)}$$
$$= \dfrac{12 - 4i}{10}$$
$$= \dfrac{12}{10} - \dfrac{4}{10}i$$
$$= \dfrac{6}{5} - \dfrac{2}{5}i$$

87. $\dfrac{2i}{5-4i} = \dfrac{2i}{5-4i} \cdot \dfrac{5+4i}{5+4i}$

$\quad = \dfrac{10i + 8i^2}{25 - 16i^2}$

$\quad = \dfrac{10i + 8(-1)}{25 - 16(-1)}$

$\quad = \dfrac{-8 + 10i}{41}$

$\quad = -\dfrac{8}{41} + \dfrac{10}{41}i$

89. $\dfrac{4}{3i} = \dfrac{4}{3i} \cdot \dfrac{-3i}{-3i}$

$\quad = \dfrac{-12i}{-9i^2}$

$\quad = \dfrac{-12i}{-9(-1)}$

$\quad = \dfrac{-12i}{9}$

$\quad = -\dfrac{4}{3}i$

91. $\dfrac{9-4i}{8i} = \dfrac{2-4i}{8i} \cdot \dfrac{-8i}{-8i}$

$\quad = \dfrac{-16i + 32i^2}{-64i^2}$

$\quad = \dfrac{-16i + 32(-1)}{-64(-1)}$

$\quad = \dfrac{-32 - 16i}{64}$

$\quad = -\dfrac{32}{64} - \dfrac{16}{64}i$

$\quad = -\dfrac{1}{2} - \dfrac{1}{4}i$

93. $\dfrac{6+3i}{6-3i} = \dfrac{6+3i}{6-3i} \cdot \dfrac{6+3i}{6+3i}$

$\quad = \dfrac{36 + 18i + 18i + 9i^2}{36 - 9i^2}$

$\quad = \dfrac{36 + 36i - 9}{36 - 9(-1)}$

$\quad = \dfrac{27 + 36i}{45}$

$\quad = \dfrac{27}{45} + \dfrac{36}{45}i$

$\quad = \dfrac{3}{5} + \dfrac{4}{5}i$

95. Substitute $1 - 2i$ for x in the equation.

$$x^2 - 2x + 5 = 0$$

$(1-2i)^2 - 2(1-2i) + 5$	0
$1 - 4i + 4i^2 - 2 + 4i + 5$	
$1 - 4i - 4 - 2 + 4i + 5$	
	0 | TRUE

$1 - 2i$ is a solution.

97. Substitute $2 + i$ for x in the equation.

$$x^2 - 4x - 5 = 0$$

$(2+i)^2 - 4(2+i) - 5$	0
$4 + 4i + i^2 - 8 - 4i - 5$	
$4 + 4i - 1 - 8 - 4i - 5$	
-10	FALSE

$2 + i$ is not a solution.

99. $\dfrac{196}{x^2 - 7x + 49} - \dfrac{2x}{x+7} = \dfrac{2058}{x^3 + 343}$

Note: $x^3 + 343 = (x+7)(x^2 - 7x + 49)$.

The LCM $= (x+7)(x^2 - 7x + 49)$.

$(x+7)(x^2 - 7x + 49)\Big(\dfrac{196}{x^2 - 7x + 49} - \dfrac{2x}{x+7}\Big) =$

$\qquad\qquad (x+7)(x^2 - 7x + 49) \cdot \dfrac{2058}{x^3 + 343}$

$196(x+7) - 2x(x^2 - 7x + 49) = 2058$

$196x + 1372 - 2x^3 + 14x^2 - 98x = 2058$

$98x - 686 - 2x^3 + 14x^2 = 0$

$49x - 343 - x^3 + 7x^2 = 0 \quad$ Dividing by 2

$49(x-7) - x^2(x-7) = 0$

$(49 - x^2)(x-7) = 0$

$(7-x)(7+x)(x-7) = 0$

$7 - x = 0 \quad$ or $\quad 7 + x = 0 \quad$ or $\quad x - 7 = 0$

$7 = x \quad$ or $\qquad x = -7 \quad$ or $\qquad x = 7$

Only 7 checks. It is the solution.

101. $\dfrac{(2i)^4 - (2i)^2}{2i - 1} = \dfrac{16i^4 - 4i^2}{-1 + 2i} = \dfrac{20}{-1 + 2i} =$

$\dfrac{20}{-1+2i} \cdot \dfrac{-1-2i}{-1-2i} = \dfrac{-20 - 40i}{5} = -4 - 8i$

103. $\dfrac{1}{8}\Big(-24 - \sqrt{-1024}\Big) = \dfrac{1}{8}(-24 - 32i) = -3 - 4i$

105. $7\sqrt{-64} - 9\sqrt{-256} = 7 \cdot 8i - 9 \cdot 16i = 56i - 144i = -88i$

107. $(1-i)^3(1+i)^3 =$

$(1-i)(1+i) \cdot (1-i)(1+i) \cdot (1-i)(1+i) =$

$(1 - i^2)(1 - i^2)(1 - i^2) = (1+1)(1+1)(1+1) =$

$2 \cdot 2 \cdot 2 = 8$

109. $\dfrac{6}{1 + \dfrac{3}{i}} = \dfrac{6}{\dfrac{i+3}{i}} = \dfrac{6i}{i+3} = \dfrac{6i}{i+3} \cdot \dfrac{-i+3}{-i+3} =$

$\dfrac{-6i^2 + 18i}{-i^2 + 9} = \dfrac{6 + 18i}{10} = \dfrac{6}{10} + \dfrac{18}{10}i = \dfrac{3}{5} + \dfrac{9}{5}i$

111. $\dfrac{i - i^{38}}{1+i} = \dfrac{i - (i^2)^{19}}{1+i} = \dfrac{i - (-1)^{19}}{1+i} = \dfrac{i - (-1)}{1+i} =$

$\dfrac{i+1}{1+i} = 1$

Chapter 8

Quadratic Equations

Exercise Set 8.1

1. $6x^2 = 30$

$\quad x^2 = 5 \qquad$ Dividing by 6

$\quad x = \sqrt{5}$ or $x = -\sqrt{5} \qquad$ Principle of square roots

Check: $\qquad \dfrac{6x^2 = 30}{}$

$\qquad \dfrac{6(\pm\sqrt{5})}{6 \cdot 5} \Big| 30$

$\qquad \qquad 30 \Big|$ TRUE

The solutions are $\sqrt{5}$ and $-\sqrt{5}$, or $\pm\sqrt{5}$.

3. $9x^2 + 25 = 0$

$\qquad 9x^2 = -25 \qquad$ Subtracting 25

$\qquad x^2 = -\dfrac{25}{9} \qquad$ Dividing by 9

$\qquad x = \sqrt{-\dfrac{25}{9}}$ or $x = -\sqrt{-\dfrac{25}{9}} \qquad$ Principle of square roots

$\qquad x = \dfrac{5}{3}i \qquad$ or $\qquad x = -\dfrac{5}{3}i \qquad$ Simplifying

Check: $\qquad \dfrac{9x^2 + 25 = 0}{}$

$\qquad \dfrac{9\left(\pm\dfrac{5}{3}i\right) + 25}{} \Big| 0$

$\qquad 9\left(-\dfrac{25}{9}\right) + 25$

$\qquad -25 + 25$

$\qquad \qquad 0 \Big|$ TRUE

The solutions are $\dfrac{5}{3}i$ and $-\dfrac{5}{3}i$, or $\pm\dfrac{5}{3}i$.

5. $2x^2 - 3 = 0$

$\qquad 2x^2 = 3$

$\qquad x^2 = \dfrac{3}{2}$

$\qquad x = \sqrt{\dfrac{3}{2}} \qquad$ or $\qquad x = -\sqrt{\dfrac{3}{2}} \qquad$ Principle of square roots

$\qquad x = \sqrt{\dfrac{3}{2} \cdot \dfrac{2}{2}} \qquad$ or $\qquad x = -\sqrt{\dfrac{3}{2} \cdot \dfrac{2}{2}} \qquad$ Rationalizing denominators

$\qquad x = \dfrac{\sqrt{6}}{2} \qquad$ or $\qquad x = -\dfrac{\sqrt{6}}{2}$

Check: $\qquad \dfrac{2x^2 - 3 = 0}{}$

$\qquad \dfrac{2\left(\pm\dfrac{\sqrt{6}}{2}\right)^2 - 3}{} \Big| 0$

$\qquad 2 \cdot \dfrac{6}{4} - 3$

$\qquad 3 - 3$

$\qquad \qquad 0 \Big|$ TRUE

The solutions are $\dfrac{\sqrt{6}}{2}$ and $-\dfrac{\sqrt{6}}{2}$, or $\pm\dfrac{\sqrt{6}}{2}$.

7. $(x+2)^2 = 49$

$\quad x + 2 = 7$ or $x + 2 = -7 \qquad$ Principle of square roots

$\quad x = 5$ or $\qquad x = -9$

The solutions are 5 and -9.

9. $(x-4)^2 = 16$

$\quad x - 4 = 4$ or $x - 4 = -4 \qquad$ Principle of square roots

$\quad x = 8$ or $\qquad x = 0$

The solutions are 8 and 0.

11. $(x-11)^2 = 7$

$\quad x - 11 = \sqrt{7} \qquad$ or $\quad x - 11 = -\sqrt{7}$

$\qquad x = 11 + \sqrt{7}$ or $\qquad x = 11 - \sqrt{7}$

The solutions are $11 + \sqrt{7}$ and $11 - \sqrt{7}$, or $11 \pm \sqrt{7}$.

13. $(x-7)^2 = -4$

$\quad x - 7 = \sqrt{-4} \qquad$ or $\quad x - 7 = -\sqrt{-4}$

$\quad x - 7 = 2i \qquad$ or $\quad x - 7 = -2i$

$\qquad x = 7 + 2i$ or $\qquad x = 7 - 2i$

The solutions are $7 + 2i$ and $7 - 2i$, or $7 \pm 2i$.

15. $(x-9)^2 = 81$

$\quad x - 9 = 9$ or $x - 9 = -9$

$\quad x = 18$ or $\qquad x = 0$

The solutions are 18 and 0.

17. $\left(x - \dfrac{3}{2}\right)^2 = \dfrac{7}{2}$

$\quad x - \dfrac{3}{2} = \sqrt{\dfrac{7}{2}} \qquad$ or $\quad x - \dfrac{3}{2} = -\sqrt{\dfrac{7}{2}}$

$\quad x - \dfrac{3}{2} = \sqrt{\dfrac{7}{2} \cdot \dfrac{2}{2}} \qquad$ or $\quad x - \dfrac{3}{2} = -\sqrt{\dfrac{7}{2} \cdot \dfrac{2}{2}}$

$\quad x - \dfrac{3}{2} = \dfrac{\sqrt{14}}{2} \qquad$ or $\quad x - \dfrac{3}{2} = -\dfrac{\sqrt{14}}{2}$

$\qquad x = \dfrac{3}{2} + \dfrac{\sqrt{14}}{2} \qquad$ or $\qquad x = \dfrac{3}{2} - \dfrac{\sqrt{14}}{2}$

$\qquad x = \dfrac{3 + \sqrt{14}}{2} \qquad$ or $\qquad x = \dfrac{3 - \sqrt{14}}{2}$

The solutions are $\dfrac{3+\sqrt{14}}{2}$ and $\dfrac{3-\sqrt{14}}{2}$, or $\dfrac{3\pm\sqrt{14}}{2}$.

19. $x^2 + 6x + 9 = 64$

$(x+3)^2 = 64$

$x + 3 = 8 \quad\text{or}\quad x + 3 = -8$

$x = 5 \quad\text{or}\qquad x = -11$

The solutions are 5 and -11.

21. $y^2 - 14y + 49 = 4$

$(y-7)^2 = 4$

$y - 7 = 2 \quad\text{or}\quad y - 7 = -2$

$y = 9 \quad\text{or}\qquad y = 5$

The solutions are 9 and 5.

23. $x^2 + 4x \quad\;\; = 2$ Original equation

$x^2 + 4x + 4 = 2 + 4$ Adding 4: $\left(\dfrac{4}{2}\right)^2 = 2^2 = 4$

$(x+2)^2 = 6$

$x + 2 = \sqrt{6} \qquad\text{or}\quad x + 2 = -\sqrt{6}$ Principle of square roots

$x = -2 + \sqrt{6} \quad\text{or}\qquad x = -2 - \sqrt{6}$

The solutions are $-2 \pm \sqrt{6}$.

25. $x^2 - 22x \qquad = 11$ Original equation

$x^2 - 22x + 121 = 11 + 121$ Adding 121: $\left(\dfrac{-22}{2}\right)^2 =$ $(-11)^2 = 121$

$(x-11)^2 = 132$

$x - 11 = \sqrt{132} \qquad\text{or}\quad x - 11 = -\sqrt{132}$

$x - 11 = 2\sqrt{33} \qquad\text{or}\quad x - 11 = -2\sqrt{33}$

$x = 11 + 2\sqrt{33} \quad\text{or}\qquad x = 11 - 2\sqrt{33}$

The solutions are $11 \pm 2\sqrt{33}$.

27. $x^2 + x \qquad = 1$

$x^2 + x + \dfrac{1}{4} = 1 + \dfrac{1}{4}$ Adding $\dfrac{1}{4}$: $\left(\dfrac{1}{2}\right)^2 = \dfrac{1}{4}$

$\left(x + \dfrac{1}{2}\right)^2 = \dfrac{5}{4}$

$x + \dfrac{1}{2} = \dfrac{\sqrt{5}}{2} \qquad\text{or}\quad x + \dfrac{1}{2} = -\dfrac{\sqrt{5}}{2}$

$x = \dfrac{-1+\sqrt{5}}{2} \quad\text{or}\quad x = \dfrac{-1-\sqrt{5}}{2}$

The solutions are $\dfrac{-1\pm\sqrt{5}}{2}$.

29. $t^2 - 5t \qquad = 7$

$t^2 - 5t + \dfrac{25}{4} = 7 + \dfrac{25}{4}$ Adding $\dfrac{25}{4}$: $\left(\dfrac{-5}{2}\right)^2 =$ $\dfrac{25}{4}$

$\left(t - \dfrac{5}{2}\right)^2 = \dfrac{53}{4}$

$t - \dfrac{5}{2} = \dfrac{\sqrt{53}}{2} \qquad\text{or}\quad t - \dfrac{5}{2} = -\dfrac{\sqrt{53}}{2}$

$t = \dfrac{5+\sqrt{53}}{2} \quad\text{or}\qquad t = \dfrac{5-\sqrt{53}}{2}$

The solutions are $\dfrac{5\pm\sqrt{53}}{2}$.

31. $x^2 + \dfrac{3}{2}x \qquad\;\; = 3$

$x^2 + \dfrac{3}{2}x + \dfrac{9}{16} = 3 + \dfrac{9}{16}$ $\left(\dfrac{1}{2}\cdot\dfrac{3}{2}\right)^2 = \left(\dfrac{3}{4}\right)^2 = \dfrac{9}{16}$

$\left(x + \dfrac{3}{4}\right)^2 = \dfrac{57}{16}$

$x + \dfrac{3}{4} = \dfrac{\sqrt{57}}{4} \qquad\text{or}\quad x + \dfrac{3}{4} = -\dfrac{\sqrt{57}}{4}$

$x = \dfrac{-3+\sqrt{57}}{4} \quad\text{or}\qquad x = \dfrac{-3-\sqrt{57}}{4}$

The solutions are $\dfrac{-3\pm\sqrt{57}}{4}$.

33. $m^2 - \dfrac{9}{2}m \qquad\;\; = \dfrac{3}{2}$ Original equation

$m^2 - \dfrac{9}{2}m + \dfrac{81}{16} = \dfrac{3}{2} + \dfrac{81}{16}$ $\left[\dfrac{1}{2}\left(-\dfrac{9}{2}\right)\right]^2 = \left(-\dfrac{9}{4}\right)^2 =$ $\dfrac{81}{16}$

$\left(m - \dfrac{9}{4}\right)^2 = \dfrac{105}{16}$

$m - \dfrac{9}{4} = \dfrac{\sqrt{105}}{4} \qquad\text{or}\quad m - \dfrac{9}{4} = -\dfrac{\sqrt{105}}{4}$

$m = \dfrac{9+\sqrt{105}}{4} \quad\text{or}\qquad m = \dfrac{9-\sqrt{105}}{4}$

The solutions are $\dfrac{9\pm\sqrt{105}}{4}$.

35. $x^2 + 6x - 16 = 0$

$x^2 + 6x \qquad = 16$ Adding 16

$x^2 + 6x + 9 = 16 + 9$ $\left(\dfrac{6}{2}\right)^2 = 3^2 = 9$

$(x+3)^2 = 25$

$x + 3 = 5 \quad\text{or}\quad x + 3 = -5$

$x = 2 \quad\text{or}\qquad x = -8$

The solutions are 2 and -8.

37. $x^2 + 22x + 102 = 0$

$x^2 + 22x \qquad = -102$ Subtracting 102

$x^2 + 22x + 121 = -102 + 121$ $\left(\dfrac{22}{2}\right)^2 = 11^2 = 121$

$(x+11)^2 = 19$

$x + 11 = \sqrt{19} \qquad\text{or}\quad x + 11 = -\sqrt{19}$

$x = -11 + \sqrt{19} \quad\text{or}\qquad x = -11 - \sqrt{19}$

The solutions are $-11 \pm \sqrt{19}$.

39. $x^2 - 10x - 4 = 0$

$x^2 - 10x = 4$ Adding 4

$x^2 - 10x + 25 = 4 + 25$ $\left(\dfrac{-10}{2}\right)^2 = (-5)^2 = 25$

$(x - 5)^2 = 29$

$x - 5 = \sqrt{29}$ or $x - 5 = -\sqrt{29}$

$x = 5 + \sqrt{29}$ or $x = 5 - \sqrt{29}$

The solutions are $5 \pm \sqrt{29}$.

41. $x^2 + 7x - 2 = 0$

$x^2 + 7x = 2$ Adding 2

$x^2 + 7x + \dfrac{49}{4} = 2 + \dfrac{49}{4}$ $\left(\dfrac{7}{2}\right)^2 = \dfrac{49}{4}$

$\left(x + \dfrac{7}{2}\right)^2 = \dfrac{57}{4}$

$x + \dfrac{7}{2} = \dfrac{\sqrt{57}}{2}$ or $x + \dfrac{7}{2} = -\dfrac{\sqrt{57}}{2}$

$x = \dfrac{-7 + \sqrt{57}}{2}$ or $x = \dfrac{-7 - \sqrt{57}}{2}$

The solutions are $\dfrac{-7 \pm \sqrt{57}}{2}$.

43. $2x^2 - 5x + 8 = 0$

$\dfrac{1}{2}(2x^2 - 5x + 8) = \dfrac{1}{2} \cdot 0$ Multiplying by $\dfrac{1}{2}$ to make the x^2-coefficient 1

$x^2 - \dfrac{5}{2}x + 4 = 0$

$x^2 - \dfrac{5}{2}x = -4$ Subtracting 4

$x^2 - \dfrac{5}{2}x + \dfrac{25}{16} = -4 + \dfrac{25}{16}$ $\left[\dfrac{1}{2}\left(-\dfrac{5}{2}\right)\right]^2 = \left(-\dfrac{5}{4}\right)^2 = \dfrac{25}{16}$

$\left(x - \dfrac{5}{4}\right)^2 = -\dfrac{64}{16} + \dfrac{25}{16}$

$\left(x - \dfrac{5}{4}\right)^2 = -\dfrac{39}{16}$

$x - \dfrac{5}{4} = \sqrt{-\dfrac{39}{16}}$ or $x - \dfrac{5}{4} = -\sqrt{-\dfrac{39}{16}}$

$x - \dfrac{5}{4} = i\sqrt{\dfrac{39}{16}}$ or $x - \dfrac{5}{4} = -i\sqrt{\dfrac{39}{16}}$

$x = \dfrac{5}{4} + i\dfrac{\sqrt{39}}{4}$ or $x = \dfrac{5}{4} - i\dfrac{\sqrt{39}}{4}$

The solutions are $\dfrac{5}{4} \pm i\dfrac{\sqrt{39}}{4}$.

45. $x^2 - \dfrac{3}{2}x - \dfrac{1}{2} = 0$

$x^2 - \dfrac{3}{2}x = \dfrac{1}{2}$

$x^2 - \dfrac{3}{2}x + \dfrac{9}{16} = \dfrac{1}{2} + \dfrac{9}{16}$ $\left[\dfrac{1}{2}\left(-\dfrac{3}{2}\right)\right]^2 = \left(-\dfrac{3}{4}\right)^2 = \dfrac{9}{16}$

$\left(x - \dfrac{3}{4}\right)^2 = \dfrac{17}{16}$

$x - \dfrac{3}{4} = \dfrac{\sqrt{17}}{4}$ or $x - \dfrac{3}{4} = -\dfrac{\sqrt{17}}{4}$

$x = \dfrac{3 + \sqrt{17}}{4}$ or $x = \dfrac{3 - \sqrt{17}}{4}$

The solutions are $\dfrac{3 \pm \sqrt{17}}{4}$.

47. $2x^2 - 3x - 17 = 0$

$\dfrac{1}{2}(2x^2 - 3x - 17) = \dfrac{1}{2} \cdot 0$ Multiplying by $\dfrac{1}{2}$ to make the x^2-coefficient 1

$x^2 - \dfrac{3}{2}x - \dfrac{17}{2} = 0$

$x^2 - \dfrac{3}{2}x = \dfrac{17}{2}$ Adding $\dfrac{17}{2}$

$x^2 - \dfrac{3}{2}x + \dfrac{9}{16} = \dfrac{17}{2} + \dfrac{9}{16}$ $\left[\dfrac{1}{2}\left(-\dfrac{3}{2}\right)\right]^2 = \left(-\dfrac{3}{4}\right)^2 = \dfrac{9}{16}$

$\left(x - \dfrac{3}{4}\right)^2 = \dfrac{145}{16}$

$x - \dfrac{3}{4} = \dfrac{\sqrt{145}}{4}$ or $x - \dfrac{3}{4} = -\dfrac{\sqrt{145}}{4}$

$x = \dfrac{3 + \sqrt{145}}{4}$ or $x = \dfrac{3 - \sqrt{145}}{4}$

The solutions are $\dfrac{3 \pm \sqrt{145}}{4}$.

49. $3x^2 - 4x - 1 = 0$

$\dfrac{1}{3}(3x^2 - 4x - 1) = \dfrac{1}{3} \cdot 0$ Multiplying to make the x^2-coefficient 1

$x^2 - \dfrac{4}{3}x - \dfrac{1}{3} = 0$

$x^2 - \dfrac{4}{3}x = \dfrac{1}{3}$ Adding $\dfrac{1}{3}$

$x^2 - \dfrac{4}{3}x + \dfrac{4}{9} = \dfrac{1}{3} + \dfrac{4}{9}$ $\left[\dfrac{1}{2}\left(-\dfrac{4}{3}\right)\right]^2 = \left(-\dfrac{2}{3}\right)^2 = \dfrac{4}{9}$

$\left(x - \dfrac{2}{3}\right)^2 = \dfrac{7}{9}$

$$x - \frac{2}{3} = \frac{\sqrt{7}}{3} \quad \text{or} \quad x - \frac{2}{3} = -\frac{\sqrt{7}}{3}$$

$$x = \frac{2 + \sqrt{7}}{3} \quad \text{or} \quad x = \frac{2 - \sqrt{7}}{3}$$

The solutions are $\dfrac{2 \pm \sqrt{7}}{3}$.

51. $x^2 + x + 2 = 0$

$$x^2 + x = -2 \qquad \text{Subtracting 2}$$

$$x^2 + x + \frac{1}{4} = -2 + \frac{1}{4} \qquad \left(\frac{1}{2}\right)^2 = \frac{1}{4}$$

$$\left(x + \frac{1}{2}\right)^2 = -\frac{7}{4}$$

$$x + \frac{1}{2} = \sqrt{-\frac{7}{4}} \qquad \text{or} \quad x + \frac{1}{2} = -\sqrt{-\frac{7}{4}}$$

$$x + \frac{1}{2} = i\sqrt{\frac{7}{4}} \qquad \text{or} \quad x + \frac{1}{2} = -i\sqrt{\frac{7}{4}}$$

$$x = -\frac{1}{2} + i\frac{\sqrt{7}}{2} \quad \text{or} \qquad x = -\frac{1}{2} - i\frac{\sqrt{7}}{2}$$

The solutions are $-\dfrac{1}{2} \pm i\dfrac{\sqrt{7}}{2}$.

53. $x^2 - 4x + 13 = 0$

$$x^2 - 4x = -13 \qquad \text{Subtracting 13}$$

$$x^2 - 4x + 4 = -13 + 4 \qquad \left(\frac{-4}{2}\right)^2 = (-2)^2 = 4$$

$$(x - 2)^2 = -9$$

$$x - 2 = \sqrt{-9} \quad \text{or} \quad x - 2 = -\sqrt{-9}$$

$$x - 2 = 3i \qquad \text{or} \quad x - 2 = -3i$$

$$x = 2 + 3i \quad \text{or} \qquad x = 2 - 3i$$

The solutions are $2 \pm 3i$.

55. $\qquad A = P(1 + r)^t$

$$1210 = 1000(1 + r)^2 \qquad \text{Substituting}$$

$$\frac{1210}{1000} = (1 + r)^2$$

$$\frac{121}{100} = (1 + r)^2$$

$$\sqrt{\frac{121}{100}} = 1 + r \quad \text{or} \quad -\sqrt{\frac{121}{100}} = 1 + r$$

$$\text{Principle of square roots}$$

$$\frac{11}{10} = 1 + r \quad \text{or} \quad -\frac{11}{10} = 1 + r$$

$$\text{Simplifying}$$

$$-\frac{10}{10} + \frac{11}{10} = r \quad \text{or} \quad -\frac{10}{10} - \frac{11}{10} = r$$

$$\frac{1}{10} = r \quad \text{or} \quad -\frac{21}{10} = r$$

Since the interest rate cannot be negative, we have

$$\frac{1}{10} = r$$

$$r = 0.1, \text{ or } 10\%.$$

The interest rate is 10%.

57. $\qquad A = P(1 + r)^t$

$$10,125 = 8000(1 + r)^2 \qquad \text{Substituting}$$

$$\frac{10,125}{8000} = (1 + r)^2$$

$$\frac{81}{64} = (1 + r)^2$$

$$\sqrt{\frac{81}{64}} = 1 + r \quad \text{or} \quad -\sqrt{\frac{81}{64}} = 1 + r$$

$$\text{Principle of square roots}$$

$$\frac{9}{8} = 1 + r \quad \text{or} \quad -\frac{9}{8} = 1 + r$$

$$\text{Simplifying}$$

$$-\frac{8}{8} + \frac{9}{8} = r \quad \text{or} \quad -\frac{8}{8} - \frac{9}{8} = r$$

$$\frac{1}{8} = r \quad \text{or} \quad -\frac{17}{8} = r$$

Since the interest rate cannot be negative, we have

$$\frac{1}{8} = r$$

$$r = 0.125, \text{ or } 12.5\%.$$

The interest rate is 12.5%.

59. $\qquad A = P(1 + r)^t$

$$6760 = 6250(1 + r)^2 \qquad \text{Substituting}$$

$$\frac{6760}{6250} = (1 + r)^2$$

$$\frac{676}{625} = (1 + r)^2$$

$$\sqrt{\frac{676}{625}} = 1 + r \quad \text{or} \quad -\sqrt{\frac{676}{625}} = 1 + r$$

$$\frac{26}{25} = 1 + r \quad \text{or} \quad -\frac{26}{25} = 1 + r$$

$$\frac{1}{25} = r \quad \text{or} \quad -\frac{51}{25} = r$$

Since the interest rate cannot be negative, we have

$$\frac{1}{25} = r$$

$$r = 0.04, \text{ or } 4\%.$$

The interest rate is 4%.

61.
$$V = 48T^2$$
$$36 = 48T^2 \quad \text{Substituting 36 for } V$$
$$\frac{36}{48} = T^2 \quad \text{Solving for } T^2$$
$$0.75 = T^2$$
$$\sqrt{0.75} = T$$
$$0.866 \approx T$$

The hang time is 0.866 sec.

63. Graph: $y = 2x + 1$

x	$\begin{array}{c}y\\y = 2x+1\end{array}$	(x,y)
-3	-5	$(-3,-5)$
0	1	$(0,1)$
2	5	$(2,5)$

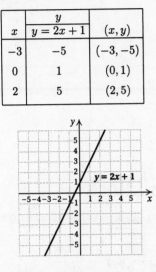

65. Graph: $x - y = -3$

To find the x-intercept, we cover up the y-term and look at the rest of the equation. We have $x = -3$. The x-intercept is $(-3, 0)$.

To find the y-intercept, we cover up the x-term and look at the rest of the equation. We have $-y = -3$, or $y = 3$. The y-intercept is $(0, 3)$.

We plot these points and draw the line

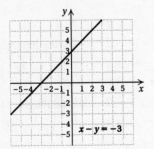

We use a third point as a check. We choose $x = 1$ and solve for y.
$$1 - y = -3$$
$$-y = -4$$
$$y = 4$$

We plot $(1, 4)$ and note that it is on the line.

67. $14 - \sqrt{88} \approx 14 - 9.3808 \approx 4.6$

69. In order for $x^2 + bx + 64$ to be a trinomial square, the following must be true:
$$\left(\frac{b}{2}\right)^2 = 64$$
$$\frac{b^2}{4} = 64$$
$$b^2 = 256$$
$$b = 16 \text{ or } b = -16$$

71.
$$x(2x^2 + 9x - 56)(3x + 10) = 0$$
$$x(2x - 7)(x + 8)(3x + 10) = 0$$
$$x = 0 \text{ or } 2x - 7 = 0 \text{ or } x + 8 = 0 \text{ or } 3x + 10 = 0$$
$$x = 0 \text{ or } \qquad x = \frac{7}{2} \text{ or } \qquad x = -8 \text{ or } \qquad x = -\frac{10}{3}$$

The solutions are -8, $-\dfrac{10}{3}$, 0, and $\dfrac{7}{2}$.

73. *Familiarize*. It is helpful to make a drawing. Let $r =$ the speed of Boat A. Then $r - 7 =$ the speed of Boat B. In 4 hr Boat A travels $4r$ km and Boat B travels $4(r - 7)$, or $4r - 28$ km.

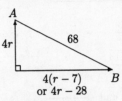

Translate. We use the Pythagorean equation:
$$a^2 + b^2 = c^2$$
$$(4r - 28)^2 + (4r)^2 = 68^2$$

Solve.
$$(4r - 28)^2 + (4r)^2 = 68^2$$
$$16r^2 - 224r + 784 + 16r^2 = 4624$$
$$32r^2 - 224r - 3840 = 0$$
$$r^2 - 7r - 120 = 0 \quad \text{Dividing by 32}$$
$$(r + 8)(r - 15) = 0$$
$$r + 8 = 0 \quad \text{or} \quad r - 15 = 0$$
$$r = -8 \quad \text{or} \qquad r = 15$$

Check. We only check 15 since the speeds of the boats cannot be negative. If the speed of Boat A is 15 km/h, then the speed of Boat B is $15 - 7$, or 8 km/h, and the distances they travel are $4 \cdot 15$, or 60 km, and $4 \cdot 8$, or 32 km.

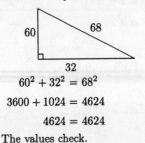

$$60^2 + 32^2 = 68^2$$
$$3600 + 1024 = 4624$$
$$4624 = 4624$$

The values check.

State. The speed of Boat A is 15 km/h, and the speed of Boat B is 8 km/h.

Exercise Set 8.2

1. $x^2 + 6x + 4 = 0$

$a = 1, \ b = 6, \ c = 4$

$x = \dfrac{-b \pm \sqrt{b^2 - 4ac}}{2a}$

$x = \dfrac{-6 \pm \sqrt{6^2 - 4 \cdot 1 \cdot 4}}{2 \cdot 1} = \dfrac{-6 \pm \sqrt{36 - 16}}{2}$

$x = \dfrac{-6 \pm \sqrt{20}}{2} = \dfrac{-6 \pm 2\sqrt{5}}{2}$

$x = \dfrac{2(-3 \pm \sqrt{5})}{2} = -3 \pm \sqrt{5}$

The solutions are $-3 + \sqrt{5}$ and $-3 - \sqrt{5}$.

3. $\qquad 3p^2 = -8p - 1$

$3p^2 + 8p + 1 = 0 \qquad$ Finding standard form

$a = 3, \ b = 8, \ c = 1$

$p = \dfrac{-b \pm \sqrt{b^2 - 4ac}}{2a}$

$p = \dfrac{-8 \pm \sqrt{8^2 - 4 \cdot 3 \cdot 1}}{2 \cdot 3} = \dfrac{-8 \pm \sqrt{64 - 12}}{6}$

$x = \dfrac{-8 \pm \sqrt{52}}{6} = \dfrac{-8 \pm 2\sqrt{13}}{6}$

$x = \dfrac{2(-4 \pm \sqrt{13})}{2 \cdot 3} = \dfrac{-4 \pm \sqrt{13}}{3}$

The solutions are $\dfrac{-4 + \sqrt{13}}{3}$ and $\dfrac{-4 - \sqrt{13}}{3}$.

5. $x^2 - x + 1 = 0$

$a = 1, \ b = -1, \ c = 1$

$x = \dfrac{-(-1) \pm \sqrt{(-1)^2 - 4 \cdot 1 \cdot 1}}{2 \cdot 1} = \dfrac{1 \pm \sqrt{1 - 4}}{2}$

$x = \dfrac{1 \pm \sqrt{-3}}{2} = \dfrac{1 \pm i\sqrt{3}}{2} = \dfrac{1}{2} \pm i\dfrac{\sqrt{3}}{2}$

The solutions are $\dfrac{1}{2} + i\dfrac{\sqrt{3}}{2}$ and $\dfrac{1}{2} - i\dfrac{\sqrt{3}}{2}$.

7. $\qquad x^2 + 13 = 4x$

$x^2 - 4x + 13 = 0 \qquad$ Finding standard form

$a = 1, \ b = -4, \ c = 13$

$x = \dfrac{-(-4) \pm \sqrt{(-4)^2 - 4 \cdot 1 \cdot 13}}{2 \cdot 1} = \dfrac{4 \pm \sqrt{16 - 52}}{2}$

$x = \dfrac{4 \pm \sqrt{-36}}{2} = \dfrac{4 \pm 6i}{2} = 2 \pm 3i$

The solutions are $2 + 3i$ and $2 - 3i$.

9. $\qquad r^2 + 3r = 8$

$r^2 + 3r - 8 = 0 \qquad$ Finding standard form

$a = 1, \ b = 3, \ c = -8$

$r = \dfrac{-3 \pm \sqrt{3^2 - 4 \cdot 1 \cdot (-8)}}{2 \cdot 1} = \dfrac{-3 \pm \sqrt{9 + 32}}{2}$

$r = \dfrac{-3 \pm \sqrt{41}}{2}$

The solutions are $\dfrac{-3 + \sqrt{41}}{2}$ and $\dfrac{-3 - \sqrt{41}}{2}$.

11. $1 + \dfrac{2}{x} + \dfrac{5}{x^2} = 0$

$x^2 + 2x + 5 = 0 \quad$ Multiplying by x^2, the LCM of the denominators

$a = 1, \ b = 2, \ c = 5$

$x = \dfrac{-2 \pm \sqrt{2^2 - 4 \cdot 1 \cdot 5}}{2 \cdot 1} = \dfrac{-2 \pm \sqrt{4 - 20}}{2}$

$x = \dfrac{-2 \pm \sqrt{-16}}{2} = \dfrac{-2 \pm 4i}{2} = -1 \pm 2i$

The solutions are $-1 + 2i$ and $-1 - 2i$.

13. $3x + x(x - 2) = 0$

$3x + x^2 - 2x = 0$

$x^2 + x = 0$

$x(x + 1) = 0$

$x = 0 \quad \text{or} \quad x + 1 = 0$

$x = 0 \quad \text{or} \qquad x = -1$

The solutions are 0 and -1.

15. $11x^2 - 3x - 5 = 0$

$a = 11, \ b = -3, \ c = -5$

$x = \dfrac{-(-3) \pm \sqrt{(-3)^2 - 4 \cdot 11 \cdot (-5)}}{2 \cdot 11}$

$x = \dfrac{3 \pm \sqrt{9 + 220}}{22} = \dfrac{3 \pm \sqrt{229}}{22}$

The solutions are $\dfrac{3 + \sqrt{229}}{22}$ and $\dfrac{3 - \sqrt{229}}{22}$.

17. $\qquad 25x^2 - 20x + 4 = 0$

$(5x - 2)(5x - 2) = 0$

$5x - 2 = 0 \quad \text{or} \quad 5x - 2 = 0$

$5x = 2 \quad \text{or} \qquad 5x = 2$

$x = \dfrac{2}{5} \quad \text{or} \qquad x = \dfrac{2}{5}$

The solution is $\dfrac{2}{5}$.

19. $4x(x - 2) - 5x(x - 1) = 2$

$4x^2 - 8x - 5x^2 + 5x = 2 \quad$ Removing parentheses

$-x^2 - 3x = 2$

$-x^2 - 3x - 2 = 0$

$x^2 + 3x + 2 = 0 \quad$ Multiplying by -1

$(x + 2)(x + 1) = 0$

$x + 2 = 0 \quad \text{or} \quad x + 1 = 0$

$x = -2 \quad \text{or} \qquad x = -1$

The solutions are -2 and -1.

21. $14(x - 4) - (x + 2) = (x + 2)(x - 4)$

$14x - 56 - x - 2 = x^2 - 2x - 8$

$13x - 58 = x^2 - 2x - 8$

$0 = x^2 - 15x + 50$

$0 = (x - 10)(x - 5)$

$$x - 10 = 0 \quad \text{or} \quad x - 5 = 0$$
$$x = 10 \quad \text{or} \quad x = 5$$

The solutions are 10 and 5.

23.
$$5x^2 = 17x - 2$$
$$5x^2 - 17x + 2 = 0$$
$$a = 5, \ b = -17, \ c = 2$$
$$x = \frac{-(-17) \pm \sqrt{(-17)^2 - 4 \cdot 5 \cdot 2}}{2 \cdot 5}$$
$$x = \frac{17 \pm \sqrt{289 - 40}}{10} = \frac{17 \pm \sqrt{249}}{10}$$

The solutions are $\dfrac{17 + \sqrt{249}}{10}$ and $\dfrac{17 - \sqrt{249}}{10}$.

25.
$$x^2 + 5 = 4x$$
$$x^2 - 4x + 5 = 0$$
$$a = 1, \ b = -4, \ c = 5$$
$$x = \frac{-(-4) \pm \sqrt{(-4)^2 - 4 \cdot 1 \cdot 5}}{2 \cdot 1} = \frac{4 \pm \sqrt{16 - 20}}{2}$$
$$x = \frac{4 \pm \sqrt{-4}}{2} = \frac{4 \pm 2i}{2} = 2 \pm i$$

The solutions are $2 + i$ and $2 - i$.

27.
$$x + \frac{1}{x} = \frac{13}{6}, \quad \text{LCM is } 6x$$
$$6x\left(x + \frac{1}{x}\right) = 6x \cdot \frac{13}{6}$$
$$6x^2 + 6 = 13x$$
$$6x^2 - 13x + 6 = 0$$
$$(2x - 3)(3x - 2) = 0$$
$$2x - 3 = 0 \quad \text{or} \quad 3x - 2 = 0$$
$$2x = 3 \quad \text{or} \quad 3x = 2$$
$$x = \frac{3}{2} \quad \text{or} \quad x = \frac{2}{3}$$

The solutions are $\dfrac{3}{2}$ and $\dfrac{2}{3}$.

29.
$$\frac{1}{y} + \frac{1}{y+2} = \frac{1}{3}, \quad \text{LCM is } 3y(y+2)$$
$$3y(y+2)\left(\frac{1}{y} + \frac{1}{y+2}\right) = 3y(y+2) \cdot \frac{1}{3}$$
$$3(y+2) + 3y = y(y+2)$$
$$3y + 6 + 3y = y^2 + 2y$$
$$6y + 6 = y^2 + 2y$$
$$0 = y^2 - 4y - 6$$
$$a = 1, \ b = -4, \ c = -6$$
$$y = \frac{-(-4) \pm \sqrt{(-4)^2 - 4 \cdot 1 \cdot (-6)}}{2 \cdot 1} = \frac{4 \pm \sqrt{16 + 24}}{2}$$
$$y = \frac{4 \pm \sqrt{40}}{2} = \frac{4 \pm 2\sqrt{10}}{2}$$
$$y = \frac{2(2 \pm \sqrt{10})}{2 \cdot 1} = 2 \pm \sqrt{10}$$

The solutions are $2 + \sqrt{10}$ and $2 - \sqrt{10}$.

31.
$$(2t - 3)^2 + 17t = 15$$
$$4t^2 - 12t + 9 + 17t = 15$$
$$4t^2 + 5t - 6 = 0$$
$$(4t - 3)(t + 2) = 0$$
$$4t - 3 = 0 \quad \text{or} \quad t + 2 = 0$$
$$t = \frac{3}{4} \quad \text{or} \quad t = -2$$

The solutions are $\dfrac{3}{4}$ and -2.

33.
$$(x - 2)^2 + (x + 1)^2 = 0$$
$$x^2 - 4x + 4 + x^2 + 2x + 1 = 0$$
$$2x^2 - 2x + 5 = 0$$
$$a = 2, \ b = -2, \ c = 5$$
$$x = \frac{-(-2) \pm \sqrt{(-2)^2 - 4 \cdot 2 \cdot 5}}{2 \cdot 2} = \frac{2 \pm \sqrt{4 - 40}}{4}$$
$$x = \frac{2 \pm \sqrt{-36}}{4} = \frac{2 \pm 6i}{4}$$
$$x = \frac{2(1 \pm 3i)}{2 \cdot 2} = \frac{1 \pm 3i}{2} = \frac{1}{2} \pm \frac{3}{2}i$$

The solutions are $\dfrac{1}{2} + \dfrac{3}{2}i$ and $\dfrac{1}{2} - \dfrac{3}{2}i$.

35.
$$x^3 - 1 = 0$$
$$(x - 1)(x^2 + x + 1) = 0$$
$$x - 1 = 0 \quad \text{or} \quad x^2 + x + 1 = 0$$
$$x = 1 \quad \text{or} \quad x = \frac{-1 \pm \sqrt{1^2 - 4 \cdot 1 \cdot 1}}{2 \cdot 1}$$
$$x = 1 \quad \text{or} \quad x = \frac{-1 \pm \sqrt{-3}}{2}$$
$$x = 1 \quad \text{or} \quad x = \frac{-1 \pm i\sqrt{3}}{2} = -\frac{1}{2} \pm i\frac{\sqrt{3}}{2}$$

The solutions are 1, $-\dfrac{1}{2} + i\dfrac{\sqrt{3}}{2}$, and $-\dfrac{1}{2} - i\dfrac{\sqrt{3}}{2}$.

37. $x^2 + 6x + 4 = 0$
$$a = 1, \ b = 6, \ c = 4$$
$$x = \frac{-6 \pm \sqrt{6^2 - 4 \cdot 1 \cdot 4}}{2 \cdot 1} = \frac{-6 \pm \sqrt{36 - 16}}{2}$$
$$x = \frac{-6 \pm \sqrt{20}}{2}$$

Using a calculator, we find that $\sqrt{20} \approx 4.472135955$.

$$\frac{-6 + \sqrt{20}}{2} \approx \frac{-6 + 4.472135955}{2}$$
$$\approx \frac{-1.527864045}{2} \qquad \text{Adding}$$
$$\approx -0.763932022 \qquad \text{Dividing}$$
$$\approx -0.8 \qquad \text{Rounding}$$

$$\frac{-6 - \sqrt{20}}{2} \approx \frac{-6 - 4.472135955}{2}$$

$$\approx \frac{-10.472135955}{2} \qquad \text{Subtracting}$$

$$\approx -5.236067978 \qquad \text{Dividing}$$

$$\approx -5.2 \qquad \text{Rounding}$$

The solutions are approximately -0.8 and -5.2.

39. $x^2 - 6x + 4 = 0$

$a = 1, \; b = -6, \; c = 4$

$$x = \frac{-(-6) \pm \sqrt{(-6)^2 - 4 \cdot 1 \cdot 4}}{2 \cdot 1} = \frac{6 \pm \sqrt{36 - 16}}{2}$$

$$x = \frac{6 \pm \sqrt{20}}{2}$$

Using a calculator, we find that $\sqrt{20} \approx 4.472135955$.

$$\frac{6 + \sqrt{20}}{2} \approx \frac{6 + 4.472135955}{2}$$

$$\approx \frac{10.472135955}{2} \qquad \text{Adding}$$

$$\approx 5.236067978 \qquad \text{Dividing}$$

$$\approx 5.2 \qquad \text{Rounding}$$

$$\frac{6 - \sqrt{20}}{2} \approx \frac{6 - 4.472135955}{2}$$

$$\approx \frac{1.527864045}{2} \qquad \text{Subtracting}$$

$$\approx 0.763932022 \qquad \text{Dividing}$$

$$\approx 0.8 \qquad \text{Rounding}$$

The solutions are approximately 5.2 and 0.8.

41. $2x^2 - 3x - 7 = 0$

$a = 2, \; b = -3, \; c = -7$

$$x = \frac{-(-3) \pm \sqrt{(-3)^2 - 4 \cdot 2 \cdot (-7)}}{2 \cdot 2} = \frac{3 \pm \sqrt{9 + 56}}{4}$$

$$x = \frac{3 \pm \sqrt{65}}{4}$$

Using a calculator, we find that $\sqrt{65} \approx 8.062257748$.

$$\frac{3 + \sqrt{65}}{4} \approx \frac{3 + 8.062257748}{4}$$

$$\approx \frac{11.062257748}{4} \qquad \text{Adding}$$

$$\approx 2.765564437 \qquad \text{Dividing}$$

$$\approx 2.8 \qquad \text{Rounding}$$

$$\frac{3 - \sqrt{65}}{4} \approx \frac{3 - 8.062257748}{4}$$

$$\approx \frac{-5.062257748}{4} \qquad \text{Subtracting}$$

$$\approx -1.265564437 \qquad \text{Dividing}$$

$$\approx -1.3 \qquad \text{Rounding}$$

The solutions are approximately 2.8 and -1.3.

43.

$$5x^2 = 3 + 8x$$

$$5x^2 - 8x - 3 = 0$$

$a = 5, \; b = -8, \; c = -3$

$$x = \frac{-(-8) \pm \sqrt{(-8)^2 - 4 \cdot 5 \cdot (-3)}}{2 \cdot 5} = \frac{8 \pm \sqrt{64 + 60}}{10}$$

$$x = \frac{8 \pm \sqrt{124}}{10}$$

Using a calculator, we find that $\sqrt{124} \approx 11.13552873$.

$$\frac{8 + \sqrt{124}}{10} \approx \frac{8 + 11.13552873}{10}$$

$$\approx \frac{19.13552873}{10} \qquad \text{Adding}$$

$$\approx 1.913552873 \qquad \text{Dividing}$$

$$\approx 1.9 \qquad \text{Rounding}$$

$$\frac{8 - \sqrt{124}}{10} \approx \frac{8 - 11.13552873}{10}$$

$$\approx \frac{-3.13552873}{10} \qquad \text{Subtracting}$$

$$\approx -0.313552873 \qquad \text{Dividing}$$

$$\approx -0.3 \qquad \text{Rounding}$$

The solutions are approximately 1.9 and -0.3.

45. *Familiarize.* Let $x =$ the number of pounds of hazelnut coffee to be used, and let $y =$ the number of pounds of French roast coffee. We organize the information in a table.

Coffee	Price per pound	Number of pounds	Total cost
Hazelnut	\$6.50	x	\$6.50x
French roast	\$7.50	y	\$7.50y
Blend	\$6.90	50	\$6.90 × 50, or \$345

Translate. From the last two columns of the table we get a system of equations.

$$x + y = 50,$$

$$6.50x + 7.50y = 345$$

Solve. Solving the system of equations, we get $(30, 20)$.

Check. The total number of pounds in the blend is $30 + 20$, or 50. The total cost of the blend is \$6.50(30) + \$7.50(20), or \$195 + \$150, or \$345. The values check.

State. The blend should consist of 30 lb of hazelnut coffee and 20 lb of French roast.

47.
$$x = \sqrt{15 - 2x}$$
$$x^2 = (\sqrt{15 - 2x})^2 \quad \text{Principle of powers}$$
$$x^2 = 15 - 2x$$
$$x^2 + 2x - 15 = 0$$
$$(x + 5)(x - 3) = 0$$
$$x + 5 = 0 \quad \text{or} \quad x - 3 = 0$$
$$x = -5 \quad \text{or} \quad x = 3$$

Only 3 checks. The solution is 3.

49.
$$\sqrt{x + 1} + 2 = \sqrt{3x + 1}$$
$$(\sqrt{x + 1} + 2)^2 = (\sqrt{3x + 1})^2 \quad \text{Principle of powers}$$
$$x + 1 + 4\sqrt{x + 1} + 4 = 3x + 1$$
$$x + 5 + 4\sqrt{x + 1} = 3x + 1 \quad \text{Collecting like terms}$$
$$4\sqrt{x + 1} = 2x - 4 \quad \text{Subtracting } x \text{ and } 5$$
$$2\sqrt{x + 1} = x - 2 \quad \text{Dividing by 2}$$
$$(2\sqrt{x + 1})^2 = (x - 2)^2 \quad \text{Principle of powers}$$
$$4(x + 1) = x^2 - 4x + 4$$
$$4x + 4 = x^2 - 4x + 4$$
$$0 = x^2 - 8x \quad \text{Subtracting } 4x \text{ and } 4$$
$$0 = x(x - 8)$$
$$x = 0 \quad \text{or} \quad x - 8 = 0$$
$$x = 0 \quad \text{or} \quad x = 8$$

Only 8 checks. The solution is 8.

51. $5.33x^2 - 8.23x - 3.24 = 0$

Since we will use a calculator, we will not clear decimals.
$$x = \frac{-(-8.23) \pm \sqrt{(-8.23)^2 - 4(5.33)(-3.24)}}{2(5.33)}$$
$$x = \frac{8.23 \pm \sqrt{136.8097}}{10.66} \approx \frac{8.23 \pm 11.69656787}{10.66}$$
$$x \approx \frac{8.23 + 11.69656787}{10.66} \approx 1.869284041$$
$$x \approx \frac{8.23 - 11.69656787}{10.66} \approx -0.325193984$$

The solutions are approximately 1.869284041 and −0.325193984.

53. $2x^2 - x - \sqrt{5} = 0$

$a = 2$, $b = -1$, $c = -\sqrt{5}$
$$x = \frac{-(-1) \pm \sqrt{(-1)^2 - 4 \cdot 2 \cdot (-\sqrt{5})}}{2 \cdot 2} = \frac{1 \pm \sqrt{1 + 8\sqrt{5}}}{4}$$

The solutions are $\dfrac{1 + \sqrt{1 + 8\sqrt{5}}}{4}$ and $\dfrac{1 - \sqrt{1 + 8\sqrt{5}}}{4}$.

55. $ix^2 - x - 1 = 0$

$a = i$, $b = -1$, $c = -1$
$$x = \frac{-(-1) \pm \sqrt{(-1)^2 - 4 \cdot i \cdot (-1)}}{2 \cdot i} = \frac{1 \pm \sqrt{1 + 4i}}{2i}$$
$$x = \frac{1 \pm \sqrt{1 + 4i}}{2i} \cdot \frac{i}{i} = \frac{i \pm i\sqrt{1 + 4i}}{2i^2} = \frac{i \pm i\sqrt{1 + 4i}}{-2}$$
$$x = \frac{-i \pm i\sqrt{1 + 4i}}{2}$$

The solutions are $\dfrac{-i + i\sqrt{1 + 4i}}{2}$ and $\dfrac{-i - i\sqrt{1 + 4i}}{2}$.

57.
$$\frac{x}{x + 1} = 4 + \frac{1}{3x^2 - 3}$$
$$\frac{x}{x + 1} = 4 + \frac{1}{3(x + 1)(x - 1)},$$
$$\text{LCM is } 3x(x + 1)(x - 1)$$
$$3(x + 1)(x - 1) \cdot \frac{x}{x + 1} =$$
$$3(x + 1)(x - 1)\left(4 + \frac{1}{3(x + 1)(x - 1)}\right)$$
$$3x(x - 1) = 12(x + 1)(x - 1) + 1$$
$$3x^2 - 3x = 12x^2 - 12 + 1$$
$$0 = 9x^2 + 3x - 11$$

$a = 9$, $b = 3$, $c = -11$
$$x = \frac{-3 \pm \sqrt{3^2 - 4 \cdot 9 \cdot (-11)}}{2 \cdot 9} = \frac{-3 \pm \sqrt{9 + 396}}{18}$$
$$x = \frac{-3 \pm \sqrt{405}}{18} = \frac{-3 \pm 9\sqrt{5}}{18}$$
$$x = \frac{3(-1 \pm 3\sqrt{5})}{3 \cdot 6} = \frac{-1 \pm 3\sqrt{5}}{6}$$

The solutions are $\dfrac{-1 + 3\sqrt{5}}{6}$ and $\dfrac{-1 - 3\sqrt{5}}{6}$.

Exercise Set 8.3

---

1. $x^2 - 8x + 16 = 0$

$a = 1$, $b = -8$, $c = 16$

We compute the discriminant.
$$b^2 - 4ac = (-8)^2 - 4 \cdot 1 \cdot 16$$
$$= 64 - 64$$
$$= 0$$

Since $b^2 - 4ac = 0$, there is just one solution, and it is a real number.

3. $x^2 + 1 = 0$

$a = 1$, $b = 0$, $c = 1$

We compute the discriminant.
$$b^2 - 4ac = 0^2 - 4 \cdot 1 \cdot 1$$
$$= -4$$

Since $b^2 - 4ac < 0$, there are two nonreal solutions.

5. $x^2 - 6 = 0$

$a = 1, \ b = 0, \ c = -6$

We compute the discriminant.

$b^2 - 4ac = 0^2 - 4 \cdot 1 \cdot (-6)$

$\qquad\qquad = 24$

Since $b^2 - 4ac > 0$, there are two real solutions.

7. $4x^2 - 12x + 9 = 0$

$a = 4, \ b = -12, \ c = 9$

We compute the discriminant.

$b^2 - 4ac = (-12)^2 - 4 \cdot 4 \cdot 9$

$\qquad\qquad = 144 - 144$

$\qquad\qquad = 0$

Since $b^2 - 4ac = 0$, there is just one solution, and it is a real number.

9. $x^2 - 2x + 4 = 0$

$a = 1, \ b = -2, \ c = 4$

We compute the discriminant.

$b^2 - 4ac = (-2)^2 - 4 \cdot 1 \cdot 4$

$\qquad\qquad = 4 - 16$

$\qquad\qquad = -12$

Since $b^2 - 4ac < 0$, there are two nonreal solutions.

11. $9t^2 - 3t = 0$

$a = 9, \ b = -3, \ c = 0$

We compute the discriminant.

$b^2 - 4ac = (-3)^2 - 4 \cdot 9 \cdot 0$

$\qquad\qquad = 9 - 0$

$\qquad\qquad = 9$

Since $b^2 - 4ac > 0$, there are two real solutions.

13. $y^2 = \dfrac{1}{2}y + \dfrac{3}{5}$

$y^2 - \dfrac{1}{2}y - \dfrac{3}{5} = 0 \qquad$ Standard form

$a = 1, \ b = -\dfrac{1}{2}, \ c = -\dfrac{3}{5}$

We compute the discriminant.

$b^2 - 4ac = \left(-\dfrac{1}{2}\right)^2 - 4 \cdot 1 \cdot \left(-\dfrac{3}{5}\right)$

$\qquad\qquad = \dfrac{1}{4} + \dfrac{12}{5}$

$\qquad\qquad = \dfrac{53}{20}$

Since $b^2 - 4ac > 0$, there are two real solutions.

15. $4x^2 - 4\sqrt{3}x + 3 = 0$

$a = 4, \ b = -4\sqrt{3}, \ c = 3$

We compute the discriminant.

$b^2 - 4ac = (-4\sqrt{3})^2 - 4 \cdot 4 \cdot 3$

$\qquad\qquad = 48 - 48$

$\qquad\qquad = 0$

Since $b^2 - 4ac = 0$, there is just one solution, and it is a real number.

17. The solutions are -4 and 4.

$x = -4 \quad$ or $\qquad x = 4$

$x + 4 = 0 \quad$ or $\quad x - 4 = 0$

$(x+4)(x-4) = 0 \qquad$ Principle of zero products

$x^2 - 16 = 0 \qquad (A+B)(A-B) = A^2 - B^2$

19. The solutions are -2 and -7.

$x = -2 \quad$ or $\qquad x = -7$

$x + 2 = 0 \quad$ or $\quad x + 7 = 0$

$(x+2)(x+7) = 0 \qquad$ Principle of zero products

$x^2 + 9x + 14 = 0 \qquad$ FOIL

21. The only solution is 8. It must be a double solution.

$x = 8 \quad$ or $\qquad x = 8$

$x - 8 = 0 \quad$ or $\quad x - 8 = 0$

$(x-8)(x-8) = 0 \qquad$ Principle of zero products

$x^2 - 16x + 64 = 0 \qquad (A-B)^2 = A^2 - 2AB + B^2$

23. The solutions are $-\dfrac{2}{5}$ and $\dfrac{6}{5}$.

$x = -\dfrac{2}{5} \quad$ or $\qquad x = \dfrac{6}{5}$

$x + \dfrac{2}{5} = 0 \quad$ or $\quad x - \dfrac{6}{5} = 0$

$5x + 2 = 0 \quad$ or $\quad 5x - 6 = 0 \qquad$ Clearing fractions

$(5x+2)(5x-6) = 0 \qquad$ Principle of zero products

$25x^2 - 20x - 12 = 0 \qquad$ FOIL

25. The solutions are $\dfrac{k}{3}$ and $\dfrac{m}{4}$.

$x = \dfrac{k}{3} \quad$ or $\qquad x = \dfrac{m}{4}$

$x - \dfrac{k}{3} = 0 \quad$ or $\quad x - \dfrac{m}{4} = 0$

$3x - k = 0 \quad$ or $\quad 4x - m = 0 \qquad$ Clearing fractions

$\qquad (3x-k)(4x-m) = 0 \qquad$ Principle of zero products

$12x^2 - 3mx - 4kx + km = 0 \qquad$ FOIL

$12x^2 - (3m + 4k)x + km = 0 \qquad$ Collecting like terms

27. The solutions are $-\sqrt{3}$ and $2\sqrt{3}$.

$$x = -\sqrt{3} \quad \text{or} \quad x = 2\sqrt{3}$$
$$x + \sqrt{3} = 0 \quad \text{or} \quad x - 2\sqrt{3} = 0$$
$$(x + \sqrt{3})(x - 2\sqrt{3}) = 0 \qquad \text{Principle of zero}$$
$$\text{products}$$
$$x^2 - 2\sqrt{3}x + \sqrt{3}x - 2(\sqrt{3})^2 = 0 \qquad \text{FOIL}$$
$$x^2 - \sqrt{3}x - 6 = 0$$

29. $\sqrt{8x} \cdot \sqrt{2x} = \sqrt{8x \cdot 2x} = \sqrt{16x^2} = \sqrt{(4x)^2} = 4x$

31. $\sqrt[4]{9a^2} \cdot \sqrt[4]{18a^3} = \sqrt[4]{9a^2 \cdot 18a^3} =$
$\sqrt[4]{3 \cdot 3 \cdot a^2 \cdot 3 \cdot 3 \cdot 2 \cdot a^2 \cdot a} = \sqrt[4]{3^4 a^4 \cdot 2a} = \sqrt[4]{3^4}\sqrt[4]{a^4}\sqrt[4]{2a} =$
$3a\sqrt[4]{2a}$

33. *Familiarize*. Let x and y represent the number of 30-sec and 60-sec commercials, respectively. Then the amount of time for the 30-sec commercials was $30x$ sec, or $\dfrac{30x}{60} = \dfrac{x}{2}$ min. The amount of time for the 60-sec commercials was $60x$ sec, or $\dfrac{60x}{60} = x$ min.

***Translate*.** Rewording, we write two equations. We will express time in minutes.

$$\underbrace{\text{Total number of commercials}}_{x + y} \underbrace{\text{is}}_{=} \underbrace{12.}_{12}$$

$$\underbrace{\begin{array}{c}\text{Time for}\\\text{30-sec}\\\text{commercials}\end{array}}_{\dfrac{x}{2}} \underbrace{\text{is}}_{=} \underbrace{\begin{array}{c}\text{total}\\\text{commercial}\\\text{time}\end{array}}_{\dfrac{x}{2} + x} \underbrace{\text{less}}_{-} \underbrace{\text{6 min.}}_{6}$$

***Solve*.** Solving the system of equations we get $(6, 6)$.

***Check*.** If there are six 30-sec and six 60-sec commercials, the total number of commercials is 12. The amount of time for six 30-sec commercials is 180 sec, or 3 min, and for six 60-sec commercials is 360 sec, or 6 min. The total commercial time is 9 min, and the amount of time for 30-sec commercials is 6 min less than this. The numbers check.

***State*.** There were six 30-sec and six 60-sec commercials.

35. a) $kx^2 - 2x + k = 0$; one solution is -3

We first find k by substituting -3 for x.
$$k(-3)^2 - 2(-3) + k = 0$$
$$9k + 6 + k = 0$$
$$10k = -6$$
$$k = -\frac{6}{10}$$
$$k = -\frac{3}{5}$$

b) $\quad -\dfrac{3}{5}x^2 - 2x + \left(-\dfrac{3}{5}\right) = 0 \quad$ Substituting $-\dfrac{3}{5}$ for k

$$3x^2 + 10x + 3 = 0 \quad \text{Multiplying by } -5$$
$$(3x + 1)(x + 3) = 0$$
$$3x + 1 = 0 \quad \text{or} \quad x + 3 = 0$$
$$3x = -1 \quad \text{or} \qquad x = -3$$
$$x = -\frac{1}{3} \quad \text{or} \qquad x = -3$$

The other solution is $-\dfrac{1}{3}$.

37. For $ax^2 + bx + c = 0$, $-\dfrac{b}{a}$ is the sum of the solutions and $\dfrac{c}{a}$ is the product of the solutions. Thus $-\dfrac{b}{a} = \sqrt{3}$ and $\dfrac{c}{a} = 8$.

$$ax^2 + bx + c = 0$$
$$x^2 + \frac{b}{a}x + \frac{c}{a} = 0 \quad \text{Multiplying by } \frac{1}{a}$$
$$x^2 - \left(-\frac{b}{a}\right)x + \frac{c}{a} = 0$$
$$x^2 - \sqrt{3}x + 8 = 0 \quad \begin{array}{l}\text{Substituting } \sqrt{3} \text{ for}\\ -\dfrac{b}{a} \text{ and 8 for } \dfrac{c}{a}\end{array}$$

Exercise Set 8.4

1. $x^4 - 6x^2 + 9 = 0$

Let $u = x^2$ and think of x^4 as $(x^2)^2$.
$$u^2 - 6u + 9 = 0 \quad \text{Substituting } u \text{ for } x^2$$
$$(u - 3)(u - 3) = 0$$
$$u - 3 = 0 \quad \text{or} \quad u - 3 = 0$$
$$u = 3 \quad \text{or} \qquad u = 3$$

Now we substitute x^2 for u and solve the equation:
$$x^2 = 3$$
$$x = \pm\sqrt{3}$$

Both $\sqrt{3}$ and $-\sqrt{3}$ check. They are the solutions.

3. $x - 10\sqrt{x} + 9 = 0$

Let $u = \sqrt{x}$ and think of x as $(\sqrt{x})^2$.
$$u^2 - 10u + 9 = 0 \quad \text{Substituting } u \text{ for } \sqrt{x}$$
$$(u - 9)(u - 1) = 0$$
$$u - 9 = 0 \quad \text{or} \quad u - 1 = 0$$
$$u = 9 \quad \text{or} \qquad u = 1$$

Now we substitute $\sqrt{x}$ for u and solve these equations:
$$\sqrt{x} = 9 \quad \text{or} \quad \sqrt{x} = 1$$
$$x = 81 \quad \text{or} \qquad x =$$

The numbers 81 and 1 both check. They are the solutions.

5. $(x^2 - 6x) - 2(x^2 - 6x) - 35 = 0$

Let $u = x^2 - 6x$.

$u^2 - 2u - 35 = 0 \quad$ Substituting u for $x^2 - 6x$

$(u - 7)(u + 5) = 0$

$u - 7 = 0 \quad$ or $\quad u + 5 = 0$

$\quad u = 7 \quad$ or $\qquad u = -5$

Now we substitute $x^2 - 6x$ for u and solve these equations:

$\quad x^2 - 6x = 7 \quad$ or $\qquad x^2 - 6x = -5$

$x^2 - 6x - 7 = 0 \quad$ or $\quad x^2 - 6x + 5 = 0$

$(x - 7)(x + 1) = 0 \quad$ or $\quad (x - 5)(x - 1) = 0$

$x = 7$ or $x = -1$ or $x = 5$ or $x = 1$

The numbers -1, 1, 5, and 7 check. They are the solutions.

7. $x^{-2} - 5^{-1} - 36 = 0$

Let $u = x^{-1}$.

$u^2 - 5u - 36 = 0 \quad$ Substituting u for x^{-1}

$(u - 9)(u + 4) = 0$

$u - 9 = 0 \quad$ or $\quad u + 4 = 0$

$\quad u = 9 \quad$ or $\qquad u = -4$

Now we substitute x^{-1} for u and solve these equations:

$x^{-1} = 9 \quad$ or $\quad x^{-1} = -4$

$\dfrac{1}{x} = 9 \quad$ or $\quad \dfrac{1}{x} = -4$

$\dfrac{1}{9} = x \quad$ or $\quad -\dfrac{1}{4} = x$

Both $\dfrac{1}{9}$ and $-\dfrac{1}{4}$ check. They are the solutions.

9. $(1 + \sqrt{x})^2 + (1 + \sqrt{x}) - 6 = 0$

Let $u = 1 + \sqrt{x}$.

$u^2 + u - 6 = 0 \quad$ Substituting u for $1 + \sqrt{x}$

$(u + 3)(u - 2) = 0$

$u + 3 = 0 \qquad$ or $\qquad u - 2 = 0$

$\quad u = -3 \qquad$ or $\qquad u = 2$

$1 + \sqrt{x} = -3 \quad$ or $\quad 1 + \sqrt{x} = 2 \quad$ Substituting $1 + \sqrt{x}$ for u

$\quad \sqrt{x} = -4 \qquad$ or $\qquad \sqrt{x} = 1$

No solution $\qquad\qquad\qquad x = 1$

The number 1 checks. It is the solution.

11. $(y^2 - 5y)^2 - 2(y^2 - 5y) - 24 = 0$

Let $u = y^2 - 5y$.

$u^2 - 2u - 24 = 0 \quad$ Substituting u for $y^2 - 5y$

$(u - 6)(u + 4) = 0$

$\quad u - 6 = 0 \quad$ or $\qquad u + 4 = 0$

$\qquad u = 6 \quad$ or $\qquad\quad u = -4$

$\quad y^2 - 5y = 6 \quad$ or $\qquad y^2 - 5y = -4$

$\qquad\qquad\qquad$ Substituting $y^2 - 5y$ for u

$y^2 - 5y - 6 = 0 \quad$ or $\quad y^2 - 5y + 4 = 0$

$(y - 6)(y + 1) = 0 \quad$ or $\quad (y - 4)(y - 1) = 0$

$y = 6$ or $y = -1$ or $y = 4$ or $y = 1$

The numbers -1, 1, 4, and 6 check. They are the solutions.

13. $t^4 - 6t^2 - 4 = 0$

Let $u = t^2$.

$u^2 - 6u - 4 = 0 \quad$ Substituting u for t^2

$u = \dfrac{-(-6) \pm \sqrt{(-6)^2 - 4 \cdot 1 \cdot (-4)}}{2 \cdot 1}$

$u = \dfrac{6 \pm \sqrt{52}}{2} = \dfrac{6 \pm 2\sqrt{13}}{2}$

$u = 3 \pm \sqrt{13}$

Now we substitute t^2 for u and solve these equations:

$t^2 = 3 + \sqrt{13} \qquad$ or $\quad t^2 = 3 - \sqrt{13}$

$t = \pm\sqrt{3 + \sqrt{13}} \quad$ or $\quad t = \pm\sqrt{3 - \sqrt{13}}$

All four numbers check. They are the solutions.

15. $2x^{-2} + x^{-1} - 1 = 0$

Let $u = x^{-1}$.

$2u^2 + u - 1 = 0 \quad$ Substituting u for x^{-1}

$(2u - 1)(u + 1) = 0$

$2u - 1 = 0 \quad$ or $\quad u + 1 = 0$

$\quad 2u = 1 \quad$ or $\qquad u = -1$

$\quad\; u = \dfrac{1}{2} \quad$ or $\qquad u = -1$

$x^{-1} = \dfrac{1}{2} \quad$ or $\quad x^{-1} = -1 \quad$ Substituting x^{-1} for u

$\dfrac{1}{x} = \dfrac{1}{2} \quad$ or $\quad \dfrac{1}{x} = -1$

$\quad x = 2 \quad$ or $\qquad x = -1$

Both 2 and -1 check. They are the solutions.

17. $6x^4 - 19x^2 + 15 = 0$

Let $u = x^2$.

$6u^2 - 19u + 15 = 0 \quad$ Substituting u for x^2

$(3u - 5)(2u - 3) = 0$

$3u - 5 = 0 \qquad$ or $\quad 2u - 3 = 0$

$\quad 3u = 5 \qquad$ or $\qquad 2u = 3$

$\quad\; u = \dfrac{5}{3} \qquad$ or $\qquad u = \dfrac{3}{2}$

$\quad x^2 = \dfrac{5}{3} \qquad$ or $\qquad x^2 = \dfrac{3}{2} \quad$ Substituting x^2 for u

$\quad x = \pm\sqrt{\dfrac{5}{3}} \qquad$ or $\qquad x = \pm\sqrt{\dfrac{3}{2}}$

$\quad x = \pm\dfrac{\sqrt{15}}{3} \qquad$ or $\qquad x = \pm\dfrac{\sqrt{6}}{2}$

$\qquad\qquad\qquad\qquad$ Rationalizing denominators

All four numbers check. They are the solutions.

19. $x^{2/3} - 4x^{1/3} - 5 = 0$

Let $u = x^{1/3}$.

$u^2 - 4u - 5 = 0$ Substituting u for $x^{1/3}$

$(u - 5)(u + 1) = 0$

$u - 5 = 0$ or $u + 1 = 0$

$u = 5$ or $u = -1$

$x^{1/3} = 5$ or $x^{1/3} = -1$ Substituting $x^{1/3}$ for u

$(x^{1/3})^3 = 5^3$ or $(x^{1/3})^3 = (-1)^3$ Principle of powers

$x = 125$ or $x = -1$

Both 125 and -1 check. They are the solutions.

21. $\left(\dfrac{x-4}{x+1}\right)^2 - 2\left(\dfrac{x-4}{x+1}\right) - 35 = 0$

Let $u = \dfrac{x-4}{x+1}$.

$u^2 - 2u - 35 = 0$ Substituting u for $\dfrac{x-4}{x+1}$

$(u - 7)(u + 5) = 0$

$u - 7 = 0$ or $u + 5 = 0$

$u = 7$ or $u = -5$

$\dfrac{x-4}{x+1} = 7$ or $\dfrac{x-4}{x+1} = -5$ Substituting $\dfrac{x-4}{x+1}$ for u

$x - 4 = 7(x+1)$ or $x - 4 = -5(x+1)$

$x - 4 = 7x + 7$ or $x - 4 = -5x - 5$

$-6x = 11$ or $6x = -1$

$x = -\dfrac{11}{6}$ or $x = -\dfrac{1}{6}$

Both $-\dfrac{11}{6}$ and $-\dfrac{1}{6}$ check. They are the solutions.

23. $9\left(\dfrac{x+2}{x+3}\right)^2 - 6\left(\dfrac{x+2}{x+3}\right) + 1 = 0$

Let $u = \dfrac{x+2}{x+3}$.

$9u^2 - 6u + 1 = 0$ Substituting u for $\dfrac{x+2}{x+3}$

$(3u - 1)(3u - 1) = 0$

$3u - 1 = 0$ or $3u - 1 = 0$

$3u = 1$ or $3u = 1$

$u = \dfrac{1}{3}$ or $u = \dfrac{1}{3}$

Now we substitute $\dfrac{x+2}{x+3}$ for u and solve the equation:

$\dfrac{x+2}{x+3} = \dfrac{1}{3}$

$3(x+2) = x+3$ Multiplying by $3(x+3)$

$3x + 6 = x + 3$

$2x = -3$

$x = -\dfrac{3}{2}$

The number $-\dfrac{3}{2}$ checks. It is the solution.

25. $\left(\dfrac{x^2-2}{x}\right)^2 - 7\left(\dfrac{x^2-2}{x}\right) - 18 = 0$

Let $u = \dfrac{x^2-2}{x}$.

$u^2 - 7u - 18 = 0$ Substituting u for $\dfrac{x^2-2}{x}$

$(u - 9)(u + 2) = 0$

$u - 9 = 0$ or $u + 2 = 0$

$u = 9$ or $u = -2$

$\dfrac{x^2-2}{x} = 9$ or $\dfrac{x^2-2}{x} = -2$

Substituting $\dfrac{x^2-2}{x}$ for u

$x^2 - 2 = 9x$ or $x^2 - 2 = -2x$

$x^2 - 9x - 2 = 0$ or $x^2 + 2x - 2 = 0$

$x = \dfrac{-(-9) \pm \sqrt{(-9)^2 - 4 \cdot 1 \cdot (-2)}}{2 \cdot 1}$

$x = \dfrac{9 \pm \sqrt{89}}{2}$

or

$x = \dfrac{-2 \pm \sqrt{2^2 - 4 \cdot 1 \cdot (-2)}}{2 \cdot 1} = \dfrac{-2 \pm \sqrt{12}}{2}$

$x = \dfrac{-2 \pm 2\sqrt{3}}{2} = -1 \pm \sqrt{3}$

All four numbers check. They are the solutions.

27. $\sqrt{3x^2}\sqrt{3x^3} = \sqrt{3x^2 \cdot 3x^3} = \sqrt{9x^5} = \sqrt{9x^4 \cdot x} = 3x^2\sqrt{x}$

29. $\dfrac{x+1}{x-1} - \dfrac{x+1}{x^2+x+1}$, LCD $= (x-1)(x^2+x+1)$

$= \dfrac{x+1}{x-1} \cdot \dfrac{x^2+x+1}{x^2+x+1} - \dfrac{x+1}{x^2+x+1} \cdot \dfrac{x-1}{x-1}$

$= \dfrac{(x^3 + 2x^2 + 2x + 1) - (x^2 - 1)}{(x-1)(x^2+x+1)}$

$= \dfrac{x^3 + x^2 + 2x + 2}{(x-1)(x^2+x+1)}$

31. $\dfrac{x+2}{x-4} + \dfrac{x-2}{x+3}$, LCD $= (x-4)(x+3)$

$= \dfrac{x+2}{x-4} \cdot \dfrac{x+3}{x+3} + \dfrac{x-2}{x+3} \cdot \dfrac{x-4}{x-4}$

$= \dfrac{(x+2)(x+3)}{(x-4)(x+3)} + \dfrac{(x-2)(x-4)}{(x+3)(x-4)}$

$= \dfrac{(x+2)(x+3) + (x-2)(x-4)}{(x-4)(x+3)}$

$= \dfrac{x^2 + 5x + 6 + x^2 - 6x + 8}{(x-4)(x+3)}$

$= \dfrac{2x^2 - x + 14}{(x-4)(x+3)}$

33. $6.75x - 35\sqrt{x} - 5.36 = 0$

Let $u = \sqrt{x}$. (Since we will use a calculator, we do not clear decimals.)

$6.75u^2 - 35u - 5.36 = 0$

$u = \dfrac{-(-35) \pm \sqrt{(-35)^2 - 4(6.75)(-5.36)}}{2(6.75)}$

$u = \dfrac{35 \pm \sqrt{1225 + 144.72}}{13.5}$

$u \approx \dfrac{35 \pm 37.01}{13.5}$

$u \approx \dfrac{72.01}{13.5}$ or $u \approx \dfrac{-2.01}{13.5}$

$u \approx 5.334$ or $u \approx -0.149$

$\sqrt{x} \approx 5.334$ or $\sqrt{x} \approx -0.149$

$x \approx 28.5$ or No solution

The number 28.5 checks, so it is the solution.

35. $\dfrac{x}{x-1} - 6\sqrt{\dfrac{x}{x-1}} - 40 = 0$

Let $u = \sqrt{\dfrac{x}{x-1}}$.

$u^2 - 6u - 40 = 0$ Substituting for $\sqrt{\dfrac{x}{x-1}}$

$(u - 10)(u + 4) = 0$

$u = 10$ or $u = -4$

$\sqrt{\dfrac{x}{x-1}} = 10$ or $\sqrt{\dfrac{x}{x-1}} = -4$

Substituting for u

$\dfrac{x}{x-1} = 100$ or No solution

$x = 100x - 100$ Multiplying by $(x-1)$

$100 = 99x$

$\dfrac{100}{99} = x$

This number checks. It is the solution.

37. $\sqrt{x-3} - \sqrt[4]{x-3} = 12$

$(x-3)^{1/2} - (x-3)^{1/4} - 12 = 0$

Let $u = (x-3)^{1/4}$.

$u^2 - u - 12 = 0$ Substituting for $(x-3)^{1/4}$

$(u - 4)(u + 3) = 0$

$u = 4$ or $u = -3$

$(x-3)^{1/4} = 4$ or $(x-3)^{1/4} = -3$

Substituting for u

$x - 3 = 4^4$ or No solution

$x - 3 = 256$

$x = 259$

This number checks. It is the solution.

39. $x^6 - 28x^3 + 27 = 0$

Let $u = x^3$.

$u^2 - 28u + 27 = 0$ Substituting for x^3

$(u - 27)(u - 1) = 0$

$u = 27$ or $u = 1$

$x^3 = 27$ or $x^3 = 1$ Substituting for u

$x = 3$ or $x = 1$

Both 3 and 1 check. They are the solutions.

Exercise Set 8.5

1. $A = 6s^2$

$\dfrac{A}{6} = s^2$ Dividing by 6

$\sqrt{\dfrac{A}{6}} = s$ Taking the positive square root

3. $F = \dfrac{Gm_1m_2}{r^2}$

$Fr^2 = Gm_1m_2$ Multiplying by r^2

$r^2 = \dfrac{Gm_1m_2}{F}$ Dividing by F

$r = \sqrt{\dfrac{Gm_1m_2}{F}}$ Taking the positive square root

5. $E = mc^2$

$\dfrac{E}{m} = c^2$ Dividing by m

$\sqrt{\dfrac{E}{m}} = c$ Taking the square root

7. $a^2 + b^2 = c^2$

$b^2 = c^2 - a^2$ Subtracting a^2

$b = \sqrt{c^2 - a^2}$ Taking the square root

9. $N = \dfrac{k^2 - 3k}{2}$

$2N = k^2 - 3k$

$0 = k^2 - 3k - 2N$ Standard form

$a = 1,\ b = -3,\ c = -2N$

$k = \dfrac{-(-3) \pm \sqrt{(-3)^3 - 4 \cdot 1 \cdot (-2N)}}{2 \cdot 1}$ Using the quadratic formula

$k = \dfrac{3 \pm \sqrt{9 + 8N}}{2}$

Since taking the negative square root would result in a negative answer, we take the positive one.

$k = \dfrac{3 + \sqrt{9 + 8N}}{2}$

11. $A = 2\pi r^2 + 2\pi rh$

$0 = 2\pi r^2 + 2\pi rh - A$ Standard form

$a = 2\pi,\ b = 2\pi h,\ c = -A$

$$r = \frac{-2\pi h \pm \sqrt{(2\pi h)^2 - 4 \cdot 2\pi \cdot (-A)}}{2 \cdot 2\pi}$$

Using the quadratic formula

$$r = \frac{-2\pi h \pm \sqrt{4\pi^2 h^2 + 8\pi A}}{4\pi}$$

$$r = \frac{-2\pi h \pm 2\sqrt{\pi^2 h^2 + 2\pi A}}{4\pi}$$

$$r = \frac{-\pi h \pm \sqrt{\pi^2 h^2 + 2\pi A}}{2\pi}$$

Since taking the negative square root would result in a negative answer, we take the positive one.

$$r = \frac{-\pi h + \sqrt{\pi^2 h^2 + 2\pi A}}{2\pi}$$

13. $$T = 2\pi\sqrt{\frac{L}{g}}$$

$$\frac{T}{2\pi} = \sqrt{\frac{L}{g}}$$ Dividing by 2π

$$\frac{T^2}{4\pi^2} = \frac{L}{g}$$ Squaring

$$gT^2 = 4\pi^2 L$$ Multiplying by $4\pi^2 g$

$$g = \frac{4\pi^2 L}{T^2}$$ Dividing by T^2

15. $$I = \frac{700W}{H^2}$$

$$H^2 I = 700W$$ Multiplying by H^2

$$H^2 = \frac{700W}{I}$$ Dividing by I

$$H = \sqrt{\frac{700W}{I}}, \text{ or}$$

$$H = \sqrt{\frac{100 \cdot 7W}{I}}, \text{ or } 10\sqrt{\frac{7W}{I}}$$

17. $$m = \frac{m_0}{\sqrt{1 - \frac{v^2}{c^2}}}$$

$$m^2 = \frac{m_0^2}{1 - \frac{v^2}{c^2}}$$ Principle of powers

$$m^2\left(1 - \frac{v^2}{c^2}\right) = m_0^2$$

$$m^2 - \frac{m^2 v^2}{c^2} = m_0^2$$

$$m^2 - m_0^2 = \frac{m^2 v^2}{c^2}$$

$$c^2(m^2 - m_0^2) = m^2 v^2$$

$$\frac{c^2(m^2 - m_0^2)}{m^2} = v^2$$

$$\sqrt{\frac{c^2(m^2 - m_0^2)}{m^2}} = v$$

$$\frac{c\sqrt{m^2 - m_0^2}}{m} = v$$

19. *Familiarize*. We make a drawing and label it. We let $x =$ the length of the rectangle. Then $x - 7 =$ the width.

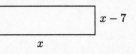

Translate. We use the formula for the area of a rectangle.

$$A = lw$$

$$18 = x(x - 7)$$ Substituting

Solve. We solve the equation.

$$18 = x^2 - 7x$$

$$0 = x^2 - 7x - 18$$

$$0 = (x - 9)(x + 2)$$

$$x - 9 = 0 \quad \text{or} \quad x + 2 = 0$$

$$x = 9 \quad \text{or} \qquad x = -2$$

Check. We only check 9 since the length cannot be negative. If $x = 9$, then $x - 7 = 9 - 7$, or 2, and the area is $9 \cdot 2$, or 18 ft². The value checks.

State. The length of 9 ft, and the width is 2 ft.

21. *Familiarize*. We make a drawing and label it. We let $x =$ the width of the rectangle. Then $2x =$ the length.

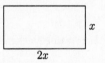

Translate.

$$A = lw$$

$$162 = 2x \cdot x$$ Substituting

Solve. We solve the equation.

$$162 = 2x^2$$

$$81 = x^2$$

$$\pm 9 = x$$

Check. We only check 9 since the width cannot be negative. If $x = 9$, then $2x = 2 \cdot 9$, or 18, and the area is $18 \cdot 9$, or 162 yd². The value checks.

State. The length is 18 yd, and the width is 9 yd.

23. *Familiarize*. We make a drawing and label it. We let $x =$ the width of the frame.

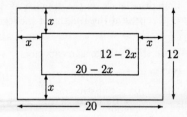

The length and width of the picture that shows are represented by $20 - 2x$ and $12 - 2x$. The area of the picture that shows is 84 cm².

Translate. Using the formula for the area of a rectangle, $A = l \cdot w$, we have

$$84 = (20 - 2x)(12 - 2x).$$

Solve. We solve the equation.

$$84 = (20 - 2x)(12 - 2x)$$
$$84 = 240 - 64x + 4x^2$$
$$0 = 156 - 64x + 4x^2$$
$$0 = 4x^2 - 64x + 156$$
$$0 = x^2 - 16x + 39 \qquad \text{Dividing by 4}$$
$$0 = (x - 13)(x - 3)$$
$$x - 13 = 0 \quad \text{or} \quad x - 3 = 0$$
$$x = 13 \quad \text{or} \qquad x = 3$$

Check. We see that 13 is not a solution, because when $x = 13$, then $20 - 2x = -6$ and $12 - 2x = -14$ and the dimensions of the frame cannot be negative. We check 3. When $x = 3$, then $20 - 2x = 14$ and $12 - 2x = 6$ and $14 \cdot 6 = 84$, the area of the picture that shows. The number 3 checks.

State. The width of the frame is 3 cm.

25. *Familiarize*. We make a drawing. We let $x =$ the length of the shorter leg. Then $x + 14 =$ the length of the longer leg.

Translate. We use the Pythagorean equation.

$$a^2 + b^2 = c^2$$
$$(x + 14)^2 + x^2 = 26^2 \qquad \text{Substituting}$$

Solve. We solve the equation.

$$x^2 + 28x + 196 + x^2 = 676$$
$$2x^2 + 28x - 480 = 0$$
$$x^2 + 14x - 240 = 0$$
$$(x - 10)(x + 24) = 0$$
$$x - 10 = 0 \quad \text{or} \quad x + 24 = 0$$
$$x = 10 \quad \text{or} \qquad x = -24$$

Check. We only check 10 since the length of a leg cannot be negative. If $x = 10$, then $x + 14 = 24$, and $10^2 + 24^2 = 676 = 26^2$. The number 10 checks.

State. The lengths of the legs are 24 ft and 10 ft.

27. *Familiarize*. Using the labels on the drawing in the text, we let x and $x + 2$ represent the lengths of the legs of the right triangle.

Translate. We use the Pythagorean equation.

$$a^2 + b^2 = c^2$$
$$x^2 + (x + 2)^2 = 10^2 \qquad \text{Substituting}$$

Solve. We solve the equation.

$$x^2 + x^2 + 4x + 4 = 100$$
$$2x^2 + 4x + 4 = 100$$
$$2x^2 + 4x - 96 = 0$$
$$x^2 + 2x - 48 = 0 \qquad \text{Dividing by 2}$$
$$(x + 8)(x - 6) = 0$$

$$x + 8 = 0 \quad \text{or} \quad x - 6 = 0$$
$$x = -8 \quad \text{or} \qquad x = 6$$

Check. We only check 6 since the length of a leg cannot be negative. When $x = 6$, then $x + 2 = 8$, and $6^2 + 8^2 = 100 = 10^2$. The number 6 checks.

State. The lengths of the legs are 6 ft and 8 ft.

29. *Familiarize*. The page numbers on facing pages are consecutive integers. Let $x =$ the number on the left-hand page. Then $x + 1 =$ the number on the right-hand page.

Translate.

$$\underbrace{\text{The product of the page numbers}}_{x(x + 1)} \underset{=}{\text{ is }} \underset{812}{812.}$$

Solve. We solve the equation.

$$x^2 + x = 812$$
$$x^2 + x - 812 = 0$$
$$(x + 29)(x - 28) = 0$$
$$x + 29 = 0 \quad \text{or} \quad x - 28 = 0$$
$$x = -29 \quad \text{or} \qquad x = 28$$

Check. We only check 28 since a page number cannot be negative. If $x = 28$, then $x + 1 = 29$ and $28 \cdot 29 = 812$. The number 28 checks.

State. The page numbers are 28 and 29.

31. *Familiarize*. We make a drawing and label it. We let $x =$ the length and $x - 4 =$ the width.

Translate. We use the formula for the area of a rectangle.

$$A = lw$$
$$10 = x(x - 4) \qquad \text{Substituting}$$

Solve. We solve the equation.

$$10 = x^2 - 4x$$
$$0 = x^2 - 4x - 10$$
$$x = \frac{-b \pm \sqrt{b^2 - 4ac}}{2a} = \frac{-(-4) \pm \sqrt{(-4)^2 - 4 \cdot 1 \cdot (-10)}}{2 \cdot 1}$$
$$x = \frac{4 \pm \sqrt{16 + 40}}{2} = \frac{4 \pm \sqrt{56}}{2} = \frac{4 \pm \sqrt{4 \cdot 14}}{2}$$
$$x = \frac{4 \pm 2\sqrt{14}}{2} = 2 \pm \sqrt{14}$$

Check. We only need to check $2 + \sqrt{14}$ since $2 - \sqrt{14}$ is negative and the length cannot be negative. If $x = 2 + \sqrt{14}$, then $x - 4 = (2 + \sqrt{14}) - 4$, or $\sqrt{14} - 2$. Using a calculator we find that the length is $2 + \sqrt{14} \approx 5.7$ ft and the width is $\sqrt{14} - 2 \approx 1.7$ ft, and $(5.7)(1.7) = 9.69 \approx 10$. Our result checks.

State. The length is $2 + \sqrt{14} \approx 5.7$ ft; the width is $\sqrt{14} - 2 \approx 1.7$ ft.

33. *Familiarize and Translate*. Using the same reasoning that we did in Exercise 21, we translate the problem to the equation

$$256 = 2x^2, \text{ or}$$
$$128 = x^2.$$

Solve. We solve the equation.

$$128 = x^2$$
$$\pm\sqrt{128} = x$$
$$\pm\sqrt{64 \cdot 2} = x$$
$$\pm 8\sqrt{2} = x$$

Check. We only check $8\sqrt{2}$ since the width cannot be negative. If $x = 8\sqrt{2}$, then $2x = 16\sqrt{2}$. Using a calculator we find that $8\sqrt{2} \approx 11.3$ and $16\sqrt{2} \approx 22.6$, and $(11.3)(22.6) = 255.38 \approx 256$. The number $8\sqrt{2}$ checks.

State. The length is $16\sqrt{2} \approx 22.6$ yd; the width is $8\sqrt{2} \approx 11.3$ yd.

35. *Familiarize*. We make a drawing and label it. We let $x =$ the width of the margin.

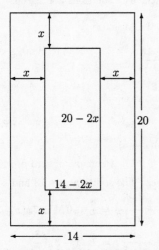

The length and width of the printed text are represented by $20 - 2x$ and $14 - 2x$. The area of the printed text is 100 in^2.

Translate. We use the formula for the area of a rectangle.

$$A = lw$$
$$100 = (20 - 2x)(14 - 2x)$$

Solve. We solve the equation.

$$100 = 280 - 68x + 4x^2$$
$$0 = 4x^2 - 68x + 180$$
$$0 = x^2 - 17x + 45 \qquad \text{Dividing by 4}$$
$$x = \frac{-b \pm \sqrt{b^2 - 4ac}}{2a} = \frac{-(-17) \pm \sqrt{(-17)^2 - 4 \cdot 1 \cdot 45}}{2 \cdot 1}$$
$$x = \frac{17 \pm \sqrt{289 - 180}}{2} = \frac{17 \pm \sqrt{109}}{2}$$
$$x \approx 13.7 \text{ or } x \approx 3.3$$

Check. If $x \approx 13.7$, then $20 - 2x \approx -7.4$ and $14 - 2x \approx -13.4$. Since the width of the margin cannot be negative, 13.7 is not a solution. If $x \approx 3.3$, then $20 - 2x \approx 13.4$ and

$14 - 2x \approx 7.4$ and $(13.4)(7.4) = 99.16 \approx 100$. The number $\frac{17 - \sqrt{109}}{2} \approx 3.3$ checks.

State. The width of the margin is $\frac{17 - \sqrt{109}}{2} \approx 3.3$ in.

37. *Familiarize*. We make a drawing. We let $x =$ the length of the shorter leg and $x + 14 =$ the length of the longer leg.

Translate. We use the Pythagorean equation.

$$a^2 + b^2 = c^2$$
$$x^2 + (x + 14)^2 = 24^2 \qquad \text{Substituting}$$

Solve. We solve the equation.

$$x^2 + x^2 + 28x + 196 = 576$$
$$2x^2 + 28x - 380 = 0$$
$$x^2 + 14x - 190 = 0 \qquad \text{Dividing by 2}$$
$$x = \frac{-b \pm \sqrt{b^2 - 4ac}}{2a} = \frac{-14 \pm \sqrt{14^2 - 4 \cdot 1 \cdot (-190)}}{2 \cdot 1}$$
$$x = \frac{-14 \pm \sqrt{196 + 760}}{2} = \frac{-14 \pm \sqrt{956}}{2} = \frac{-14 \pm \sqrt{4 \cdot 239}}{2}$$
$$x = \frac{-14 \pm 2\sqrt{239}}{2} = -7 \pm \sqrt{239}$$
$$x \approx 8.5 \text{ or } x \approx -22.5$$

Check. Since the length of a leg cannot be negative, we only need to check 8.5. If $x = -7 + \sqrt{239} \approx 8.5$, then $x + 14 = -7 + \sqrt{239} + 14 = 7 + \sqrt{239} \approx 22.5$ and $(8.5)^2 + (22.5)^2 = 578.5 \approx 576 = 24^2$. The number $-7 + \sqrt{239} \approx 8.5$ checks.

State. The lengths of the legs are $-7 + \sqrt{239} \approx 8.5$ ft and $7 + \sqrt{239} \approx 22.5$ ft.

39. $\sqrt{-20} = \sqrt{-1 \cdot 4 \cdot 5} = i \cdot 2 \cdot \sqrt{5} = 2\sqrt{5}i$

41. $\sqrt{x^2} = -20$

There is no solution since $\sqrt{x^2}$ is always nonnegative.

43.
$$\sqrt{x + 5} + \sqrt{x - 3} = 4$$
$$\sqrt{x + 5} = 4 - \sqrt{x - 3} \qquad \text{Isolating one radical}$$
$$(\sqrt{x + 5})^2 = (4 - \sqrt{x - 3})^2 \qquad \begin{array}{l}\text{Principle of} \\ \text{powers}\end{array}$$
$$x + 5 = 16 - 8\sqrt{x - 3} + x - 3$$
$$x + 5 = 13 + x - 8\sqrt{x - 3}$$
$$-8 = -8\sqrt{x - 3} \qquad \text{Isolating the radical}$$
$$1 = \sqrt{x - 3} \qquad \text{Dividing by } -8$$
$$1^2 = (\sqrt{x - 3})^2$$
$$1 = x - 3$$
$$4 = x$$

The number 4 checks and is the solution.

Exercise Set 8.6

1. $\dfrac{1}{x} = \dfrac{x-3}{40}$, LCM is $40x$

$$40x \cdot \frac{1}{x} = 40x \cdot \frac{x-3}{40} \qquad \text{Multiplying by the LCM}$$

$$40 = x(x-3)$$
$$40 = x^2 - 3x$$
$$0 = x^2 - 3x - 40 \qquad \text{Standard form}$$
$$0 = (x-8)(x+5)$$

$x = 8$ or $x = -5$ Principle of zero products

Check:

For 8:
$$\begin{array}{c|c} \dfrac{1}{x} = \dfrac{x-3}{40} \\[2mm] \hline \dfrac{1}{8} & \dfrac{8-3}{40} \\[3mm] & \dfrac{5}{40} \\[3mm] & \dfrac{1}{8} \quad \text{TRUE} \end{array}$$

For -5:
$$\begin{array}{c|c} \dfrac{1}{x} = \dfrac{x-3}{40} \\[2mm] \hline \dfrac{1}{-5} & \dfrac{-5-3}{40} \\[3mm] -\dfrac{1}{5} & \dfrac{-8}{40} \\[3mm] & -\dfrac{1}{5} \quad \text{TRUE} \end{array}$$

The solutions are 8 and -5.

3. $\dfrac{1}{4-x} - \dfrac{1}{2+x} = \dfrac{1}{4}$,

$\qquad\qquad\qquad$ LCM is $4(4-x)(2+x)$

$$4(4-x)(2+x) \cdot \left(\frac{1}{4-x} - \frac{1}{2+x} \right) = 4(4-x)(2+x) \cdot \frac{1}{4}$$

$$4(2+x) - 4(4-x) = (4-x)(2+x)$$
$$8 + 4x - 16 + 4x = 8 + 2x - x^2$$
$$8x - 8 = 8 + 2x - x^2$$
$$x^2 + 6x - 16 = 0$$
$$(x+8)(x-2) = 0$$

$x = -8$ or $x = 2$ Principle of zero products

Both numbers check. The solutions are -8 and 2.

5. $\dfrac{50}{x} - \dfrac{50}{x-5} = -\dfrac{1}{2}$, LCM is $2x(x-5)$

$$2x(x-5) \cdot \left(\frac{50}{x} - \frac{50}{x-5} \right) = 2x(x-5) \cdot \left(-\frac{1}{2} \right)$$

$$100(x-5) - 100x = -x(x-5)$$
$$100x - 500 - 100x = -x^2 + 5x$$
$$x^2 - 5x - 500 = 0$$
$$(x-25)(x+20) = 0$$

$x = 25$ or $x = -20$ Principle of zero products

Both numbers check. The solutions are 25 and -20.

7. $\dfrac{x+2}{x} = \dfrac{x-1}{2}$, LCM is $2x$

$$2x \cdot \frac{x+2}{x} = 2x \cdot \frac{x-1}{2}$$
$$2(x+2) = x(x-1)$$
$$2x + 4 = x^2 - x$$
$$0 = x^2 - 3x - 4$$
$$0 = (x-4)(x+1)$$

$x = 4$ or $x = -1$ Principle of zero products

Both numbers check. The solutions are 4 and -1.

9. $x - 8 = \dfrac{1}{x+8}$, LCM is $x+8$

$$(x+8)(x-8) = (x+8) \cdot \frac{1}{x+8}$$
$$x^2 - 64 = 1$$
$$x^2 = 65$$
$$x = \pm\sqrt{65}$$

Both numbers check. The solutions are $\pm\sqrt{65}$.

11. $\dfrac{6}{x} = \dfrac{x-15}{9}$, LCM is $9x$

$$9x \cdot \frac{6}{x} = 9x \cdot \frac{x-15}{9}$$
$$54 = x(x-15)$$
$$54 = x^2 - 15x$$
$$0 = x^2 - 15x - 54$$
$$0 = (x-18)(x+3)$$

$x = 18$ or $x = -3$ Principle of zero products

Both numbers check. The solutions are 18 and -3.

13. $\dfrac{40}{x} - \dfrac{20}{x-3} = \dfrac{8}{7}$, LCM is $7x(x-3)$

$$7x(x-3) \cdot \left(\frac{40}{x} - \frac{20}{x-3} \right) = 7x(x-3) \cdot \frac{8}{7}$$

$$280(x-3) - 140x = 8x(x-3)$$
$$280x - 840 - 140x = 8x^2 - 24x$$
$$0 = 8x^2 - 164x + 840$$
$$0 = 2x^2 - 41x + 210$$
$$0 = (2x-21)(x-10)$$

$2x - 21 = 0$ or $x - 10 = 0$

$x = \dfrac{21}{2}$ or $x = 10$

Both numbers check. The solutions are $\dfrac{21}{2}$ and 10.

15.
$$\frac{x+1}{3x+2} = \frac{2x-3}{3x-2} - 1 - \frac{36}{4-9x^2}$$
$$\frac{x+1}{3x+2} = \frac{2x-3}{3x-2} - 1 - \frac{36}{4-9x^2} \cdot \frac{-1}{-1}$$
$$\frac{x+1}{3x+2} = \frac{2x-3}{3x-2} - 1 - \frac{-36}{9x^2-4},$$
$$\text{LCM is } (3x+2)(3x-2)$$
$$(3x+2)(3x-2) \cdot \frac{x+1}{3x+2} =$$
$$(3x+2)(3x-2)\left(\frac{2x-3}{3x-2} - 1 - \frac{-36}{(3x+2)(3x-2)}\right)$$
$$(3x-2)(x+1) =$$
$$(3x+2)(2x-3) - (3x+2)(3x-2) - (-36)$$
$$3x^2 + x - 2 = 6x^2 - 5x - 6 - (9x^2-4) + 36$$
$$3x^2 + x - 2 = 6x^2 - 5x - 6 - 9x^2 + 4 + 36$$
$$6x^2 + 6x - 36 = 0$$
$$x^2 + x - 6 = 0 \qquad \text{Dividing by 6}$$
$$(x+3)(x-2) = 0$$
$$x+3 = 0 \quad \text{or} \quad x-2 = 0$$
$$x = -3 \quad \text{or} \quad x = 2$$
Both numbers check. The solutions are -3 and 2.

17.
$$\frac{x-2}{x+2} + \frac{x+2}{x-2} = \frac{10x-8}{4-x^2}$$
$$\frac{x-2}{x+2} + \frac{x+2}{x-2} = \frac{10x-8}{4-x^2} \cdot \frac{-1}{-1}$$
$$\frac{x-2}{x+2} + \frac{x+2}{x-2} = \frac{8-10x}{x^2-4},$$
$$\text{LCM is } (x+2)(x-2)$$
$$(x+2)(x-2) \cdot \left(\frac{x-2}{x+2} + \frac{x+2}{x-2}\right) =$$
$$(x+2)(x-2) \cdot \frac{8-10x}{(x+2)(x-2)}$$
$$(x-2)(x-2) + (x+2)(x+2) = 8 - 10x$$
$$x^2 - 4x + 4 + x^2 + 4x + 4 = 8 - 10x$$
$$2x^2 + 10x = 0$$
$$2x(x+5) = 0$$
$$2x = 0 \quad \text{or} \quad x+5 = 0$$
$$x = 0 \quad \text{or} \quad x = -5$$
Both numbers check. The solutions are 0 and -5.

19.
$$\frac{13}{7x-5} + \frac{11x}{3} = \frac{39}{21x-15},$$
$$\text{LCM is } 3(7x-5)$$
$$3(7x-5)\left(\frac{13}{7x-5} + \frac{11x}{3}\right) = 3(7x-5) \cdot \frac{39}{3(7x-5)}$$
$$3 \cdot 13 + (7x-5)(11x) = 39$$
$$39 + 77x^2 - 55x = 39$$
$$77x^2 - 55x = 0$$
$$11x(7x-5) = 0$$
$$11x = 0 \quad \text{or} \quad 7x - 5 = 0$$
$$x = 0 \quad \text{or} \quad 7x = 5$$
$$x = 0 \quad \text{or} \quad x = \frac{5}{7}$$

The number 0 checks, but $\frac{5}{7}$ produces a zero denominator $\left(7 \cdot \frac{5}{7} - 5 = 5 - 5 = 0\right)$. The solution is 0.

21.
$$\frac{12}{x^2-9} = 1 + \frac{3}{x-3},$$
$$\text{LCM is } (x+3)(x-3)$$
$$(x+3)(x-3) \cdot \frac{12}{(x+3)(x-3)} = (x+3)(x-3)\left(1 + \frac{3}{x-3}\right)$$
$$12 = (x+3)(x-3) + 3(x+3)$$
$$12 \quad x^2 - 9 + 3x + 9$$
$$0 = x^2 + 3x - 12$$
$$x = \frac{-b \pm \sqrt{b^2-4ac}}{2a} = \frac{-3 \pm \sqrt{3^2 - 4 \cdot 1 \cdot (-12)}}{2 \cdot 1}$$
$$x = \frac{-3 \pm \sqrt{9+48}}{2} = \frac{-3 \pm \sqrt{57}}{2}$$

Both numbers check. The solutions are $\frac{-3+\sqrt{57}}{2}$ and $\frac{-3-\sqrt{57}}{2}$, or $\frac{-3\pm\sqrt{57}}{2}$.

23. *Familiarize*. We first make a drawing. We can also organize the information in a table. We let r = the speed for the first part of the trip. Then $r - 5$ = the speed for the second part of the trip. We let t_1 and t_2 represent the time for the first and second parts of the trip, respectively.

r mph t_1 hr $r-5$ mph t_2 hr
_____ • _____
       80 mi   35 mi

Trip	Distance	Speed	Time
1st part	80	r	t_1
2nd part	35	$r-5$	t_2
	Total time		3

Translate. Since the total time is 3 hr, we can write one equation:
$$t_1 + t_2 = 3$$
Using $t = d/r$ in the first two rows of the table, we have
$$t_1 = \frac{80}{r} \text{ and } t_2 = \frac{35}{r-5}.$$

Substituting into the first equation, we get

$$\frac{80}{r} + \frac{35}{r-5} = 3.$$

Solve. The LCM is $r(r-5)$.

$$r(r-5)\left(\frac{80}{r} + \frac{35}{r-5}\right) = r(r-5) \cdot 3$$

Multiplying by the LCM

$$80(r-5) + 35r = 3r^2 - 15r$$
$$80r - 400 + 35r = 3r^2 - 15r$$
$$115r - 400 = 3r^2 - 15r$$
$$0 = 3r^2 - 130r + 400$$
$$0 = (3r - 10)(r - 40)$$

$r = \frac{10}{3}$ or $r = 40$ Principle of zero products

Check. $\frac{10}{3}$ is not a solution, because it would make the speed on the second part of the trip negative: $\frac{10}{3} - 5 = -\frac{5}{3}$.

If $r = 40$, then the time for the first part of the trip was 80/40, or 2 hr. The speed on the second part of the trip was $40 - 5$, or 35 mph and the time was 35/35, or 1 hr. The total time was 2 hr + 1 hr, or 3 hr. The result checks.

State. The speed on the first part of the trip was 40 mph; the speed on the second part of the trip was 35 mph.

25. **Familiarize.** We first make a drawing. We also organize the information in a table. We let $r =$ the speed and $t =$ the time of the slower trip.

280 mi r mph t hr

280 mi $r + 5$ mph $t - 1$ hr

Trip	Distance	Speed	Time
Slower	280	r	t
Faster	280	$r + 5$	$t - 1$

Translate. Using $t = d/r$, we get two equations from the table:

$$t = \frac{280}{r} \text{ and } t - 1 = \frac{280}{r+5}$$

Solve. We substitute $\frac{280}{r}$ for t in the second equation and solve for r.

$$\frac{280}{r} - 1 = \frac{280}{r+5}, \text{ LCM is } r(r+5)$$
$$r(r+5)\left(\frac{280}{r} - 1\right) = r(r+5) \cdot \frac{280}{r+5}$$
$$280(r+5) - r(r+5) = 280r$$
$$280r + 1400 - r^2 - 5r = 280r$$
$$0 = r^2 + 5r - 1400$$
$$0 = (r - 35)(r + 40)$$

$r = 35$ or $r = -40$

Check. Since negative speed has no meaning in this problem, we only check 35. If $r = 35$, then the time for the

slow trip is $\frac{280}{35}$, or 8 hours. If $r = 35$, then $r + 5 = 40$ and the time for the fast trip is $\frac{280}{40}$, or 7 hours. This is 1 hour less time than the slow trip took, so we have an answer to the problem.

State. The speed is 35 mph.

27. **Familiarize.** We make a drawing and then organize the information in a table. We let $r =$ the speed and $t =$ the time of plane A.

2800 km r km/h t hr

2000 km $r + 50$ km/h $t - 3$ hr

Plane	Distance	Speed	Time
A	2800	r	t
B	2000	$r + 50$	$t - 3$

Translate. Using $r = d/t$, we get two equations from the table:

$$r = \frac{2800}{t} \text{ and } r + 50 = \frac{2000}{t-3}$$

Solve. We substitute $\frac{2800}{t}$ for r in the second equation and solve for t.

$$\frac{2800}{t} + 50 = \frac{2000}{t-3}, \text{ LCM is } t(t-3)$$
$$t(t-3)\left(\frac{2800}{t} + 50\right) = t(t-3) \cdot \frac{2000}{t-3}$$
$$2800(t-3) + 50t(t-3) = 2000t$$
$$2800t - 8400 + 50t^2 - 150t = 2000t$$
$$50t^2 + 650t - 8400 = 0$$
$$t^2 + 13t - 168 = 0$$
$$(t + 21)(t - 8) = 0$$

$t = -21$ or $t = 8$

Check. Since negative time has no meaning in this problem, we only check 8 hours. If $t = 8$, then $t - 3 = 5$. The speed of plane A is $\frac{2800}{8}$, or 350 km/h. The speed of plane B is $\frac{2000}{5}$, or 400 km/h. Since the speed of plane B is 50 km/h faster than the speed of plane A, the value checks.

State. The speed of plane A is 350 km/h; the speed of plane B is 400 km/h.

29. **Familiarize.** Let $x =$ the time it takes the smaller pipe to fill the tank. Then $x - 3 =$ the time it takes the larger pipe to fill the tank.

Translate. Using the work formula, we write an equation.

$$\frac{2}{x} + \frac{2}{x-3} = 1$$

Solve. We solve the equation.

We multiply by the LCM, $x(x-3)$.

$$x(x-3)\left(\frac{2}{x} + \frac{2}{x-3}\right) = x(x-3)\cdot 1$$

$$2(x-3) + 2x = x(x-3)$$

$$2x - 6 + 2x = x^2 - 3x$$

$$0 = x^2 - 7x + 6$$

$$0 = (x-6)(x-1)$$

$x = 6$ or $x = 1$

Check. Since negative time has no meaning in this problem, 1 is not a solution $(1 - 3 = -2)$. We only check 6 hr. This is the time it would take the smaller pipe working alone. Then the larger pipe would take $6 - 3$, or 3 hr working alone. The larger pipe would fill $2\left(\frac{1}{3}\right)$, or $\frac{2}{3}$, of the tank in 2 hr, and the smaller pipe would fill $2\left(\frac{1}{6}\right)$, or $\frac{1}{3}$, of the tank in 2 hr. Thus in 2 hr they would fill $\frac{2}{3} + \frac{1}{3}$ of the tank. This is all of it, so the numbers check.

State. It takes the smaller pipe, working alone, 6 hr to fill the tank.

31. Familiarize. We make a drawing and then organize the information in a table. We let r = the speed of the boat in still water. Then $r - 2$ = the speed upstream and $r + 2$ = the speed downstream.

1 km $r - 2$ km/h → Upstream

Downstream ← 1 km $r + 2$ km/h

Trip	Distance	Speed	Time
Upstream	1	$r-2$	t_1
Downstream	1	$r+2$	t_2
	Total time		1

Translate. Since the total time is 1 hr, we can write one equation:

$$t_1 + t_2 = 1$$

Using $t = d/r$ in the first two rows of the table, we have

$$t_1 = \frac{1}{r-2} \text{ and } t_2 = \frac{1}{r+2}.$$

Substituting into the first equation, we get

$$\frac{1}{r-2} + \frac{1}{r+2} = 1.$$

Solve. We solve the equation. We multiply by the LCM, $(r-2)(r+2)$.

$$(r-2)(r+2)\left(\frac{1}{r-2} + \frac{1}{r+2}\right) = (r-2)(r+2)\cdot 1$$

$$(r+2) + (r-2) = (r-2)(r+2)$$

$$2r = r^2 - 4$$

$$0 = r^2 - 2r - 4$$

$$r = \frac{-(-2) \pm \sqrt{(-2)^2 - 4\cdot 1\cdot(-4)}}{2\cdot 1}$$

$$r = \frac{2 \pm \sqrt{4 + 16}}{2} = \frac{2 \pm \sqrt{20}}{2}$$

$$r = \frac{2 \pm 2\sqrt{5}}{2} = 1 \pm \sqrt{5}$$

$$1 + \sqrt{5} \approx 1 + 2.236 \approx 3.24$$

$$1 - \sqrt{5} \approx 1 - 2.236 \approx -1.24$$

Check. Since negative speed has no meaning in this problem, we only check 3.24 km/h. If $r \approx 3.24$, then $r - 2 \approx 1.24$ and $r + 2 \approx 5.24$. The time it takes to travel upstream is approximately $\frac{1}{1.24}$, or 0.806 hr, and the time it takes to travel downstream is approximately $\frac{1}{5.24}$, or 0.191 hr. The total time is 0.997 which is approximately 1 hour. The value checks.

State. The speed of the boat in still water is approximately 3.24 km/h.

33. Familiarize. We make a drawing and then organize the information in a table. We let r = the speed on the first part of the trip. Then $r - 5$ = the speed on the second part of the trip.

r km/h $r - 5$ km/h

80 mi   25 mi

Trip	Distance	Speed	Time
First part	80	r	t_1
Second part	25	$r-5$	t_2
	Total time		3

Translate. Since the total time is 3 hr, we can write one equation:

$$t_1 + t_2 = 3$$

Using $t = d/r$ in the first two rows of the table, we have

$$t_1 = \frac{80}{r} \text{ and } t_2 = \frac{25}{r-5}.$$

Substituting into the first equation, we get

$$\frac{80}{r} + \frac{25}{r-5} = 3.$$

Solve. We solve the equation. We multiply by the LCM, $r(r-5)$.

$$r(r-5)\left(\frac{80}{r} + \frac{25}{r-5}\right) = r(r-5)\cdot 3$$

$$80(r-5) + 25r = 3r(r-5)$$

$$80r - 400 + 25r = 3r^2 - 15r$$

$$0 = 3r^2 - 120r + 400$$

$$r = \frac{-(-120) \pm \sqrt{(-120)^2 - 4\cdot 3\cdot 400}}{2\cdot 3}$$

$$r = \frac{120 \pm \sqrt{14,400 - 4800}}{6} = \frac{120 \pm \sqrt{9600}}{6}$$

$$r = \frac{120 \pm 40\sqrt{6}}{6} = \frac{60 \pm 20\sqrt{6}}{3}$$

$$\frac{60 + 20\sqrt{6}}{3} \approx \frac{60 + 20(2.449)}{3} \approx 36.33$$

$$\frac{60 - 20\sqrt{6}}{3} \approx \frac{60 - 20(2.449)}{3} \approx 3.67$$

Check. If $r \approx 3.67$, then $r - 5 \approx -1.33$. Since negative time has no meaning in this problem, 3.67 is not a solution. If $r \approx 36.33$, then $r - 5 \approx 36.33 - 5 \approx 31.33$. The time for the first part of the trip is $\dfrac{80}{36.33} \approx 2.2$ hr, and the time for the second part of the trip is $\dfrac{25}{31.33} \approx 0.8$ hr. The total time is $2.2 + 0.8$, or 3 hr. The value checks.

State. The speed on the first part of the trip was approximately 36.33 mph, and the speed on the second part of the trip was approximately 31.33 mph.

35.
$$x = \sqrt{x + 6}$$
$$x^2 = (\sqrt{x + 6})^2 \qquad \text{Principle of powers}$$
$$x^2 = x + 6$$
$$x^2 - x - 6 = 0$$
$$(x - 3)(x + 2) = 0$$
$$x - 3 = 0 \quad \text{or} \quad x + 2 = 0$$
$$x = 3 \quad \text{or} \qquad x = -2$$

Only 3 checks. It is the solution.

37. $\sqrt{2x - 7} = -9$

This equation has no solution, because the principle square root is always nonnegative.

39.
$$= \frac{1}{x - 1} + \frac{1}{x^2 - 3x + 2}$$
$$= \frac{1}{x - 1} + \frac{1}{(x - 2)(x - 1)}, \text{ LCM is } (x - 1)(x - 2)$$
$$= \frac{1}{x - 1} \cdot \frac{x - 2}{x - 2} + \frac{1}{(x - 2)(x - 1)}$$
$$= \frac{x - 2}{(x - 1)(x - 2)} + \frac{1}{(x - 1)(x - 2)}$$
$$= \frac{x - 2 + 1}{(x - 1)(x - 2)}$$
$$= \frac{x - 1}{(x - 1)(x - 2)}$$
$$= \frac{(x - 1)(1)}{(x - 1)(x - 2)}$$
$$= \frac{1}{x - 2}$$

41. Familiarize. Let $a =$ the time it takes smokestack A to generate the given amount of pollution. Then $2.13a =$ the time it takes smokestack B to generate the same amount of pollution.

Translate. We use the work formula.
$$\frac{16.3}{a} + \frac{16.3}{2.13a} = 1$$

Solve. We solve the equation. We multiply by the LCM, $2.13a$.
$$2.13a\left(\frac{16.3}{a} + \frac{16.3}{2.13a}\right) = 2.13a \cdot 1$$
$$(2.13)(16.3) + 16.3 = 2.13a$$
$$34.719 + 16.3 = 2.13a$$
$$51.019 = 2.13a$$
$$24.0 \approx a$$

Check. If it takes smokestack A about 24.0 hr to generate the given amount of pollution, then it takes smokestack B about 2.13(24.0), or about 51.0 hr, to generate the same amount of pollution. In about 16.3 hr smokestack A generates about $\dfrac{16.3}{24.0} \approx 0.68$ of the given amount of pollution. In 16.3 hr smokestack B generates about $\dfrac{16.3}{51.0} \approx 0.32$ of the given amount of pollution. Together they generate $0.68 + 0.32$, or the entire amount of pollution. The result checks.

State. It would take smokestack A about 24.0 hr and smokestack B about 51.0 hr to generate the given amount of pollution.

43.
$$\frac{4}{2x + i} - \frac{1}{x - i} = \frac{2}{x + i},$$
$$\text{LCM is } (2x + i)(x - i)(x + i)$$
$$(2x + i)(x - i)(x + i)\left(\frac{4}{2x + i} - \frac{1}{x - i}\right) =$$
$$(2x + i)(x - i)(x + i) \cdot \frac{2}{x + i}$$
$$4(x - i)(x + i) - (2x + i)(x + i) = 2(2x + i)(x - i)$$
$$4(x^2 - i^2) - (2x^2 + 3ix + i^2) = 2(2x^2 - ix - i^2)$$
$$4(x^2 + 1) - (2x^2 + 3ix - 1) = 2(2x^2 - ix + 1)$$
$$4x^2 + 4 - 2x^2 - 3ix + 1 = 4x^2 - 2ix + 2$$
$$2x^2 - 3ix + 5 = 4x^2 - 2ix + 2$$
$$0 = 2x^2 + ix - 3$$

$a = 2$, $b = i$, $c = -3$
$$x = \frac{-i \pm \sqrt{i^2 - 4 \cdot 2 \cdot (-3)}}{2 \cdot 2} = \frac{-i \pm \sqrt{-1 + 24}}{4}$$
$$x = \frac{-i \pm \sqrt{23}}{4}$$

Both numbers check. The solutions are $\dfrac{-i \pm \sqrt{23}}{4}$.

Exercise Set 8.7

1. $y = 4x^2$

a), b) $y = 4x^2$ is of the form $y = ax^2$. Thus we know that the vertex is $(0, 0)$ and $x = 0$ is the line of symmetry.

c) We know that $y = 0$ when $x = 0$ since the vertex is $(0, 0)$.

For $x = 1$, $y = 4x^2 = 4 \cdot 1^2 = 4$.

For $x = -1$, $y = 4x^2 = 4 \cdot (-1)^2 = 4$.

For $x = 2$, $y = 4x^2 = 4 \cdot 2^2 = 16$.

For $x = -2$, $y = 4x^2 = 4 \cdot (-2)^2 = 16$.

We complete the table.

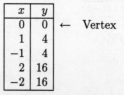

x	y	
0	0	$\leftarrow$ Vertex
1	4	
-1	4	
2	16	
-2	16	

d) We plot the ordered pairs (x, y) from the table and connect them with a smooth curve.

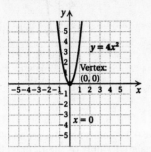

3. $y = \frac{1}{3}x^2$ is of the form $y = ax^2$. Thus we know that the vertex is $(0, 0)$ and $x = 0$ is the line of symmetry. We choose some numbers for x and find the corresponding values for y. Then we plot the ordered pairs (x, y) and connect them with a smooth curve.

For $x = 3$, $y = \frac{1}{3}x^2 = \frac{1}{3} \cdot 3^2 = 3$.

For $x = -3$, $y = \frac{1}{3}x^2 = \frac{1}{3} \cdot (-3)^2 = 3$.

For $x = 6$, $y = \frac{1}{3}x^2 = \frac{1}{3} \cdot 6^2 = 12$.

For $x = -6$, $y = \frac{1}{3}x^2 = \frac{1}{3} \cdot (-6)^2 = 12$.

x	y	
0	0	← Vertex
3	3	
-3	3	
6	12	
-6	12	

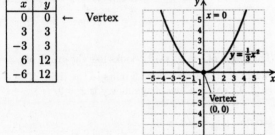

5. $y = -\frac{1}{2}x^2$ is of the form $y = ax^2$. Thus we know that the vertex is $(0, 0)$ and $x = 0$ is the line of symmetry. We choose some numbers for x and find the corresponding values for y. Then we plot the ordered pairs (x, y) and connect them with a smooth curve.

For $x = 2$, $y = -\frac{1}{2}x^2 = -\frac{1}{2} \cdot 2^2 = -2$.

For $x = -2$, $y = -\frac{1}{2}x^2 = -\frac{1}{2} \cdot (-2)^2 = -2$.

For $x = 4$, $y = -\frac{1}{2}x^2 = -\frac{1}{2} \cdot 4^2 = -8$.

For $x = -4$, $y = -\frac{1}{2}x^2 = -\frac{1}{2} \cdot (-4)^2 = -8$.

x	y	
0	0	← Vertex
2	-2	
-2	-2	
4	-8	
-4	-8	

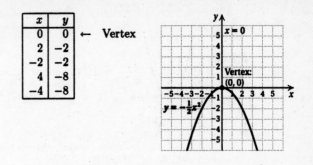

7. $y = -4x^2$ is of the form $y = ax^2$. Thus we know that the vertex is $(0, 0)$ and $x = 0$ is the line of symmetry. We choose some numbers for x and find the corresponding values for y. Then we plot the ordered pairs (x, y) and connect them with a smooth curve.

For $x = 1$, $y = -4x^2 = -4 \cdot 1^2 = -4$.

For $x = -1$, $y = -4x^2 = -4 \cdot (-1)^2 = -4$.

For $x = 2$, $y = -4x^2 = -4 \cdot 2^2 = -16$.

For $x = -2$, $y = -4x^2 = -4 \cdot (-2)^2 = -16$.

x	y	
0	0	← Vertex
1	-4	
-1	-4	
2	-16	
-2	-16	

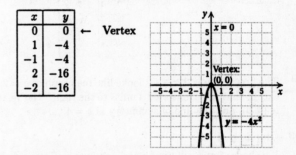

9. $y = (x - 3)^2$

a) For $x = 3$, $y = (3 - 3)^2 = 0^2 = 0$.

For $x = 4$, $y = (4 - 3)^2 = 1^2 = 1$.

For $x = 2$, $y = (2 - 3)^2 = (-1)^2 = 1$.

For $x = 5$, $y = (5 - 3)^2 = 2^2 = 4$.

For $x = 1$, $y = (1 - 3)^2 = (-2)^2 = 4$.

For $x = 0$, $y = (0 - 3)^2 = (-3)^2 = 9$.

We complete the table.

x	y	
3	0	← Vertex
4	1	
2	1	
5	4	
1	4	
0	9	← y-intercept

b) Since $y = (x - 3)^2$ is of the form $y = a(x - h)^2$, we know that the vertex is $(3, 0)$ and the line of symmetry is $x = 3$. We plot the points found in part(a) and draw the graph.

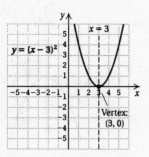

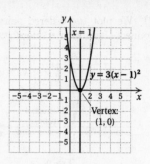

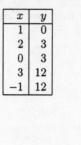

x	y
1	0
2	3
0	3
3	12
-1	12

11. Graph: $y = 2(x-4)^2$

We choose some values of x and compute y. Then we plot these ordered pairs and connect them with a smooth curve.

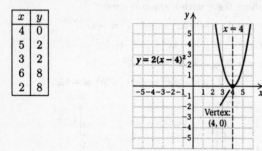

x	y
4	0
5	2
3	2
6	8
2	8

The graph of $y = 2(x-4)^2$ looks like the graph of $y = 2x^2$ except that it is translated 4 units to the right. The vertex is $(4, 0)$, and the line of symmetry is $x = 4$.

13. Graph: $y = -2(x+2)^2$

We choose some values of x and compute y. Then we plot these ordered pairs and connect them with a smooth curve.

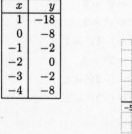

x	y
1	-18
0	-8
-1	-2
-2	0
-3	-2
-4	-8

We can express the equation in the equivalent form $y = -2[x - (-2)]^2$. Then we know that the graph looks like the graph of $y = -2x^2$ translated 2 units to the left. The vertex is $(-2, 0)$, and the line of symmetry is $x = -2$.

15. Graph: $y = 3(x-1)^2$

We choose some values of x and compute y. Then we plot these ordered pairs and connect them with a smooth curve.

The graph of $y = 3(x-1)^2$ looks like the graph of $y = 3x^2$ except that it is translated 1 unit to the right. The vertex is $(1, 0)$, and the line of symmetry is $x = 1$.

17. Graph: $y = \dfrac{1}{3}(x-3)^2$

We choose some values of x and compute y. Then we plot these ordered pairs and connect them with a smooth curve.

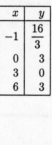

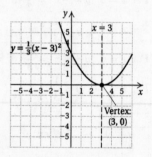

x	y
-1	$\dfrac{16}{3}$
0	3
3	0
6	3

The graph of $y = \dfrac{1}{3}(x-3)^2$ looks like the graph of $y = \dfrac{1}{3}x^2$ except that it is translated 3 units to the right. The vertex is $(3, 0)$, and the line of symmetry is $x = 3$.

19. Graph: $y = -\dfrac{3}{2}(x+2)^2$

We choose some values of x and compute y. Then we plot these ordered pairs and connect them with a smooth curve.

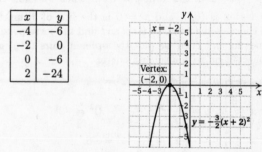

x	y
-4	-6
-2	0
0	-6
2	-24

We can express the equation in the equivalent form $y = -\dfrac{3}{2}[x - (-2)]^2$. Then we know that the graph looks like the graph of $y = -\dfrac{3}{2}x^2$ translated 2 units to the left. The vertex is $(-2, 0)$, and the line of symmetry is $x = -2$.

21. Graph: $y = (x-3)^2 + 1$

We choose some values of x and compute y. Then we plot these ordered pairs and connect them with a smooth curve.

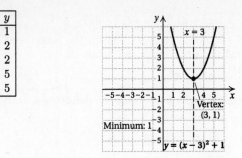

x	y
3	1
4	2
2	2
5	5
1	5

The graph of $y = (x-3)^2 + 1$ looks like the graph of $y = x^2$ except that it is translated 3 units right and 1 unit up. The vertex is $(3, 1)$, and the line of symmetry is $x = 3$. The equation is of the form $y = a(x-h)^2 + k$ with $a = 1$. Since $1 > 0$, we know that 1 is the minimum value.

23. Graph: $y = -3(x+4)^2 + 1$
    $y = -3[x - (-4)]^2 + 1$

We choose some values of x and compute y. Then we plot these ordered pairs and connect them with a smooth curve.

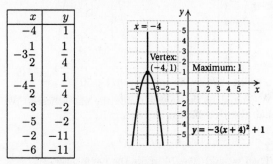

x	y
-4	1
$-3\frac{1}{2}$	$\frac{1}{4}$
$-4\frac{1}{2}$	$\frac{1}{4}$
-3	-2
-5	-2
-2	-11
-6	-11

The graph of $y = -3(x+4)^2 + 1$ looks like the graph of $y = 3x^2$ except that it is translated 4 units left and 1 unit up and opens downward. The vertex is $(-4, 1)$, and the line of symmetry is $x = -4$. Since $-3 < 0$, we know that 1 is the maximum value.

25. Graph: $y = \frac{1}{2}(x+1)^2 + 4$

$y = \frac{1}{2}[x - (-1)]^2 + 4$

We choose some values of x and compute y. Then we plot these ordered pairs and connect them with a smooth curve.

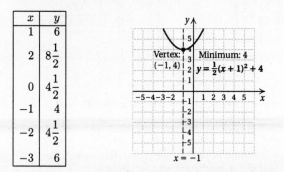

x	y
1	6
2	$8\frac{1}{2}$
0	$4\frac{1}{2}$
-1	4
-2	$4\frac{1}{2}$
-3	6

The graph of $y = \frac{1}{2}(x+1)^2 + 4$ looks like the graph of $y = \frac{1}{2}x^2$ except that it is translated 1 unit left and 4 units

up. The vertex is $(-1, 4)$, and the line of symmetry is $x = -1$. Since $\frac{1}{2} > 0$, we know that 4 is the minimum value.

27. Graph: $y = -2(x+2)^2 - 3$
    $y = -2[x - (-2)]^2 + (-3)$

We choose some values of x and compute y. Then we plot these ordered pairs and connect them with a smooth curve.

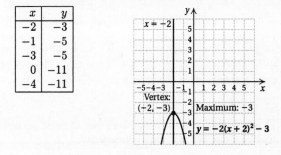

x	y
-2	-3
-1	-5
-3	-5
0	-11
-4	-11

The graph of $y = -2(x+2)^2 - 3$ looks like the graph of $y = 2x^2$ except that it is translated 2 units left and 3 units down and opens downward. The vertex is $(-2, -3)$, and the line of symmetry is $x = -2$. Since $-2 < 0$, we know that -3 is the maximum value.

29. Graph: $y = -(x+1)^2 - 2$
    $y = -[x - (-1)]^2 + (-2)$

We choose some values of x and compute y. Then we plot these ordered pairs and connect them with a smooth curve.

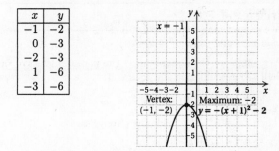

x	y
-1	-2
0	-3
-2	-3
1	-6
-3	-6

The graph of $y = -(x+1)^2 - 2$ looks like the graph of $y = x^2$ except that it is translated 1 unit left and 2 units down and opens downward. The vertex is $(-1, -2)$, and the line of symmetry is $x = -1$. Since $-1 < 0$, we know that -2 is the maximum value.

31. $\sqrt{x} = -7$

The equation has no solution, because the principal square root of a number is always nonnegative.

33.
$$x - 5 = \sqrt{x + 7}$$
$$(x - 5)^2 = (\sqrt{x + 7})^2 \quad \text{Principle of powers}$$
$$x^2 - 10x + 25 = x + 7$$
$$x^2 - 11x + 18 = 0$$
$$(x - 9)(x - 2) = 0$$
$$x - 9 = 0 \quad \text{or} \quad x - 2 = 0$$
$$x = 9 \quad \text{or} \quad x = 2$$

Check: For 9:

$$x - 5 = \sqrt{x+7}$$

$9 - 5$	$\sqrt{9+7}$
4	$\sqrt{16}$
	4 TRUE

For 2:

$$x - 5 = \sqrt{x+7}$$

$2 - 5$	$\sqrt{2+7}$
-3	$\sqrt{9}$
	3 FALSE

Only 9 checks. It is the solution.

35. $\sqrt{x+4} = -11$

The equation has no solution, because the principal square root of a number is always nonnegative.

37. We observe that the points $(0,2), (-2,0), (1,0), (2,4)$, and $(-1,4)$ are on the graph. We complete the table.

x	y
0	2
-2	0
1	0
2	4
-1	4

Exercise Set 8.8

1. $y = x^2 - 2x - 3 = (x^2 - 2x) - 3$

We complete the square inside the parentheses. We take half the x-coefficient and square it.

$\frac{1}{2} \cdot (-2) = -1$ and $(-1)^2 = 1$

Then we add $1 - 1$ inside the parentheses.

$$y = (x^2 - 2x + 1 - 1) - 3$$
$$= (x^2 - 2x + 1) - 1 - 3$$
$$= (x - 1)^2 - 4$$
$$= (x - 1)^2 + (-4)$$

Vertex: $(1, -4)$

Line of symmetry: $x = 1$

We plot a few points and draw the curve.

x	y
1	-4
2	-3
0	-3
3	0
-1	0
4	5
-2	5

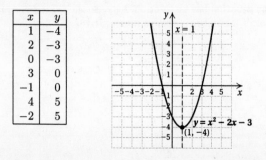

3. $y = -x^2 + 4x + 1 = -(x^2 - 4x) + 1$

We complete the square inside the parentheses. We take half the x-coefficient and square it.

$\frac{1}{2} \cdot (-4) = -2$ and $(-2)^2 = 4$

Then we add $4 - 4$ inside the parentheses.

$$y = -(x^2 - 4x + 4 - 4) + 1$$
$$= -(x^2 - 4x + 4) + 4 + 1$$
$$= -(x - 2)^2 + 5$$

Vertex: $(2, 5)$

Line of symmetry: $x = 2$

We plot a few points and draw the curve.

x	y
2	5
3	4
1	4
4	1
0	1
5	-4
-1	-4

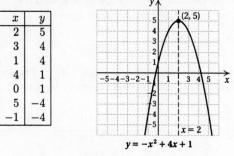

5. $y = 3x^2 - 24x + 50 = 3(x^2 - 8x) + 50$

We complete the square inside the parentheses. We take half the x-coefficient and square it.

$\frac{1}{2} \cdot (-8) = -4$ and $(-4)^2 = 16$

Then we add $16 - 16$ inside the parentheses.

$$y = 3(x^2 - 8x + 16 - 16) + 50$$
$$= 3(x^2 - 8x + 16) - 48 + 50$$
$$= 3(x - 4)^2 + 2$$

Vertex: $(4, 2)$

Line of symmetry: $x = 4$

We plot a few points and draw the curve.

x	y
4	2
5	5
3	5
6	14
2	14

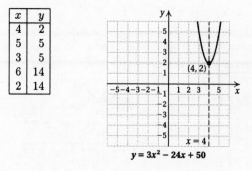

7. $y = -2x^2 + 2x + 1 = -2(x^2 - x) + 1$

We complete the square inside the parentheses. We take half the x-coefficient and square it.

$\frac{1}{2} \cdot (-1) = -\frac{1}{2}$ and $\left(-\frac{1}{2}\right)^2 = \frac{1}{4}$

Then we add $\frac{1}{4} - \frac{1}{4}$ inside the parentheses.

$$y = -2\left(x^2 - x + \frac{1}{4} - \frac{1}{4}\right) + 1$$

$$= -2\left(x^2 - x + \frac{1}{4}\right) + \frac{1}{2} + 1$$

$$= -2\left(x - \frac{1}{2}\right)^2 + \frac{3}{2}$$

Vertex: $\left(\frac{1}{2}, \frac{3}{2}\right)$

Line of symmetry: $x = \frac{1}{2}$

We plot a few points and draw the curve.

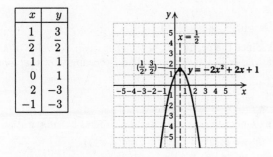

x	y
$\frac{1}{2}$	$\frac{3}{2}$
1	1
0	1
2	-3
-1	-3

9. $y = 5 - x^2 = -x^2 + 5 = -(x - 0)^2 + 5$

Vertex: $(0, 5)$

Line of symmetry: $x = 0$

We plot a few points and draw the curve.

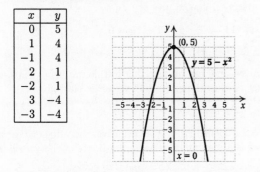

x	y
0	5
1	4
-1	4
2	1
-2	1
3	-4
-3	-4

11. $y = x^2 + 2x + 12$

We solve the equation $x^2 + 2x + 12 = 0$ using the quadratic formula.

$$x = \frac{-2 \pm \sqrt{2^2 - 4 \cdot 1 \cdot 12}}{2 \cdot 1}$$

$$x = \frac{-2 \pm \sqrt{4 - 48}}{2} = \frac{-2 \pm \sqrt{-44}}{2}$$

$$x = \frac{-2 \pm \sqrt{-1 \cdot 4 \cdot 11}}{2} = \frac{-2 \pm 2i\sqrt{11}}{2}$$

$$x = -1 \pm i\sqrt{11}$$

Since the equation has no real solutions, there are no x-intercepts.

13. $y = -x^2 + 5x + 24$

We solve the equation $-x^2 + 5x + 24 = 0$.

$$-x^2 + 5x + 24 = 0$$

$$x^2 - 5x - 24 = 0 \quad \text{Multiplying by } -1$$

$$(x - 8)(x + 3) = 0$$

$x = 8$ or $x = -3$ Principle of zero products

The x-intercepts are $(-3, 0)$ and $(8, 0)$.

15. $y = 3x^2 - 6x + 1$

We solve the equation $3x^2 - 6x + 1 = 0$ using the quadratic formula.

$$x = \frac{-(-6) \pm \sqrt{(-6)^2 - 4 \cdot 3 \cdot 1}}{2 \cdot 3}$$

$$x = \frac{6 \pm \sqrt{36 - 12}}{6} = \frac{6 \pm \sqrt{24}}{6}$$

$$x = \frac{6 \pm 2\sqrt{6}}{6} = \frac{2(3 \pm \sqrt{6})}{2 \cdot 3}$$

$$x = \frac{3 \pm \sqrt{6}}{3}$$

The x-intercepts are $\left(\frac{3 - \sqrt{6}}{3}, 0\right)$ and $\left(\frac{3 + \sqrt{6}}{3}, 0\right)$.

17. $y = 2x^2 + 4x - 1$

We solve the equation $2x^2 + 4x - 1 = 0$ using the quadratic formula.

$$x = \frac{-4 \pm \sqrt{4^2 - 4 \cdot 2 \cdot (-1)}}{2 \cdot 2}$$

$$x = \frac{-4 \pm \sqrt{16 + 8}}{4} = \frac{-4 \pm \sqrt{24}}{4}$$

$$x = \frac{-4 \pm 2\sqrt{6}}{4} = \frac{2(-2 \pm \sqrt{6})}{2 \cdot 2}$$

$$x = \frac{-2 \pm \sqrt{6}}{2}$$

The x-intercepts are $\left(\frac{-2 - \sqrt{6}}{2}, 0\right)$ and $\left(\frac{-2 + \sqrt{6}}{2}, 0\right)$.

19. We make a drawing and label it.

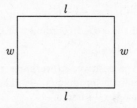

The perimeter must be 88 ft, so we have

$$2l + 2w = 88. \quad (1)$$

We wish to find the maximum area. We know

$$A = lw. \quad (2)$$

Solving (1) for l, we get $l = 44 - w$. Substituting in (2), we get a quadratic equation:

$$A = (44 - w)w = -w^2 + 44w$$

The w-coordinate of the vertex is

$$w = -\frac{b}{2a} = -\frac{44}{2(-1)} = 22.$$

Substituting into the equation, we find the A-coordinate of the vertex:

$$A = -(22)^2 + 44(22) = 484$$

The coefficient of w^2 is negative, so we know that 484 is a maximum. The maximum area occurs when the width is 22 ft and the length is $44 - 22$, or 22 ft. It is 484 ft^2.

21. Let x and y represent the numbers. Their sum is 45, so we have

$$x + y = 45. \qquad (1)$$

We wish to find the maximum product P. We know

$$P = xy. \qquad (2)$$

Solving (1) for y, we get $y = 45 - x$. Substituting in (2), we get a quadratic equation:

$$P = x(45 - x) = -x^2 + 45x$$

The first coordinate of the vertex is

$$x = -\frac{b}{2a} = -\frac{45}{2(-1)} = 22.5.$$

Note that when $x = 22.5$, $y = 45 - 22.5 = 22.5$.

We substitute to find the second coordinate of the vertex:

$$P = -(22.5)^2 + 45(22.5) = 506.25$$

The coefficient of x^2 is negative, so we know that 506.25 is a maximum. The maximum product is 506.25. The numbers 22.5 and 22.5 yield this product.

23. Let x and y represent the numbers. Their difference is 6, so we have

$$x - y = 6. \qquad (1)$$

We wish to find the minimum product P. We know

$$P = xy. \qquad (2)$$

Solving (1) for x, we get $x = y + 6$. Substituting in (2), we get a quadratic equation:

$$P = (y + 6)y = y^2 + 6y$$

The first coordinate of the vertex is

$$y = -\frac{b}{2a} = -\frac{6}{2 \cdot 1} = -3.$$

Note that when $y = -3$, $x = -3 + 6 = 3$.

We substitute to find the second coordinate of the vertex:

$$P = (-3)^2 + 6(-3) = -9$$

The coefficient of y^2 is positive, so we know that -9 is a minimum. The minimum product is -9. The numbers that yield this product are -3 and 3.

25. Let x and y represent the numbers. Their difference is 7, so we have

$$x - y = 7. \qquad (1)$$

We wish to find the minimum product P. We know

$$P = xy. \qquad (2)$$

Solving (1) for x, we get $x = y + 7$. Substituting in (2), we get a quadratic equation:

$$P = (y + 7)y = y^2 + 7y$$

The first coordinate of the vertex is

$$y = -\frac{b}{2a} = -\frac{7}{2 \cdot 1} = -\frac{7}{2}.$$

Note that when $y = -\frac{7}{2}$, $x = -\frac{7}{2} + 7 = \frac{7}{2}$.

We substitute to find the second coordinate of the vertex:

$$P = \left(-\frac{7}{2}\right)^2 + 7\left(-\frac{7}{2}\right) = -\frac{49}{4}$$

The coefficient of y^2 is positive, so we know that $-\frac{49}{4}$ is a minimum. The minimum product is $-\frac{49}{4}$. The numbers that yield this product are $-\frac{7}{2}$ and $\frac{7}{2}$.

27. $h = -16t^2 + 64t + 2240$

a) Since the coefficient of t^2 is negative, the value of h at the vertex is a maximum. The t-coordinate of the vertex is

$$t = -\frac{b}{2a} = -\frac{64}{2(-16)} = 2.$$

The h-coordinate of the vertex is found by substitution:

$$h = -16(2)^2 + 64(2) + 2240 = 2304$$

Thus, the rocket attains its maximum height of 2304 ft 2 sec after blastoff.

b) When the rocket reaches the ground, $h = 0$.

$$\begin{aligned}
0 &= -16t^2 + 64t + 2240 \\
0 &= t^2 - 4t - 140 \qquad \text{Dividing by } -16 \\
0 &= (t - 14)(t + 10) \\
t - 14 &= 0 \quad \text{or} \quad t + 10 = 0 \\
t &= 14 \quad \text{or} \qquad t = -10
\end{aligned}$$

Negative time has no meaning in this problem. Thus, the rocket reaches the ground 14 sec after blastoff.

29. $P = -x^2 + 980x - 3000$

Since the coefficient of x^2 is negative, the value of P at the vertex is a maximum. The x-coordinate of the vertex is

$$x = -\frac{b}{2a} = -\frac{980}{2(-1)} = 490.$$

Thus, 490 doors should be produced and sold in order to maximize profit.

31. We look for an equation of the form $y = ax^2 + bx + c$. We use the three data points to find a, b, and c.

$$\begin{aligned}
4 &= a(1)^2 + b(1) + c, \\
6 &= a(-1)^2 + b(-1) + c, \\
16 &= a(-2)^2 + b(-2) + c, \text{ or}
\end{aligned}$$

$$\begin{aligned}
4 &= a + b + c, \\
6 &= a - b + c, \\
16 &= 4a - 2b + c
\end{aligned}$$

Solving, we get $a = 3$, $b = -1$, and $c = 2$. Then the equation is $y = 3x^2 - x + 2$.

33. We look for an equation of the form $y = ax^2 + bx + c$. We use the three data points to find a, b, and c.

$$-4 = a(1)^2 + b(1) + c,$$
$$-6 = a(2)^2 + b(2) + c,$$
$$-6 = a(3)^2 + b(3) + c, \text{ or}$$

$$-4 = a + b + c,$$
$$-6 = 4a + 2b + c,$$
$$-6 = 9a + 3b + c$$

Solving, we get $a = 1$, $b = -5$, and $c = 0$. Then the equation is $y = x^2 - 5x$.

35. a) We look for an equation of the form $y = ax^2 + bx + c$. We use the three data points to find a, b, and c.

$$1000 = a(1)^2 + b(1) + c,$$
$$2000 = a(2)^2 + b(2) + c,$$
$$8000 = a(3)^2 + b(3) + c, \text{ or}$$

$$1000 = a + b + c,$$
$$2000 = 4a + 2b + c,$$
$$8000 = 9a + 3b + c$$

Solving, we get $a = 2500$, $b = -6500$, and $c = 5000$. Then the equation is $y = 2500x^2 - 6500x + 5000$.

b) To predict the earnings for the fourth month, we substitute 4 for x and compute y:

$$y = 2500(4)^2 - 6500(4) + 5000 = 19,000$$

Thus, the earnings are $19,000.

37. a) We look for an equation of the form $y = ax^2 + bx + c$. We use the data points $(60, 200)$, $(80, 130)$, and $(100, 100)$ to find a, b, and c.

$$200 = a(60)^2 + b(60) + c,$$
$$130 = a(80)^2 + b(80) + c,$$
$$100 = a(100)^2 + b(100) + c, \text{ or}$$

$$200 = 3600a + 60b + c,$$
$$130 = 6400a + 80b + c,$$
$$100 = 10,000a + 100b + c$$

Solving, we get $a = 0.05$, $b = -10.5$, and $c = 650$. Then the equation is $y = 0.05x^2 - 10.5x + 650$.

b) We substitute 50 for x and compute y:

$$y = 0.05(50)^2 - 10.5(50) + 650 = 250$$

Thus, 250 daytime accidents occur at 50 km/h.

39. $\sqrt{9a^3}\sqrt{16ab^4} = \sqrt{144a^4b^4} = 12a^2b^2$

41.
$$\sqrt{5x - 4} + \sqrt{13 - x} = 7$$
$$\sqrt{5x - 4} = 7 - \sqrt{13 - x}$$
$$\text{Isolating one radical}$$
$$(\sqrt{5x - 4})^2 = (7 - \sqrt{13 - x})^2$$
$$\text{Principle of powers}$$
$$5x - 4 = 49 - 14\sqrt{13 - x} + (13 - x)$$
$$6x - 66 = -14\sqrt{13 - x}$$
$$3x - 33 = -7\sqrt{13 - x} \quad \text{Dividing by 2}$$
$$(3x - 33)^2 = (-7\sqrt{13 - x})^2$$
$$9x^2 - 198x + 1089 = 49(13 - x)$$
$$9x^2 - 198x + 1089 = 637 - 49x$$
$$9x^2 - 149x + 452 = 0$$
$$(9x - 113)(x - 4) = 0$$
$$9x - 113 = 0 \quad \text{ or } \quad x - 4 = 0$$
$$9x = 113 \quad \text{ or } \quad x = 4$$
$$x = \frac{113}{9} \quad \text{ or } \quad x = 4$$

The number 4 checks, but $\frac{113}{9}$ does not. The solution is 4.

43.
$$\sqrt{7x - 5} = \sqrt{4x + 7}$$
$$(\sqrt{7x - 5})^2 = (\sqrt{4x + 7})^2 \quad \text{Principle of powers}$$
$$7x - 5 = 4x + 7$$
$$3x = 12 \quad \text{Adding 5 and subtracting } 4x$$
$$x = 4$$

The number 4 checks and is the solution.

45. Graph $y = |x^2 + 6x + 4|$

$$y = |(x^2 + 6x + 9 - 9) + 4|$$
$$y = |(x^2 + 6x + 9) - 9 + 4|$$
$$y = |(x + 3)^2 - 5|$$

The graph will lie entirely on or above the x-axis since absolute value must be nonnegative. For values of x for which $(x + 3)^2 - 5 \geq 0$, the graph will be the same as the graph of $y = (x + 3)^2 - 5$. For values of x for which $(x + 3)^2 - 5 < 0$, the graph will be the reflection of the graph of $y = (x + 3)^2 - 5$ across the x-axis. We first graph $y = (x + 3)^2 - 5$:

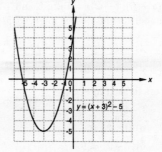

Then, reflecting the points below the axis across the x-axis, we get the graph of $y = |(x+3)^2 - 5|$, or $y = |x^2 + 6x + 4|$:

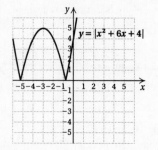

47. We make a drawing.

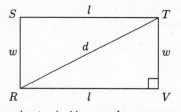

The perimeter is 44, so we have

$$2l + 2w = 44. \qquad (1)$$

Then we use the Pythagorean equation:

$$d^2 = l^2 + w^2. \qquad (2)$$

Solving (1) for w, we get $w = 22 - l$. Substituting in (2), we get

$$d^2 = l^2 + (22 - l)^2$$
$$d^2 = l^2 + 484 - 44l + l^2$$
$$d^2 = 2l^2 - 44l + 484$$

Since the coefficient of l^2 is positive, we know the value of d^2 at the vertex is a minimum. The l-coordinate of the vertex is

$$l = -\frac{b}{2a} = -\frac{-44}{2 \cdot 2} = 11.$$

We substitute to find the second coordinate of the vertex:

$$d^2 = 2(11)^2 - 44(11) + 484 = 242$$

The minimum value of d^2 is 242, so the minimum value of $d = \sqrt{242} = 11\sqrt{2}$ ft.

Exercise Set 8.9

1. $(x - 6)(x + 2) > 0$

The solutions of $(x - 6)(x + 2) = 0$ are 6 and -2. They divide the real-number line into three intervals as shown:

We try test numbers in each interval.

A: Test -3, $y = (-3 - 6)(-3 + 2) = 9$

B: Test 0, $y = (0 - 6)(0 + 2) = -12$

C: Test 7, $y = (7 - 6)(7 + 2) = 9$

The expression is positive for all values of x in intervals A and C. The solution set is $\{x | x < -2 \text{ or } x > 6\}$.

3. $3(x + 1)(x - 4) \leq 0$

The solutions of $3(x + 1)(x - 4) = 0$ are -1 and 4. They divide the real-number line into three intervals as shown:

We try test numbers in each interval.

A: Test -2, $y = 3(-2 + 1)(-2 - 4) = 18$

B: Test 0, $y = 3(0 + 1)(0 - 4) = -12$

C: Test 5, $y = 3(5 + 1)(5 - 4) = 18$

The expression is negative for all numbers in interval B. The inequality symbol is $\leq$, so we need to include the intercepts. The solution set is $\{x | -1 \leq x \leq 4\}$.

5. $x^2 - x - 2 < 0$

$(x + 1)(x - 2) < 0$ Factoring

The solutions of $(x + 1)(x - 2) = 0$ are -1 and 2. They divide the real-number line into three intervals as shown:

We try test numbers in each interval.

A: Test -2, $y = (-2 + 1)(-2 - 2) = 4$

B: Test 0, $y = (0 + 1)(0 - 2) = -2$

C: Test 3, $y = (3 + 1)(3 - 2) = 4$

The expression is negative for all numbers in interval B. The solution set is $\{x | -1 < x < 2\}$.

7. $9 - x^2 \leq 0$

$(3 + x)(3 - x) \leq 0$

The solutions of $(3 + x)(3 - x) = 0$ are -3 and 3. They divide the real-number line into three intervals as shown:

We try test numbers in each interval.

A: Test -4, $y = 9 - (-4)^2 = -7$

B: Test 0, $y = 9 - (0)^2 = 9$

C: Test 4, $y = 9 - (4)^2 = -7$

The expression is negative for all numbers in intervals A and C. The inequality symbol is $\leq$, so we need to include the intercepts. The solution set is $\{x | x \leq -3 \text{ or } x \geq 3\}$.

9. $x^2 - 2x + 1 \geq 0$

$(x - 1)^2 \geq 0$

The solution of $(x - 1)^2 = 0$ is 1. For all real-number values of x except 1, $(x - 1)^2$ will be positive. Thus the solution set is $\{x | x \text{ is a real number}\}$.

11.
$$x^2 + 8 < 6x$$
$$x^2 - 6x + 8 < 0$$
$$(x - 4)(x - 2) < 0$$

The solutions of $(x - 4)(x - 2) = 0$ are 4 and 2. They divide the real-number line into three intervals as shown:

We try test numbers in each interval.

A: Test 0, $y = (0 - 4)(0 - 2) = 8$
B: Test 3, $y = (3 - 4)(3 - 2) = -1$
C: Test 5, $y = (5 - 4)(5 - 2) = 3$

The expression is negative for all numbers in interval B. The solution set is $\{x | 2 < x < 4\}$.

13. $3x(x + 2)(x - 2) < 0$

The solutions of $3x(x + 2)(x - 2) = 0$ are 0, -2, and 2. They divide the real-number line into four intervals as shown:

We try test numbers in each interval.

A: Test -3, $y = 3(-3)(-3 + 2)(-3 - 2) = -45$
B: Test -1, $y = 3(-1)(-1 + 2)(-1 - 2) = 9$
C: Test 1, $y = 3(1)(1 + 2)(1 - 2) = -9$
D: Test 3, $y = 3(3)(3 + 2)(3 - 2) = 45$

The expression is negative for all numbers in intervals A and C. The solution set is $\{x | x < -2 \text{ or } 0 < x < 2\}$.

15. $(x + 9)(x - 4)(x + 1) > 0$

The solutions of $(x + 9)(x - 4)(x + 1) = 0$ are -9, 4, and -1. They divide the real-number line into four intervals as shown:

We try test numbers in each interval.

A: Test -10, $y = (-10 + 9)(-10 - 4)(-10 + 1) = -126$
B: Test -2, $y = (-2 + 9)(-2 - 4)(-2 + 1) = 42$
C: Test 0, $y = (0 + 9)(0 - 4)(0 + 1) = -36$
D: Test 5, $y = (5 + 9)(5 - 4)(5 + 1) = 84$

The expression is positive for all values of x in intervals B and D. The solution set is $\{x | -9 < x < -1 \text{ or } x > 4\}$.

17. $(x + 3)(x + 2)(x - 1) < 0$

The solutions of $(x + 3)(x + 2)(x - 1) = 0$ are -3, -2, and 1. They divide the real-number line into four intervals as shown:

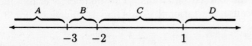

We try test numbers in each interval.

A: Test -4, $y = (-4 + 3)(-4 + 2)(-4 - 1) = -10$
B: Test $-\dfrac{5}{2}$, $y = \left(-\dfrac{5}{2} + 3\right)\left(-\dfrac{5}{2} + 2\right)\left(-\dfrac{5}{2} - 1\right) = \dfrac{7}{8}$
C: Test 0, $y = (0 + 3)(0 + 2)(0 - 1) = -6$
D: Test 2, $y = (2 + 3)(2 + 2)(2 - 1) = 20$

The expression is negative for all numbers in intervals A and C. The solution set is $\{x | x < -3 \text{ or } -2 < x < 1\}$.

19. $\dfrac{1}{x - 6} < 0$

We write the related equation by changing the $<$ symbol to $=$:

$$\frac{1}{x - 6} = 0$$

We solve the related equation.

$$(x - 6) \cdot \frac{1}{x - 6} = (x - 6) \cdot 0$$
$$1 = 0$$

We get a false equation, so the related equation has no solution.

Next we find the numbers for which the rational expression is undefined by setting the denominator equal to 0 and solving:

$$x - 6 = 0$$
$$x = 6$$

We use 6 to divide the number line into two intervals as shown:

We try test numbers in each interval.

A: Test 0,
$$\begin{array}{c|c} \dfrac{1}{x - 6} < 0 & \\ \hline \dfrac{1}{0 - 6} & 0 \\ \hline -\dfrac{1}{6} & \text{TRUE} \end{array}$$

The number 0 is a solution of the inequality, so the interval A is part of the solution set.

B: Test 7,
$$\begin{array}{c|c} \dfrac{1}{x - 6} < 0 & \\ \hline \dfrac{1}{7 - 6} & 0 \\ \hline 1 & \text{FALSE} \end{array}$$

The number 7 is not a solution of the inequality, so the interval B is not part of the solution set. The solution set is $\{x | x < 6\}$.

21. $\dfrac{x+1}{x-3} > 0$

Solve the related equation.

$$\frac{x+1}{x-3} = 0$$

$$x + 1 = 0$$

$$x = -1$$

Find the numbers for which the rational expression is undefined.

$$x - 3 = 0$$

$$x = 3$$

Use the numbers -1 and 3 to divide the number line into intervals as shown:

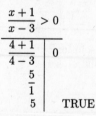

Try test numbers in each interval.

A: Test -2,

$$\frac{x+1}{x-3} > 0$$

$$\frac{-2+1}{-2-3} \;\Big|\; 0$$

$$\frac{-1}{-5}$$

$$\frac{1}{5} \;\Big|\; \text{TRUE}$$

The number -2 is a solution of the inequality, so the interval A is part of the solution set.

B: Test 0,

$$\frac{x+1}{x-3} > 0$$

$$\frac{0+1}{0-3} \;\Big|\; 0$$

$$-\frac{1}{3} \;\Big|\; \text{FALSE}$$

The number 0 is not a solution of the inequality, so the interval B is not part of the solution set.

C: Test 4,

$$\frac{x+1}{x-3} > 0$$

$$\frac{4+1}{4-3} \;\Big|\; 0$$

$$\frac{5}{1}$$

$$5 \;\Big|\; \text{TRUE}$$

The number 4 is a solution of the inequality, so the interval C is part of the solution set. The solution set is

$\{x \,|\, x < -1 \ or \ x > 3\}.$

23. $\dfrac{3x+2}{x-3} \le 0$

Solve the related equation.

$$\frac{3x+2}{x-3} = 0$$

$$3x + 2 = 0$$

$$3x = -2$$

$$x = -\frac{2}{3}$$

Find the numbers for which the rational expression is undefined.

$$x - 3 = 0$$

$$x = 3$$

Use the numbers $-\dfrac{2}{3}$ and 3 to divide the number line into intervals as shown:

Try test numbers in each interval.

A: Test -1,

$$\frac{3x+2}{x-3} \le 0$$

$$\frac{3(-1)+2}{-1-3} \;\Big|\; 0$$

$$\frac{-1}{-4}$$

$$\frac{1}{4} \;\Big|\; \text{FALSE}$$

The number -1 is not a solution of the inequality, so the interval A is not part of the solution set.

B: Test 0,

$$\frac{3x+2}{x-3} \le 0$$

$$\frac{3\cdot 0+2}{0-3} \;\Big|\; 0$$

$$\frac{2}{-3}$$

$$-\frac{2}{3} \;\Big|\; \text{TRUE}$$

The number 0 is a solution of the inequality, so the interval B is part of the solution set.

C: Test 4,

$$\frac{3x+2}{x-3} \le 0$$

$$\frac{3\cdot 4+2}{4-3} \;\Big|\; 0$$

$$14 \;\Big|\; \text{FALSE}$$

The number 4 is not a solution of the inequality, so the interval C is not part of the solution set. The solution set includes the interval B. The number $-\dfrac{2}{3}$ is also included since the inequality symbol is $\le$ and $-\dfrac{2}{3}$ is the solution of the related equation. The number 3 is not included because the rational expression is undefined for 3. The solution set is $\left\{x \,\Big|\, -\dfrac{2}{3} \le x < 3\right\}.$

25. $\dfrac{x-1}{x-2} > 3$

Solve the related equation.

$$\frac{x-1}{x-2} = 3$$
$$x - 1 = 3(x - 2)$$
$$x - 1 = 3x - 6$$
$$5 = 2x$$
$$\frac{5}{2} = x$$

Find the numbers for which the rational expression is undefined.

$$x - 2 = 0$$
$$x = 2$$

Use the numbers $\dfrac{5}{2}$ and 2 to divide the number line into intervals as shown:

$$\begin{array}{c} \overset{A}{\rule{0pt}{0pt}} \qquad \overset{B}{\rule{0pt}{0pt}} \qquad \overset{C}{\rule{0pt}{0pt}} \\ 2 \qquad \frac{5}{2} \end{array}$$

Try test numbers in each interval.

A: Test 0, $\dfrac{x-1}{x-2} > 3$

$$\begin{array}{c|c} \dfrac{0-1}{0-2} & 3 \\[2mm] \dfrac{1}{2} & \text{FALSE} \end{array}$$

The number 0 is not a solution of the inequality, so the interval A is not part of the solution set.

B: Test $\dfrac{9}{4}$, $\dfrac{x-1}{x-2} > 3$

$$\begin{array}{c|c} \dfrac{\frac{9}{4}-1}{\frac{9}{4}-2} & 3 \\[3mm] \dfrac{\frac{5}{4}}{\frac{1}{4}} & \\[3mm] 5 & \text{TRUE} \end{array}$$

The number $\dfrac{9}{4}$ is a solution of the inequality, so the interval B is part of the solution set.

C: Test 3, $\dfrac{x-1}{x-2} > 3$

$$\begin{array}{c|c} \dfrac{3-1}{3-2} & 3 \\[2mm] 2 & \text{FALSE} \end{array}$$

The number 3 is not a solution of the inequality, so the interval C is not part of the solution set. The solution set is $\left\{ x \middle| 2 < x < \dfrac{5}{2} \right\}$.

27. $\dfrac{(x-2)(x+1)}{x-5} < 0$

Solve the related equation.

$$\frac{(x-2)(x+1)}{x-5} = 0$$
$$(x-2)(x+1) = 0$$
$$x = 2 \text{ or } x = -1$$

Find the numbers for which the rational expression is undefined.

$$x - 5 = 0$$
$$x = 5$$

Use the numbers 2, -1, and 5 to divide the number line into intervals as shown:

$$\begin{array}{c} \overset{A}{\rule{0pt}{0pt}} \quad \overset{B}{\rule{0pt}{0pt}} \quad \overset{C}{\rule{0pt}{0pt}} \quad \overset{D}{\rule{0pt}{0pt}} \\ -1 \qquad 2 \qquad 5 \end{array}$$

Try test numbers in each interval.

A: Test -2, $\dfrac{(x-2)(x+1)}{x-5} < 0$

$$\begin{array}{c|c} \dfrac{(-2-2)(-2+1)}{-2-5} & 0 \\[3mm] \dfrac{-4(-1)}{-7} & \\[3mm] -\dfrac{4}{7} & \text{TRUE} \end{array}$$

Interval A is part of the solution set.

B: Test 0, $\dfrac{(x-2)(x+1)}{x-5} < 0$

$$\begin{array}{c|c} \dfrac{(0-2)(0+1)}{0-5} & 0 \\[3mm] \dfrac{-2 \cdot 1}{-5} & \\[3mm] \dfrac{2}{5} & \text{FALSE} \end{array}$$

Interval B is not part of the solution set.

C: Test 3, $\dfrac{(x-2)(x+1)}{x-5} < 0$

$$\begin{array}{c|c} \dfrac{(3-2)(3+1)}{3-5} & 0 \\[3mm] \dfrac{1 \cdot 4}{-2} & \\[3mm] -2 & \text{TRUE} \end{array}$$

Interval C is part of the solution set.

D: Test 6, $\dfrac{(x-2)(x+1)}{x-5} < 0$

$$\begin{array}{c|c} \dfrac{(6-2)(6+1)}{6-5} & 0 \\[3mm] \dfrac{4 \cdot 7}{1} & \\[3mm] 28 & \text{FALSE} \end{array}$$

Interval D is not part of the solution set.

The solution set is $\{x | x < -1 \text{ or } 2 < x < 5\}$.

29. $\dfrac{x+3}{x} \le 0$

Solve the related equation.

$$\dfrac{x+3}{x} = 0$$

$$x + 3 = 0$$

$$x = -3$$

Find the numbers for which the rational expression is undefined.

$$x = 0$$

Use the numbers -3 and 0 to divide the number line into intervals as shown:

Try test numbers in each interval.

A: Test -4,

$$\dfrac{x+3}{x} \le 0$$

$$\begin{array}{c|c} \dfrac{-4+3}{-4} & 0 \\ \hline \dfrac{1}{4} & \text{FALSE} \end{array}$$

Interval A is not part of the solution set.

B: Test -1,

$$\dfrac{x+3}{x} \le 0$$

$$\begin{array}{c|c} \dfrac{-1+3}{-1} & 0 \\ \hline -2 & \text{TRUE} \end{array}$$

Interval B is part of the solution set.

C: Test 1,

$$\dfrac{x+3}{x} \le 0$$

$$\begin{array}{c|c} \dfrac{1+3}{1} & 0 \\ \hline 4 & \text{FALSE} \end{array}$$

Interval C is not part of the solution set.

The solution set includes the interval B. The number -3 is also included since the inequality symbol is $\le$ and -3 is a solution of the related equation. The number 0 is not included because the rational expression is undefined for 0. The solution set is $\{x| -3 \le x < 0\}$.

31. $\dfrac{x}{x-1} > 2$

Solve the related equation.

$$\dfrac{x}{x-1} = 2$$

$$x = 2x - 2$$

$$2 = x$$

Find the numbers for which the rational expression is undefined.

$$x - 1 = 0$$

$$x = 1$$

Use the numbers 1 and 2 to divide the number line into intervals as shown:

Try test numbers in each interval.

A: Test 0,

$$\dfrac{x}{x-1} > 2$$

$$\begin{array}{c|c} \dfrac{0}{0-1} & 2 \\ \hline 0 & \text{FALSE} \end{array}$$

Interval A is not part of the solution set.

B: Test $\dfrac{3}{2}$,

$$\dfrac{x}{x-1} > 2$$

$$\begin{array}{c|c} \dfrac{\dfrac{3}{2}}{\dfrac{3}{2}-1} & 2 \\[2ex] \hline \dfrac{\dfrac{3}{2}}{\dfrac{1}{2}} & \\[1ex] 3 & \text{TRUE} \end{array}$$

Interval B is part of the solution set.

C: Test 3,

$$\dfrac{x}{x-1} > 2$$

$$\begin{array}{c|c} \dfrac{3}{3-1} & 2 \\ \hline \dfrac{3}{2} & \text{FALSE} \end{array}$$

Interval C is not part of the solution set.

The solution set is $\{x| 1 < x < 2\}$.

33. $\dfrac{x-1}{(x-3)(x+4)} < 0$

Solve the related equation.

$$\dfrac{x-1}{(x-3)(x+4)} = 0$$

$$x - 1 = 0$$

$$x = 1$$

Find the numbers for which the rational expression is undefined.

$$(x-3)(x+4) = 0$$

$$x = 3 \text{ or } x = -4$$

Use the numbers 1, 3, and -4 to divide the number line into intervals as shown:

Try test numbers in each interval.

A: Test −5,

$$\frac{x-1}{(x-3)(x+4)} < 0$$

$$\frac{-5-1}{(-5-3)(-5+4)} \bigg| 0$$

$$\frac{-6}{-8(-1)}$$

$$-\frac{3}{4} \bigg|\ \text{TRUE}$$

Interval *A* is part of the solution set.

B: Test 0,

$$\frac{x-1}{(x-3)(x+4)} < 0$$

$$\frac{0-1}{(0-3)(0+4)} \bigg| 0$$

$$\frac{-1}{-3 \cdot 4}$$

$$\frac{1}{12} \bigg|\ \text{FALSE}$$

Interval *B* is not part of the solution set.

C: Test 2,

$$\frac{x-1}{(x-3)(x+4)} < 0$$

$$\frac{2-1}{(2-3)(2+4)} \bigg| 0$$

$$\frac{1}{-1 \cdot 6}$$

$$-\frac{1}{6} \bigg|\ \text{TRUE}$$

Interval *C* is part of the solution set.

D: Test 4,

$$\frac{x-1}{(x-3)(x+4)} < 0$$

$$\frac{4-1}{(4-3)(4+4)} \bigg| 0$$

$$\frac{3}{1 \cdot 8}$$

$$\frac{3}{8} \bigg|\ \text{FALSE}$$

Interval *D* is not part of the solution set.

The solution set is $\{x | x < -4 \ or \ 1 < x < 3\}$.

35. $3 < \dfrac{1}{x}$

Solve the related equation.

$$3 = \frac{1}{x}$$

$$x = \frac{1}{3}$$

Find the numbers for which the rational expression is undefined.

$$x = 0$$

Use the numbers 0 and $\dfrac{1}{3}$ to divide the number line into intervals as shown:

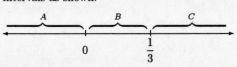

Try test numbers in each interval.

A: Test −1,

$$3 < \frac{1}{x}$$

$$3 \bigg| \frac{1}{-1}$$

$$\bigg| -1 \quad \text{FALSE}$$

Interval *A* is not part of the solution set.

B: Test $\dfrac{1}{6}$,

$$3 < \frac{1}{x}$$

$$3 \Bigg| \frac{1}{\frac{1}{6}}$$

$$\bigg| 6 \quad \text{TRUE}$$

Interval *B* is part of the solution set.

C: Test 1,

$$3 < \frac{1}{x}$$

$$3 \bigg| \frac{1}{1}$$

$$\bigg| 1 \quad \text{FALSE}$$

Interval *C* is not part of the solution set.

The solution set is $\left\{x \middle| 0 < x < \dfrac{1}{3}\right\}$.

37. $\dfrac{(x-1)(x+2)}{(x+3)(x-4)} > 0$

Solve the related equation.

$$\frac{(x-1)(x+2)}{(x+3)(x-4)} = 0$$

$$(x-1)(x+2) = 0$$

$$x = 1 \text{ or } x = -2$$

Find the numbers for which the rational expression is undefined.

$$(x+3)(x-4) = 0$$

$$x = -3 \text{ or } x = 4$$

Use the numbers 1, −2, −3, and 4 to divide the number line into intervals as shown:

Try test numbers in each interval.

A: Test −4,

$$\frac{(x-1)(x+2)}{(x+3)(x-4)} > 0$$

$$\frac{(-4-1)(-4+2)}{(-4+3)(-4-4)} \bigg| 0$$

$$\frac{-5(-2)}{-1(-8)}$$

$$\frac{5}{4} \bigg|\ \text{TRUE}$$

Interval *A* is part of the solution set.

B: Test $-\dfrac{5}{2}$,

$$\frac{(x-1)(x+2)}{(x+3)(x-4)} > 0$$

$$\begin{array}{c|c} \dfrac{\left(-\dfrac{5}{2}-1\right)\left(-\dfrac{5}{2}+2\right)}{\left(-\dfrac{5}{2}+3\right)\left(-\dfrac{5}{2}-4\right)} & 0 \\ \dfrac{-\dfrac{7}{2}\left(-\dfrac{1}{2}\right)}{\dfrac{1}{2}\left(-\dfrac{13}{2}\right)} & \\ -\dfrac{7}{13} & \text{FALSE} \end{array}$$

Interval B is not part of the solution set.

C: Test 1,

$$\frac{(x-1)(x+2)}{(x+3)(x-4)} > 0$$

$$\begin{array}{c|c} \dfrac{(0-1)(0+2)}{(0+3)(0-4)} & 0 \\ \dfrac{-1\cdot 2}{3(-4)} & \\ \dfrac{1}{6} & \text{TRUE} \end{array}$$

Interval C is part of the solution set.

D: Test 2,

$$\frac{(x-1)(x+2)}{(x+3)(x-4)} > 0$$

$$\begin{array}{c|c} \dfrac{(2-1)(2+2)}{(2+3)(2-4)} & 0 \\ \dfrac{1\cdot 4}{5(-2)} & \\ -\dfrac{2}{5} & \text{FALSE} \end{array}$$

Interval D is not part of the solution set.

E: Test 5,

$$\frac{(x-1)(x+2)}{(x+3)(x-4)} > 0$$

$$\begin{array}{c|c} \dfrac{(5-1)(5+2)}{(5+3)(5-4)} & 0 \\ \dfrac{4\cdot 7}{8\cdot 1} & \\ \dfrac{7}{2} & \text{TRUE} \end{array}$$

Interval E is part of the solution set.

The solution set is $\{x|x<-3\ or\ -2<x<1\ or\ x>4\}$.

39. $\dfrac{x^2+3x-10}{x^2-x-56} \le 0$

$\dfrac{(x+5)(x-2)}{(x-8)(x+7)} \le 0$

Solve the related equation.

$\dfrac{(x+5)(x-2)}{(x-8)(x+7)} = 0$

$(x+5)(x-2) = 0$

$x = -5$ or $x = 2$

Find the numbers for which the rational expression is undefined.

$(x-8)(x+7) = 0$

$x = 8$ or $x = -7$

Use the numbers -5, 2, 8, and -7 to divide the number line into intervals as shown:

Try test numbers in each interval.

A: Test -8,

$$\frac{(x+5)(x-2)}{(x-8)(x+7)} \le 0$$

$$\begin{array}{c|c} \dfrac{(-8+5)(-8-2)}{(-8-8)(-8+7)} & 0 \\ \dfrac{-3(-10)}{-16(-1)} & \\ \dfrac{15}{8} & \text{FALSE} \end{array}$$

Interval A is not part of the solution set.

B: Test -6,

$$\frac{(x+5)(x-2)}{(x-8)(x+7)} \le 0$$

$$\begin{array}{c|c} \dfrac{(-6+5)(-6-2)}{(-6-8)(-6+7)} & 0 \\ \dfrac{-1(-8)}{-14\cdot 1} & \\ -\dfrac{4}{7} & \text{TRUE} \end{array}$$

Interval B is part of the solution set.

C: Test 0,

$$\frac{(x+5)(x-2)}{(x-8)(x+7)} \le 0$$

$$\begin{array}{c|c} \dfrac{(0+5)(0-2)}{(0-8)(0+7)} & 0 \\ \dfrac{5(-2)}{-8\cdot 7} & \\ \dfrac{5}{28} & \text{FALSE} \end{array}$$

Interval C is not part of the solution set.

D: Test 3,

$$\frac{(x+5)(x-2)}{(x-8)(x+7)} \le 0$$

$$\begin{array}{c|c} \dfrac{(3+5)(3-2)}{(3-8)(3+7)} & 0 \\ \dfrac{8\cdot 1}{-5\cdot 10} & \\ -\dfrac{4}{25} & \text{TRUE} \end{array}$$

Interval D is part of the solution set.

E: Test 9,

$$\frac{(x+5)(x-2)}{(x-8)(x+7)} \le 0$$

$$\frac{(9+5)(9-2)}{(9-8)(9+7)} \,\Big|\, 0$$

$$\frac{14 \cdot 7}{1 \cdot 16}$$

$$\frac{49}{8} \quad \text{FALSE}$$

Interval E is not part of the solution set.

The solution set includes intervals B and D. The numbers -5 and 2 are also included since the inequality symbol is $\le$ and -5 and 2 are solutions of the related equation. The numbers -7 and 8 are not included because the rational expression is undefined for -7 and 8. The solution set is $\{x| -7 < x \le -5 \text{ or } 2 \le x < 8\}$.

41.
$$x^2 - 2x \le 2$$
$$x^2 - 2x - 2 \le 0$$

The solutions of $x^2 - 2x - 2 = 0$ are found using the quadratic formula. They are $1 \pm \sqrt{3}$, or about 2.7 and -0.7. These numbers divide the number line into three intervals as shown:

We try test numbers in each interval.

A: Test -1, $y = (-1)^2 - 2(-1) - 2 = 1$

B: Test 0, $y = 0^2 - 2 \cdot 0 - 2 = -2$

C: Test 3, $y = 3^2 - 2 \cdot 3 - 2 = 1$

The expression is negative for all values of x in interval B. The inequality symbol is $\le$, so we must also include the intercepts. The solution set is $\{x|1 - \sqrt{3} \le x \le 1 + \sqrt{3}\}$.

43.
$$x^4 + 2x^2 > 0$$
$$x^2(x^2 + 2) > 0$$

$x^2 > 0$ for all $x \ne 0$, and $x^2 + 2 > 0$ for all values of x. Then $x^2(x^2 + 2) > 0$ for all $x \ne 0$. The solution set is $\{x|x \ne 0\}$, or the set of all real numbers except 0.

45.
$$\left|\frac{x+2}{x-1}\right| < 3$$
$$-3 < \frac{x+2}{x-1} < 3$$

We rewrite the inequality using "and."

$$-3 < \frac{x+2}{x-1} \text{ and } \frac{x+2}{x-1} < 3$$

We will solve each inequality and then find the intersection of their solution sets.

Solve: $-3 < \dfrac{x+2}{x-1}$

Solve the related equation.

$$-3 = \frac{x+2}{x-1}$$
$$-3x + 3 = x + 2$$
$$1 = 4x$$
$$\frac{1}{4} = x$$

Find the numbers for which the rational expression is undefined.

$$x - 1 = 0$$
$$x = 1$$

Use the numbers $\dfrac{1}{4}$ and 1 to divide the number line into intervals as shown:

Try test numbers in each interval.

A: Test 0,

$$-3 < \frac{x+2}{x-1}$$

$$-3 \,\Big|\, \frac{0+2}{0-1}$$

$$-2 \quad \text{TRUE}$$

Interval A is part of the solution set.

B: Test $\dfrac{1}{2}$,

$$-3 < \frac{x+2}{x-1}$$

$$-3 \,\Bigg|\, \frac{\frac{1}{2}+2}{\frac{1}{2}-1}$$

$$\frac{\frac{5}{2}}{-\frac{1}{2}}$$

$$-5 \quad \text{FALSE}$$

Interval B is not part of the solution set.

C: Test 2,

$$-3 < \frac{x+2}{x-1}$$

$$-3 \,\Big|\, \frac{2+2}{2-1}$$

$$4 \quad \text{TRUE}$$

Interval C is part of the solution set.

The solution set of $-3 < \dfrac{x+2}{x-1}$ is $\left\{x\Big|x < \dfrac{1}{4} \text{ or } x > 1\right\}$.

Solve: $\dfrac{x+2}{x-1} > 3$

Solve the related equation.

$$\frac{x+2}{x-1} = 3$$

$$x + 2 = 3x - 3$$

$$5 = 2x$$

$$\frac{5}{2} = x$$

From our work above we know that the rational expression is undefined for 1.

Use the numbers $\dfrac{5}{2}$ and 1 to divide the number line into intervals as shown:

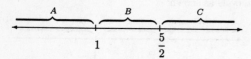

Try test numbers in each interval.

A: Test 0,
$$\frac{x+2}{x-1} < 3$$
$$\begin{array}{c|c} \dfrac{0+2}{0-1} & 3 \\ \hline -2 & \text{TRUE} \end{array}$$

Interval A is part of the solution set.

B: Test 2,
$$\frac{x+2}{x-1} < 3$$
$$\begin{array}{c|c} \dfrac{2+2}{2-1} & 3 \\ \hline 4 & \text{FALSE} \end{array}$$

Interval B is not part of the solution set.

C: Test 3,
$$\frac{x+2}{x-1} < 3$$
$$\begin{array}{c|c} \dfrac{3+2}{3-1} & 3 \\ \hline \dfrac{5}{2} & \text{TRUE} \end{array}$$

Interval C is part of the solution set.

The solution set of $\dfrac{x+2}{x-1} < 3$ is $\left\{ x \middle| x < 1 \text{ or } x > \dfrac{5}{2} \right\}$.

The solution set of the original inequality is
$$\left\{ x \middle| x < \frac{1}{4} \text{ or } x > 1 \right\} \cap \left\{ x \middle| x < 1 \text{ or } x > \frac{5}{2} \right\}, \text{ or}$$
$$\left\{ x \middle| x < \frac{1}{4} \text{ or } x > \frac{5}{2} \right\}.$$

47. a) Solve: $-16t^2 + 32t + 1920 > 1920$

$$-16t^2 + 32t > 0$$

$$t^2 - 2t < 0$$

$$t(t-2) < 0$$

The solutions of $t(t-2) = 0$ are 0 and 2. They divide the number line into three intervals as shown:

Try test numbers in each interval.

A: Test -1, $y = -1(-1-2) = 3$

B: Test 1, $y = 1(1-2) = -1$

C: Test 3, $y = 3(3-2) = 3$

The expression is negative for all values of t in interval B. The solution set is $\{t | 0 < t < 2\}$.

b) Solve: $-16t^2 + 32t + 1920 < 640$

$$-16t^2 + 32t + 1280 < 0$$

$$t^2 - 2t - 80 > 0$$

$$(t-10)(t+8) > 0$$

The solutions of $(t-10)(t+8) = 0$ are 10 and -8. They divide the number line into three intervals as shown:

Try test numbers in each interval.

A: Test -10, $y = (-10-10)(-10+8) = 40$

B: Test 0, $y = (0-10)(0+8) = -80$

C: Test 20, $y = (20-10)(20+8) = 80 = 280$

The expression is positive for all values of t in intervals A and C. However, since negative values of t have no meaning in this problem, we disregard interval A. Thus, the solution set is $\{t | t > 10\}$.

Chapter 9

Conic Sections, Relations, and Functions

1. Graph: $y = x^2$

The graph is a parabola. The vertex is $(0,0)$; the line of symmetry is $x = 0$. The curve opens upward. We choose some x-values on both sides of the vertex and compute the corresponding y-values. Then we plot the points and graph the parabola.

x	y
0	0
1	1
-1	1
2	4
-2	4

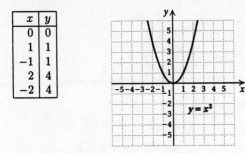

3. Graph: $x = y^2 + 4y + 1$

We complete the square.

$$x = (y^2 + 4y + 4 - 4) + 1$$
$$= (y^2 + 4y + 4) - 4 + 1$$
$$= (y + 2)^2 - 3, \text{ or}$$
$$= [y - (-2)]^2 + (-3)$$

The graph is a parabola. The vertex is $(-3, -2)$; the line of symmetry is $y = -2$. The curve opens to the right.

x	y
-3	-2
-2	-3
-2	-1
1	-4
1	0

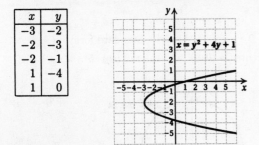

5. Graph: $y = -x^2 + 4x - 5$

We use the formula to find the first coordinate of the vertex:

$$x = -\frac{b}{2a} = -\frac{4}{2(-1)} = 2$$

Then $y = -x^2 + 4x - 5 = -(2)^2 + 4(2) - 5 = -1$.

The vertex is $(2, -1)$; the line of symmetry is $x = 2$. The curve opens downward.

x	y
2	-1
1	-2
3	-2
0	-5
4	-5

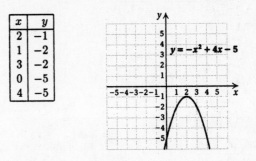

7. Graph: $x = -3y^2 - 6y - 1$

We complete the square.

$$x = -3(y^2 + 2y) - 1$$
$$= -3(y^2 + 2y + 1 - 1) - 1$$
$$= -3(y^2 + 2y + 1) + 3 - 1$$
$$= -3(y + 1)^2 + 2$$
$$= -3[y - (-1)]^2 + 2$$

The graph is a parabola. The vertex is $(2, -1)$; the line of symmetry is $y = -1$. The curve opens to the left.

x	y
2	-1
-1	-2
-1	0
-10	-3
-10	1

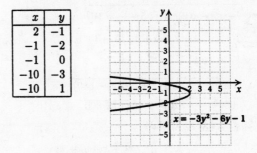

9. $d = \sqrt{(x_2 - x_1)^2 + (y_2 - y_1)^2}$

$d = \sqrt{(2 - 6)^2 + [-7 - (-4)]^2}$ Substituting

$= \sqrt{(-4)^2 + (-3)^2}$

$= \sqrt{25} = 5$

11. $d = \sqrt{(x_2 - x_1)^2 + (y_2 - y_1)^2}$

$d = \sqrt{(5 - 0)^2 + [-6 - (-4)]^2}$

$= \sqrt{5^2 + (-2)^2}$

$= \sqrt{29} \approx 5.385$

13. $d = \sqrt{(x_2 - x_1)^2 + (y_2 - y_1)^2}$

$d = \sqrt{(-9 - 9)^2 + (-9 - 9)^2}$

$= \sqrt{(-18)^2 + (-18)^2}$

$= \sqrt{648} = \sqrt{324 \cdot 2}$

$= 18\sqrt{2} \approx 25.456$

15. $d = \sqrt{(x_2 - x_1)^2 + (y_2 - y_1)^2}$

$d = \sqrt{(-4.3 - 2.8)^2 + [-3.5 - (-3.5)]^2}$

$= \sqrt{(-7.1)^2 + 0^2} = \sqrt{(-7.1)^2}$

$= 7.1$

17. $d = \sqrt{(x_2 - x_1)^2 + (y_2 - y_1)^2}$

$d = \sqrt{\left(\dfrac{5}{7} - \dfrac{1}{7}\right)^2 + \left(\dfrac{1}{14} - \dfrac{11}{14}\right)^2}$

$= \sqrt{\left(\dfrac{4}{7}\right)^2 + \left(-\dfrac{5}{7}\right)^2}$

$= \sqrt{\dfrac{16}{49} + \dfrac{25}{49}}$

$= \sqrt{\dfrac{41}{49}}$

$= \dfrac{\sqrt{41}}{7} \approx 0.915$

19. $d = \sqrt{[56 - (-23)]^2 + (-17 - 10)^2}$

$= \sqrt{79^2 + (-27)^2} = \sqrt{6970} \approx 83.487$

21. $d = \sqrt{(a - 0)^2 + (b - 0)^2}$

$= \sqrt{a^2 + b^2}$

23. $d = \sqrt{(-\sqrt{7} - \sqrt{2})^2 + [\sqrt{5} - (-\sqrt{3})]^2}$

$= \sqrt{7 + 2\sqrt{14} + 2 + 5 + 2\sqrt{15} + 3}$

$= \sqrt{17 + 2\sqrt{14} + 2\sqrt{15}} \approx 5.677$

25. $d = \sqrt{[1000 - (-2000)]^2 + (-240 - 580)^2}$

$= \sqrt{3000^2 + (-820)^2} = \sqrt{9,672,400}$

$= 20\sqrt{24,181} \approx 3110.048$

27. Using the midpoint formula $\left(\dfrac{x_1 + x_2}{2}, \dfrac{y_1 + y_2}{2}\right)$, we obtain

$\left(\dfrac{-1 + 4}{2}, \dfrac{9 + (-2)}{2}\right)$, or $\left(\dfrac{3}{2}, \dfrac{7}{2}\right)$.

29. Using the midpoint formula $\left(\dfrac{x_1 + x_2}{2}, \dfrac{y_1 + y_2}{2}\right)$, we obtain

$\left(\dfrac{3 + (-3)}{2}, \dfrac{5 + 6}{2}\right)$, or $\left(\dfrac{0}{2}, \dfrac{11}{2}\right)$, or $\left(0, \dfrac{11}{2}\right)$.

31. Using the midpoint formula $\left(\dfrac{x_1 + x_2}{2}, \dfrac{y_1 + y_2}{2}\right)$, we obtain

$\left(\dfrac{-10 + 8}{2}, \dfrac{-13 + (-4)}{2}\right)$, or $\left(\dfrac{-2}{2}, \dfrac{-17}{2}\right)$, or $\left(-1, -\dfrac{17}{2}\right)$.

33. Using the midpoint formula $\left(\dfrac{x_1 + x_2}{2}, \dfrac{y_1 + y_2}{2}\right)$, we obtain

$\left(\dfrac{-3.4 + 2.9}{2}, \dfrac{8.1 + (-8.7)}{2}\right)$, or $\left(\dfrac{-0.5}{2}, \dfrac{-0.6}{2}\right)$, or $(-0.25, -0.3)$.

35. Using the midpoint formula $\left(\dfrac{x_1 + x_2}{2}, \dfrac{y_1 + y_2}{2}\right)$, we obtain

$\left(\dfrac{\frac{1}{6} + \left(-\frac{1}{3}\right)}{2}, \dfrac{-\frac{3}{4} + \frac{5}{6}}{2}\right)$, or $\left(\dfrac{-\frac{1}{6}}{2}, \dfrac{\frac{1}{12}}{2}\right)$, or $\left(-\dfrac{1}{12}, \dfrac{1}{24}\right)$.

37. Using the midpoint formula $\left(\dfrac{x_1 + x_2}{2}, \dfrac{y_1 + y_2}{2}\right)$, we obtain

$\left(\dfrac{\sqrt{2} + \sqrt{3}}{2}, \dfrac{-1 + 4}{2}\right)$, or $\left(\dfrac{\sqrt{2} + \sqrt{3}}{2}, \dfrac{3}{2}\right)$.

39. $\quad (x + 1)^2 + (y + 3)^2 = 4$

$[x - (-1)]^2 + [y - (-3)]^2 = 2^2 \quad$ Standard form

The center is $(-1, -3)$, and the radius is 2.

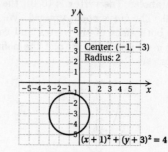

41. $\quad (x - 3)^2 + y^2 = 2$

$(x - 3)^2 + (y - 0)^2 = (\sqrt{2})^2 \quad$ Standard form

The center is $(3, 0)$, and the radius is $\sqrt{2}$.

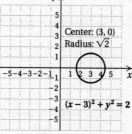

43. $\quad x^2 + y^2 = 25$

$(x - 0)^2 + (y - 0)^2 = 5^2 \quad$ Standard form

The center is $(0, 0)$, and the radius is 5.

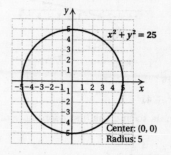

45. $(x - h)^2 + (y - k)^2 = r^2 \quad$ Standard form

$(x - 0)^2 + (y - 0)^2 = 7^2 \quad$ Substituting

$\qquad x^2 + y^2 = 49$

47. $(x - h)^2 + (y - k)^2 = r^2 \qquad$ Standard form

$[x - (-5)]^2 + (y - 3)^2 = (\sqrt{7})^2 \quad$ Substituting

$\qquad (x + 5)^2 + (y - 3)^2 = 7$

49.
$$x^2 + y^2 + 8x - 6y - 15 = 0$$
$$(x^2 + 8x) + (y^2 - 6y) - 15 = 0$$
Regrouping
$$(x^2 + 8x + 16 - 16) + (y^2 - 6y + 9 - 9) - 15 = 0$$
Completing the square twice
$$(x^2 + 8x + 16) + (y^2 - 6y + 9) - 16 - 9 - 15 = 0$$
$$(x + 4)^2 + (y - 3)^2 = 40$$
$$[x - (-4)]^2 + (y - 3)^2 = (\sqrt{40})^2$$
$$[x - (-4)]^2 + (y - 3)^2 = (2\sqrt{10})^2$$
The center is $(-4, 3)$, and the radius is $2\sqrt{10}$.

51.
$$x^2 + y^2 - 8x + 2y + 13 = 0$$
$$(x^2 - 8x) + (y^2 + 2y) + 13 = 0$$
Regrouping
$$(x^2 - 8x + 16 - 16) + (y^2 + 2y + 1 - 1) + 13 = 0$$
Completing the square twice
$$(x^2 - 8x + 16) + (y^2 + 2y + 1) - 16 - 1 + 13 = 0$$
$$(x - 4)^2 + (y + 1)^2 = 4$$
$$(x - 4)^2 + [y - (-1)]^2 = 2^2$$
The center is $(4, -1)$, and the radius is 2.

53.
$$x^2 + y^2 - 4x = 0$$
$$(x^2 - 4x) + y^2 = 0$$
$$(x^2 - 4x + 4 - 4) + y^2 = 0$$
$$(x^2 - 4x + 4) + y^2 - 4 = 0$$
$$(x - 2)^2 + y^2 = 4$$
$$(x - 2)^2 + (y - 0)^2 = 2^2$$
The center is $(2, 0)$, and the radius is 2.

55. We use the elimination method.
$$\begin{array}{rl} x + y = & 8 \quad (1) \\ x - y = & -24 \quad (2) \\ \hline 2x = & -16 \quad \text{Adding} \\ x = & -8 \end{array}$$

Substitute -8 for x in one of the original equations and solve for y.
$$x + y = 8 \quad (1)$$
$$-8 + y = 8 \quad \text{Substituting}$$
$$y = 16$$
The solution is $(-8, 16)$.

57. $2x + 3y = 8, \quad (1)$
$\quad\quad x - 2y = -3 \quad (2)$
We use the elimination method.
$$\begin{array}{rl} 2x + 3y = & 8 \\ -2x + 4y = & 6 \quad \text{Multiplying (2) by } -2 \\ \hline 7y = & 14 \quad \text{Adding} \\ y = & 2 \end{array}$$

Substitute 2 for y in either of the original equations and solve for x.
$$x - 2y = -3 \quad (2)$$
$$x - 2 \cdot 2 = -3 \quad \text{Substituting}$$
$$x - 4 = -3$$
$$x = 1$$
The solution is $(1, 2)$.

59. $-4x + 12y = -9, \quad (1)$
$\quad\quad x - 3y = 2 \quad (2)$
We use the elimination method.
$$\begin{array}{rl} -4x + 12y = & -9 \quad (1) \\ -4x + 12y = & -8 \quad \text{Multiplying (2) by } -4 \\ \hline 0 = & -17 \quad \text{Adding} \end{array}$$

We get a false equation. The system of equations has no solution.

61. We first find the length of the radius, which is the distance between $(0, 0)$ and $\left(\frac{1}{4}, \frac{\sqrt{31}}{4}\right)$.
$$r = \sqrt{\left(\frac{1}{4} - 0\right)^2 + \left(\frac{\sqrt{31}}{4} - 0\right)^2}$$
$$= \sqrt{\left(\frac{1}{4}\right)^2 + \left(\frac{\sqrt{31}}{4}\right)^2}$$
$$= \sqrt{\frac{1}{16} + \frac{31}{16}}$$
$$= \sqrt{\frac{32}{16}}$$
$$= \sqrt{2}$$
$$(x - h)^2 + (y - k)^2 = r^2 \quad \text{Standard form}$$
$$(x - 0)^2 + (y - 0)^2 = (\sqrt{2})^2$$
Substituting $(0, 0)$ for the center and $\sqrt{2}$ for the radius
$$x^2 + y^2 = 2$$

63.

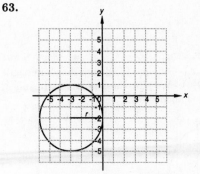

The center is $(-3, -2)$ and the radius is 3.
$$(x - h)^2 + (y - k)^2 = r^2 \quad \text{Standard form}$$
$$[x - (-3)]^2 + [y - (-2)]^2 = 3^2 \quad \text{Substituting}$$
$$(x + 3)^2 + (y + 2)^2 = 9$$

65. $d = \sqrt{(x_2 - x_1)^2 + (y_2 - y_1)^2}$

$ = \sqrt{(-1 - 6)^2 + (3k - 2k)^2}$ Substituting

$ = \sqrt{(-7)^2 + (k)^2}$

$ = \sqrt{49 + k^2}$

67. $d = \sqrt{(x_2 - x_1)^2 + (y_2 - y_1)^2}$

$ = \sqrt{[6m - (-2m)]^2 + (-7n - n)^2}$ Substituting

$ = \sqrt{(8m)^2 + (-8n)^2}$

$ = \sqrt{64m^2 + 64n^2}$

$ = \sqrt{64(m^2 + n^2)}$

$ = 8\sqrt{m^2 + n^2}$

69. $d = \sqrt{(x_2 - x_1)^2 + (y_2 - y_1)^2}$

$ = \sqrt{(-3\sqrt{3} - \sqrt{3})^2 + (-\sqrt{6} - \sqrt{6})^2}$ Substituting

$ = \sqrt{(-4\sqrt{3})^2 + (-2\sqrt{6})^2}$

$ = \sqrt{48 + 24}$

$ = \sqrt{72}$

$ = \sqrt{36 \cdot 2}$

$ = 6\sqrt{2}$

71. The distance between $(-8, -5)$ and $(6, 1)$ is

$\sqrt{(-8 - 6)^2 + (-5 - 1)^2} = \sqrt{196 + 36} = \sqrt{232}$.

The distance between $(6, 1)$ and $(-4, 5)$ is

$\sqrt{[6 - (-4)]^2 + (1 - 5)^2} = \sqrt{100 + 16} = \sqrt{116}$.

The distance between $(-4, 5)$ and $(-8, -5)$ is

$\sqrt{[-4 - (-8)]^2 + [5 - (-5)]^2} = \sqrt{16 + 100} = \sqrt{116}$.

Since $(\sqrt{116})^2 + (\sqrt{116})^2 = (\sqrt{232})^2$, the points are vertices of a right triangle.

73. $\left(\dfrac{(2 - \sqrt{3}) + (2 + \sqrt{3})}{2}, \dfrac{5\sqrt{2} + 3\sqrt{2}}{2}\right)$, or $\left(\dfrac{4}{2}, \dfrac{8\sqrt{2}}{2}\right)$, or

$(2, 4\sqrt{2})$

Exercise Set 9.2

1. $\dfrac{x^2}{9} + \dfrac{y^2}{36} = 1$

$\dfrac{x^2}{3^2} + \dfrac{y^2}{6^2} = 1$

The x-intercepts are $(-3, 0)$ and $(3, 0)$, and the y-intercepts are $(0, -6)$ and $(0, 6)$. We plot these points and connect them with an oval-shaped curve.

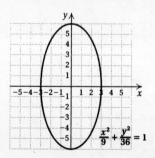

3. $\dfrac{x^2}{1} + \dfrac{y^2}{4} = 1$

$\dfrac{x^2}{1^2} + \dfrac{y^2}{2^2} = 1$

The x-intercepts are $(-1, 0)$ and $(1, 0)$, and the y-intercepts are $(0, -2)$ and $(0, 2)$. We plot these points and connect them with an oval-shaped curve.

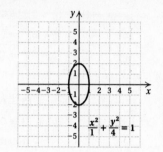

5. $4x^2 + 9y^2 = 36$

$\dfrac{x^2}{9} + \dfrac{y^2}{4} = 1$ Dividing by 36

$\dfrac{x^2}{3^2} + \dfrac{y^2}{2^2} = 1$

The x-intercepts are $(-3, 0)$ and $(3, 0)$, and the y-intercepts are $(0, -2)$ and $(0, 2)$. We plot these points and connect them with an oval-shaped curve.

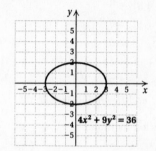

7. $x^2 + 4y^2 = 4$

$\dfrac{x^2}{4} + \dfrac{y^2}{1} = 1$ Dividing by 4

$\dfrac{x^2}{2^2} + \dfrac{y^2}{1^2} = 1$

The x-intercepts are $(-2, 0)$ and $(2, 0)$, and the y-intercepts are $(0, -1)$ and $(0, 1)$. We plot these points and connect them with an oval-shaped curve.

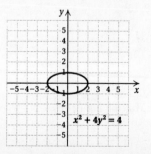

9. $\dfrac{x^2}{16} - \dfrac{y^2}{16} = 1$

$\dfrac{x^2}{4^2} - \dfrac{y^2}{4^2} = 1$

a) $a = 4$ and $b = 4$, so the asymptotes are $y = \dfrac{4}{4}x$ and
$y = -\dfrac{4}{4}x$, or $y = x$ and $y = -x$. We sketch them.

b) If we let $y = 0$, we get $x = \pm 4$, so the intercepts are
$(4, 0)$ and $(-4, 0)$.

c) We plot the intercepts and draw smooth curves through
them that approach the asymptotes.

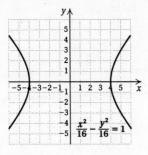

11. $\dfrac{y^2}{16} - \dfrac{x^2}{9} = 1$

$\dfrac{y^2}{4^2} - \dfrac{x^2}{3^2} = 1$

a) $a = 3$ and $b = 4$, so the asymptotes are $y = \dfrac{4}{3}x$ and
$y = -\dfrac{4}{3}x$. We sketch them.

b) If we let $x = 0$, we get $y = \pm 4$, so the intercepts are
$(0, 4)$ and $(0, -4)$.

c) We plot the intercepts and draw smooth curves through
them that approach the asymptotes.

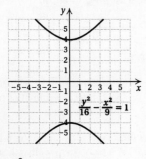

13. $\dfrac{x^2}{25} - \dfrac{y^2}{36} = 1$

$\dfrac{x^2}{5^2} - \dfrac{y^2}{6^2} = 1$

a) $a = 5$ and $b = 6$, so the asymptotes are $y = \dfrac{6}{5}x$ and
$y = -\dfrac{6}{5}x$. We sketch them.

b) If we let $y = 0$, we get $x = \pm 5$, so the intercepts are
$(5, 0)$ and $(-5, 0)$.

c) We plot the intercepts and draw smooth curves through
them that approach the asymptotes.

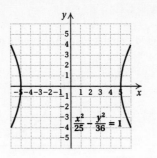

15. $x^2 - y^2 = 4$

$\dfrac{x^2}{4} - \dfrac{y^2}{4} = 1$ Dividing by 4

$\dfrac{x^2}{2^2} - \dfrac{y^2}{2^2} = 1$

a) $a = 2$ and $b = 2$, so the asymptotes are $y = \dfrac{2}{2}x$ and
$y = -\dfrac{2}{2}x$, or $y = x$ and $y = -x$. We sketch them.

b) If we let $y = 0$, we get $x = \pm 2$, so the intercepts are
$(2, 0)$ and $(-2, 0)$.

c) We plot the intercepts and draw smooth curves through
them that approach the asymptotes.

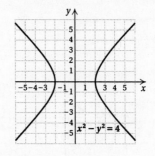

17. $\sqrt[3]{216s^{18}} = \sqrt[3]{216}\,\sqrt[3]{s^{18}} = 6s^6$

19. $3x^2 + 21 = 0$

$3x^2 = -21$

$x^2 = -7$

$x = \pm\sqrt{-7}$

$x = \pm i\sqrt{7}$

21. $\dfrac{2\sqrt{3} - 4\sqrt{2}}{\sqrt{2} - 5\sqrt{3}} = \dfrac{2\sqrt{3} - 4\sqrt{2}}{\sqrt{2} - 5\sqrt{3}} \cdot \dfrac{\sqrt{2} + 5\sqrt{3}}{\sqrt{2} + 5\sqrt{3}}$

$= \dfrac{2\sqrt{6} + 10 \cdot 3 - 4 \cdot 2 - 20\sqrt{6}}{2 - 25 \cdot 3}$

$= \dfrac{2\sqrt{6} + 30 - 8 - 20\sqrt{6}}{2 - 75}$

$= \dfrac{22 - 18\sqrt{6}}{-73}$

23. $y = \dfrac{6}{x}$

We find some ordered pairs, keeping the results in a table.

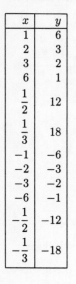

x	y
1	6
2	3
3	2
6	1
$\frac{1}{2}$	12
$\frac{1}{3}$	18
-1	-6
-2	-3
-3	-2
-6	-1
$-\frac{1}{2}$	-12
$-\frac{1}{3}$	-18

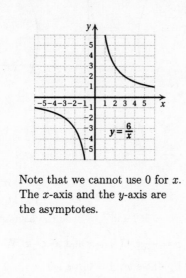

Note that we cannot use 0 for x. The x-axis and the y-axis are the asymptotes.

25. $y = -\dfrac{1}{x}$

x	y
1	-1
2	$-\frac{1}{2}$
4	$-\frac{1}{4}$
$\frac{1}{2}$	-2
$\frac{1}{4}$	-4
-1	1
-2	$\frac{1}{2}$
-4	$\frac{1}{4}$
$-\frac{1}{2}$	2
$-\frac{1}{4}$	4

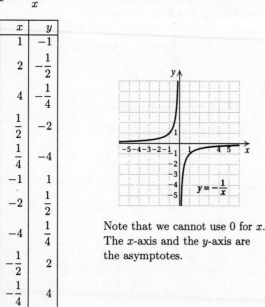

Note that we cannot use 0 for x. The x-axis and the y-axis are the asymptotes.

27.
$$16x^2 + y^2 + 96x - 8y + 144 = 0$$
$$(16x^2 + 96x) + (y^2 - 8y) + 144 = 0$$
$$16(x^2 + 6x) + (y^2 - 8y) + 144 = 0$$
$$16(x^2 + 6x + 9 - 9) + (y^2 - 8y + 16 - 16) + 144 = 0$$
$$16(x^2 + 6x + 9) + (y^2 - 8y + 16) - 144 - 16 + 144 = 0$$
$$16(x + 3)^2 + (y - 4)^2 = 16$$
$$\frac{(x + 3)^2}{1} + \frac{(y - 4)^2}{16} = 1$$
$$\frac{[x - (-3)]^2}{1^2} + \frac{(y - 4)^2}{4^2} = 1$$

Center: $(-3, 4)$

Vertices: $[1 + (-3), 4]$, $[-1 + (-3), 4]$, $(-3, 4 + 4)$,

$(-3, -4 + 4)$

or

$(-2, 4)$, $(-4, 4)$, $(-3, 8)$, $(-3, 0)$

$16x^2 + y^2 + 96x - 8y + 144 = 0$

29.
$$x^2 + y^2 - 10x + 8y - 40 = 0$$
$$x^2 - 10x + y^2 + 8y = 40$$
$$(x^2 - 10x + 25) + (y^2 + 8y + 16) = 40 + 25 + 16$$
$$(x - 5)^2 + (y + 4)^2 = 81$$

The graph is a circle.

31. $1 - 3y = 2y^2 - x$
$$x = 2y^2 + 3y - 1$$

The graph is a parabola.

33.
$$4x^2 + 25y^2 - 8x - 100y + 4 = 0$$
$$4x^2 - 8x + 25y^2 - 100y = -4$$
$$4(x^2 - 2x + 1) + 25(y^2 - 4y + 4) = -4 + 4 + 100$$
$$4(x - 1)^2 + 25(y - 2)^2 = 100$$
$$\frac{(x - 1)^2}{25} + \frac{(y - 2)^2}{4} = 1$$

The graph is an ellipse.

35. $y = ax^2 + bx + c$ Standard form of a parabola with line of symmetry parallel to the y-axis

We substitute values for x and y.

$3 = a(0)^2 + b(0) + c$ Substituting 0 for x and 3 for y

$6 = a(-1)^2 + b(-1) + c$ Substituting -1 for x and 6 for y

$9 = a(2)^2 + b(2) + c$ Substituting 2 for x and 9 for y

or

$3 = c$

$6 = a - b + c$

$9 = 4a + 2b + c$

Next we substitute 3 for c in the second and third equations and solve the resulting system for a and b.

$$6 = a - b + 3, \quad \text{or} \quad a - b = 3,$$
$$9 = 4a + 2b + 3 \qquad 4a + 2b = 6$$

Solving this system we get $a = 2$ and $b = -1$,. Thus the equation of the parabola whose line of symmetry is parallel to the y-axis and which passes through $(0,3)$, $(-1,6)$, and $(2,9)$ is $y = 2x^2 - x + 3$.

Exercise Set 9.3

1. $x^2 + y^2 = 100,$ (1)

   $y - x = 2$ (2)

First solve Equation (2) for y.

    $y = x + 2$ (3)

Then substitute $x + 2$ for y in Equation (1) and solve for x.

$$x^2 + y^2 = 100$$
$$x^2 + (x+2)^2 = 100$$
$$x^2 + x^2 + 4x + 4 = 100$$
$$2x^2 + 4x - 96 = 0$$
$$x^2 + 2x - 48 = 0 \qquad \text{Multiplying by } \frac{1}{2}$$
$$(x+8)(x-6) = 0$$

$x + 8 = 0 \quad$ or $\quad x - 6 = 0 \quad$ Principle of zero products

      $x = -8 \quad$ or $\qquad x = 6$

Now substitute these numbers into Equation (3) and solve for y.

$y = -8 + 2 = -6$

$y = 6 + 2 = 8$

The pairs $(-8, -6)$ and $(6, 8)$ check, so they are the solutions.

3. $9x^2 + 4y^2 = 36,$ (1)

   $3x + 2y = 6$ (2)

First solve Equation (2) for x.

   $3x = 6 - 2y$

    $x = 2 - \dfrac{2}{3}y$ (3)

Then substitute $2 - \dfrac{2}{3}y$ for x in Equation (1) and solve for y.

$$9x^2 + 4y^2 = 36$$
$$9\left(2 - \frac{2}{3}y\right)^2 + 4y^2 = 36$$
$$9\left(4 - \frac{8}{3}y + \frac{4}{9}y^2\right) + 4y^2 = 36$$
$$36 - 24y + 4y^2 + 4y^2 = 36$$
$$8y^2 - 24y = 0$$
$$y^2 - 3y = 0$$
$$y(y - 3) = 0$$

$y = 0 \quad$ or $\quad y - 3 = 0 \quad$ Principle of zero products

$y = 0 \quad$ or $\qquad y = 3$

Now substitute these numbers into Equation (3) and solve for x.

$$x = 2 - \frac{2}{3} \cdot 0 = 2$$
$$x = 2 - \frac{2}{3} \cdot 3 = 2 - 2 = 0$$

The pairs $(2, 0)$ and $(0, 3)$ check, so they are the solutions.

5. $y^2 = x + 3,$ (1)

   $2y = x + 4$ (2)

First solve Equation (2) for x.

   $2y - 4 = x$ (3)

Then substitute $y - 4$ for x in Equation (1) and solve for y.

$$y^2 = x + 3$$
$$y^2 = (2y - 4) + 3$$
$$y^2 = 2y - 1$$
$$y^2 - 2y + 1 = 0$$
$$(y - 1)(y - 1) = 0$$

$y - 1 = 0 \quad$ or $\quad y - 1 = 0$

$\quad y = 1 \quad$ or $\qquad y = 1$

Now substitute 1 for y in Equation (3) and solve for x.

$2 \cdot 1 - 4 = x$

$\quad -2 = x$

The pair $(-2, 1)$ checks. It is the solution.

7. $x^2 - xy + 3y^2 = 27,$ (1)

   $x - y = 2$ (2)

First solve Equation (2) for y.

   $x - 2 = y$ (3)

Then substitute $x - 2$ for y in Equation (1) and solve for x.

$$x^2 - xy + 3y^2 = 27$$
$$x^2 - x(x-2) + 3(x-2)^2 = 27$$
$$x^2 - x^2 + 2x + 3x^2 - 12x + 12 = 27$$
$$3x^2 - 10x - 15 = 0$$
$$x = \frac{-(-10) \pm \sqrt{(-10)^2 - 4(3)(-15)}}{2 \cdot 3}$$
$$x = \frac{10 \pm \sqrt{100 + 180}}{6} = \frac{10 \pm \sqrt{280}}{6}$$
$$x = \frac{10 \pm 2\sqrt{70}}{6} = \frac{5 \pm \sqrt{70}}{3}$$

Now substitute these numbers in Equation (3) and solve for y.

$$y = \frac{5 + \sqrt{70}}{3} - 2 = \frac{-1 + \sqrt{70}}{3}$$
$$y = \frac{5 - \sqrt{70}}{3} - 2 = \frac{-1 - \sqrt{70}}{3}$$

The pairs $\left(\dfrac{5+\sqrt{70}}{3}, \dfrac{-1+\sqrt{70}}{3}\right)$ and

$\left(\dfrac{5-\sqrt{70}}{3}, \dfrac{-1-\sqrt{70}}{3}\right)$ check, so they are the solutions.

9. $x^2 - xy + 3y^2 = 5,$ (1)

   $x - y = 2$ (2)

First solve Equation (2) for y.

   $x - 2 = y$ (3)

Then substitute $x - 2$ for y in Equation (1) and solve for x.

$$x^2 - xy + 3y^2 = 5$$
$$x^2 - x(x-2) + 3(x-2)^2 = 5$$
$$x^2 - x^2 + 2x + 3x^2 - 12x + 12 = 5$$
$$3x^2 - 10x + 7 = 0$$
$$(3x - 7)(x - 1) = 0$$

$3x - 7 = 0$ or $x - 1 = 0$

    $x = \dfrac{7}{3}$ or $x = 1$

Now substitute these numbers in Equation (3) and solve for y.

$y = \dfrac{7}{3} - 2 = \dfrac{1}{3}$

$y = 1 - 2 = -1$

The pairs $\left(\dfrac{7}{3}, \dfrac{1}{3}\right)$ and $(1, -1)$ check, so they are the solutions.

11. $a + b = -6,$ (1)

   $ab = -7$ (2)

First solve Equation (1) for a.

   $a = -b - 6$ (3)

Then substitute $-b - 6$ for a in Equation (2) and solve for b.

$$(-b - 6)b = -7$$
$$-b^2 - 6b = -7$$
$$0 = b^2 + 6b - 7$$
$$0 = (b + 7)(b - 1)$$

$b + 7 = 0$ or $b - 1 = 0$

    $b = -7$ or $b = 1$

Now substitute these numbers in Equation (3) and solve for a.

$a = -(-7) - 6 = 1$

$a = -1 - 6 = -7$

The pairs $(-7, 1)$ and $(1, -7)$ check, so they are the solutions.

13. $2a + b = 1,$ (1)

   $b = 4 - a^2$ (2)

Equation (2) is already solved for b. Substitute $4 - a^2$ for b in Equation (1) and solve for a.

$$2a + 4 - a^2 = 1$$
$$0 = a^2 - 2a - 3$$
$$0 = (a - 3)(a + 1)$$

$a - 3 = 0$ or $a + 1 = 0$

   $a = 3$ or $a = -1$

Substitute these numbers in Equation (2) and solve for b.

$b = 4 - 3^2 = -5$

$b = 4 - (-1)^2 = 3$

The pairs $(3, -5)$ and $(-1, 3)$ check.

15. $x^2 + y^2 = 5,$ (1)

   $x - y = 8$ (2)

First solve Equation (2) for x.

   $x = y + 8$ (3)

Then substitute $y + 8$ for x in Equation (1) and solve for y.

$$(y + 8)^2 + y^2 = 5$$
$$y^2 + 16y + 64 + y^2 = 5$$
$$2y^2 + 16y + 59 = 0$$
$$y = \frac{-16 \pm \sqrt{(16)^2 - 4(2)(59)}}{2 \cdot 2}$$
$$y = \frac{-16 \pm \sqrt{-216}}{4} = \frac{-16 \pm 6i\sqrt{6}}{4}$$
$$y = \frac{-8 \pm 3i\sqrt{6}}{2}, \text{ or } -4 \pm \frac{3}{2}i\sqrt{6}$$

Now substitute these numbers in Equation (3) and solve for x.

$x = -4 + \dfrac{3}{2}i\sqrt{6} + 8 = 4 + \dfrac{3}{2}i\sqrt{6}, \text{ or } \dfrac{8 + 3i\sqrt{6}}{2}$

$x = -4 - \dfrac{3}{2}i\sqrt{6} + 8 = 4 - \dfrac{3}{2}i\sqrt{6}, \text{ or } \dfrac{8 - 3i\sqrt{6}}{2}$

The pairs $\left(4 + \dfrac{3}{2}i\sqrt{6}, -4 + \dfrac{3}{2}i\sqrt{6}\right)$ and

$\left(4 - \dfrac{3}{2}i\sqrt{6}, -4 - \dfrac{3}{2}i\sqrt{6}\right)$, or $\left(\dfrac{8 + 3i\sqrt{6}}{2}, \dfrac{-8 + 3i\sqrt{6}}{2}\right)$ and

$\left(\dfrac{8 - 3i\sqrt{6}}{2}, \dfrac{-8 - 3i\sqrt{6}}{2}\right)$ check.

17. $x^2 + y^2 = 25,$ (1)

   $y^2 = x + 5$ (2)

We substitute $x + 5$ for y^2 in Equation (1) and solve for x.

$$x^2 + y^2 = 25$$
$$x^2 + (x + 5) = 25$$
$$x^2 + x - 20 = 0$$
$$(x + 5)(x - 4) = 0$$

$x + 5 = 0$ or $x - 4 = 0$

   $x = -5$ or $x = 4$

We substitute these numbers for x in either Equation (1) or Equation (2) and solve for y. Here we use Equation (2).

$y^2 = -5 + 5 = 0$ and $y = 0$.

$y^2 = 4 + 5 = 9$ and $y = \pm 3$.

The pairs $(-5, 0)$, $(4, 3)$ and $(4, -3)$ check.

19. $\quad x^2 + y^2 = 9,$ $\qquad$ (1)

$\quad x^2 - y^2 = 9$ $\qquad$ (2)

Here we use the elimination method.

$$\begin{array}{ll} x^2 + y^2 = 9 & (1) \\ \underline{x^2 - y^2 = 9} & (2) \\ 2x^2 \quad\quad = 18 & \text{Adding} \\ x^2 \quad = 9 \\ x = \pm 3 \end{array}$$

If $x = 3$, $x^2 = 9$, and if $x = -3$, $x^2 = 9$, so substituting 3 or -3 in Equation (1) gives us

$$x^2 + y^2 = 9$$
$$9 + y^2 = 9$$
$$y^2 = 0$$
$$y = 0.$$

The pairs $(3, 0)$ and $(-3, 0)$ check.

21. $\quad x^2 + y^2 = 20,$ $\qquad$ (1)

$\quad xy = 8$ $\qquad$ (2)

First we solve Equation (2) for y.

$$y = \frac{8}{x}$$

Then substitute $\frac{8}{x}$ for y in Equation (1) and solve for x.

$$x^2 + \left(\frac{8}{x}\right)^2 = 20$$
$$x^2 + \frac{64}{x^2} = 20$$
$$x^4 + 64 = 20x^2 \quad \text{Multiplying by } x^2$$
$$x^4 - 20x^2 + 64 = 0$$
$$u^2 - 20u + 64 = 0 \quad \text{Letting } u = x^2$$
$$(u - 4)(u - 16) = 0$$
$$u - 4 = 0 \quad \text{or} \quad u - 16 = 0$$
$$u = 4 \quad \text{or} \quad u = 16$$

We now substitute x^2 for u and solve for x.

$$x^2 = 4 \quad \text{or} \quad x^2 = 16$$
$$x = \pm 2 \quad \text{or} \quad x = \pm 4$$

Since $y = \frac{8}{x}$, if $x = 2$, $y = 4$; if $x = -2$, $y = -4$; if $x = 4$, $y = 2$; if $x = -4$, $y = -2$. The pairs $(2, 4)$, $(-2, -4)$, $(4, 2)$, and $(-4, -2)$ check. They are the solutions.

23. $\quad x^2 + y^2 = 13,$ $\qquad$ (1)

$\quad xy = 6$ $\qquad$ (2)

First we solve Equation (2) for y.

$$y = \frac{6}{x}$$

Then substitute $\frac{6}{x}$ for y in Equation (1) and solve for x.

$$x^2 + \left(\frac{6}{x}\right)^2 = 13$$
$$x^2 + \frac{36}{x^2} = 13$$
$$x^4 + 36 = 13x^2 \quad \text{Multiplying by } x^2$$
$$x^4 - 13x^2 + 36 = 0$$
$$u^2 - 13u + 36 = 0 \quad \text{Letting } u = x^2$$
$$(u - 9)(u - 4) = 0$$
$$u - 9 = 0 \quad \text{or} \quad u - 4 = 0$$
$$u = 9 \quad \text{or} \quad u = 4$$

We now substitute x^2 for u and solve for x.

$$x^2 = 9 \quad \text{or} \quad x^2 = 4$$
$$x = \pm 3 \quad \text{or} \quad x = \pm 2$$

Since $y = \frac{6}{x}$, if $x = 3$, $y = 2$; if $x = -3$, $y = -2$; if $x = 2$, $y = 3$; if $x = -2$, $y = -3$. The pairs $(3, 2)$, $(-3, -2)$, $(2, 3)$, and $(-2, -3)$ check. They are the solutions.

25. $\quad 2xy + 3y^2 = 7,$ $\qquad$ (1)

$\quad 3xy - 2y^2 = 4$ $\qquad$ (2)

$$\begin{array}{ll} 6xy + 9y^2 = 21 & \text{Multiplying (1) by 3} \\ \underline{-6xy + 4y^2 = -8} & \text{Multiplying (2) by } -2 \\ 13y^2 = 13 \\ y^2 = 1 \\ y = \pm 1 \end{array}$$

Substitute for y in Equation (1) and solve for x.

When $y = 1$: $\quad 2 \cdot x \cdot 1 + 3 \cdot 1^2 = 7$

$$2x = 4$$
$$x = 2$$

When $y = -1$: $\quad 2 \cdot x \cdot (-1) + 3 \cdot (-1)^2 = 7$

$$-2x = 4$$
$$x = -2$$

The pairs $(2, 1)$ and $(-2, -1)$ check. They are the solutions.

27. $\quad 4a^2 - 25b^2 = 0,$ $\qquad$ (1)

$\quad 2a^2 - 10b^2 = 3b + 4$ $\qquad$ (2)

$$\begin{array}{ll} 4a^2 - 25b^2 = 0 \\ \underline{-4a^2 + 20b^2 = -6b - 8} & \text{Multiplying (2) by } -2 \\ -5b^2 = -6b - 8 \\ 0 = 5b^2 - 6b - 8 \\ 0 = (5b + 4)(b - 2) \end{array}$$

$$5b + 4 = 0 \quad \text{or} \quad b - 2 = 0$$
$$b = -\frac{4}{5} \quad \text{or} \quad b = 2$$

Substitute for b in Equation (1) and solve for a.

When $b = -\dfrac{4}{5}$: $4a^2 - 25\left(-\dfrac{4}{5}\right)^2 = 0$

$$4a^2 = 16$$
$$a^2 = 4$$
$$a = \pm 2$$

When $b = 2$: $4a^2 - 25(2)^2 = 0$
$$4a^2 = 100$$
$$a^2 = 25$$
$$a = \pm 5$$

The pairs $\left(2, -\dfrac{4}{5}\right)$, $\left(-2, -\dfrac{4}{5}\right)$, $(5, 2)$ and $(-5, 2)$ check. They are the solutions.

29. $ab -\ b^2 = -4,$ (1)
$ab - 2b^2 = -6$ (2)

$$
\begin{array}{ll}
ab -\ b^2 = \quad -4 & \\
\underline{-ab + 2b^2 = \quad\ \ 6} & \text{Multiplying (2) by } -1 \\
\qquad\quad b^2 = \quad\ \ 2 & \\
\qquad\ \ b = \pm\sqrt{2} &
\end{array}
$$

Substitute for b in Equation (1) and solve for a.

When $b = \sqrt{2}$: $a(\sqrt{2}) - (\sqrt{2})^2 = -4$

$$a\sqrt{2} = -2$$
$$a = -\dfrac{2}{\sqrt{2}} = -\sqrt{2}$$

When $b = -\sqrt{2}$: $a(-\sqrt{2}) - (-\sqrt{2})^2 = -4$

$$-a\sqrt{2} = -2$$
$$a = \dfrac{2}{\sqrt{2}} = \sqrt{2}$$

The pairs $(-\sqrt{2}, \sqrt{2})$ and $(\sqrt{2}, -\sqrt{2})$ check. They are the solutions.

31. $x^2 +\ y^2 = 25,$ (1)
$9x^2 + 4y^2 = 36$ (2)

$$
\begin{array}{ll}
-4x^2 - 4y^2 = -100 & \text{Multiplying (1) by } -4 \\
\underline{\ 9x^2 + 4y^2 = \quad\ 36} & \\
\ 5x^2 \qquad\quad = -64 &
\end{array}
$$

$$x^2 = -\dfrac{64}{5}$$

$$x = \pm\sqrt{-\dfrac{64}{5}} = \pm\dfrac{8i}{\sqrt{5}}$$

$$x = \pm\dfrac{8i\sqrt{5}}{5} \qquad \text{Rationalizing the denominator}$$

Substituting $\dfrac{8i\sqrt{5}}{5}$ or $-\dfrac{8i\sqrt{5}}{5}$ for x in Equation (1) and solving for y gives us

$$-\dfrac{64}{5} + y^2 = 25$$

$$y^2 = \dfrac{189}{5}$$

$$y = \pm\sqrt{\dfrac{189}{5}} = \pm 3\sqrt{\dfrac{21}{5}}$$

$$y = \pm\dfrac{3\sqrt{105}}{5}. \qquad \text{Rationalizing the denominator}$$

The pairs $\left(\dfrac{8i\sqrt{5}}{5}, \dfrac{3\sqrt{105}}{5}\right)$, $\left(-\dfrac{8i\sqrt{5}}{5}, \dfrac{3\sqrt{105}}{5}\right)$, $\left(\dfrac{8i\sqrt{5}}{5}, -\dfrac{3\sqrt{105}}{5}\right)$, and $\left(-\dfrac{8i\sqrt{5}}{5}, -\dfrac{3\sqrt{105}}{5}\right)$, check. They are the solutions.

33. *Familiarize.* Using the labels on the drawing in the text we let $l =$ the length of the tile and $w =$ the width.

Translate. The perimeter is 6 m, so we have $2l + 2w = 6$. We use the Pythagorean equation to write another equation: $l^2 + w^2 = (\sqrt{5})^2$, or $l^2 + w^2 = 5$.

Solve. We solve the system.

$$2l + 2w = 6, \qquad (1)$$
$$l^2 + w^2 = 5 \qquad (2)$$

First solve Equation (1) for w.

$$w = 3 - l \qquad (3)$$

Then substitute $3 - l$ for w in Equation (2) and solve for l.

$$l^2 + w^2 = 5$$
$$l^2 + (3 - l)^2 = 5$$
$$l^2 + 9 - 6l + l^2 = 5$$
$$2l^2 - 6l + 4 = 0$$
$$l^2 - 3l + 2 = 0 \qquad \text{Multiplying by } \dfrac{1}{2}$$
$$(l - 2)(l - 1) = 0$$
$$l - 2 = 0 \quad \text{or} \quad l - 1 = 0$$
$$l = 2 \quad \text{or} \qquad l = 1$$

If $l = 2$, then $w = 3 - 2$, or 1. If $l = 1$, then $w = 3 - 1$, or 2. Since length is usually considered to be longer than the width, we have the solution $l = 2$ and $w = 1$, or $(2, 1)$.

Check. If $l = 2$ and $w = 1$, then the perimeter is $2 \cdot 2 + 2 \cdot 1$, or 6. The length of a diagonal is $\sqrt{2^2 + 1^2}$, or $\sqrt{5}$. The numbers check.

State. The length is 2 m, and the width is 1 m.

35. *Familiarize.* We first make a drawing. We let $l =$ the length and $w =$ the width of the rectangle.

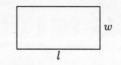

Translate.

Area: $lw = 14$

Perimeter: $2l + 2w = 18$, or $l + w = 9$

Solve. We solve the system.

Solve the second equation for l: $\quad l = 9 - w$

Substitute $9 - w$ for l in the first equation and solve for w.

$$(9 - w)w = 14$$
$$9w - w^2 = 14$$
$$0 = w^2 - 9w + 14$$
$$0 = (w - 7)(w - 2)$$
$$w - 7 = 0 \quad \text{or} \quad w - 2 = 0$$
$$w = 7 \quad \text{or} \qquad w = 2$$

If $w = 7$, then $l = 9 - 7$, or 2. If $w = 2$, then $l = 9 - 2$, or 7. Since length is usually considered to be longer than the width, we have the solution $l = 7$ and $w = 2$, or $(7, 2)$.

Check. If $l = 7$ and $w = 2$, the area is $7 \cdot 2$, or 14. The perimeter is $2 \cdot 7 + 2 \cdot 2$, or 18. The numbers check.

State. The length is 7 in., and the width is 2 in.

37. *Familiarize.* Let $l =$ the length and $w =$ the width of the rectangle. Then the length of a diagonal is $\sqrt{l^2 + w^2}$.

Translate.

The diagonal   is   1 ft   longer than   the length.
$\underbrace{\sqrt{l^2 + w^2}}_{} \quad \underset{=}{\downarrow} \quad \underset{1}{\downarrow} \quad \underset{+}{\downarrow} \quad \underset{l}{\downarrow}$

The diagonal   is   3 ft   longer than   twice the width.
$\underbrace{\sqrt{l^2 + w^2}}_{} \quad \underset{=}{\downarrow} \quad \underset{3}{\downarrow} \quad \underset{+}{\downarrow} \quad \underset{2 \cdot}{\downarrow} \quad \underset{w}{\downarrow}$

Solve. We solve the system of equations.

$$\sqrt{l^2 + w^2} = 1 + l, \qquad (1)$$
$$\sqrt{l^2 + w^2} = 3 + 2w \qquad (2)$$

We substitute $3 + 2w$ for $\sqrt{l^2 + w^2}$ in Equation (1) and solve for l.

$$3 + 2w = 1 + l$$
$$2 + 2w = l \qquad (3)$$

Now we substitute $2 + 2w$ for l in Equation (2).

$$\sqrt{(2 + 2w)^2 + w^2} = 3 + 2w$$
$$\sqrt{4 + 8w + 4w^2 + w^2} = 3 + 2w$$
$$\sqrt{5w^2 + 8w + 4} = 3 + 2w$$
$$(\sqrt{5w^2 + 8w + 4})^2 = (3 + 2w)^2$$
$$5w^2 + 8w + 4 = 9 + 12w + 4w^2$$
$$w^2 - 4w - 5 = 0$$
$$(w - 5)(w + 1) = 0$$
$$w - 5 = 0 \quad \text{or} \quad w + 1 = 0$$
$$w = 5 \quad \text{or} \qquad w = -1$$

Since the width cannot be negative, we consider only 5. We substitute 5 for w in Equation (3) and solve for l.

$$l = 2 + 2 \cdot 5 = 12$$

Check. The length of a diagonal of a 12 ft by 5 ft rectangle is $\sqrt{12^2 + 5^2} = \sqrt{169} = 13$. This is 1 ft longer than the

length, 12, and 3 ft longer than twice the width, $2 \cdot 5$, or 10. The numbers check.

State. The length is 12 ft, and the width is 5 ft.

39. *Familiarize.* We make a drawing and label it. Let x and y represent the lengths of the legs of the triangle.

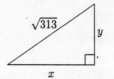

Translate. The product of the lengths of the legs is 156, so we have:

$$xy = 156$$

We use the Pythagorean theorem to get a second equation:

$$x^2 + y^2 = (\sqrt{313})^2, \text{ or } x^2 + y^2 = 313$$

Solve. We solve the system of equations.

$$xy = 156, \qquad (1)$$
$$x^2 + y^2 = 313 \qquad (2)$$

First solve Equation (1) for y.

$$y = \frac{156}{x}$$

Then we substitute $\dfrac{156}{x}$ for y in Equation (2) and solve for x.

$$x^2 + y^2 = 313 \qquad (2)$$
$$x^2 + \left(\frac{156}{x}\right)^2 = 313$$
$$x^2 + \frac{24,336}{x^2} = 313$$
$$x^4 + 24,336 = 313x^2$$
$$x^4 - 313x^2 + 24,336 = 0$$
$$u^2 - 313u + 24,336 = 0 \qquad \text{Letting } u = x^2$$
$$(u - 169)(u - 144) = 0$$
$$u - 169 = 0 \quad \text{or} \quad u - 144 = 0$$
$$u = 169 \quad \text{or} \qquad u = 144$$

We now substitute x^2 for u and solve for x.

$$x^2 = 169 \quad \text{or} \quad x^2 = 144$$
$$x = \pm 13 \quad \text{or} \qquad x = \pm 12$$

Since $y = 156/x$, if $x = 13$, $y = 12$; if $x = -13$, $y = -12$; if $x = 12$, $y = 13$; and if $x = -12$, $y = -13$. The possible solutions are $(13, 12)$, $(-13, -12)$, $(12, 13)$, and $(-12, -13)$.

Check. Since negative lengths do not make sense in this problem, we consider only $(13, 12)$ and $(12, 13)$. Since both possible solutions give the same pair of legs, we only need to check $(13, 12)$. If $x = 13$ and $y = 12$, their product is 156. Also, $\sqrt{13^2 + 12^2} = \sqrt{313}$. The numbers check.

State. The lengths of the legs are 13 and 12.

41. *Familiarize*. We let x = the length of a side of one broccoli bed and y = the length of a side of the other broccoli bed. Make a drawing.

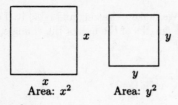

Area: x^2 Area: y^2

Translate.

The sum of the areas is 832 ft^2.

$$x^2 + y^2 = 832$$

The difference of the areas is 320 ft^2.

$$x^2 - y^2 = 320$$

Solve. We solve the system of equations.

$$
\begin{aligned}
x^2 + y^2 &= 832 \\
x^2 - y^2 &= 320 \\
\hline
2x^2 &= 1152 \quad \text{Adding} \\
x^2 &= 576 \\
x &= \pm 24
\end{aligned}
$$

Since length cannot be negative, we consider only $x = 24$. Substitute 24 for x in the first equation and solve for y.

$$
\begin{aligned}
24^2 + y^2 &= 832 \\
576 + y^2 &= 832 \\
y^2 &= 256 \\
y &= \pm 16
\end{aligned}
$$

Again, we consider only the positive value, 16. The possible solution is $(24, 16)$.

Check. The areas of the beds are 24^2, or 576, and 16^2, or 256. The sum of the areas is $576 + 256$, or 832. The difference of the areas is $576 - 256$, or 320. The values check.

State. The lengths of the beds are 24 ft and 16 ft.

43. *Familiarize*. We first make a drawing. We let l = the length and w = the width.

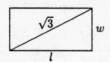

Translate.

Area: $\sqrt{2} = lw$ (1)

Using the Pythagorean equation: $l^2 + w^2 = (\sqrt{3})^2$, or

$$l^2 + w^2 = 3 \qquad (2)$$

Solve. We solve the system of equations. First we solve Equation (1) for w.

$$\frac{\sqrt{2}}{l} = w$$

Then we substitute $\dfrac{\sqrt{2}}{l}$ for w in Equation (2) and solve for l.

$$
\begin{aligned}
l^2 + \left(\frac{\sqrt{2}}{l}\right)^2 &= 3 \\
l^2 + \frac{2}{l^2} &= 3 \\
l^4 + 2 &= 3l^2 \quad \text{Multiplying by } l^2 \\
l^4 - 3l^2 + 2 &= 0 \\
u^2 - 3u + 2 &= 0 \quad \text{Letting } u = l^2 \\
(u - 2)(u - 1) &= 0
\end{aligned}
$$

$$
\begin{array}{lll}
u - 2 = 0 & \text{or} & u - 1 = 0 \\
u = 2 & \text{or} & u = 1 \\
l^2 = 2 & \text{or} & l^2 = 1 \quad \text{Substituting } l^2 \text{ for } u \\
l = \pm\sqrt{2} & \text{or} & l = \pm 1
\end{array}
$$

Length cannot be negative, we only need to consider $l = \sqrt{2}$ or $l = 1$. Since $w = \sqrt{2}/l$, if $l = \sqrt{2}$, $w = 1$ and if $l = 1$, $w = \sqrt{2}$. Length is usually considered to be longer than width, so we have the solution $l = \sqrt{2}$ and $w = 1$, or $(\sqrt{2}, 1)$.

Check. If $l = \sqrt{2}$ and $w = 1$, the area is $\sqrt{2} \cdot 1$, or $\sqrt{2}$. Also $(\sqrt{2})^2 + 1^2 = 2 + 1 = 3 = (\sqrt{3})^2$. The numbers check.

State. The length is $\sqrt{2}$ m, and the width is 1 m.

45.
$$
\begin{aligned}
2x^2 + 1 &= x \\
2x^2 - x + 1 &= 0 \\
x &= \frac{-(-1) \pm \sqrt{(-1)^2 - 4 \cdot 2 \cdot 1}}{2 \cdot 2} = \frac{1 \pm \sqrt{1 - 8}}{4} \\
x &= \frac{1 \pm \sqrt{-7}}{4} = \frac{1 \pm i\sqrt{7}}{4} \\
x &= \frac{1}{4} \pm i\frac{\sqrt{7}}{4}
\end{aligned}
$$

47. $\sqrt{80} = \sqrt{16 \cdot 5} = \sqrt{16}\sqrt{5} = 4\sqrt{5}$

49. $\dfrac{\sqrt{x} - \sqrt{h}}{\sqrt{x} + \sqrt{h}} = \dfrac{\sqrt{x} - \sqrt{h}}{\sqrt{x} + \sqrt{h}} \cdot \dfrac{\sqrt{x} + \sqrt{h}}{\sqrt{x} + \sqrt{h}} = \dfrac{x - h}{x + 2\sqrt{xh} + h}$

51. *Familiarize*. Let r = the speed of the boat in still water and t = the time of the trip upstream. Organize the information in a table.

	Speed	Time	Distance
Upstream	$r - 2$	t	4
Downstream	$r + 2$	$3 - t$	4

Recall that $rt = d$, or $t = d/r$.

Translate. From the first line of the table we obtain $t = \dfrac{4}{r - 2}$. From the second line we obtain $3 - t = \dfrac{4}{r + 2}$.

Solve. Substitute $\dfrac{4}{r - 2}$ for t in the second equation and solve for r.

$$3 - \frac{4}{r-2} = \frac{4}{r+2},$$

$$\text{LCM is } (r-2)(r+2)$$

$$(r-2)(r+2)\left(3 - \frac{4}{r-2}\right) = (r-2)(r+2) \cdot \frac{4}{r+2}$$

$$3(r-2)(r+2) - 4(r+2) = 4(r-2)$$

$$3(r^2 - 4) - 4r - 8 = 4r - 8$$

$$3r^2 - 12 - 4r - 8 = 4r - 8$$

$$3r^2 - 8r - 12 = 0$$

$$r = \frac{-(-8) \pm \sqrt{(-8)^2 - 4(3)(-12)}}{2 \cdot 3}$$

$$r = \frac{8 \pm \sqrt{208}}{6} = \frac{8 \pm 4\sqrt{13}}{6}$$

$$r = \frac{4 \pm 2\sqrt{13}}{3}$$

Since negative speed has no meaning in this problem, we consider only the positive square root.

$$r = \frac{4 + 2\sqrt{13}}{3} \approx 3.7$$

Check. The value checks. The check is left to the student.

State. The speed of the boat in still water is approximately 3.7 mph.

53. It is helpful to draw a picture.

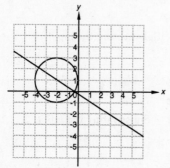

Let (h, k) represent the point on the line $5x + 8y = -2$ which is the center of a circle that passes through the points $(-2, 3)$ and $(-4, 1)$. The distance between (h, k) and $(-2, 3)$ is the same as the distance between (h, k) and $(-4, 1)$. This gives us one equation:

$$\sqrt{[h-(-2)]^2 + (k-3)^2} = \sqrt{[h-(-4)]^2 + (k-1)^2}$$

$$(h+2)^2 + (k-3)^2 = (h+4)^2 + (k-1)^2$$

$$h^2 + 4h + 4 + k^2 - 6k + 9 = h^2 + 8h + 16 + k^2 - 2k + 1$$

$$4h - 6k + 13 = 8h - 2k + 17$$

$$-4h - 4k = 4$$

$$h + k = -1$$

We get a second equation by substituting (h, k) in $5x + 8y = -2$.

$$5h + 8k = -2$$

We now solve the following system:

$$h + k = -1,$$
$$5h + 8k = -2$$

The solution, which is the center of the circle, is $(-2, 1)$.

Next we find the length of the radius. We can find the distance between either $(-2, 3)$ or $(-4, 1)$ and the center $(-2, 1)$. We use $(-2, 3)$.

$$r = \sqrt{[-2-(-2)]^2 + (1-3)^2}$$

$$r = \sqrt{0^2 + (-2)^2}$$

$$r = \sqrt{4} = 2$$

We now have an equation knowing that the center is $(-2, 1)$ and the radius is 2.

$$(x-h)^2 + (y-k)^2 = r^2$$

$$[x-(-2)]^2 + (y-1)^2 = 2^2$$

$$(x+2)^2 + (y-1)^2 = 4$$

55. $\dfrac{x^2}{a^2} + \dfrac{y^2}{b^2} = 1$ \qquad Standard form

Substitute the coordinates of the given points:

$$\frac{2^2}{a^2} + \frac{(-3)^2}{b^2} = 1,$$

$$\frac{1^2}{a^2} + \frac{(\sqrt{13})^2}{b^2} = 1, \text{ or}$$

$$\frac{4}{a^2} + \frac{9}{b^2} = 1, \qquad (1)$$

$$\frac{1}{a^2} + \frac{13}{b^2} = 1 \qquad (2)$$

Solve Equation (2) for $\dfrac{1}{a^2}$:

$$\frac{1}{a^2} = 1 - \frac{13}{b^2}$$

$$\frac{1}{a^2} = \frac{b^2 - 13}{b^2} \qquad (3)$$

Substitute $\dfrac{b^2 - 13}{b^2}$ for $\dfrac{1}{a^2}$ in Equation (1) and solve for b^2.

$$4\left(\frac{b^2 - 13}{b^2}\right) + \frac{9}{b^2} = 1$$

$$\frac{4b^2 - 52}{b^2} + \frac{9}{b^2} = 1$$

$$4b^2 - 52 + 9 = b^2$$

$$3b^2 = 43$$

$$b^2 = \frac{43}{3}$$

Substitute $\dfrac{43}{3}$ for b^2 in Equation (3) and solve for a^2.

$$\frac{1}{a^2} = \frac{\frac{43}{3} - 13}{\frac{43}{3}} = \frac{\frac{43}{3} - 13}{\frac{43}{3}} \cdot \frac{3}{3}$$

$$\frac{1}{a^2} = \frac{43 - 3 \cdot 13}{43} = \frac{43 - 39}{43}$$

$$\frac{1}{a^2} = \frac{4}{43}$$

$$a^2 = \frac{43}{4}$$

The equation of the ellipse is

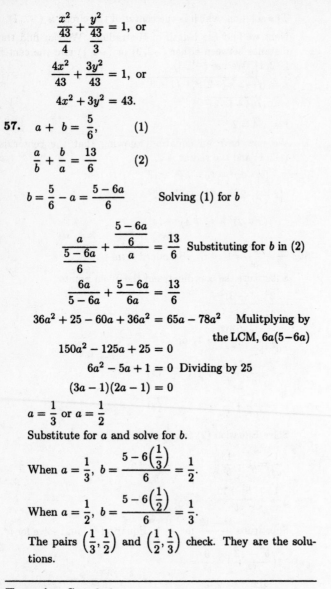

$$\frac{x^2}{\frac{43}{4}} + \frac{y^2}{\frac{43}{3}} = 1, \text{ or}$$

$$\frac{4x^2}{43} + \frac{3y^2}{43} = 1, \text{ or}$$

$$4x^2 + 3y^2 = 43.$$

57. $a + b = \dfrac{5}{6},$ (1)

$\dfrac{a}{b} + \dfrac{b}{a} = \dfrac{13}{6}$ (2)

$b = \dfrac{5}{6} - a = \dfrac{5 - 6a}{6}$ Solving (1) for b

$\dfrac{a}{\dfrac{5-6a}{6}} + \dfrac{\dfrac{5-6a}{6}}{a} = \dfrac{13}{6}$ Substituting for b in (2)

$\dfrac{6a}{5-6a} + \dfrac{5-6a}{6a} = \dfrac{13}{6}$

$36a^2 + 25 - 60a + 36a^2 = 65a - 78a^2$ Multiplying by
$\qquad\qquad\qquad\qquad$ the LCM, $6a(5-6a)$

$150a^2 - 125a + 25 = 0$

$6a^2 - 5a + 1 = 0$ Dividing by 25

$(3a - 1)(2a - 1) = 0$

$a = \dfrac{1}{3} \text{ or } a = \dfrac{1}{2}$

Substitute for a and solve for b.

When $a = \dfrac{1}{3}$, $b = \dfrac{5 - 6\left(\dfrac{1}{3}\right)}{6} = \dfrac{1}{2}$.

When $a = \dfrac{1}{2}$, $b = \dfrac{5 - 6\left(\dfrac{1}{2}\right)}{6} = \dfrac{1}{3}$.

The pairs $\left(\dfrac{1}{3}, \dfrac{1}{2}\right)$ and $\left(\dfrac{1}{2}, \dfrac{1}{3}\right)$ check. They are the solutions.

Exercise Set 9.4

1. Each arrow gives an ordered pair. Thus we have the following:

$d = \{$(New York, Mets), (New York, Yankees), (Los Angeles, Rams), (Los Angeles, Raiders), (St. Louis, Cardinals)$\}$

3. Each arrow gives an ordered pair. Thus we have the following:

$f = \{(-4, 5), (-5, 5), (-6, 5), (-7, 2)\}$

5. The relation is a function because each member of the domain is matched to only one member of the range.

7. The relation is not a function because the member of the domain 20 is matched to more than one member of the range.

9. The domain is the set of all first numbers of ordered pairs in the relation. The domain of $s = \{8, -6, -1, 2\}$.

The range is the set of all second members of ordered pairs in the relation. The range of $s = \{3, -4, -1\}$.

11. Domain of $t = \{20, 21, 22, 23\}$

Range of $t = \{7, -5, 0\}$

13. $g(x) = -2x - 4$

$g(-2) = -2(-2) - 4$ Substituting the input, -2

$\qquad = 4 - 4$

$\qquad = 0$

15. $g(x) = -2x - 4$

$g\left(-\dfrac{1}{4}\right) = -2\left(-\dfrac{1}{4}\right) - 4$ Substituting the input, $-\dfrac{1}{4}$

$\qquad = \dfrac{1}{2} - 4$

$\qquad = -\dfrac{7}{2}$

17. $h(x) = 3x^2$

$h(-1) = 3(-1)^2$ Substituting the input, -1

$\qquad = 3 \cdot 1$

$\qquad = 3$

19. $h(x) = 3x^2$

$h(2) = 3 \cdot 2^2$ Substituting the input, 2

$\qquad = 3 \cdot 4$

$\qquad = 12$

21. $P(x) = x^3 - x$

$P(2) = 2^3 - 2$ Substituting the input, 2

$\qquad = 8 - 2$

$\qquad = 6$

23. $P(x) = x^3 - x$

$P(-1) = (-1)^3 - (-1)$ Substituting the input, -1

$\qquad = -1 + 1$

$\qquad = 0$

25. $Q(x) = x^4 - 2x^3 + x^2 - x + 2$

$Q(-1) = (-1)^4 - 2(-1)^3 + (-1)^2 - (-1) + 2$

$\qquad\qquad\qquad\qquad$ Substituting the input, -1

$\qquad = 1 + 2 + 1 + 1 + 2$

$\qquad = 7$

27. $Q(x) = x^4 - 2x^3 + x^2 - x + 2$

$Q(2) = 2^4 - 2 \cdot 2^3 + 2^2 - 2 + 2$

$\qquad\qquad\qquad\qquad$ Substituting the input, 2

$\qquad = 16 - 16 + 4 - 2 + 2$

$\qquad = 4$

29. $P(d) = 1 + \dfrac{d}{33}$

$P(20) = 1 + \dfrac{20}{33} = \dfrac{53}{33}$, or $1\dfrac{20}{33}$ atm

$P(30) = 1 + \dfrac{30}{33} = \dfrac{63}{33} = \dfrac{21}{11}$, or $1\dfrac{10}{11}$ atm

$P(100) = 1 + \dfrac{100}{33} = \dfrac{133}{33}$, or $4\dfrac{1}{33}$ atm

31. $S(p) = 7p - 5.9$

$S(1) = 7 \cdot 1 - 5.9 = 7 - 5.9 = 1.1$ million

$S(2) = 7 \cdot 2 - 5.9 = 14 - 5.9 = 8.1$ million

$S(3) = 7 \cdot 3 - 5.9 = 21 - 5.9 = 15.1$ million

$S(4) = 7 \cdot 4 - 5.9 = 28 - 5.9 = 22.1$ million

$S(5) = 7 \cdot 5 - 5.9 = 35 - 5.9 = 29.1$ million

33. Graph: $f(x) = -2x - 3$

We compute some function values.

$f(0) = -2 \cdot 0 - 3 = -3$

$f(-2) = -2(-2) - 3 = 1$

$f(1) = -2 \cdot 1 - 3 = -5$

A list of function values is shown in this table.

x	$f(x) = -2x - 3$
0	-3
-2	1
1	-5

We plot the points and connect them. The graph is a straight line.

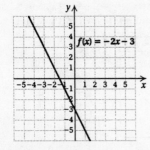

35. Graph: $h(x) = |x|$

We compute some function values.

$h(-4) = |-4| = 4$

$h(-1) = |-1| = 1$

$h(0) = |0| = 0$

$h(2) = |2| = 2$

$h(5) = |5| = 5$

A list of function values is shown in this table.

x	$h(x) = |x|$
-4	4
-1	1
0	0
2	2
5	5

We plot the points and connect them. The graph is V-shaped.

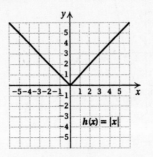

37. Graph: $f(x) = -5$

We compute some function values.

$f(-4) = -5$

$f(0) = -5$

$f(3) = -5$

A list of function values is shown in this table.

x	$f(x) = -5$
-4	-5
0	-5
3	-5

We plot the points and connect them. The graph is a straight line.

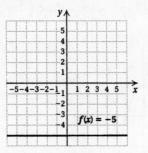

39. Graph: $f(x) = x^2 - 4$

We compute some function values.

$f(-3) = (-3)^2 - 4 = 5$

$f(-1) = (-1)^2 - 4 = -3$

$f(0) = 0^2 - 4 = -4$

$f(1) = 1^2 - 4 = -3$

$f(2) = 2^2 - 4 = 0$

A list of function values is shown in this table.

x	$f(x) = x^2 - 4$
-3	5
-1	-3
0	-4
1	-3
2	0

We plot the points and connect them. The graph is a parabola.

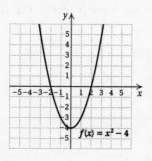

41. Graph: $f(x) = 3 - x^2$

We compute some function values.

$f(-3) = 3 - (-3)^2 = -6$

$f(-1) = 3 - (-1)^2 = 2$

$f(0) = 3 - 0^2 = 3$

$f(2) = 3 - 2^2 = -1$

$f(3) = 3 - 3^2 = -6$

A list of function values is shown in this table.

x	$f(x) = 3 - x^2$
-3	-6
-1	2
0	3
2	-1
3	-6

We plot the points and connect them. The graph is a parabola.

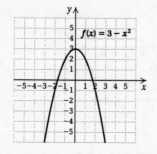

43. Graph: $g(x) = -\dfrac{4}{x}$

We compute some function values. Note that 0 cannot be an input.

$g(-4) = -\dfrac{4}{-4} = 1$

$g(-2) = -\dfrac{4}{-2} = 2$

$g(-1) = -\dfrac{4}{-1} = 4$

$g(1) = -\dfrac{4}{1} = -4$

$g(2) = -\dfrac{4}{2} = -2$

$g(4) = -\dfrac{4}{4} = -1$

A list of function values is shown in this table.

x	$g(x) = -\dfrac{4}{x}$
-4	1
-2	2
-1	4
1	-4
2	-2
4	-1

We plot the points and connect them. The graph is a hyperbola.

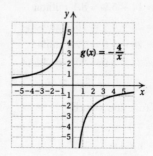

45. Graph: $f(x) = 3 - |x|$

We compute some function values.

$f(-5) = 3 - |-5| = -2$

$f(-3) = 3 - |-3| = 0$

$f(-1) = 3 - |-1| = 2$

$f(0) = 3 - |0| = 3$

$f(2) = 3 - |2| = 1$

$f(4) = 3 - |4| = -1$

A list of function values is shown in this table.

x	$f(x) = 3 - |x|$
-5	-2
-3	0
-1	2
0	3
2	1
4	-1

We plot the points and connect them. The graph is V-shaped.

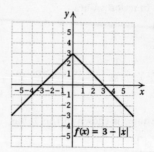

47.

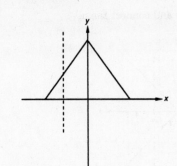

The graph is the graph of a function. It passes the vertical-line test. No vertical line intersects the graph more than once.

49.

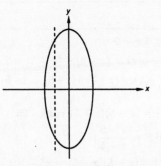

The graph is not the graph of a function. It does not pass the vertical-line test. It is possible for a vertical line to intersect the graph more than once.

51. $p(x) = x^4 - x^3 + x - 10$

There are no restrictions on the numbers we can substitute into this formula. The domain is the set of all real numbers.

53. $f(x) = \dfrac{7}{x^2 - 25}$

This formula is meaningful for all replacements that do not make the denominator 0. To find those replacements that do make the denominator 0, we solve:

$$x^2 - 25 = 0$$
$$(x + 5)(x - 5) = -0$$
$$x = -5 \text{ or } x = 5$$

The domain is $\{x | x \neq -5 \text{ or } x \neq 5\}$.

55. $g(x) = \dfrac{x}{x^2 + 8x + 15}$

We find the replacements that make the denominator 0. Then the formula is meaningful for all other replacements.

$$x^2 + 8x + 15 = 0$$
$$(x + 3)(x + 5) = -0$$
$$x = -3 \text{ or } x = -5$$

The domain is $\{x | x \neq -3 \text{ or } x \neq -5\}$.

57. $f(x) = \sqrt{6 + 8x}$

The formula is meaningful for all real numbers x for which the following is true:

$$6 + 8x \geq 0$$
$$8x \geq -6$$
$$x \geq -\frac{6}{8}, \text{ or } -\frac{3}{4}$$

The domain is $\left\{x \Big| x \geq -\dfrac{3}{4}\right\}$.

59. $g(x) = \dfrac{4}{5 - 2x}$

We find the replacements that make the denominator 0. Then the formula is meaningful for all other replacements.

$$5 - 2x = 0$$
$$5 = 2x$$
$$\frac{5}{2} = x$$

The domain is $\left\{x \Big| x \neq \dfrac{5}{2}\right\}$.

61. $f(x) = 3x + 5$

The domain of $f = \{0, 1, 2, 3\}$. We find function values by substituting into the formula. The range is the resulting set of function values.

$$f(0) = 3 \cdot 0 + 5 = 0 + 5 = 5$$
$$f(1) = 3 \cdot 1 + 5 = 3 + 5 = 8$$
$$f(2) = 3 \cdot 2 + 5 = 6 + 5 = 11$$
$$f(3) = 3 \cdot 3 + 5 = 9 + 5 = 14$$

The range of $f = \{5, 8, 11, 14\}$.

63. Graph: $f(x) = \dfrac{|x|}{x}$

We compute some function values. Note that 0 cannot be an input.

$$f(-5) = \frac{|-5|}{-5} = -1$$
$$f(-3) = \frac{|-3|}{-3} = -1$$
$$f(-1) = \frac{|-1|}{-1} = -1$$
$$f(1) = \frac{|1|}{1} = 1$$
$$f(2) = \frac{|2|}{2} = 1$$
$$f(4) = \frac{|4|}{4} = 1$$

A list of function values is shown in this table.

x	$f(x) = \dfrac{|x|}{x}$
-5	-1
-3	-1
-1	-1
1	1
2	1
4	1

We plot the points and connect them. The open circles indicate that neither the point $(0, -1)$ nor the point $(0, 1)$ is on the graph.

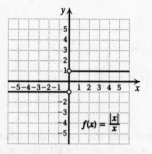

65. Graph: $h(x) = |x| - x$

We compute some function values.

$h(-3) = |-3| - (-3) = 6$

$h(-1) = |-1| - (-1) = 2$

$h(0) = |0| - 0 = 0$

$h(1) = |1| - 1 = 0$

$h(2) = |2| - 2 = 0$

$h(4) = |4| - 4 = 0$

A list of function values is shown in this table.

x	$h(x) = |x| - x$
-3	6
-1	2
0	0
1	0
2	0
4	0

We plot the points and connect them.

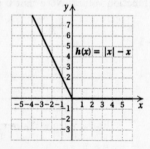

67. Graph: $|y| = x$

We choose some values for y and find the corresponding x-values.

When $y = -5$, $x = |y| = |-5| = 5$.

When $y = -3$, $x = |y| = |-3| = 3$.

When $y = 0$, $x = |y| = |0| = 0$.

When $y = 1$, $x = |y| = |1| = 1$.

When $y = 4$, $x = |y| = |4| = 4$.

A list of function values is shown in this table.

$x = |y|$	y
5	-5
3	-3
0	0
1	1
4	4

We plot the points and connect them.

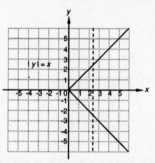

Since the graph does not pass the vertical-line test, it is not the graph of a function.

Exercise Set 9.5

1. $f(x) = x - 5$, $g(x) = x + 5$

$(f + g)(x) = f(x) + g(x) = (x - 5) + (x + 5) = 2x$

$(f - g)(x) = f(x) - g(x) = (x-5)-(x+5) = x-5-x-5 = -10$

$fg(x) = f(x) \cdot g(x) = (x - 5)(x + 5) = x^2 - 25$

$(f/g)(x) = f(x)/g(x) = \dfrac{x - 5}{x + 5}$

$ff(x) = f(x) \cdot f(x) = (x - 5)(x - 5) = x^2 - 10x - 25$

3. $f(x) = 3x^2 - 2x + 1$, $g(x) = x^3$

$(f + g)(x) = f(x) + g(x) = (3x^2 - 2x + 1) + (x^3) = x^3 + 3x^2 - 2x + 1$

$(f - g)(x) = f(x) - g(x) = (3x^2 - 2x + 1) - (x^3) = -x^3 + 3x^2 - 2x + 1$

$fg(x) = f(x) \cdot g(x) = (3x^2 - 2x + 1)(x^3) = 3x^5 - 2x^4 + x^3$

$(f/g)(x) = f(x)/g(x) = \dfrac{3x^2 - 2x + 1}{x^3}$

$ff(x) = f(x) \cdot f(x) = (3x^2 - 2x + 1)(3x^2 - 2x + 1) = 9x^4 - 6x^3 + 3x^2 - 6x^3 + 4x^2 - 2x + 3x^2 - 2x + 1 = 9x^4 - 12x^3 + 10x^2 - 4x + 1$

5. $f(x) = -5x^2$, $g(x) = 4x^3$

$(f + g)(x) = f(x) + g(x) = -5x^2 + 4x^3 = 4x^3 - 5x^2$

$(f - g)(x) = f(x) - g(x) = -5x^2 - 4x^3 = -4x^3 - 5x^2$

$fg(x) = f(x) \cdot g(x) = -5x^2 \cdot 4x^3 = -20x^5$

$(f/g)(x) = f(x)/g(x) = \dfrac{-5x^2}{4x^3} = \dfrac{-5}{4x}$, or $-\dfrac{5}{4x}$

$ff(x) = f(x) \cdot f(x) = -5x^2(-5x^2) = 25x^4$

7. $f(x) = 10$, $g(x) = -3$

$(f + g)(x) = f(x) + g(x) = 10 + (-3) = 7$

$(f - g)(x) = f(x) - g(x) = 10 - (-3) = 13$

$fg(x) = f(x) \cdot g(x) = 10(-3) = -30$

$(f/g)(x) = f(x)/g(x) = \dfrac{10}{-3}$, or $-\dfrac{10}{3}$

$ff(x) = f(x) \cdot f(x) = 10 \cdot 10 = 100$

9. $f \circ g(x) = f(g(x)) = f(6 - 4x) = 2(6 - 4x) - 3 =$
$12 - 8x - 3 = -8x + 9$

$g \circ f(x) = g(f(x)) = g(2x - 3) = 6 - 4(2x - 3) =$
$6 - 8x + 12 = -8x + 18$

11. $f \circ g(x) = f(g(x)) = f(2x - 1) = 3(2x - 1)^2 + 2 =$
$3(4x^2 - 4x + 1) + 2 = 12x^2 - 12x + 3 + 2 =$
$12x^2 - 12x + 5$

$g \circ f(x) = g(f(x)) = g(3x^2 + 2) = 2(3x^2 + 2) - 1 =$
$6x^2 + 4 - 1 = 6x^2 + 3$

13. $f \circ g(x) = f(g(x)) = f\left(\dfrac{2}{x}\right) = 4\left(\dfrac{2}{x}\right)^2 - 1 =$
$4\left(\dfrac{4}{x^2}\right) - 1 = \dfrac{16}{x^2} - 1$

$g \circ f(x) = g(f(x)) = g(4x^2 - 1) = \dfrac{2}{4x^2 - 1}$

15. $f \circ g(x) = f(g(x)) = f(x^2 - 5) = (x^2 - 5)^2 + 5 =$
$x^4 - 10x^2 + 25 + 5 = x^4 - 10x^2 + 30$

$g \circ f(x) = g(f(x)) = g(x^2 + 5) = (x^2 + 5)^2 - 5 =$
$x^4 + 10x^2 + 25 - 5 = x^4 + 10x^2 + 20$

17. $h(x) = (5 - 3x)^2$

This is $5 - 3x$ raised to the second power, so the two most obvious functions are $f(x) = x^2$ and $g(x) = 5 - 3x$.

19. $h(x) = \sqrt{5x + 2}$

This is the square root of $5x + 2$, so the two most obvious functions are $f(x) = \sqrt{x}$ and $g(x) = 5x + 2$.

21. $h(x) = \dfrac{1}{x - 1}$

This is the reciprocal of $x - 1$, so the two most obvious functions are $f(x) = \dfrac{1}{x}$ and $g(x) = x - 1$.

23. $h(x) = \dfrac{1}{\sqrt{7x + 2}}$

This is the reciprocal of the square root of $7x + 2$. Two functions that can be used are $f(x) = \dfrac{1}{\sqrt{x}}$ and $g(x) = 7x + 2$.

25. $h(x) = (\sqrt{x} + 5)^4$

This is $\sqrt{x} + 5$ raised to the fourth power, so the two most obvious functions are $f(x) = x^4$ and $g(x) = \sqrt{x} + 5$.

27. $\sqrt{24x^3 y^5} = \sqrt{4 \cdot 6 \cdot x^2 \cdot x \cdot y^4 \cdot y} = \sqrt{4x^2 y^4}\sqrt{6xy} = 2xy^2\sqrt{6xy}$

29. $\sqrt[3]{27x^6 y^{12}} = 3x^2 y^4$

31. $2x^2 - 7x + 4 = 0$

$a = 2,\ b = -7,\ c = 4$

$x = \dfrac{-b \pm \sqrt{b^2 - 4ac}}{2a} = \dfrac{-(-7) \pm \sqrt{(-7)^2 - 4 \cdot 2 \cdot 4}}{2 \cdot 2}$

$x = \dfrac{7 \pm \sqrt{49 - 32}}{4} = \dfrac{7 \pm \sqrt{17}}{4}$

33. $\dfrac{f(x + h) - f(x)}{h} = \dfrac{3(x + h) + 7 - (3x + 7)}{h}$

$= \dfrac{3x + 3h + 7 - 3x - 7}{h}$

$= \dfrac{3h}{h} = 3$

35. $\dfrac{f(x + h) - f(x)}{h} = \dfrac{-7 - (-7)}{h} = \dfrac{-7 + 7}{h} = 0$

37. $\dfrac{f(x + h) - f(x)}{h} = \dfrac{(x + h)^2 - x^2}{h}$

$= \dfrac{x^2 + 2xh + h^2 - x^2}{h}$

$= \dfrac{2xh + h^2}{h} = \dfrac{h(2x + h)}{h}$

$= 2x + h$

39. $\dfrac{f(x + h) - f(x)}{h} = \dfrac{\dfrac{1}{x + h} - \dfrac{1}{x}}{h} = \dfrac{\dfrac{x - (x + h)}{x(x + h)}}{h}$

$= \dfrac{\dfrac{x - x - h}{x(x + h)}}{h} = \dfrac{\dfrac{-h}{x(x + h)}}{h}$

$= \dfrac{-h}{x(x + h)} \cdot \dfrac{1}{h} = \dfrac{-1}{x(x + h)}$

41. $\dfrac{f(x + h) - f(x)}{h} = \dfrac{\sqrt{x + h} - \sqrt{x}}{h}$

$= \dfrac{\sqrt{x + h} - \sqrt{x}}{h} \cdot \dfrac{\sqrt{x + h} + \sqrt{x}}{\sqrt{x + h} + \sqrt{x}}$

Rationalizing the numerator

$= \dfrac{x + h - x}{h(\sqrt{x + h} + \sqrt{x})}$

$= \dfrac{h}{h(\sqrt{x + h} + \sqrt{x})}$

$= \dfrac{1}{\sqrt{x + h} + \sqrt{x}}$

43. If $C(x)$ is the cost per person, then $x \cdot C(x)$ is the total cost of the bus for x people.

$x \cdot C(x) = x \cdot \dfrac{100 + 5x}{x} = 100 + 5x$

$40 \cdot C(40) = 100 + 5 \cdot 40 = \$300$

$80 \cdot C(80) = 100 + 5 \cdot 80 = \$500$

Exercise Set 9.6

1.

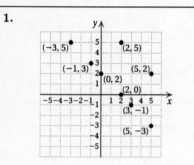

The inverse relation is $\{(-1,3),(2,5),(-3,5),(0,2)\}$.

3.

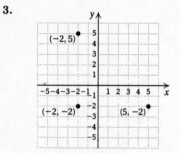

The inverse relation is $\{(-2,-2),(5,-2)\}$.

5. The graph of $f(x) = 5x - 2$ is shown below. It passes the horizontal-line test, so it is one-to-one.

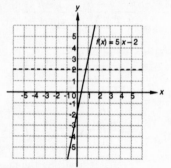

We find a formula for the inverse.

1. Replace $f(x)$ by y : $y = 5x - 2$

2. Interchange x and y : $x = 5y - 2$

3. Solve for y : $x + 2 = 5y$

$$\frac{x+2}{5} = y$$

4. Replace y by $f^{-1}(x)$: $f^{-1}(x) = \dfrac{x+2}{5}$

7. The graph of $f(x) = x^2 - 2$ is shown below. There are many horizontal lines that cross the graph more than once, so the function is not one-to-one.

9. The graph of $g(x) = |x| - 3$ is shown below. There are many horizontal lines that cross the graph more than once, so the function is not one-to-one.

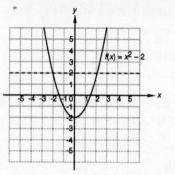

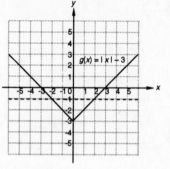

11. The graph of $f(x) = |x + 3|$ is shown below. There are many horizontal lines that cross the graph more than once, so the function is not one-to-one.

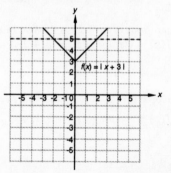

13. The graph of $g(x) = \dfrac{-2}{x}$ is shown below. It passes the horizontal-line test, so it is one-to-one.

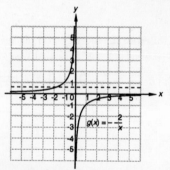

We find a formula for the inverse.

1. Replace $g(x)$ by y: $y = \dfrac{-2}{x}$

2. Interchange x and y: $x = \dfrac{-2}{y}$

3. Solve for y: $y = \dfrac{-2}{x}$

4. Replace y by $g^{-1}(x)$: $g^{-1}(x) = \dfrac{-2}{x}$

15. $f(x) = x + 9$

  1. Replace $f(x)$ by y: $y = x + 9$

  2. Interchange x and y: $x = y + 9$

  3. Solve for y: $x - 9 = y$

  4. Replace y by $f^{-1}(x)$: $f^{-1}(x) = x - 9$

17. $f(x) = 1 - x$

  1. Replace $f(x)$ by y: $y = 1 - x$

  2. Interchange x and y: $x = 1 - y$

  3. Solve for y: $y = 1 - x$

  4. Replace y by $f^{-1}(x)$: $f^{-1}(x) = 1 - x$

19. $g(x) = x - 12$

  1. Replace $g(x)$ by y: $y = x - 12$

  2. Interchange x and y: $x = y - 12$

  3. Solve for y: $x + 12 = y$

  4. Replace y by $g^{-1}(x)$: $g^{-1}(x) = x + 12$

21. $f(x) = 3x$

  1. Replace $f(x)$ by y: $y = 3x$

  2. Interchange x and y: $x = 3y$

  3. Solve for y: $\dfrac{x}{3} = y$

  4. Replace y by $f^{-1}(x)$: $f^{-1}(x) = \dfrac{x}{3}$

23. $g(x) = 3x + 2$

  1. Replace $g(x)$ by y: $y = 3x + 2$

  2. Interchange x and y: $x = 3y + 2$

  3. Solve for y: $x - 2 = 3y$

$$\dfrac{x - 2}{3} = y$$

  4. Replace y by $g^{-1}(x)$: $g^{-1}(x) = \dfrac{x - 2}{3}$

25. $h(x) = \dfrac{2}{x + 5}$

  1. Replace $h(x)$ by y: $y = \dfrac{2}{x + 5}$

  2. Interchange x and y: $x = \dfrac{2}{y + 5}$

  3. Solve for y: $x(y + 5) = 2$

$$y + 5 = \dfrac{2}{x}$$

$$y = \dfrac{2}{x} - 5$$

  4. Replace y by $h^{-1}(x)$: $h^{-1}(x) = \dfrac{2}{x} - 5$

27. $f(x) = \dfrac{2x + 1}{5x + 3}$

  1. Replace $f(x)$ by y: $y = \dfrac{2x + 1}{5x + 3}$

  2. Interchange x and y: $x = \dfrac{2y + 1}{5y + 3}$

  3. Solve for y: $5xy + 3x = 2y + 1$

$$5xy - 2y = 1 - 3x$$

$$y(5x - 2) = 1 - 3x$$

$$y = \dfrac{1 - 3x}{5x - 2}$$

  4. Replace y by $f^{-1}(x)$: $f^{-1}(x) = \dfrac{1 - 3x}{5x - 2}$

29. $g(x) = \dfrac{x - 3}{x + 4}$

  1. Replace $g(x)$ by y: $y = \dfrac{x - 3}{x + 4}$

  2. Interchange x and y: $x = \dfrac{y - 3}{y + 4}$

  3. Solve for y: $xy + 4x = y - 3$

$$4x + 3 = y - xy$$

$$4x + 3 = y(1 - x)$$

$$\dfrac{4x + 3}{1 - x} = y$$

  4. Replace y by $g^{-1}(x)$: $g^{-1}(x) = \dfrac{4x + 3}{1 - x}$

31. $f(x) = x^3 - 1$

  1. Replace $f(x)$ by y: $y = x^3 - 1$

  2. Interchange x and y: $x = y^3 - 1$

  3. Solve for y: $x + 1 = y^3$

$$\sqrt[3]{x + 1} = y$$

  4. Replace y by $f^{-1}(x)$: $f^{-1}(x) = \sqrt[3]{x + 1}$

33. $G(x) = (x - 2)^3$

  1. Replace $G(x)$ by y: $y = (x - 2)^3$

  2. Interchange x and y: $x = (y - 2)^3$

  3. Solve for y: $\sqrt[3]{x} = y - 2$

$$\sqrt[3]{x} + 2 = y$$

  4. Replace y by $G^{-1}(x)$: $G^{-1}(x) = \sqrt[3]{x} + 2$

35. $f(x) = \sqrt[3]{x}$

  1. Replace $f(x)$ by y: $y = \sqrt[3]{x}$

  2. Interchange x and y: $x = \sqrt[3]{y}$

  3. Solve for y: $x^3 = y$

  4. Replace y by $f^{-1}(x)$: $f^{-1}(x) = x^3$

37. Relation: $y = -\dfrac{1}{2}x + 2$

Inverse: $x = -\dfrac{1}{2}y + 2$

We graph the relation and its inverse:

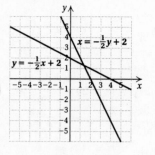

39. Relation: $y = x^2 - 3$

Inverse: $x = y^2 - 3$

We graph the relation and its inverse:

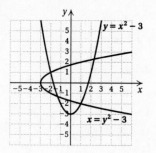

41. We first graph $f(x) = \frac{1}{2}x - 3$. The graph of f^{-1} can be obtained by reflecting the graph of f across the line $y = x$.

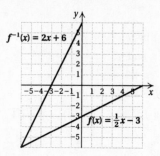

43. We first graph $f(x) = x^3$. The graph of f^{-1} can be obtained by reflecting the graph of f across the line $y = x$.

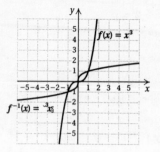

45. a) $f(8) = 2(8 + 12) = 2 \cdot 20 = 40$

Size 40 in Italy corresponds to size 8 in the United States.

$$f(10) = 2(10 + 12) = 2 \cdot 22 = 44$$

Size 44 in Italy corresponds to size 10 in the United States.

$$f(14) = 2(14 + 12) = 2 \cdot 26 = 52$$

Size 52 in Italy corresponds to size 14 in the United States.

$$f(18) = 2(18 + 12) = 2 \cdot 30 = 60$$

Size 60 in Italy corresponds to size 18 in the United States.

b) The graph of $f(x) = 2(x+12)$ is shown below. It passes the horizontal-line test, so the function is one-to-one and, hence, has an inverse that is a function.

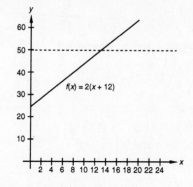

We find a formula for the inverse.

1. Replace $f(x)$ by y : $y = 2(x + 12)$

2. Interchange x and y : $x = 2(y + 12)$

3. Solve for y : $\dfrac{x}{2} = y + 12$

$$\frac{x}{2} - 12 = y$$

4. Replace y by $f^{-1}(x)$: $f^{-1}(x) = \dfrac{x}{2} - 12$

c) $f^{-1}(40) = \dfrac{40}{2} - 12 = 20 - 12 = 8$

Size 8 in the United States corresponds to size 40 in Italy.

$$f^{-1}(44) = \frac{44}{2} - 12 = 22 - 12 = 10$$

Size 10 in the United States corresponds to size 44 in Italy.

$$f^{-1}(52) = \frac{52}{2} - 12 = 26 - 12 = 14$$

Size 14 in the United States corresponds to size 52 in Italy.

$$f^{-1}(60) = \frac{60}{2} - 12 = 30 - 12 = 18$$

Size 18 in the United States corresponds to size 60 in Italy.

47. The graph of $f(x) = 4$ is shown below. Since the horizontal line $y = 4$ crosses the graph in more than one place, the function does not have an inverse that is a function.

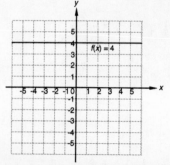

49. $C(x) = \dfrac{100 + 5x}{x}$

    1. Replace $C(x)$ by y : $y = \dfrac{100 + 5x}{x}$

    2. Interchange x and y : $x = \dfrac{100 + 5y}{y}$

    3. Solve for y : $xy = 100 + 5y$

$$xy - 5y = 100$$
$$y(x - 5) = 100$$
$$y = \frac{100}{x - 5}$$

    4. Replace y by $C^{-1}(x)$: $C^{-1}(x) = \dfrac{100}{x - 5}$

$C^{-1}(x)$ gives the number of people in the group, where x is the cost per person (in dollars) of chartering a bus.

51. $f \circ f^{-1}(x) = f(f^{-1}(x)) = f(3x - 7) = \dfrac{(3x - 7) + 7}{3} = \dfrac{3x}{3} = x$

$f^{-1} \circ f(x) = f^{-1}(f(x)) = 3\left(\dfrac{x + 7}{3}\right) - 7 = x + 7 - 7 = x$

53. $f \circ f^{-1}(x) = f(f^{-1}(x)) = (\sqrt[3]{x + 5})^3 - 5 = x + 5 - 5 = x$

$f^{-1} \circ f(x) = f^{-1}(f(x)) = \sqrt[3]{(x^3 - 5) + 5} = \sqrt[3]{x^3} = x$

55. $f \circ g(x) = f(g(x)) = \sqrt[5]{x^5} = x$

$g \circ f(x) = g(f(x)) = (\sqrt[5]{x})^5 = x$

f and g are inverses of each other.

57. $f \circ g(x) = f(g(x)) = \dfrac{2\left(\dfrac{7x - 4}{3x + 2}\right) - 3}{4\left(\dfrac{7x - 4}{3x + 2}\right) + 7}$

$$= \frac{\dfrac{14x - 8}{3x + 2} - 3}{\dfrac{28x - 16}{3x + 2} + 7}$$

$$= \frac{\dfrac{14x - 8}{3x + 2} - 3}{\dfrac{28x - 16}{3x + 2} + 7} \cdot \frac{3x + 2}{3x + 2}$$

$$= \frac{14x - 8 - 3(3x + 2)}{28x - 16 + 7(3x + 2)}$$

$$= \frac{14x - 8 - 9x - 6}{28x - 16 + 21x + 14}$$

$$= \frac{5x - 14}{49x - 2}$$

Since $f \circ g(x) \neq x$, f and g are not inverses of each other.

Chapter 10

Exponential and Logarithmic Functions

---

1. Graph: $y = 2^x$

We compute some function values, thinking of y as $f(x)$, and keep the results in a table.

$f(0) = 2^0 = 1$

$f(1) = 2^1 = 2$

$f(2) = 2^2 = 4$

$f(-1) = 2^{-1} = \dfrac{1}{2^1} = \dfrac{1}{2}$

$f(-2) = 2^{-2} = \dfrac{1}{2^2} = \dfrac{1}{4}$

x	y, or $f(x)$
0	1
1	2
2	4
-1	$\dfrac{1}{2}$
-2	$\dfrac{1}{4}$

Next we plot these points and connect them with a smooth curve.

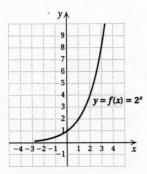

3. Graph: $y = 5^x$

We compute some function values, thinking of y as $f(x)$, and keep the results in a table.

$f(0) = 5^0 = 1$

$f(1) = 5^1 = 5$

$f(2) = 5^2 = 25$

$f(-1) = 5^{-1} = \dfrac{1}{5^1} = \dfrac{1}{5}$

$f(-2) = 5^{-2} = \dfrac{1}{5^2} = \dfrac{1}{25}$

x	y, or $f(x)$
0	1
1	5
2	25
-1	$\dfrac{1}{5}$
-2	$\dfrac{1}{25}$

Next we plot these points and connect them with a smooth curve.

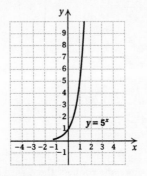

5. Graph: $y = 2^{x+1}$

We compute some function values, thinking of y as $f(x)$, and keep the results in a table.

$f(0) = 2^{0+1} = 2^1 = 2$

$f(-1) = 2^{-1+1} = 2^0 = 1$

$f(-2) = 2^{-2+1} = 2^{-1} = \dfrac{1}{2^1} = \dfrac{1}{2}$

$f(-3) = 2^{-3+1} = 2^{-2} = \dfrac{1}{2^2} = \dfrac{1}{4}$

$f(1) = 2^{1+1} = 2^2 = 4$

$f(2) = 2^{2+1} = 2^3 = 8$

x	y, or $f(x)$
0	2
-1	1
-2	$\dfrac{1}{2}$
-3	$\dfrac{1}{4}$
1	4
2	8

Next we plot these points and connect them with a smooth curve.

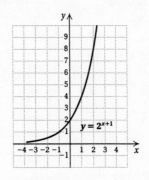

7. Graph: $y = 3^{x-2}$

We compute some function values, thinking of y as $f(x)$, and keep the results in a table.

$f(0) = 3^{0-2} = 3^{-2} = \dfrac{1}{3^2} = \dfrac{1}{9}$

$f(1) = 3^{1-2} = 3^{-1} = \dfrac{1}{3^1} = \dfrac{1}{3}$

$f(2) = 3^{2-2} = 3^0 = 1$

$f(3) = 3^{3-2} = 3^1 = 3$

$f(4) = 3^{4-2} = 3^2 = 9$

$f(-1) = 3^{-1-2} = 3^{-3} = \dfrac{1}{3^3} = \dfrac{1}{27}$

$f(-2) = 3^{-2-2} = 3^{-4} = \dfrac{1}{3^4} = \dfrac{1}{81}$

x	y, or $f(x)$
0	$\dfrac{1}{9}$
1	$\dfrac{1}{3}$
2	1
3	3
4	9
-1	$\dfrac{1}{27}$
-2	$\dfrac{1}{81}$

Next we plot these points and connect them with a smooth curve.

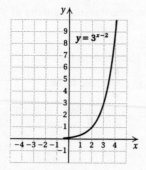

9. Graph: $y = 2^x - 3$

We construct a table of values, thinking of y as $f(x)$. Then we plot the points and connect them with a smooth curve.

$f(0) = 2^0 - 3 = 1 - 3 = -2$

$f(1) = 2^1 - 3 = 2 - 3 = -1$

$f(2) = 2^2 - 3 = 4 - 3 = 1$

$f(3) = 2^3 - 3 = 8 - 3 = 5$

$f(-1) = 2^{-1} - 3 = \dfrac{1}{2} - 3 = -\dfrac{5}{2}$

$f(-2) = 2^{-2} - 3 = \dfrac{1}{4} - 3 = -\dfrac{11}{4}$

x	y, or $f(x)$
0	-2
1	-1
2	1
3	5
-1	$-\dfrac{5}{2}$
-2	$-\dfrac{11}{4}$

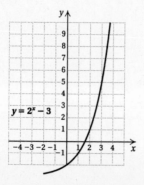

11. Graph: $y = 5^{x+3}$

We construct a table of values, thinking of y as $f(x)$. Then we plot the points and connect them with a smooth curve.

$f(0) = 5^{0+3} = 5^3 = 125$

$f(-1) = 5^{-1+3} = 5^2 = 25$

$f(-2) = 5^{-2+3} = 5^1 = 5$

$f(-3) = 5^{-3+3} = 5^0 = 1$

$f(-4) = 5^{-4+3} = 5^{-1} = \dfrac{1}{5}$

$f(-5) = 5^{-5+3} = 5^{-2} = \dfrac{1}{25}$

x	y, or $f(x)$
0	125
-1	25
-2	5
-3	1
-4	$\dfrac{1}{5}$
-5	$\dfrac{1}{25}$

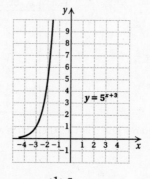

13. Graph: $y = \left(\dfrac{1}{2}\right)^x$

We construct a table of values, thinking of y as $f(x)$. Then we plot the points and connect them with a smooth curve.

$f(0) = \left(\dfrac{1}{2}\right)^0 = 1$

$f(1) = \left(\dfrac{1}{2}\right)^1 = \dfrac{1}{2}$

$f(2) = \left(\dfrac{1}{2}\right)^2 = \dfrac{1}{4}$

$f(3) = \left(\dfrac{1}{2}\right)^3 = \dfrac{1}{8}$

$f(-1) = \left(\dfrac{1}{2}\right)^{-1} = \dfrac{1}{\left(\dfrac{1}{2}\right)^1} = \dfrac{1}{\dfrac{1}{2}} = 2$

$f(-2) = \left(\dfrac{1}{2}\right)^{-2} = \dfrac{1}{\left(\dfrac{1}{2}\right)^2} = \dfrac{1}{\dfrac{1}{4}} = 4$

$f(-3) = \left(\dfrac{1}{2}\right)^{-3} = \dfrac{1}{\left(\dfrac{1}{2}\right)^3} = \dfrac{1}{\dfrac{1}{8}} = 8$

x	y, or $f(x)$
0	1
1	$\dfrac{1}{2}$
2	$\dfrac{1}{4}$
3	$\dfrac{1}{8}$
-1	2
-2	4
-3	8

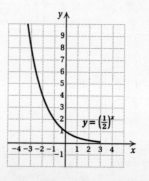

15. Graph: $y = \left(\frac{1}{5}\right)^x$

We construct a table of values, thinking of y as $f(x)$. Then we plot the points and connect them with a smooth curve.

$f(0) = \left(\frac{1}{5}\right)^0 = 1$

$f(1) = \left(\frac{1}{5}\right)^1 = \frac{1}{5}$

$f(2) = \left(\frac{1}{5}\right)^2 = \frac{1}{25}$

$f(-1) = \left(\frac{1}{5}\right)^{-1} = \frac{1}{\frac{1}{5}} = 5$

$f(-2) = \left(\frac{1}{5}\right)^{-2} = \frac{1}{\frac{1}{25}} = 25$

x	y, or $f(x)$
0	1
1	$\frac{1}{5}$
2	$\frac{1}{25}$
-1	5
-2	25

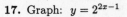

17. Graph: $y = 2^{2x-1}$

We construct a table of values, thinking of y as $f(x)$. Then we plot the points and connect them with a smooth curve.

$f(0) = 2^{2 \cdot 0 - 1} = 2^{-1} = \frac{1}{2}$

$f(1) = 2^{2 \cdot 1 - 1} = 2^1 = 2$

$f(2) = 2^{2 \cdot 2 - 1} = 2^3 = 8$

$f(-1) = 2^{2(-1)-1} = 2^{-3} = \frac{1}{8}$

$f(-2) = 2^{2(-2)-1} = 2^{-5} = \frac{1}{32}$

x	y, or $f(x)$
0	$\frac{1}{2}$
1	2
2	8
-1	$\frac{1}{8}$
-2	$\frac{1}{32}$

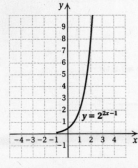

19. Graph: $x = 2^y$

We can find ordered pairs by choosing values for y and then computing values for x.

For $y = 0$, $x = 2^0 = 1$.

For $y = 1$, $x = 2^1 = 2$.

For $y = 2$, $x = 2^2 = 4$.

For $y = 3$, $x = 2^3 = 8$.

For $y = -1$, $x = 2^{-1} = \frac{1}{2^1} = \frac{1}{2}$.

For $y = -2$, $x = 2^{-2} = \frac{1}{2^2} = \frac{1}{4}$.

For $y = -3$, $x = 2^{-3} = \frac{1}{2^3} = \frac{1}{8}$.

x	y
1	0
2	1
4	2
8	3
$\frac{1}{2}$	-1
$\frac{1}{4}$	-2
$\frac{1}{8}$	-3

⌐ (1) Choose values for y.

└ (2) Compute values for x.

We plot these points and connect them with a smooth curve.

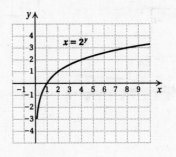

21. Graph: $x = \left(\frac{1}{2}\right)^y$

We can find ordered pairs by choosing values for y and then computing values for x. Then we plot these points and connect them with a smooth curve.

For $y = 0$, $x = \left(\frac{1}{2}\right)^0 = 1$.

For $y = 1$, $x = \left(\frac{1}{2}\right)^1 = \frac{1}{2}$.

For $y = 2$, $x = \left(\frac{1}{2}\right)^2 = \frac{1}{4}$.

For $y = 3$, $x = \left(\frac{1}{2}\right)^3 = \frac{1}{8}$.

For $y = -1$, $x = \left(\frac{1}{2}\right)^{-1} = \frac{1}{\frac{1}{2}} = 2$.

For $y = -2$, $x = \left(\frac{1}{2}\right)^{-2} = \frac{1}{\frac{1}{4}} = 4$.

For $y = -3$, $x = \left(\frac{1}{2}\right)^{-3} = \frac{1}{\frac{1}{8}} = 8$.

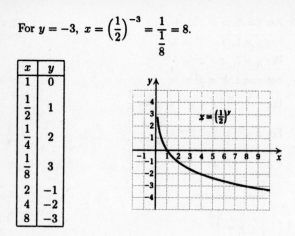

x	y
1	0
$\frac{1}{2}$	1
$\frac{1}{4}$	2
$\frac{1}{8}$	3
2	-1
4	-2
8	-3

23. Graph: $x = 5^y$

We can find ordered pairs by choosing values for y and then computing values for x. Then we plot these points and connect them with a smooth curve.

For $y = 0$, $x = 5^0 = 1$.

For $y = 1$, $x = 5^1 = 5$.

For $y = 2$, $x = 5^2 = 25$.

For $y = -1$, $x = 5^{-1} = \frac{1}{5}$.

For $y = -2$, $x = 5^{-2} = \frac{1}{25}$.

x	y
1	0
5	1
25	2
$\frac{1}{5}$	-1
$\frac{1}{25}$	-2

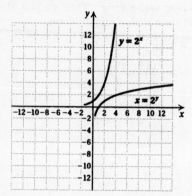

25. Graph $y = 2^x$ (see Exercise 1) and $x = 2^y$ (see Exercise 19) using the same set of axes.

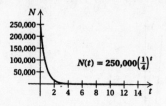

27. a) $N(t) = 250,000\left(\frac{1}{4}\right)^t$

Substitute for t and do the computations.

$N(0) = 250,000\left(\frac{1}{4}\right)^0 = 250,000 \cdot 1 = 250,000$

$N(1) = 250,000\left(\frac{1}{4}\right)^1 = 250,000 \cdot \frac{1}{4} = 62,500$

$N(4) = 250,000\left(\frac{1}{4}\right)^4 = 250,000 \cdot \frac{1}{256} \approx 976$

(We round down since we are not considering partial cans.)

$N(10) = 250,000\left(\frac{1}{4}\right)^{10} = 250,000 \cdot \frac{1}{1,048,576} \approx 0$

b) We use the function values computed in part (a) with others, if we wish, and draw the graph. Note that the axes are scaled differently because of the large function values.

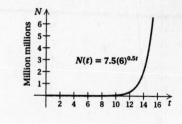

29. a) $N(t) = 7.5(6)^{0.5t}$

Substitute for t and do the computations.

$N(1) = 7.5(6)^{0.5(1)} = 7.5(6)^{0.5} \approx 18.4$ million

In 1987, $t = 1987 - 1985$, or 2.

$N(2) = 7.5(6)^{0.5(2)} = 7.5(6) = 45$ million

In 1988, $t = 1988 - 1985$, or 3.

$N(3) = 7.5(6)^{0.5(3)} = 7.5(6)^{1.5} \approx 110.2$ million

In 1990, $t = 1990 - 1985$, or 5.

$N(5) = 7.5(6)^{0.5(5)} = 7.5(6)^{2.5} \approx 661.4$ million

In 1995, $t = 1995 - 1985$, or 10.

$N(10) = 7.5(6)^{0.5(10)} = 7.5(6)^5 \approx 58,320$ million, or 58.32 billion

In 2000, $t = 2000 - 1985$, or 15.

$N(15) = 7.5(6)^{0.5(15)} = 7.5(6)^{7.5} \approx 5,142,753$ million, or 5.1 trillion

b) We use the function values computed in part (a) with others, if we wish, and draw the graph.

31. $x^{-5} \cdot x^3 = x^{-5+3} = x^{-2} = \frac{1}{x^2}$

33. $9^0 = 1$ (For any nonzero number a, $a^0 = 1$.)

35. $\frac{x^{-3}}{x^4} = x^{-3-4} = x^{-7} = \frac{1}{x^7}$

37. Graph: $y = 2^x + 2^{-x}$

Construct a table of values, thinking of y as $f(x)$. Then plot these points and connect them with a curve.

$f(0) = 2^0 + 2^{-0} = 1 + 1 = 2$

$f(1) = 2^1 + 2^{-1} = 2 + \frac{1}{2} = 2\frac{1}{2}$

$f(2) = 2^2 + 2^{-2} = 4 + \frac{1}{4} = 4\frac{1}{4}$

$f(3) = 2^3 + 2^{-3} = 8 + \frac{1}{8} = 8\frac{1}{8}$

$f(-1) = 2^{-1} + 2^{-(-1)} = \frac{1}{2} + 2 = 2\frac{1}{2}$

$f(-2) = 2^{-2} + 2^{-(-2)} = \frac{1}{4} + 4 = 4\frac{1}{4}$

$f(-3) = 2^{-3} + 2^{-(-3)} = \frac{1}{8} + 8 = 8\frac{1}{8}$

x	y, or $f(x)$
0	2
1	$2\frac{1}{2}$
2	$4\frac{1}{4}$
3	$8\frac{1}{8}$
-1	$2\frac{1}{2}$
-2	$4\frac{1}{4}$
-3	$8\frac{1}{8}$

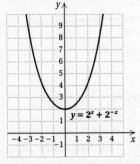

39. Graph: $y = |2^x - 2|$

Construct a table of values, thinking of y as $f(x)$. Then plot these points and connect them with a curve.

$f(-4) = |2^{-4} - 2| = \left|\frac{1}{16} - 2\right| = \left|-1\frac{15}{16}\right| = 1\frac{15}{16}$

$f(-2) = |2^{-2} - 2| = \left|\frac{1}{4} - 2\right| = \left|-1\frac{3}{4}\right| = 1\frac{3}{4}$

$f(0) = |2^0 - 2| = |1 - 2| = |-1| = 1$

$f(1) = |2^1 - 2| = |2 - 2| = |0| = 0$

$f(2) = |2^2 - 2| = |4 - 2| = |2| = 2$

$f(3) = |2^3 - 2| = |8 - 2| = |6| = 6$

$f(4) = |2^4 - 2| = |16 - 2| = |14| = 14$

x	y, or $f(x)$
-4	$1\frac{15}{16}$
-2	$1\frac{3}{4}$
0	1
1	0
2	2
3	6
4	14

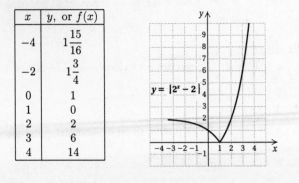

1. Graph: $y = \log_2 x$

The equation $y = \log_2 x$ is equivalent to $2^y = x$. We can find ordered pairs by choosing values for y and computing the corresponding x-values.

For $y = 0$, $x = 2^0 = 1$.

For $y = 1$, $x = 2^1 = 2$.

For $y = 2$, $x = 2^2 = 4$.

For $y = 3$, $x = 2^3 = 8$.

For $y = -1$, $x = 2^{-1} = \frac{1}{2}$.

For $y = -2$, $x = 2^{-2} = \frac{1}{4}$.

x, or 2^y	y
1	0
2	1
4	2
8	3
$\frac{1}{2}$	-1
$\frac{1}{4}$	-2

(1) Select y.

(2) Compute x.

We plot the set of ordered pairs and connect the points with a smooth curve.

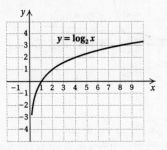

3. Graph: $y = \log_{1/3} x$

The equation $y = \log_{1/3} x$ is equivalent to $\left(\frac{1}{3}\right)^y = x$. We can find ordered pairs by choosing values for y and computing the corresponding x-values.

For $y = 0$, $x = \left(\frac{1}{3}\right)^0 = 1$.

For $y = 1$, $x = \left(\frac{1}{3}\right)^1 = \frac{1}{3}$.

For $y = 2$, $x = \left(\frac{1}{3}\right)^2 = \frac{1}{9}$.

For $y = -1$, $x = \left(\frac{1}{3}\right)^{-1} = 3$.

For $y = -2$, $x = \left(\frac{1}{3}\right)^{-2} = 9$.

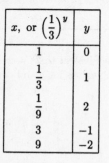

x, or $\left(\frac{1}{3}\right)^y$	y
1	0
$\frac{1}{3}$	1
$\frac{1}{9}$	2
3	−1
9	−2

We plot the set of ordered pairs and connect the points with a smooth curve.

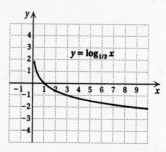

5. Graph $f(x) = 3^x$ (see Exercise Set 10.1, Exercise 2) and $f^{-1}(x) = \log_3 x$ on the same set of axes. We can obtain the graph of f^{-1} by reflecting the graph of f across the line $y = x$.

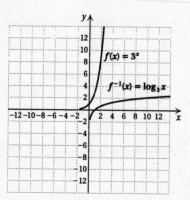

7.   The exponent is the logarithm.
$$10^3 = 1000 \Rightarrow 3 = \log_{10} 1000$$
   The base remains the same.

9.   The exponent is the logarithm.
$$5^{-3} = \frac{1}{125} \Rightarrow -3 = \log_5 \frac{1}{125}$$
   The base remains the same.

11. $8^{1/3} = 2 \Rightarrow \frac{1}{3} = \log_8 2$

13. $10^{0.3010} = 2 \Rightarrow 0.3010 = \log_{10} 2$

15. $e^2 = t \Rightarrow 2 = \log_e t$

17. $Q^t = x \Rightarrow t = \log_Q x$

19. $e^2 = 7.3891 \Rightarrow 2 = \log_e 7.3891$

21. $e^{-2} = 0.1353 \Rightarrow -2 = \log_e 0.1353$

23.   The logarithm is the exponent.
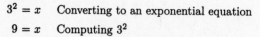
$$w = \log_4 10 \Rightarrow 4^w = 10$$
   The base remains the same.

25.   The logarithm is the exponent.
$$\log_6 36 = 2 \Rightarrow 6^2 = 36$$
   The base remains the same.

27. $\log_{10} 0.01 = -2 \Rightarrow 10^{-2} = 0.01$

29. $\log_{10} 8 = 0.9031 \Rightarrow 10^{0.9031} = 8$

31. $\log_e 100 = 4.6052 \Rightarrow e^{4.6052} = 100$

33. $\log_t Q = k \Rightarrow t^k = Q$

35. $\log_3 x = 2$
   $3^2 = x$ Converting to an exponential equation
   $9 = x$ Computing 3^2

37. $\log_x 16 = 2$
   $x^2 = 16$ Converting to an exponential equation

   $x = 4$ or $x = -4$ Principle of square roots

$\log_4 16 = 2$ because $4^2 = 16$. Thus, 4 is a solution. Since all logarithm bases must be positive, $\log_{-4} 16$ is not defined and -4 is not a solution.

39. $\log_2 16 = x$
   $2^x = 16$ Converting to an exponential equation
   $2^x = 2^4$
   $x = 4$ The exponents are the same.

41. $\log_3 27 = x$
   $3^x = 27$ Converting to an exponential equation
   $3^x = 3^3$
   $x = 3$ The exponents are the same.

43. $\log_x 25 = 1$
   $x^1 = 25$ Converting to an exponential equation
   $x = 25$

45. $\log_3 x = 0$
   $3^0 = x$ Converting to an exponential equation
   $1 = x$

47. $\log_2 x = -1$
   $2^{-1} = x$ Converting to an exponential equation
   $\frac{1}{2} = x$ Simplifying

49. $\log_8 x = \frac{1}{3}$

$8^{1/3} = x$

$2 = x$

51. Let $\log_{10} 100 = x$. Then

$10^x = 100$

$10^x = 10^2$

$x = 2$

Thus, $\log_{10} 100 = 2$.

53. Let $\log_{10} 0.1 = x$. Then

$10^x = 0.1 = \frac{1}{10}$

$10^x = 10^{-1}$

$x = -1$

Thus, $\log_{10} 0.1 = -1$.

55. Let $\log_{10} 1 = x$. Then

$10^x = 1$

$10^x = 10^0 \qquad (10^0 = 1)$

$x = 0$

Thus, $\log_{10} 1 = 0$.

57. Let $\log_5 625 = x$. Then

$5^x = 625$

$5^x = 5^4$

$x = 4$

Thus, $\log_5 625 = 4$.

59. Think of the meaning of $\log_7 49$. It is the exponent to which you raise 7 to get 49. That exponent is 2. Therefore, $\log_7 49 = 2$.

61. Think of the meaning of $\log_2 8$. It is the exponent to which you raise 2 to get 8. That exponent is 3. Therefore, $\log_2 8 = 3$.

63. Let $\log_9 \frac{1}{81} = x$. Then

$9^x = \frac{1}{81}$

$9^x = 9^{-2}$

$x = -2$

Thus, $\log_9 81 = -2$.

65. Let $\log_8 1 = x$. Then

$8^x = 1$

$8^x = 8^0 \qquad (8^0 = 1)$

$x = 0$

Thus, $\log_8 1 = 0$.

67. Let $\log_e e = x$. Then

$e^x = e$

$e^x = e^1$

$x = 1$

Thus, $\log_e e = 1$.

69. Let $\log_{27} 9 = x$. Then

$27^x = 9$

$(3^3)^x = 3^2$

$3^{3x} = 3^2$

$3x = 2$

$x = \frac{2}{3}$

Thus, $\log_{27} 9 = \frac{2}{3}$.

71.
$E = mc^2$

$\frac{E}{m} = c^2 \qquad$ Dividing by m

$\sqrt{\frac{E}{m}} = c \qquad$ Taking the positive square root

73.
$P = ab^2$

$\frac{P}{a} = b^2 \qquad$ Dividing by a

$\sqrt{\frac{P}{a}} = b \qquad$ Taking the positive square root

75.
$A = \sqrt{3ab}$

$A^2 = (\sqrt{3ab})^2 \qquad$ Principle of powers

$A^2 = 3ab$

$\frac{A^2}{3a} = b \qquad$ Dividing by $3a$

77. Graph: $y = \left(\frac{3}{2}\right)^x$ $\qquad$ Graph: $y = \log_{3/2} x$, or

$$x = \left(\frac{3}{2}\right)^y$$

x	y, or $\left(\frac{3}{2}\right)^x$
0	1
1	$\frac{3}{2}$
2	$\frac{9}{4}$
3	$\frac{27}{8}$
-1	$\frac{2}{3}$
-2	$\frac{4}{9}$

x, or $\left(\frac{3}{2}\right)^y$	y
1	0
$\frac{3}{2}$	1
$\frac{9}{4}$	2
$\frac{27}{8}$	3
$\frac{2}{3}$	-1
$\frac{4}{9}$	-2

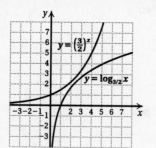

79. Graph: $y = \log_3 |x + 1|$

x	y
0	0
2	1
8	2
-2	0
-4	1
-9	2

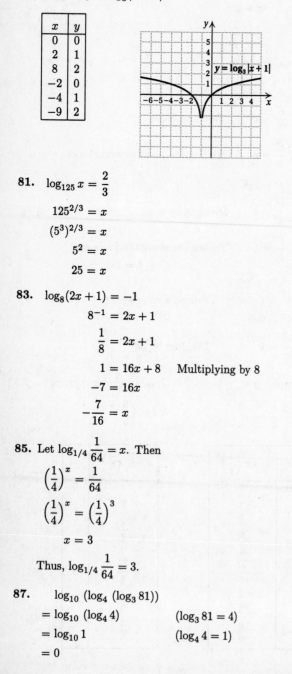

81. $\log_{125} x = \dfrac{2}{3}$

$125^{2/3} = x$

$(5^3)^{2/3} = x$

$5^2 = x$

$25 = x$

83. $\log_8(2x + 1) = -1$

$8^{-1} = 2x + 1$

$\dfrac{1}{8} = 2x + 1$

$1 = 16x + 8$ Multiplying by 8

$-7 = 16x$

$-\dfrac{7}{16} = x$

85. Let $\log_{1/4} \dfrac{1}{64} = x$. Then

$\left(\dfrac{1}{4}\right)^x = \dfrac{1}{64}$

$\left(\dfrac{1}{4}\right)^x = \left(\dfrac{1}{4}\right)^3$

$x = 3$

Thus, $\log_{1/4} \dfrac{1}{64} = 3$.

87. $\log_{10}(\log_4(\log_3 81))$

$= \log_{10}(\log_4 4)$ $(\log_3 81 = 4)$

$= \log_{10} 1$ $(\log_4 4 = 1)$

$= 0$

89. Let $\log_{1/5} 25 = x$. Then

$\left(\dfrac{1}{5}\right)^x = 25$

$(5^{-1})^x = 25$

$5^{-x} = 5^2$

$-x = 2$

$x = -2$

Thus, $\log_{1/5} 25 = -2$.

Exercise Set 10.3

1. $\log_2(32 \cdot 8) = \log_2 32 + \log_2 8$ Property 1

3. $\log_4(64 \cdot 16) = \log_4 64 + \log_4 16$ Property 1

5. $\log_a Qx = \log_a Q + \log_a x$ Property 1

7. $\log_b 3 + \log_b 84 = \log_b(3 \cdot 84)$ Property 1

$= \log_b 252$

9. $\log_c K + \log_c y = \log_c K \cdot y$ Property 1

$= \log_c Ky$

11. $\log_c y^4 = 4 \log_c y$ Property 2

13. $\log_b t^6 = 6 \log_b t$ Property 2

15. $\log_b C^{-3} = -3 \log_b C$ Property 2

17. $\log_a \dfrac{67}{5} = \log_a 67 - \log_a 5$ Property 3

19. $\log_b \dfrac{2}{5} = \log_b 2 - \log_b 5$ Property 3

21. $\log_c 22 - \log_c 3 = \log_c \dfrac{22}{3}$ Property 3

23. $\log_a x^2 y^3 z$

$= \log_a x^2 + \log_a y^3 + \log_a z$ Property 1

$= 2 \log_a x + 3 \log_a y + \log_a z$ Property 2

25. $\log_b \dfrac{xy^2}{z^3}$

$= \log_b xy^2 - \log_b z^3$ Property 3

$= \log_b x + \log_b y^2 - \log_b z^3$ Property 1

$= \log_b x + 2 \log_b y - 3 \log_b z$ Property 2

27. $\log_c \sqrt[3]{\dfrac{x^4}{y^3 z^2}}$

$= \log_c \left(\dfrac{x^4}{y^3 z^2}\right)^{1/3}$

$= \dfrac{1}{3}\, \log_c \dfrac{x^4}{y^3 z^2}$ Property 2

$= \dfrac{1}{3}(\log_c x^4 - \log_c y^3 z^2)$ Property 3

$= \dfrac{1}{3}[\log_c x^4 - (\log_c y^3 + \log_c z^2)]$ Property 1

$= \dfrac{1}{3}(\log_c x^4 - \log_c y^3 - \log_c z^2)$ Removing parentheses

$= \dfrac{1}{3}(4\,\log_c x - 3\,\log_c y - 2\,\log_c z)$ Property 2

$= \dfrac{4}{3}\,\log_c x - \log_c y - \dfrac{2}{3}\log_c z$

29. $\log_a \sqrt[4]{\dfrac{m^8 n^{12}}{a^3 b^5}}$

$= \log_a \left(\dfrac{m^8 n^{12}}{a^3 b^5}\right)^{1/4}$

$= \dfrac{1}{4}\,\log_a \dfrac{m^8 n^{12}}{a^3 b^5}$ Property 2

$= \dfrac{1}{4}(\log_a m^8 n^{12} - \log_a a^3 b^5)$ Property 3

$= \dfrac{1}{4}[\log_a m^8 + \log_a n^{12} - (\log_a a^3 + \log_a b^5)]$ Property 1

$= \dfrac{1}{4}(\log_a m^8 + \log_a n^{12} - \log_a a^3 - \log_a b^5)$

     Removing parentheses

$= \dfrac{1}{4}(\log_a m^8 + \log_a n^{12} - 3 - \log_a b^5)$ Property 4

$= \dfrac{1}{4}(8\,\log_a m + 12\,\log_a n - 3 - 5\,\log_a b)$ Property 2

$= 2\,\log_a m + 3\,\log_a n - \dfrac{3}{4} - \dfrac{5}{4}\log_a b$

31. $\dfrac{2}{3}\,\log_a x - \dfrac{1}{2}\,\log_a y$

$= \log_a x^{2/3} - \log_a y^{1/2}$ Property 2

$= \log_a \dfrac{x^{2/3}}{y^{1/2}}$, or   Property 3

$\log_a \dfrac{\sqrt[3]{x^2}}{\sqrt{y}}$

33. $\log_a 2x + 3(\log_a x - \log_a y)$

$= \log_a 2x + 3\,\log_a x - 3\,\log_a y$

$= \log_a 2x + \log_a x^3 - \log_a y^3$ Property 2

$= \log_a 2x^4 - \log_a y^3$ Property 1

$= \log_a \dfrac{2x^4}{y^3}$ Property 3

35. $\log_a \dfrac{a}{\sqrt{x}} - \log_a \sqrt{ax}$

$= \log_a ax^{-1/2} - \log_a a^{1/2}x^{1/2}$

$= \log_a \dfrac{ax^{-1/2}}{a^{1/2}x^{1/2}}$ Property 3

$= \log_a \dfrac{a^{1/2}}{x}$, or

$\log_a \dfrac{\sqrt{a}}{x}$

37. $\log_b 15 = \log_b (3 \cdot 5)$

$= \log_b 3 + \log_b 5$ Property 1

$= 1.099 + 1.609$

$= 2.708$

39. $\log_b \dfrac{5}{3} = \log_b 5 - \log_b 3$ Property 3

$= 1.609 - 1.099$

$= 0.51$

41. $\log_b \dfrac{1}{5} = \log_b 1 - \log_b 5$ Property 3

$= 0 - 1.609$ $(\log_b 1 = 0)$

$= -1.609$

43. $\log_b \sqrt{b^3} = \log_b b^{3/2} = \dfrac{3}{2}$ Property 4

45. $\log_b 5b = \log_b 5 + \log_b b$ Property 1

$= 1.609 + 1$ $(\log_b b = 1)$

$= 2.609$

47. $\log_e e^t = t$ Property 4

49. $\log_p p^5 = 5$ Property 4

51. $\log_2 2^7 = x$

$7 = x$ Property 4

53. $\log_e e^x = -7$

$x = -7$ Property 4

55. $i^{29} = i^{28} \cdot i = (i^2)^{14} \cdot i = (-1)^{14} \cdot i = 1 \cdot i = i$

57. $\dfrac{2+i}{2-i} = \dfrac{2+i}{2-i} \cdot \dfrac{2+i}{2+i} = \dfrac{4+4i+i^2}{4-i^2} = \dfrac{4+4i-1}{4-(-1)} = \dfrac{3+4i}{5} = \dfrac{3}{5} + \dfrac{4}{5}i$

59. $2i^2 \cdot 5i^3 = 10i^5 = 10 \cdot (i^2)^2 \cdot i = 10 \cdot (-1)^2 \cdot i = 10i$

61. $\log_a (x^8 - y^8) - \log_a (x^2 + y^2)$

$= \log_a \dfrac{x^8 - y^8}{x^2 + y^2}$ Property 3

$= \log_a \dfrac{(x^4 + y^4)(x^2 + y^2)(x + y)(x - y)}{x^2 + y^2}$ Factoring

$= \log_a [(x^4 + y^4)(x + y)(x - y)]$ Simplifying

$= \log_a (x^6 - x^4 y^2 + x^2 y^4 - y^6)$ Multiplying

63. $\log_a \sqrt{1 - s^2}$

$= \log_a (1 - s^2)^{1/2}$

$= \dfrac{1}{2} \log_a (1 - s^2)$

$= \dfrac{1}{2} \log_a [(1 - s)(1 + s)]$

$= \dfrac{1}{2} \log_a (1 - s) + \dfrac{1}{2} \log_a (1 + s)$

65. False. For example, let $a = 10$, $P = 100$, and $Q = 10$.

$\dfrac{\log 100}{\log 10} = \dfrac{2}{1} = 2$, but

$\log \dfrac{100}{10} = \log 10 = 1.$

67. True, by Property 1

69. False. For example, let $a = 2$, $P = 1$, and $Q = 1$.

$\log_2(1 + 1) = \log_2 2 = 1$, but

$\log_2 1 + \log_2 1 = 0 + 0 = 0.$

Exercise Set 10.4

1. 0.3010

3. 0.9031

5. 0.7185

7. 1.7226

9. 2.5159

11. 4.0864

13. -0.2441

15. -1.2840

17. -2.2069

19. 1000

21. 31,622.777

23. 3

25. 0.2841

27. 0.01494

29. 0.6931

31. 2.0794

33. 3.9703

35. -5.0832

37. 36.7890

39. 0.0023

41. 1.0057

43. 1.190×10^{10}

45. 8.1490

47. -3.6420

49. 1637.9488

51. 7.6331

53. We will use common logarithms for the conversion. Let $a = 10$, $b = 6$, and $M = 100$ and substitute in the change-of-base formula.

$\log_b M = \dfrac{\log_a M}{\log_a b}$

$\log_6 100 = \dfrac{\log_{10} 100}{\log_{10} 6}$

$\approx \dfrac{2}{0.7782}$

≈ 2.5702

55. We will use common logarithms for the conversion. Let $a = 10$, $b = 2$, and $M = 10$ and substitute in the change-of-base formula.

$\log_2 10 = \dfrac{\log_{10} 10}{\log_{10} 2}$

$\approx \dfrac{1}{0.3010}$

≈ 3.3219

57. We will use natural logarithms for the conversion. Let $a = e$, $b = 200$, and $M = 30$ and substitute in the change-of-base formula.

$\log_{200} 30 = \dfrac{\ln 30}{\ln 200}$

$\approx \dfrac{3.4012}{5.2983}$

≈ 0.6419

59. We will use natural logarithms for the conversion. Let $a = e$, $b = 0.5$, and $M = 5$ and substitute in the change-of-base formula.

$\log_{0.5} 5 = \dfrac{\ln 5}{\ln 0.5}$

$\approx \dfrac{1.6094}{-0.6931}$

≈ -2.3219

61. We will use common logarithms for the conversion. Let $a = 10$, $b = 2$, and $M = 0.2$ and substitute in the change-of-base formula.

$\log_2 0.2 = \dfrac{\log_{10} 0.2}{\log_{10} 2}$

$\approx \dfrac{-0.6990}{0.3010}$

≈ -2.3219

63. We will use natural logarithms for the conversion. Let $a = e$, $b = \pi$, and $M = 58$ and substitute in the change-of-base formula.

$$\log_\pi 58 = \frac{\ln 58}{\ln \pi}$$

$$\approx \frac{4.0604}{1.1447}$$

$$\approx 3.5471$$

65. $ax^2 - b = 0$

$ax^2 = b$

$x^2 = \dfrac{b}{a}$

$x = \pm\sqrt{\dfrac{b}{a}}$

The solution is $\pm\sqrt{\dfrac{b}{a}}$.

67. $x^{1/2} - 6x^{1/4} + 8 = 0$

Let $u = x^{1/4}$.

$u^2 - 6u + 8 = 0$ Substituting

$(u - 4)(u - 2) = 0$

$u = 4$ or $u = 2$

$x^{1/4} = 4$ or $x^{1/4} = 2$

$x = 256$ or $x = 16$ Raising both sides to the fourth power

Both numbers check. The solutions are 256 and 16.

69. $x - 18\sqrt{x} + 77 = 0$

Let $u = \sqrt{x}$.

$u^2 - 18u + 77 = 0$ Substituting

$(u - 7)(u - 11) = 0$

$u = 7$ or $u = 11$

$\sqrt{x} = 7$ or $\sqrt{x} = 11$

$x = 49$ or $x = 121$ Squaring both sides

Both numbers check. The solutions are 49 and 121.

71. Use the change-of-base formula with $a = e$, and $b = 10$. We obtain

$$\log M = \frac{\ln M}{\ln 10}$$

$$\log M = 0.4343 \ln M.$$

73.

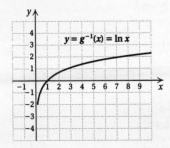

75. $\dfrac{\log_2 47}{\log_2 16} = \log_{16} 47$ Change-of-base formula

≈ 1.3886

77. $\dfrac{4.31}{\ln x} = \dfrac{28}{3.01}$

$\dfrac{4.31(3.01)}{28} = \ln x$

$0.463325 = \ln x$

$1.5893 \approx x$

Exercise Set 10.5

1. $2^x = 8$

$2^x = 2^3$

$x = 3$ The exponents are the same.

3. $4^x = 256$

$4^x = 4^4$

$x = 4$ The exponents are the same.

5. $2^{2x} = 32$

$2^{2x} = 2^5$

$2x = 5$

$x = \dfrac{5}{2}$

7. $3^{5x} = 27$

$3^{5x} = 3^3$

$5x = 3$

$x = \dfrac{3}{5}$

9. $2^x = 11$

$\log 2^x = \log 11$ Taking the common logarithm on both sides

$x \log 2 = \log 11$ Property 2

$x = \dfrac{\log 11}{\log 2}$

$x \approx 3.4594$

11. $2^x = 43$

$\log 2^x = \log 43$ Taking the common logarithm on both sides

$x \log 2 = \log 43$ Property 2

$x = \dfrac{\log 43}{\log 2}$

$x \approx 5.4263$

13. $5^{4x-7} = 125$

$5^{4x-7} = 5^3$

$4x - 7 = 3$ The exponents are the same.

$4x = 10$

$x = \dfrac{10}{4}$, or $\dfrac{5}{2}$

15. $3^{x^2} \cdot 3^{4x} = \dfrac{1}{27}$

$3^{x^2+4x} = 3^{-3}$

$x^2 + 4x = -3$

$x^2 + 4x + 3 = 0$

$(x+3)(x+1) = 0$

$x = -3 \text{ or } x = -1$

17. $4^x = 8$

$(2^2)^x = 2^3$

$2^{2x} = 2^3$

$2x = 3$ The exponents are the same.

$x = \dfrac{3}{2}$

19. $e^t = 100$

$\ln e^t = \ln 100$ Taking ln on both sides

$t = \ln 100$ Property 4

$t \approx 4.6052$ Using a calculator

21. $e^{-t} = 0.1$

$\ln e^{-t} = \ln 0.1$ Taking ln on both sides

$-t = \ln 0.1$ Property 4

$-t \approx -2.3026$

$t \approx 2.3026$

23. $e^{-0.02t} = 0.06$

$\ln e^{-0.02t} = \ln 0.06$ Taking ln on both sides

$-0.02t = \ln 0.06$ Property 4

$t = \dfrac{\ln 0.06}{-0.02}$

$t \approx \dfrac{-2.8134}{-0.02}$

$t \approx 140.6705$

25. $2^x = 3^{x-1}$

$\log 2^x = \log 3^{x-1}$

$x \log 2 = (x-1) \log 3$

$x \log 2 = x \log 3 - \log 3$

$\log 3 = x \log 3 - x \log 2$

$\log 3 = x(\log 3 - \log 2)$

$\dfrac{\log 3}{\log 3 - \log 2} = x$

$\dfrac{0.4771}{0.4771 - 0.3010} \approx x$

$-2.7095 \approx x$

27. $(3.6)^x = 62$

$\log (3.6)^x = \log 62$

$x \log 3.6 = \log 62$

$x = \dfrac{\log 62}{\log 3.6}$

$x \approx 3.2220$

29. $\log_4 x = 4$

$x = 4^4$ Writing an equivalent exponential equation

$x = 256$

31. $\log_2 x = -5$

$x = 2^{-5}$ Writing an equivalent exponential equation

$x = \dfrac{1}{32}$

33. $\log x = 1$ The base is 10.

$x = 10^1$

$x = 10$

35. $\log x = -2$ The base is 10.

$x = 10^{-2}$

$x = \dfrac{1}{100}$

37. $\ln x = 2$

$x = e^2 \approx 7.3891$

39. $\ln x = -1$

$x = e^{-1}$

$x = \dfrac{1}{e} \approx 0.3679$

41. $\log_3 (2x+1) = 5$

$2x + 1 = 3^5$ Writing an equivalent exponential equation

$2x + 1 = 243$

$2x = 242$

$x = 121$

43. $\log x + \log (x-9) = 1$ The base is 10.

$\log_{10} [x(x-9)] = 1$ Property 1

$x(x-9) = 10^1$

$x^2 - 9x = 10$

$x^2 - 9x - 10 = 0$

$(x-10)(x+1) = 0$

$x = 10 \text{ or } x = -1$

Check: For 10:

$$\frac{\log x + \log (x-9) = 1}{\log 10 + \log (10-9) \mid 1}$$

$\log 10 + \log 1$

$1 + 0$

$1 \mid$ TRUE

For −1:

$$\frac{\log x + \log (x-9) = 1}{\log(-1) + \log (-1-9) \mid 1} \quad \text{FALSE}$$

The number −1 does not check, because negative numbers do not have logarithms. The solution is 10.

45. $\log x - \log (x+3) = -1$ The base is 10.

$$\log_{10} \frac{x}{x+3} = -1 \quad \text{Property 3}$$

$$\frac{x}{x+3} = 10^{-1}$$

$$\frac{x}{x+3} = \frac{1}{10}$$

$$10x = x + 3$$

$$9x = 3$$

$$x = \frac{1}{3}$$

The answer checks. The solution is $\frac{1}{3}$.

47. $\log_2 (x+1) + \log_2 (x-1) = 3$

$$\log_2[(x+1)(x-1)] = 3 \quad \text{Property 1}$$

$$(x+1)(x-1) = 2^3$$

$$x^2 - 1 = 8$$

$$x^2 = 9$$

$$x = \pm 3$$

The number 3 checks, but −3 does not. The solution is 3.

49. $\log_4 (x+6) - \log_4 x = 2$

$$\log_4 \frac{x+6}{x} = 2 \quad \text{Property 3}$$

$$\frac{x+6}{x} = 4^2$$

$$\frac{x+6}{x} = 16$$

$$x + 6 = 16x$$

$$6 = 15x$$

$$\frac{2}{5} = x$$

The answer checks. The solution is $\frac{2}{5}$.

51. $\log_4 (x+3) + \log_4 (x-3) = 2$

$$\log_4[(x+3)(x-3)] = 2 \quad \text{Property 1}$$

$$(x+3)(x-3) = 4^2$$

$$x^2 - 9 = 16$$

$$x^2 = 25$$

$$x = \pm 5$$

The number 5 checks, but −5 does not. The solution is 5.

53. $\log_3 (2x-6) - \log_3 (x+4) = 2$

$$\log_3 \frac{2x-6}{x+4} = 2 \quad \text{Property 3}$$

$$\frac{2x-6}{x+4} = 3^2$$

$$\frac{2x-6}{x+4} = 9$$

$$2x - 6 = 9x + 36$$
$$\text{Multiplying by } (x+4)$$

$$-42 = 7x$$

$$-6 = x$$

Check:

$$\frac{\log_3 (2x-6) - \log_3 (x+4) = 2}{\log_3 [2(-6) - 6] - \log_3 (-6+4) \mid 2}$$
$$\log_3 (-18) - \log_3 (-2) \mid \quad \text{FALSE}$$

The number −6 does not check, because negative numbers do not have logarithms. There is no solution.

55. $x^2 + y^2 = 25,$ (1)

     $y - x = 1$ (2)

First solve Equation (2) for y.

$y = x + 1$

Then substitute $x + 1$ for y in Equation (1) and solve for x.

$$x^2 + y^2 = 25$$

$$x^2 + (x+1)^2 = 25 \quad \text{Substituting}$$

$$x^2 + x^2 + 2x + 1 = 25$$

$$2x^2 + 2x - 24 = 0$$

$$x^2 + x - 12 = 0$$

$$(x+4)(x-3) = 0$$

$$x + 4 = 0 \quad \text{or} \quad x - 3 = 0$$

$$x = -4 \quad \text{or} \quad x = 3$$

We now substitute these numbers for x in Equation (2) and solve for y.

When $x = -4$, $y = -4 + 1$, or −3.

When $x = 3$, $y = 3 + 1$, or 4.

The ordered pairs $(-4, -3)$ and $(3, 4)$ check and are the solutions.

57. $2x^2 + 1 = y^2,$ or $2x^2 - y^2 = -1,$ (1)

   $2y^2 + x^2 = 22$ $x^2 + 2y^2 = 22$ (2)

Solve the system using the elimination method.

$$4x^2 - 2y^2 = -2 \quad \text{Multiplying (1) by 2}$$

$$\underline{x^2 + 2y^2 = 22}$$

$$5x^2 \qquad\; = 20 \quad \text{Adding}$$

$$x^2 = 4$$

$$x = \pm 2$$

Substitute these values for x in either of the original equations and solve for y. Here we use the first equation.

When $x = 2$, $y^2 = 2 \cdot 2^2 + 1$, or 9. Thus $y = \pm 3$.

When $x = -2$, $y^2 = 2(-2)^2 + 1$, or 9. Thus $y = \pm 3$.

The ordered pairs $(2, 3)$, $(2, -3)$, $(-2, 3)$, and $(-2, -3)$ check and are the solutions.

59.
$$8^x = 16^{3x+9}$$
$$(2^3)^x = (2^4)^{3x+9}$$
$$2^{3x} = 2^{12x+36}$$
$$3x = 12x + 36$$
$$-36 = 9x$$
$$-4 = x$$

61.
$$\log_6 (\log_2 x) = 0$$
$$\log_2 x = 6^0$$
$$\log_2 x = 1$$
$$x = 2^1$$
$$x = 2$$

63.
$$\log_5 \sqrt{x^2 - 9} = 1$$
$$\sqrt{x^2 - 9} = 5^1$$
$$x^2 - 9 = 25 \qquad \text{Squaring both sides}$$
$$x^2 = 34$$
$$x = \pm\sqrt{34}$$

Both numbers check. The solutions are $\pm\sqrt{34}$.

65.
$$\log (\log x) = 5 \qquad \text{The base is 10.}$$
$$\log x = 10^5$$
$$\log x = 100,000$$
$$x = 10^{100,000}$$

The number checks. The solution is $10^{100,000}$.

67.
$$\log x^2 = (\log x)^2$$
$$2 \log x = (\log x)^2$$
$$0 = (\log x)^2 - 2 \log x$$
let $u = \log x$.
$$0 = u^2 - 2u$$
$$0 = u(u - 2)$$

$$u = 0 \qquad \text{or} \qquad u = 2$$
$$\log x = 0 \qquad \text{or} \qquad \log x = 2$$
$$x = 10^0 \qquad \text{or} \qquad x = 10^2$$
$$x = 1 \qquad \text{or} \qquad x = 100$$

Both numbers check. The solutions are 1 and 100.

69.
$$\log_a a^{x^2+4x} = 21$$
$$x^2 + 4x = 21 \qquad \text{Property 4}$$
$$x^2 + 4x - 21 = 0$$
$$(x + 7)(x - 3) = 0$$
$$x = -7 \text{ or } x = 3$$

Both numbers check. The solutions are $= -7$ and 3.

71.
$$3^{2x} - 8 \cdot 3^x + 15 = 0$$
Let $u = 3^x$ and substitute.
$$u^2 - 8u + 15 = 0$$
$$(u - 5)(u - 3) = 0$$

$$u = 5 \qquad \text{or} \qquad u = 3$$
$$3^x = 5 \qquad \text{or} \qquad 3^x = 3 \qquad \text{Substituting } 3^x \text{ for } u$$
$$\log 3^x = \log 5 \qquad \text{or} \qquad 3^x = 3^1$$
$$x \log 3 = \log 5 \qquad \text{or} \qquad x = 1$$
$$x = \frac{\log 5}{\log 3} \qquad \text{or} \qquad x = 1, \text{ or}$$
$$x \approx 1.4650 \qquad \text{or} \qquad x = 1$$

Both numbers check. Note that we can also express $\dfrac{\log 5}{\log 3}$ as $\log_3 5$ using the change-of-base formula.

73. $\log_5 125 = 3$ and $\log_{125} 5 = \dfrac{1}{3}$, so $x = (log_{125}5)^{log_5 125}$ is equivalent to $x = \left(\dfrac{1}{3}\right)^3 = \dfrac{1}{27}$. Then $\log_3 x = \log_3 \dfrac{1}{27} = -3$.

Exercise Set 10.6

1. a) We set $A(t) = \$450,000$ and solve for t:
$$450,000 = 50,000(1.06)^t$$
$$\frac{450,000}{50,000} = (1.06)^t$$
$$9 = (1.06)^t$$
$$\log 9 = \log (1.06)^t \qquad \begin{array}{l}\text{Taking the common}\\ \text{logarithm on both sides}\end{array}$$
$$\log 9 = t \log 1.06 \qquad \text{Property 2}$$
$$t = \frac{\log 9}{\log 1.06} \approx \frac{0.95424}{0.02531} \approx 37.7$$

It will take about 37.7 years for the $50,000 to grow to $450,000.

b) We set $A(t) = \$100,000$ and solve for t:
$$100,000 = 50,000(1.06)^t$$
$$\frac{100,000}{50,000} = (1.06)^t$$
$$2 = (1.06)^t$$
$$\log 2 = \log (1.06)^t \qquad \begin{array}{l}\text{Taking the common}\\ \text{logarithm on both sides}\end{array}$$
$$\log 2 = t \log 1.06 \qquad \text{Property 2}$$
$$t = \frac{\log 2}{\log 1.06} \approx \frac{0.30103}{0.02531} \approx 11.9$$

The doubling time is about 11.9 years.

3. a) We substitute 5 for t:
$$N(t) = 100 \cdot 2^{2t}$$
$$N(5) = 100 \cdot 2^{2 \cdot 5}$$
$$= 100 \cdot 2^{10}$$
$$= 102,400$$

There will be 102,400 bacteria present after 5 hr.

b) We set $N(t) = 128,000$ and solve for t:

$$128,000 = 100 \cdot 2^{2t}$$

$$1280 = 2^{2t} \qquad \text{Dividing by 100}$$

$$\log 1280 = \log 2^{2t}$$

$$\log 1280 = 2t \, \log 2$$

$$\frac{\log 1280}{2 \, \log 2} = t$$

$$\frac{3.10721}{2(0.30103)} \approx t$$

$$5.16 \approx t$$

About 5.16 hours will be required to obtain 128,000 bacteria.

5. a) We substitute 1 for a:

$$N(a) = 1000 + 200 \, \log a, \ a \geq 1$$

$$N(1) = 1000 + 200 \, \log 1$$

$$= 1000 + 200 \cdot 0$$

$$= 1000$$

1000 units were sold after spending \$1000.

b) We substitute 5 for a since a is in thousands.

$$N(5) = 1000 + 200 \, \log 5$$

$$\approx 1000 + 200(0.69897)$$

$$\approx 1140$$

1140 units were sold after spending \$5000.

c) We set $N(a) = 1276$ and solve for a.

$$1276 = 1000 + 200 \, \log a$$

$$276 = 200 \, \log a \qquad \text{Subtracting 1000}$$

$$1.38 = \log a \qquad \text{Dividing by 200}$$

$$24 \approx a \qquad \text{Finding the antilogarithm}$$

About \$24,000 must be spent in order to sell 1276 units.

7. We substitute 223.018 for P, since P is in thousands.

$$R(223.018) = 0.37 \, \ln 223.018 + 0.05$$

$$\approx 0.37(5.40725) + 0.05$$

$$\approx 2.1$$

The average walking speed of people living in Akron, Ohio, is about 2.1 ft/sec.

9. We substitute 42.045 for P, since P is in thousands.

$$R(42.045) = 0.37 \, \ln 42.045 + 0.05$$

$$\approx 0.37(3.73874) + 0.05$$

$$\approx 1.4$$

The average walking speed of people living in Ocala, Florida, is about 1.4 ft/sec.

11. a) The equation $P(t) = P_0 e^{kt}$ can be used to model population growth. At $t = 0$ (1987), the population was 5.0 billion. We substitute 5 for P_0 and 2.8%, or 0.028, for k to obtain the exponential growth function:

$P(t) = 5e^{0.028t}$, where t is the number of years after 1987 and P is in billions

b) In 1996, $t = 1996 - 1987$, or 9. To find the population in 1996, we substitute 9 for t:

$$P(9) = 5e^{0.028(9)}$$

$$= 5e^{0.252}$$

$$\approx 5(1.2866)$$

$$\approx 6.4$$

We can predict that the population will be about 6.4 billion in 1996.

In 2000, $t = 2000 - 1987$, or 13. To find the population in 2000, we substitute 13 for t:

$$P(13) = 5e^{0.028(13)}$$

$$= 5e^{0.364}$$

$$\approx 5(1.4391)$$

$$\approx 7.2$$

We can predict that the population will be about 7.2 billion in 2000.

c) We set $P(t) = 8$ and solve for t.

$$8 = 5e^{0.028t}$$

$$1.6 = e^{0.028t}$$

$$\ln 1.6 = \ln e^{0.028t}$$

$$\ln 1.6 = 0.028t$$

$$\frac{\ln 1.6}{0.028} = t$$

$$\frac{0.4700}{0.028} \approx t$$

$$17 \approx t$$

The population will be 8.0 billion about 17 years after 1987, or in 2004.

13. a) The exponential growth function is $C(t) = C_0 e^{kt}$. We assume that $t = 0$ corresponds to 1967 and $C_0 = \$80$ thousand. Then the growth function is $C(t) = 80e^{kt}$. To find k, we can use the fact that at $t = 21$ (1988), the cost was \$1,350,000. Then we substitute 21 for t and 1350 for $C(t)$ (since $C(t)$ is in thousands of dollars) and solve for k.

$$1350 = 80e^{k(21)}$$

$$\frac{1350}{80} = e^{21k}$$

$$\ln \frac{1350}{80} = \ln e^{21k}$$

$$\ln 16.875 = 21k$$

$$\frac{2.82583}{21} \approx k$$

$$0.135 \approx k$$

Thus, the exponential growth function is $C(t) = 80e^{0.135t}$, where t is the number of years after 1967 and $C(t)$ is in thousands of dollars.

b) In 1995, $t = 1995 - 1967$, or 28. We substitute 28 for t.

$$C(28) = 80e^{0.135(28)} = 80e^{3.78} \approx 3505$$

The cost will be about $3505 thousand, or $3,505,000 in 1995.

c) We set $C(t) = 8000$ (8,000,000 = 8000 thousand) and solve for t.

$$8000 = 80e^{0.135t}$$
$$100 = e^{0.135t} \qquad \text{Dividing by 80}$$
$$\ln 100 = \ln \ e^{0.135t}$$
$$\ln 100 = 0.135t$$
$$4.60517 \approx 0.135t$$
$$34.1 \approx t$$

The cost will be $8,000,000 about 34.1 yr after 1967, or in 2002.

d) To find the doubling time, we set $C(t) = 160$ and solve for t.

$$160 = 80e^{0.135t}$$
$$2 = e^{0.135t} \qquad \text{Dividing by 80}$$
$$\ln 2 = \ln \ e^{0.135t}$$
$$\ln 2 = 0.135t$$
$$0.69315 \approx 0.135t$$
$$5.1 \approx t$$

The cost will double in about 5.1 yr.

15. a) We substitute 0.09 for k.

$$P(t) = P_0 e^{kt}$$
$$P(t) = P_0 e^{0.09t}$$

b) We substitute 1000 for P_0 and 1 for t.

$$P(1) = 1000e^{0.09(1)}$$
$$= 1000e^{0.09}$$
$$\approx 1000(1.09417)$$
$$\approx 1094.17$$

After 1 yr the balance will be $1094.17.

We substitute 1000 for P_0 and 2 for t.

$$P(2) = 1000e^{0.09(2)}$$
$$= 1000e^{0.18}$$
$$\approx 1000(1.19722)$$
$$\approx 1197.22$$

After 2 yr the balance will be $1197.22.

c) We set $P_0 = 1000$ and $P(t) = 2000$ and solve for t.

$$2000 = 1000e^{0.09t}$$
$$2 = e^{0.09t}$$
$$\ln 2 = \ln \ e^{0.09t}$$
$$\ln 2 = 0.09t$$
$$0.69135 \approx 0.09t$$
$$7.7 \approx t$$

An investment of $1000 will double itself in about 7.7 yr.

17. We will use the function derived in Example 6:

$$P(t) = P_0 e^{-0.00012t}$$

If the tusk has lost 20% of its carbon-14 from an initial amount P_0, then 80%(P_0) is the amount present. To find the age of the tusk t, we substitute 80%(P_0), or $0.8P_0$, for $P(t)$ in the function above and solve for t.

$$0.8P_0 = P_0 e^{-0.00012t}$$
$$0.8 = e^{-0.00012t}$$
$$\ln 0.8 = \ln \ e^{-0.00012t}$$
$$-0.2231 \approx -0.00012t$$
$$t \approx \frac{-0.2231}{-0.00012} \approx 1860$$

The tusk is about 1860 years old.

19. a) We substitute 0 for t.

$$V(0) = 28,000e^{-0} = 28,000 \cdot 1 = \$28,000$$

b) We substitute 2 for t.

$$V(2) = 28,000e^{-2} = 28,000(0.13534) \approx \$3789$$

21. a) The exponential growth function is $V(t) = V_0 e^{kt}$. Since $V_0 = 84,000$, we have $V(t) = 84,000e^{kt}$, where t is the number of years after 1947. To find k we use the fact that at $t = 40$ (1987), the value of the painting was $53,900,000. We substitute and solve for k:

$$53,900,000 = 84,000e^{k(40)}$$
$$\frac{1925}{3} = e^{40k}$$
$$\ln \frac{1925}{3} = \ln \ e^{40k}$$
$$\ln \frac{1925}{3} = 40k$$
$$\frac{\ln \dfrac{1925}{3}}{40} = k$$
$$\frac{6.4641}{40} \approx k$$
$$0.1616 \approx k$$

The exponential growth rate is about 0.1616, or 16.16%. The exponential growth function is $V(t) = 84,000e^{0.1616t}$, where t is the number of years after 1947.

b) We find $V(50)$:

$$V(50) = 84,000e^{0.1616(50)}$$
$$= 84,000e^{8.08}$$
$$\approx 84,000(3229.2332)$$
$$\approx \$271,000,000$$

c) We substitute 2(84,000), or 168,000 for V and solve for t:

$$168,000 = 84,000e^{0.1616t}$$

$$2 = e^{0.1616t}$$

$$\ln 2 = \ln e^{0.1616t}$$

$$\ln 2 = 0.1616t$$

$$\frac{\ln 2}{0.1616} = t$$

$$\frac{0.6931}{0.1616} \approx t$$

$$4.3 \approx t$$

The doubling time is about 4.3 yr.

d) We substitute 1,000,000,000 for V and solve for t.

$$1,000,000,000 = 84,000e^{0.1616t}$$

$$\frac{250,000}{21} = e^{0.1616t}$$

$$\ln \frac{250,000}{21} = \ln e^{0.1616t}$$

$$\ln \frac{250,000}{21} = 0.1616t$$

$$\frac{\ln \dfrac{250,000}{21}}{0.1616} = t$$

$$\frac{9.3847}{0.1616} \approx t$$

$$58 \approx t$$

The value of the painting will be \$1 billion after about 58 years.

23. $i^{46} = (i^2)^{23} = (-1)^{23} = -1$

25. $i^{14} + i^{15} = (i^2)^7 + (i^2)^7 \cdot i = (-1)^7 + (-1)^7 \cdot i = -1 - i$

27. $(5 - 4i)(5 + 4i) = 25 - 16i^2 = 25 + 16 = 41$